LINEAR ALGEBRA
for
MATHEMATICS, SCIENCE, AND ENGINEERING

LINEAR ALGEBRA
for
MATHEMATICS, SCIENCE, AND ENGINEERING

Edward M. Landesman
University of California, Santa Cruz

Magnus R. Hestenes
University of California, Los Angeles

PRENTICE HALL
Englewood Cliffs, New Jersey 07632

Library of Congress Cataloging-in-Publication Data

Landesman, Edward M.
   Linear algebra for mathematics, science, and
engineering / Edward M. Landesman, Magnus R. Hestenes.
      p.   cm.
   Includes index.
   ISBN 0–13–529561–0
   1. Algebras, Linear.   I. Hestenes, Magnus Rudolph, 1906–91.
II. Title.
QA184.L364   1991
512′.5—dc20   91–24266
                        CIP

Acquisition editor: Steve Conmy
Production: Nicholas Romanelli
Interior design: Lee Goldstein
Cover design: Patricia A. McGowan
Prepress buyer: Paula Massenaro
Manufacturing buyer: Lori Bulwin

 © 1992 by Prentice-Hall, Inc.
A Simon & Schuster Company
Englewood Cliffs, New Jersey 07632

Printed in the United States of America
10 9 8 7 6 5 4 3 2 1

ISBN 0-13-529561-0

Prentice-Hall International (UK) Limited, *London*
Prentice-Hall of Australia Pty. Limited, *Sydney*
Prentice-Hall Canada, Inc., *Toronto*
Prentice-Hall Hispanoamericana, S.A., *Mexico*
Prentice-Hall of India Private Limited, *New Delhi*
Prentice-Hall of Japan, Inc., *Tokyo*
Simon & Schuster Asia Pte. Ltd., *Singapore*
Editor Prentice-Hall do Brasil, Ltda., *Rio de Janeiro*
Prentice Hall, *Englewood Cliffs, New Jersey*

To Miriam Landesman and Carolyn Hestenes

Contents

4 Eigenvalues and Eigenvectors of a Square Matrix   205

5 Linear Spaces   283

6 Linear Transformations and Their Properties   331

Preface

This text evolved from many years of our teaching and discussing linear algebra. Our purpose is to present the fundamentals of the subject in a way that will be comprehensible to various levels of students.

Most colleges offer such a course traditionally after a year or two of calculus. Although few courses in linear algebra include much calculus, it is usually maintained that students need some mathematical maturity before they can master the more abstract theory encountered in linear algebra. Today, knowledge of linear algebra is needed by those working in the natural sciences, including mathematicians, physicists, engineers, as well as social scientists, including economists, psychologists, and sociologists.

We ease the student into learning the abstract notions of a vector space and a linear transformation by first exemplifying these concepts via the theory of matrices, this being not a compromise in rigor but a learning tool. The algebra of matrices is linked to the algebra of real numbers, and the similarities and differences of the two systems are explored. Examples of applications including circuit theory, graph theory, and chemistry are used to indicate how matrices play a problem-solving role in many fields.

In Chapter 1 we introduce the student to the notions of "linear combinations," "linear independence," and "linear dependence." The presentation of $AX = H$, both as a system of linear equations and as a linear combination of the columns of A being equal to the column vector H, is used to advantage in preparing the reader for the more abstract concepts of rank and of a vector space.

Chapter 2 introduces the idea of nonsingularity and Gaussian elimination for the computation of inverses. A detailed explanation that includes several examples provides the reader with a thorough treatment of systems of equations.

Chapter 3 defines determinants and elucidates the basic theory.

In Chapter 4, we break with tradition and present a comprehensive explanation of eigenvalues and eigenvectors before the concepts of a vector space and a linear transformation are introduced. Again, via matrices, the student can comprehend the basic theory without the abstract formality.

This approach is our response to many who contend that a basic course in linear algebra often includes insufficient (if any) time for briefing students on the fundamental theory of eigenvalues and eigenvectors.

After the student becomes familiar with the working of matrices, the formal definition of a vector space and its consequences is introduced in Chapter 5.

The concepts of a linear transformation are introduced in Chapter 6 and are tied in with the theory of matrices. Section 6.3 terminates the equivalent of a first course in linear algebra.

Beginning with Section 6.4, more advanced topics are presented. The material is treated more concisely, being intended for the reader who has completed the earlier part of the book.

Many topics can be chosen by the instructor to complement the previous part of the course, provided time permits. Topics less often encountered in linear algebra texts, but of importance to many disciplines, include orthogonal projections, least square solutions, pseudoinverses, and Rayleigh quotients, all being fairly independent of one another.

Chapter 7 contains the basic theory of quadratic forms and their geometric interpretation. Emphasis is placed on the properties of ellipsoids.

Finally, in Chapter 8 we introduce some computational and numerical techniques, including the power method, a conjugate gradient routine, and a Gauss–Seidel method for solving $A\mathbf{x} = \mathbf{h}$.

An Appendix covering some basic theory of polynomials is included that will be useful to students needing a review of that topic before proceeding to Chapter 4.

Answers to selected exercises also appear at the end of the book, following the appendix.

We offer the following schedule for a first course in linear algebra. This version has been used for many years by one of the authors at the University of California, Santa Cruz.

Quarter System (30 meetings)

1. Organization, §§ 1.0, 1.1	11. Exam	21. § 4.4
2. § 1.2	12. § 3.1	22. Exam
3. § 1.3	13. § 3.2	23. § 5.1
4. §§ 1.3, 1.4	14. § 3.3	24. §§ 5.1, 5.2
5. § 1.5	15. §§ 3.3, 3.4	25. §§ 5.2, 5.3
6. § 2.1	16. § 3.4	26. § 5.3
7. § 2.2	17. § 4.1	27. § 6.1
8. § 2.3	18. § 4.2	28. § 6.2
9. §§ 2.3, 2.4	19. § 4.3	29. 6.2
10. § 2.4	20. § 4.3	30. § 6.3

If a semester system is used and there are 45 class sessions, the instructor may wish to devote more time to each of the sections cited above, and may also wish to include §§ 1.6, 4.5, and 5.5. Time permitting, one can then choose several of the remaining sections in Chapter 6 as well as topics from Chapters 7 and 8. For example, in a more theoretically oriented course you could include §§ 6.6, 6.7, 7.1, and 7.2, whereas in a more computational or numerically oriented course you might wish to cover §§ 8.1, 8.3, 8.4, and 8.5.

To the following colleagues, who reviewed the manuscript, we extend our thanks for their many helpful comments and suggestions:

> Terry L. Herdman, Virginia Polytechnic Institute & State University
> Stephen D. Smith, University of Illinois, Chicago
> William Rundell, Texas A & M
> James Northrup, Colby College
> Adolfo J. Rumbos, University of Utah
> Ivie Stein, Jr., University of Toledo
> Herbert E. Kasube, Bradley University

We are also grateful to many colleagues for offering their advice and for class-testing various versions of the manuscript. In particular, we thank Professors David Bao, Nick Burgoyne, Svetlana Katok, Al Kelley, and Andrey Todorov. Professor Gerhard Ringel drew the original sketches from which the illustrations were prepared. Many students, graduate and undergraduate, also participated. Our special thanks to Robert Curtis for his useful suggestions and to Ozlem Imamoglu for writing the solutions manual. Ms Dorothy Hollinger typed the original manuscript with patience and precision. For their highly competent proofreading and for their compilation of the index, we thank Laura Peterson and Kim Viviani. Finally, to Steve Conmy and Nicholas Romanelli, our editors at Prentice Hall, we extend our appreciation for their support and cooperation prior to and during the production stage of the book.

E.M.L. / M.R.H.

LINEAR ALGEBRA
for
MATHEMATICS, SCIENCE, AND ENGINEERING

1

Matrices:
Basic Operations

Linear algebra is often referred to as the first course in mathematics in which a student is introduced to some very abstract mathematical concepts. These concepts include the notions of a linear space and a linear transformation, geometric applications, and many of the ideas that follow. There are many reasons for studying linear algebra. Certainly, one of the more practical reasons for studying the subject is because of the interest in solving systems of linear equations. The study of systems of linear equations is of great importance not only to mathematicians but also to scientists, engineers, and social scientists. Such equations arise in describing various kinds of physical phenomena, economics, and behavioral theory. Even in situations where highly complex "nonlinear" behavior is studied, one often approximates the nonlinear behavior or first needs to understand what is occurring in the linear situation. This often reduces to the study of systems of linear equations.

As an example, consider the pair of linear equations

$$2x - 3y = 1$$

$$x + 5y = 7.$$

In high school you probably learned how to solve such a system of equations by first eliminating one of the variables, say x, and then solving for the other variable y. Substituting this value of y in either of the original equations then gave the value for x. In this example, $x = 2$ and $y = 1$ is the solution.

There is, however, another way of solving this system of equations, a method that is far more practical, especially when dealing with large systems. The method involves the use of blocks of numbers. You already know how important the real numbers are in mathematics and its application to other fields. Experience has shown that in advanced mathematics and its applications, it is often advantageous to consider a block of numbers as a single entity.

To this end, we can view the left side of the system of equations above as the block

$$\begin{bmatrix} 2 & -3 \\ 1 & 5 \end{bmatrix}$$

of numbers describing the coefficients of x and y in these equations. On the right side of these equations, we have another block of numbers, the column

$$\begin{bmatrix} 1 \\ 7 \end{bmatrix}.$$

From these blocks of numbers we can form still another block of numbers, the block

$$\begin{bmatrix} 2 & -3 & \vdots & 1 \\ 1 & 5 & \vdots & 7 \end{bmatrix}.$$

We observe that this block is made up of the two original blocks above. The dotted line separates the block that represents the coefficients of x and y and the block that represents the numbers on the right side of the system of equations.

In the book, we shall learn how to operate on the block

$$\begin{bmatrix} 2 & -3 & \vdots & 1 \\ 1 & 5 & \vdots & 7 \end{bmatrix}$$

by certain rules that will not alter the nature of the original system of equations but will transform this block into another block, which looks like

$$\begin{bmatrix} 1 & 0 & \vdots & 2 \\ 0 & 1 & \vdots & 1 \end{bmatrix}.$$

Just as the original system of equations

$$2x - 3y = 1$$

$$x + 5y = 7$$

had associated with it the block

$$\begin{bmatrix} 2 & -3 & \vdots & 1 \\ 1 & 5 & \vdots & 7 \end{bmatrix},$$

so does the block

$$\begin{bmatrix} 1 & 0 & \vdots & 2 \\ 0 & 1 & \vdots & 1 \end{bmatrix}$$

have associated with it the system of equations

$$x + 0y = 2$$

$$0x + y = 1,$$

i.e., $x = 2$ and $y = 1$, the solution to our original system of equations. If the object was to solve this original system on a computer, we would actually use the blocks described above in order to get x and y. While it is highly unlikely that we would need to use a computer to solve such a simple system of equations, it is very much the case that a computer would be used to solve large systems of this type. Our blocks of numbers would of course also be large.

We proceed by presenting two examples in which systems of equations arise.

■ **EXAMPLE 1** ────────────────────────────────

In calculus, we often evaluate integrals in which partial fractions are employed. In order to complete the evaluation of the integral, a system of linear equations must be solved. Consider, for example,

$$\int \frac{2x^2 - 4x + 3}{x^3 + x^2 + x}\, dx$$

Since the denominator can be factored into $x(x^2 + x + 1)$, our original rational expression becomes

$$\frac{2x^2 - 4x + 3}{x(x^2 + x + 1)},$$

whose partial fraction decomposition can be written as

$$\frac{2x^2 - 4x + 3}{x(x^2 + x + 1)} = \frac{a}{x} + \frac{bx + c}{x^2 + x + 1}.$$

The right side of this equation, after some simple algebra, can be rewritten as

$$\frac{(a + b)x^2 + (a + c)x + a}{x(x^2 + x + 1)}.$$

Equating coefficients of like powers of x on the left and right sides of the equation, we get the system of linear equations

$$
\begin{aligned}
a + b \quad\;\;\;\; &= \quad 2 \\
a \quad\;\; + c &= -4 \\
a \quad\quad\;\;\;\; &= \quad 3.
\end{aligned}
$$

We can view this system of equations as a block of numbers:

$$
\begin{bmatrix}
1 & 1 & 0 & \vdots & 2 \\
1 & 0 & 1 & \vdots & -4 \\
1 & 0 & 0 & \vdots & 3
\end{bmatrix}.
$$

■ **EXAMPLE 2** ────────────────────────────────

In chemistry, a chemical equation is balanced whenever each side of the equation has the same number of atoms of every element involved. For example, consider the unbalanced "equation"

$$H_2SO_3 + HBrO_3 \longrightarrow H_2SO_4 + Br_2 + H_2O.$$

The "equation" above will be balanced if we can find positive integers $x, y, z, w,$ and u so that

$$x\text{H}_2\text{SO}_3 + y\text{HBrO}_3 \longrightarrow z\text{H}_2\text{SO}_4 + w\text{Br}_2 + u\text{H}_2\text{O}.$$

For the number of hydrogen atoms to be the same, we see that

$$2x + y = 2z + 2u.$$

For sulfur, we have

$$x = z.$$

Continuing, for oxygen

$$3x + 3y = 4z + u.$$

Finally, for bromine

$$y = 2w.$$

This system of equations is equivalent to the homogeneous system of equations

$$
\begin{aligned}
2x + y - 2z \quad\quad\; - 2u &= 0 \\
x \quad\quad - z \quad\quad\quad\quad &= 0 \\
3x + 3y - 4z \quad\quad\; - u &= 0 \\
y \quad - 2w \quad\quad &= 0.
\end{aligned}
$$

The block of numbers now is

$$
\begin{bmatrix}
2 & 1 & -2 & 0 & -2 & \vdots & 0 \\
1 & 0 & -1 & 0 & 0 & \vdots & 0 \\
3 & 3 & -4 & 0 & -1 & \vdots & 0 \\
0 & 1 & 0 & -2 & 0 & \vdots & 0
\end{bmatrix}
$$

Rectangular blocks of the type described above are called *matrices*. Matrices can be viewed to be generalized numbers because to a great extent, they behave like numbers. We shall see that every matrix can be multiplied by a number to obtain a new matrix. We will also learn how to add matrices of the same shape and learn the rules for multiplying matrices of related shapes so as to obtain new matrices. We will also see that there is a generalization of the notion of division when applied to matrices.

It is for the reasons above as well as for many other reasons that the idea of a matrix is introduced. In fact, sometimes the emphasis is not on solving a system of linear equations but on describing a process that occurs in everyday life. In elementary school when you studied multiplication tables, you were, without realizing it, using the idea of a matrix. In this case, however, you were dealing with only one matrix and the only operation used was a "look-up." In manufacturing, one often seeks information about a product that is being produced in many factories at different locations. A matrix can be extremely useful for this purpose. Another example that should interest you as a student is your instructor's gradebook. Before your course in linear algebra is complete, that gradebook will probably contain the names of all students in the course, in what year they are at the school, their majors, and their scores on exams, quizzes, and homework. Once again, a matrix can play a valuable role in this example.

Beginning in the next section, you will learn some of the fundamental properties of matrices and how useful they are in solving systems of linear equations as well as other applications. Later, in Chapters 5 and 6, we will see that some of the fundamental properties that are shared by matrices lead to the more abstract notions of a linear space and a linear transformation. But for now, in the first few chapters, we will be content with defining and studying the basic properties of matrices and some of their applications.

Exercises

1. Consider the following systems of linear equations:

   (i) $2x + 3y = 7$ (ii) $9x + y = 19$ (iii) $2x + 3y = \ \ 7$
   $\ \ \ 5x + 8y = 4$ $-x + y = \ \ 1$ $6x + 9y = 20$

   (iv) $\ \ \ x + 7y = 8$ (v) $4x + 8y = 9$ (vi) $-9x + 2y = 3$
   $\ \ \ 12x - 3y = 4$ $2x - 5y = 2$ $36x - 8y = 1$

   (a) Using the systems above, write down the blocks of numbers (matrices) that describe the coefficients of the variables in each system.

   (b) Using the systems above, write down the blocks of numbers (matrices) that describe the numbers on the right-hand side of these systems.

   (c) Using the systems above, write down the blocks of numbers (matrices) that are made up of the two blocks in parts (a) and (b) for each system.

2. Consider the following systems of linear equations:

   (i) $2x + 5y + z = 1$ (ii) $x + y - z = 2$ (iii) $\ \ \ x + y + z = 1$
   $\ \ \ x - y \ \ \ \ \ \ \ = 1$ $\ \ \ x \ \ \ \ \ \ + z = 1$ $\ \ \ x - y - z = 4$
   $\ \ \ x + \ y + z = 2$ $\ \ \ \ \ \ \ y - z = 0$ $3x + y + z = 2$

   (iv) $\ \ x + \ y + z = 1$ (v) $\ \ \ x + 2y + 3z = 1$ (vi) $\ \ x + \ y \ \ \ \ \ \ = 0$
   $\ \ \ x - \ y + z = 4$ $-x + 9y + 4z = 2$ $\ \ \ x - \ y + z = 1$
   $\ \ 2x + 3y - z = 2$ $\ \ 2x + \ y - 3z = 3$ $6x - 4y - z = 2$

   Repeat parts (a), (b), and (c) in Exercise 1 for the systems above.

3. Write down the system of linear equations that is associated with each of the following blocks of numbers (matrices).

(a) $\begin{bmatrix} 2 & -3 & : & 4 \\ 6 & 9 & : & -3 \end{bmatrix}$ (b) $\begin{bmatrix} 1 & 3 & : & 5 \\ 2 & 0 & : & -6 \end{bmatrix}$

(c) $\begin{bmatrix} 2 & 3 & 0 & : & -2 \\ 1 & 0 & 0 & : & 8 \end{bmatrix}$ (d) $\begin{bmatrix} 3 & 0 & 0 & : & -6 \\ 0 & 1 & -2 & : & 5 \end{bmatrix}$

1.1   INTRODUCTION TO MATRICES

As we have indicated in our preliminary remarks, one of the fundamental concepts in the study of linear algebra is the idea of a matrix. A *matrix* is a rectangular array of numbers, such as

$$\begin{bmatrix} 1 & 2 & 4 \\ -3 & 5 & 6 \end{bmatrix}, \quad \begin{bmatrix} 2 & 1 \\ -1 & 2 \end{bmatrix}, \quad \begin{bmatrix} 0 & 1 \\ 1 & 0 \\ 0 & 1 \end{bmatrix},$$

$$\begin{bmatrix} 7 & -1 & 2 & 1 \\ 6 & -1 & 4 & 4 \\ 2 & 3 & 3 & 2 \end{bmatrix}, \quad \begin{bmatrix} \sqrt{2} & \pi & 0 \\ 1 & 0.8 & 3/2 \\ -1 & 3 & 5 \end{bmatrix}, \quad \begin{bmatrix} 6 \\ -1 \\ 5 \end{bmatrix}.$$

The numbers which appear in a matrix are called *entries* or *elements* of the matrix. We shall use the terms *entry* and *element* interchangeably. An entry may be a whole number (integer), a fraction, or a decimal. For convenience in computation, we use fractions and integers as entries. In applications, the entries of a matrix are usually decimals. Normally, we will label matrices by capital letters. For example, we could label the matrices listed above as follows:

$$A = \begin{bmatrix} 1 & 2 & 4 \\ -3 & 5 & 6 \end{bmatrix}, \quad B = \begin{bmatrix} 2 & 1 \\ -1 & 2 \end{bmatrix}, \quad E = \begin{bmatrix} 1 & 0 \\ 0 & 1 \\ 1 & 0 \end{bmatrix},$$

$$G = \begin{bmatrix} 7 & -1 & 2 & 1 \\ 6 & -1 & 4 & 4 \\ 2 & 3 & 3 & 2 \end{bmatrix}, \quad U = \begin{bmatrix} \sqrt{2} & \pi & 0 \\ 1 & 0.8 & 3/2 \\ -2 & 3 & 5 \end{bmatrix}, \quad X = \begin{bmatrix} 6 \\ -1 \\ 5 \end{bmatrix}.$$

Of course, any other six capital letters could be used. However, in a given problem, when we have labeled a particular matrix with a given letter, we will retain that letter for the matrix until we have finished the problem.

We observe that matrices have *rows* and *columns*. A row is displayed horizontally and a column vertically. For example, the matrix

$$A = \begin{bmatrix} 1 & 2 & 4 \\ -3 & 5 & 6 \end{bmatrix}$$

has two rows and three columns. Its two rows, R_1 and R_2, are

$$R_1 = [1 \quad 2 \quad 4], \qquad R_2 = [-3 \quad 5 \quad 6].$$

Likewise, the three columns C_1, C_2, and C_3 of A are

$$C_1 = \begin{bmatrix} 1 \\ -3 \end{bmatrix}, \qquad C_2 = \begin{bmatrix} 2 \\ 5 \end{bmatrix}, \qquad C_3 = \begin{bmatrix} 4 \\ 6 \end{bmatrix}.$$

Observe that A can be viewed as a column

$$A = \begin{bmatrix} R_1 \\ R_2 \end{bmatrix}$$

of its two rows R_1 and R_2. Similarly, A can be represented as a row

$$A = [C_1 \quad C_2 \quad C_3]$$

of its three columns C_1, C_2, and C_3.

As another example, we note that the matrix

$$M = \begin{bmatrix} 7 & -1 & 2 & 1 \\ 6 & -1 & 4 & 4 \\ 2 & 3 & 3 & 2 \end{bmatrix}$$

has the representation

$$M = \begin{bmatrix} R_1 \\ R_2 \\ R_3 \end{bmatrix}$$

as a column of three rows

$$R_1 = [7 \quad -1 \quad 2 \quad 1], \quad R_2 = [6 \quad -1 \quad 4 \quad 4], \quad R_3 = [2 \quad 3 \quad 3 \quad 2].$$

Similarly, M has the representation

$$M = [C_1 \quad C_2 \quad C_3 \quad C_4],$$

as a row of the four columns

$$C_1 = \begin{bmatrix} 7 \\ 6 \\ 2 \end{bmatrix}, \qquad C_2 = \begin{bmatrix} -1 \\ -1 \\ 3 \end{bmatrix}, \qquad C_3 = \begin{bmatrix} 2 \\ 4 \\ 3 \end{bmatrix}, \qquad C_4 = \begin{bmatrix} 1 \\ 4 \\ 2 \end{bmatrix}.$$

In general, a matrix A having m rows $R_1, R_2, \ldots, R_m$ and n columns C_1, $C_2, \ldots, C_n$ has the representations

$$A = \begin{bmatrix} R_1 \\ R_2 \\ \vdots \\ R_m \end{bmatrix} \qquad \text{and} \qquad A = [C_1 \quad C_2 \quad \cdots \quad C_n].$$

The first representation of the matrix *A* is a column of *m* rows and the second representation of the matrix *A* is a row of *n* columns. Such a matrix *A* is called an ***m* × *n-matrix*** or an ***m* × *n-dimensional matrix***. The symbol "*m* × *n*" is read "*m* by *n*." For example, a 3 × 2-matrix has three rows and two columns. A 100 × 200-matrix has one hundred rows and two hundred columns. A *j* × 1-matrix has *j* rows and one column. A 1 × 1-matrix is a single number.

■ **EXAMPLE 1** _____

In electronic circuit theory, irrigation systems, and the flow of traffic, the theory of networks or graphs *G* comes into play. In many practical situations such as the ones above, the graphs *G* which we consider consist of a finite number of edges and vertices. We assume further that there is at most one edge between any two vertices. Such graphs *G* are called *simple graphs*. The two graphs below are simple graphs.

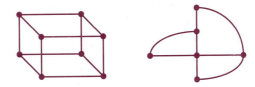

It should be clear that the following graph is not simple:

Matrices play a role in graph theory. One such matrix *A* is called the *adjacency matrix* which describes how each edge is related to each vertex in a simple graph *G*. To illustrate this, consider the following simple graph *G*.

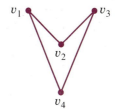

We can construct a table that provides us with all of the information about G described in the illustration above. In the table, the letters T and F denote the words *true* and *false*, respectively. For example, the T in the row containing v_2 and the column containing v_1 tells us that it is true that there is an edge joining v_2 to v_1. Similarly, the F in the row containing v_3 and the column containing v_1 tells us that there is not an edge joining v_3 to v_1.

v_1	F	T	F	T
v_2	T	F	T	F
v_3	F	T	F	T
v_4	T	F	T	F
	v_1	v_2	v_3	v_4

We can see that column 1 of the table conveys the information that the vertex v_1 is joined to vertices v_2 and v_4. Column 2 indicates that the vertex v_2 is also joined to the vertex v_3, while column 3 shows that the vertex v_3 is also joined to the vertex v_4. By symmetry, we see that since v_2 joins v_1, so v_1 joins v_2 and since v_4 joins v_1, so v_1 joins v_4, and so on. Similarly, since a vertex does not have an edge joining to itself, it follows that v_1 is not joined to v_1, v_2 is not joined to v_2, and so on. While the table above can be written as a 4×4-matrix, we usually represent it by replacing the T's and F's with 1's and 0's respectively. We write

$$A = \begin{bmatrix} 0 & 1 & 0 & 1 \\ 1 & 0 & 1 & 0 \\ 0 & 1 & 0 & 1 \\ 1 & 0 & 1 & 0 \end{bmatrix}.$$

This 4×4-matrix A is the ***adjacency matrix*** for the simple graph G above.

We return to the study of a general matrix. The *size* of a matrix is determined by the number of rows and columns in the matrix. Two matrices are of the same size if they have the same number of rows and columns. In mathematics, we often use the term ***dimension*** or ***order*** instead of the term *size*. For example, the matrices

$$\begin{bmatrix} 1 & 2 & 4 \\ -3 & 5 & 6 \end{bmatrix} \quad \text{and} \quad \begin{bmatrix} 2 & -1 & 5 \\ 6 & 0 & 1 \end{bmatrix}$$

are of the same size (or dimension or order), since each has two rows and three columns. Whereas the matrices

$$\begin{bmatrix} 1 & 2 & 4 \\ -3 & 5 & 6 \end{bmatrix} \quad \text{and} \quad \begin{bmatrix} 2 & 6 \\ -1 & 0 \\ 5 & 1 \end{bmatrix}$$

are of different size.

Two matrices are said to be *equal* if they are of the same size and if corresponding entries are equal. As an example, the equation

$$\begin{bmatrix} a & b & c \\ d & e & f \end{bmatrix} = \begin{bmatrix} 1 & 2 & 4 \\ -3 & 5 & 6 \end{bmatrix}$$

holds if and only if $a = 1$, $b = 2$, $c = 4$, $d = -3$, $e = 5$, and $f = 6$.

In applications, the concept of a square matrix plays an important role. A matrix is *square* if it has the same number of rows as columns. The matrices

$$\begin{bmatrix} 2 & 9 \\ -1 & 6 \end{bmatrix}, \quad \begin{bmatrix} 1 & 0 & 1 \\ -1 & 5 & 6 \\ 7 & -8 & 9 \end{bmatrix}, \quad \begin{bmatrix} 1 & 7 & 1 & -8 \\ -1 & 4 & -1 & 1 \\ -2 & 3 & 1 & 9 \\ 1 & 6 & -1 & 5 \end{bmatrix}$$

are square matrices. Each has the same number of rows as columns.

A matrix having n rows and n columns is called an $n \times n$-*matrix* or simply an *n-square matrix.* In the three examples above, the first is a 2×2-matrix, the second is a 3×3-matrix, and the third is a 4×4-matrix.

Because of their importance, we give special names to matrices having a single row or a single column. A matrix having a single row is called a *row vector.* Similarly, a matrix having a single column is called a *column vector.* For example, the matrices

$$[3 \quad 4], \quad [5 \quad -6 \quad 7], \quad [-1 \quad 0 \quad 2 \quad -3]$$

are row vectors of dimensions two, three, and four respectively, while the matrices

$$\begin{bmatrix} 3 \\ 4 \end{bmatrix}, \quad \begin{bmatrix} 5 \\ -6 \\ 7 \end{bmatrix}, \quad \begin{bmatrix} -1 \\ 0 \\ 2 \\ -3 \end{bmatrix}$$

are column vectors of dimensions two, three, and four respectively. An entry of a row or a column vector is also called a *component* of the row or column vector. For the column vector

$$\begin{bmatrix} 5 \\ -6 \\ 7 \end{bmatrix},$$

we see that its first component is 5, its second component is -6, and its third component is 7. In general, a row vector or a column vector with n components is a $1 \times n$-matrix or an $n \times 1$-matrix respectively. We will often use symbols such as

$$[a \quad b], \quad [u \quad v], \quad [x \quad y], \quad [a \quad b \quad c], \quad [x \quad y \quad z]$$

and

$$\begin{bmatrix} a \\ b \end{bmatrix}, \quad \begin{bmatrix} u \\ v \end{bmatrix}, \quad \begin{bmatrix} x \\ y \end{bmatrix}, \quad \begin{bmatrix} a \\ b \\ c \end{bmatrix} \quad \begin{bmatrix} x \\ y \\ z \end{bmatrix}$$

to denote row and column vectors of small dimensions. For general row and column vectors, we use expressions such as

$$[a_1 \quad a_2 \quad \cdots \quad a_n], \quad [x_1 \quad x_2 \quad \cdots \quad x_n], \quad \begin{bmatrix} a_1 \\ a_2 \\ \vdots \\ a_n \end{bmatrix}, \quad \begin{bmatrix} x_1 \\ x_2 \\ \vdots \\ x_n \end{bmatrix}.$$

Of course, any other letters besides a and x can be used. The subscripted letters $a_1, a_2, \ldots$ are read "a one," "a two," and so on. This is helpful in that the subscript tells you the component with which you are dealing.

Matrices that appear in applications are frequently very large. It is not unusual to encounter a matrix having 10,000 rows and 5000 columns. Fortunately, the basic properties of matrices are exhibited by relatively small matrices. As a result, we begin our study of matrices by considering matrices that have either two or three rows or columns. In working with these smaller matrices we shall use special notations, such as

$$\begin{bmatrix} a & b \\ c & d \end{bmatrix}, \quad \begin{bmatrix} a_1 & b_1 & c_1 \\ a_2 & b_2 & c_2 \end{bmatrix}, \quad \begin{bmatrix} a_1 & b_1 & c_1 \\ a_2 & b_2 & c_2 \\ a_3 & b_3 & c_3 \end{bmatrix}.$$

We will also use the more conventional notations

$$\begin{bmatrix} a_{11} & a_{12} \\ a_{21} & a_{22} \end{bmatrix}, \quad \begin{bmatrix} a_{11} & a_{12} & a_{13} \\ a_{21} & a_{22} & a_{23} \end{bmatrix}, \quad \begin{bmatrix} a_{11} & a_{12} & a_{13} \\ a_{21} & a_{22} & a_{23} \\ a_{31} & a_{32} & a_{33} \end{bmatrix}.$$

Here the subscripts 1 and 2 on a_{12} (read "a one two") signify that a_{12} is in the first row and second column. Similarly, a_{23} is the entry in the second row and third column. In general, the first subscript i of a_{ij} denotes the row, while the second subscript j of a_{ij} denotes the column in which a_{ij} belongs.

For a general $m \times n$-matrix A, we use the notation

$$A = \begin{bmatrix} a_{11} & a_{12} & \cdots & a_{1j} & \cdots & a_{1n} \\ a_{21} & a_{22} & \cdots & a_{2j} & \cdots & a_{2n} \\ \cdot & \cdot & \cdots & \cdot & \cdots & \cdot \\ \cdot & \cdot & \cdots & \cdot & \cdots & \cdot \\ \cdot & \cdot & \cdots & \cdot & \cdots & \cdot \\ a_{i1} & a_{i2} & \cdots & a_{ij} & \cdots & a_{in} \\ \cdot & \cdot & \cdots & \cdot & \cdots & \cdot \\ \cdot & \cdot & \cdots & \cdot & \cdots & \cdot \\ \cdot & \cdot & \cdots & \cdot & \cdots & \cdot \\ a_{m1} & a_{m2} & \cdots & a_{mj} & \cdots & a_{mn} \end{bmatrix}.$$

The element a_{ij} is in the ith row,

$$R_i = [a_{i1} \quad a_{i2} \quad \cdots \quad a_{ij} \quad \cdots \quad a_{in}]$$

and in the jth column,

$$C_j = \begin{bmatrix} a_{1j} \\ a_{2j} \\ \cdot \\ \cdot \\ \cdot \\ a_{ij} \\ \cdot \\ \cdot \\ \cdot \\ a_{mj} \end{bmatrix}$$

of A. The element a_{ij} is called the **ijth entry** or the **(i, j)th entry** of A. This suggests the following abbreviation for the matrix:

$$A = [a_{ij}], \quad i = 1, \ldots, m; \quad j = 1, \ldots, n$$

or simply the notation

$$A = [a_{ij}].$$

This notation should be viewed as the "general form" of the entries of the matrix. For example, as above,

$$[a_{ij}] = \begin{bmatrix} a_{11} & a_{12} & \cdots & a_{1n} \\ a_{21} & a_{22} & \cdots & a_{2n} \\ \vdots & \vdots & & \vdots \\ a_{m1} & a_{m2} & \cdots & a_{mn} \end{bmatrix}.$$

As further examples,

$$[3 + \pi a_{ij}] = \begin{bmatrix} 3 + \pi a_{11} & 3 + \pi a_{12} & \cdots & 3 + \pi a_{1n} \\ 3 + \pi a_{21} & 3 + \pi a_{22} & \cdots & 3 + \pi a_{2n} \\ \cdots & \cdots & \cdots & \cdots \\ 3 + \pi a_{m1} & 3 + \pi a_{m2} & \cdots & 3 + \pi a_{mn} \end{bmatrix}$$

$$[\sin a_{ij}] = \begin{bmatrix} \sin a_{11} & \sin a_{12} & \cdots & \sin a_{1n} \\ \sin a_{21} & \sin a_{22} & \cdots & \sin a_{2n} \\ \cdots & \cdots & \cdots & \cdots \\ \sin a_{m1} & \sin a_{m2} & \cdots & \sin a_{mn} \end{bmatrix}.$$

If the dimension of the matrix A is known in advance, we simply omit writing the range of the subscripts i and j. When i and j have the same range, the matrix is square. In this case we often designate this situation by writing

$$A = [a_{ij}]; \qquad i, j = 1, \ldots, n.$$

The elements $a_{11}, a_{22}, \ldots, a_{nn}$ form the **main diagonal** of a **square** matrix $A = [a_{ij}]$. In particular, a_{11} and a_{22} are the main diagonal elements of the first of the following two matrices:

$$A = \begin{bmatrix} a_{11} & a_{12} \\ a_{21} & a_{22} \end{bmatrix}, \qquad B = \begin{bmatrix} b_{11} & b_{12} & b_{13} \\ b_{21} & b_{22} & b_{23} \\ b_{31} & b_{32} & b_{33} \end{bmatrix}.$$

The entries b_{11}, b_{22}, and b_{33} are the main diagonal elements of B.

■ **EXAMPLE 2** _____

Consider the 6×6-matrix given by

$$A = [a_{ij}] = i + j, \qquad i, j = 1, 2, \ldots, 6.$$

We have

$$A = \begin{bmatrix} 2 & 3 & 4 & 5 & 6 & 7 \\ 3 & 4 & 5 & 6 & 7 & 8 \\ 4 & 5 & 6 & 7 & 8 & 9 \\ 5 & 6 & 7 & 8 & 9 & 10 \\ 6 & 7 & 8 & 9 & 10 & 11 \\ 7 & 8 & 9 & 10 & 11 & 12 \end{bmatrix}.$$

A dice player should be interested in this matrix since each entry $a_{ij} = i + j$ represents a possible sum when rolling dice. For example, the 3 in the $(1, 2)$ position represents the sum 3 when the first die shows 1 and the second die shows 2. Similarly, the 3 in the $(2, 1)$ position is also the sum 3 when the first die shows 2 and the second die shows 1. One sees that the sum 7 (which appears in the diagonal from upper right to lower left of *A*) occurs more often than any other sum. In fact, 7 occurs six times out of the possible 36 sums. It follows that the probability that a sum of 7 will occur when rolling dice is 6/36 or 1/6. By looking at other diagonals of the matrix *A*, one easily determines the probability of other sums. The main diagonal of *A* represents those sums for which each die has the same value. A dice player refers to such sums as doubles. One sees that the probability of rolling doubles is also 6/36 or 1/6.

In an analogous approach to the algebra of real and complex numbers, we will be studying the algebra of matrices. We can think of matrices as being a generalization of ordinary numbers, and we will begin to see how our *algebra of real numbers* works when we simply replace the real numbers with matrices. In our real number system, we can add, subtract, multiply, and divide (except by 0) any two numbers. As we shall see, we can do the same with matrices, under certain restrictions. We will also investigate the validity of the *operational laws,* such as associativity, commutativity, and distributivity of matrices, as well as seeking solutions of *matrix equations,* such as $AX = B$ (where *A, X,* and *B* are all matrices). All of the above notions are quite familiar when dealing with real or complex numbers (1 × 1-matrices), but new questions arise when one begins to study the matrix number system.

You are familiar with the fact that in a number system, the numbers 0 and 1 play a fundamental role. In matrix theory, a matrix, all of whose entries are zero, is called a *zero matrix.* It is designated by 0 irrespective of its size. Thus

$$0 = [0 \quad 0], \qquad 0 = \begin{bmatrix} 0 & 0 & 0 \\ 0 & 0 & 0 \end{bmatrix}, \qquad 0 = \begin{bmatrix} 0 & 0 \\ 0 & 0 \\ 0 & 0 \end{bmatrix},$$

and so on. In each case, the 0 on the left represents the *zero matrix.* There is obviously a unique zero matrix of every size.

As we shall see, *square* matrices of the form

$$\begin{bmatrix} 1 & 0 \\ 0 & 1 \end{bmatrix}, \qquad \begin{bmatrix} 1 & 0 & 0 \\ 0 & 1 & 0 \\ 0 & 0 & 1 \end{bmatrix}, \qquad \begin{bmatrix} 1 & 0 & 0 & 0 \\ 0 & 1 & 0 & 0 \\ 0 & 0 & 1 & 0 \\ 0 & 0 & 0 & 1 \end{bmatrix}$$

play a role in matrix theory similar to the number 1. These matrices are called *identity matrices* and are usually designated by the capital letter *I*, irrespective of their size. When we wish to designate the size of the identity

I, we use a subscript. Thus I_2 is the 2×2 identity, I_3 is the 3×3 identity, and so on.

We shall have occasion to use submatrices of a matrix. A *submatrix* is a matrix within a matrix. It can be obtained by deleting rows and columns. As an example, consider the matrix

$$A = \begin{bmatrix} 5 & -3 & 0 \\ 0 & 1 & 0 \\ 2 & 4 & 6 \end{bmatrix}.$$

If we delete row 3 and column 3 from the matrix A, we are left with the matrix

$$B = \begin{bmatrix} 5 & -3 \\ 0 & 1 \end{bmatrix}.$$

The matrix B is a submatrix of the matrix A. Similarly, if we delete column 2 of the matrix A, we are left with the matrix

$$C = \begin{bmatrix} 5 & 0 \\ 0 & 0 \\ 2 & 6 \end{bmatrix}.$$

The matrix C is also a submatrix of the matrix A. The matrices B and C are just two of many submatrices of A.

As a further example, consider the matrix

$$A = \begin{bmatrix} 0 & 0 & -1 & 5 & 1 & -2 \\ 1 & 4 & 0 & 1 & -1 & 2 \\ 0 & -7 & 6 & 5 & -4 & -3 \\ 1 & 9 & 3 & -2 & 1 & -5 \end{bmatrix}.$$

The matrices

$$M = \begin{bmatrix} 4 & 1 \\ 9 & -2 \end{bmatrix}, \quad N = \begin{bmatrix} -1 & 2 \\ 1 & -5 \end{bmatrix}, \quad P = \begin{bmatrix} 0 & 5 & 1 \\ 9 & -2 & 1 \end{bmatrix},$$

$$R = \begin{bmatrix} 0 & 5 & 1 \\ 4 & 1 & -1 \\ 9 & -2 & 1 \end{bmatrix}, \quad S = \begin{bmatrix} 0 & -1 & 5 & 1 \\ 4 & 0 & 1 & -1 \\ -7 & 6 & 5 & -4 \\ 9 & 3 & -2 & 1 \end{bmatrix}$$

are submatrices of A. For example one sees that the matrix

$$M = \begin{bmatrix} 4 & 1 \\ 9 & -2 \end{bmatrix}$$

is obtained from A by deleting rows 1 and 3 and columns 1, 3, 5, and 6 of A. The reader should verify the other examples.

Unless otherwise stated or implied, it will be understood that the entries in our matrices are real numbers. Such matrices are called **real matrices.** Occasionally, we shall consider **complex matrices,** that is, matrices whose entries are complex numbers. Throughout the book we often use the term **scalar** instead of **real number.** This is to facilitate the generalization of the concepts developed in this book to matrices other than those with real entries. For example, when we permit the entries to be complex numbers, the scalars are complex numbers.

Finally, we make use of the concept of **block matrices.** These are matrices whose entries are matrices of appropriate sizes. For example, consider the matrix

$$M = \left[\begin{array}{ccc:cc} 2 & 3 & -1 & 4 & 5 \\ 0 & 3 & 0 & 1 & 1 \\ \hdashline 5 & 6 & -1 & 7 & -2 \\ 3 & 4 & 6 & 7 & 5 \end{array}\right].$$

Draw vertical and horizontal lines as indicated. These lines break the matrix up into four submatrices

$$A = \begin{bmatrix} 2 & 3 & -1 \\ 0 & 3 & 0 \end{bmatrix}, \quad B = \begin{bmatrix} 4 & 5 \\ 1 & 1 \end{bmatrix}, \quad C = \begin{bmatrix} 5 & 6 & -1 \\ 3 & 4 & 6 \end{bmatrix},$$

$$D = \begin{bmatrix} 7 & -2 \\ 7 & 5 \end{bmatrix}.$$

It follows that M has the block representation

$$M = \begin{bmatrix} A & B \\ C & D \end{bmatrix}$$

where A, B, C, D are again matrices.

Of course the matrix M can be divided into blocks in other ways, such as

$$M = \left[\begin{array}{ccc:cc} 2 & 3 & -1 & 4 & 5 \\ 0 & 3 & 0 & 1 & 1 \\ 5 & 6 & -1 & 7 & -2 \\ \hdashline 3 & 4 & 6 & 7 & 5 \end{array}\right].$$

In this case M is still of the form

$$M = \begin{bmatrix} A & B \\ C & D \end{bmatrix}$$

where now

$$A = \begin{bmatrix} 2 & 3 & -1 \\ 0 & 3 & 0 \\ 5 & 6 & -1 \end{bmatrix}, \qquad B = \begin{bmatrix} 4 & 5 \\ 1 & 1 \\ 7 & -2 \end{bmatrix},$$

$$C = \begin{bmatrix} 3 & 4 & 6 \end{bmatrix}, \qquad D = \begin{bmatrix} 7 & 5 \end{bmatrix}$$

Still another division of M into blocks is

$$M = \left[\begin{array}{ccc|cc} 2 & 3 & -1 & 4 & 5 \\ \hline 0 & 3 & 0 & 1 & 1 \\ 5 & 6 & -1 & 7 & -2 \\ \hline 3 & 4 & 6 & 7 & 5 \end{array} \right].$$

Then M is in the block form,

$$M = \begin{bmatrix} A_1 & A_2 & A_3 \\ B_1 & B_2 & B_3 \\ C_1 & C_2 & C_3 \end{bmatrix},$$

where

$$A_1 = [2], \qquad A_2 = \begin{bmatrix} 3 & -1 \end{bmatrix}, \qquad A_3 = \begin{bmatrix} 4 & 5 \end{bmatrix}$$

$$B_1 = \begin{bmatrix} 0 \\ 5 \end{bmatrix}, \qquad B_2 = \begin{bmatrix} 3 & 0 \\ 6 & -1 \end{bmatrix}, \qquad B_3 = \begin{bmatrix} 1 & 1 \\ 7 & -2 \end{bmatrix}$$

$$C_1 = [3], \qquad C_2 = \begin{bmatrix} 4 & 6 \end{bmatrix}, \qquad C_3 = \begin{bmatrix} 7 & 5 \end{bmatrix}.$$

If we draw lines between each row and each column,

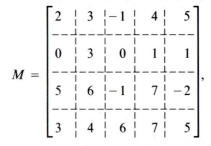

the blocks are 1×1-matrices and so are the entries of the matrix itself.

Thus a matrix can be considered to be a block matrix whose blocks are 1×1-matrices.

Another example of a block matrix is

$$J = \begin{bmatrix} J_1 & 0 & 0 \\ 0 & J_2 & 0 \\ 0 & 0 & J_3 \end{bmatrix}$$

where

$$J_1 = \begin{bmatrix} 2 & 1 & 0 \\ 0 & 2 & 1 \\ 0 & 0 & 2 \end{bmatrix}, \qquad J_2 = \begin{bmatrix} 2 & 1 \\ 0 & 2 \end{bmatrix}, \qquad J_3 = [2]$$

The zero-blocks in this matrix have various sizes. What are their sizes? A matrix of this type is called a ***block diagonal matrix***. We shall encounter matrices of this type in Section 6.8.

Exercises

1. The size of matrix A in the following list is 2×3. What are the sizes of the other matrices?

$$A = \begin{bmatrix} 2 & 5 & 6 \\ 0 & -1 & 7 \end{bmatrix}, \qquad B = \begin{bmatrix} 4 & 6 & 8 & 7 \\ 5 & 7 & 9 & 1 \end{bmatrix}, \qquad C = \begin{bmatrix} 1 & -1 \\ 2 & -2 \\ 3 & -3 \end{bmatrix},$$

$$D = \begin{bmatrix} 1 & 4 & 5 & -6 \\ -2 & 5 & -4 & 7 \\ 3 & 6 & 0 & 8 \end{bmatrix}, \qquad E = \begin{bmatrix} 5 \\ 6 \\ 7 \end{bmatrix}, \qquad F = [0 \quad 1 \quad 2 \quad 3 \quad 4],$$

$$G = \begin{bmatrix} g_{11} & g_{12} & \cdots & g_{19} \\ g_{21} & g_{22} & \cdots & g_{29} \\ \cdots & \cdots & \cdots & \cdots \\ g_{81} & g_{82} & \cdots & g_{89} \end{bmatrix}, \qquad H = \begin{bmatrix} h_1 & k_1 & m_1 & n_1 \\ h_2 & k_2 & m_2 & n_2 \\ h_3 & k_3 & m_3 & n_3 \end{bmatrix}.$$

2. Referring to Exercise 1, what is the second column of each of the matrices A, B, C, F, and H? What is the third row of each of the matrices D, E, G, and H? For the matrices A, B, C, and D in Exercise 1, write their rows R_i and their columns C_j.

3. Referring to the matrices in Exercise 1:

   (a) Construct a new matrix from A by adding 10 to its (1, 3)-entry.

   (b) Construct a second new matrix from A by multiplying each entry in its second column by 2.

   (c) Construct a new matrix from C by adding -1 to each entry in its third row.

   (d) Construct a second new matrix from C by multiplying each entry in its third row by -1.

(e) Construct a third new matrix from C by adding each entry in its first column to the corresponding entry in its second column. Note that the entries in the second column of the new matrix are all 0.

(f) Construct the matrix $[C \quad E]$. It has three columns.

4. Expressing the matrix D in Exercise 1 in the form

$$D = [d_{ij}], \quad i = 1, 2, 3; \quad d = 1, 2, 3, 4,$$

give the values of the elements (entries) d_{12}, d_{23}, d_{34}, and d_{32}. For what values of i and j is $d_{ij} = 0$?

5. Referring to the matrices in Exercise 1:

(a) Find the sum of the entries in A.

(b) Find the sum of the absolute values of the entries in A.

(c) Find the sum of the entries in the second row of D.

(d) Find the sum of the absolute values of the entries in the second row of D.

(e) Find the sum of the entries in the first and last columns of B.

(f) Find the sum of the (1, 2)-, (2, 3)-, and (3, 3)-entries in D.

6. What is the

(a) Third component of the row vector $[1 \quad 2 \quad -4 \quad 3]$?

(b) First component of the column vector $\begin{bmatrix} 6 \\ 27 \\ -3 \\ \sqrt{5} \end{bmatrix}$?

(c) Second component of the row vector $[\sqrt{10} \quad \pi \quad 3.1415 \quad 22/7]$?

7. Consider the matrices

$$L = \begin{bmatrix} 1 & 0 & 1 \\ 0 & 1 & 1 \end{bmatrix}, \quad M = \begin{bmatrix} 1 & b & 1 \\ 0 & 1 & 1 \end{bmatrix}, \quad N = \begin{bmatrix} 1 & b & 1 \\ a & 1 & c \end{bmatrix}$$

(a) Determine b so that L and M are equal.

(b) Determine a, b, and c so that L and N are equal.

(c) Find the sum of the entries in L.

(d) Choose b in M so that the sum of its entries are 0.

(e) Choose b in M so that its last two columns are identical.

8. Suppose that the entry a_{ij} in the matrix $A = [a_{ij}]$ is given by the formula $a_{ij} = 2i + 3j$. Write out matrix A in full when i has the values 1, 2, and 3 and j has the values 1 and 2.

9. Suppose that the entry b_{ij} in the matrix $B = [b_{ij}]$ is given by the formula

$$b_{ij} = 5i^2 + 3ij + 4j^2 \quad \text{so that} \quad B = [5i^2 + 3ij + 4j^2]$$

Write out the matrix B in full when i has the values 1 and 2 and j has the values 1, 2, 3, and 4.

10. Suppose that the entry c_{ij} in $C = [c_{ij}]$ is given by the formula

$$c_{ij} = \sin i\pi + \cos j\pi + ij\sqrt{10} \quad \text{so that} \quad C = [\sin i\pi + \cos j\pi + ij\sqrt{10}]$$

Write out the matrix C in full when $i = 1, 2, 3$ and $j = 1, 2, 3, 4$.

11. The first index is the row index and the second index is the column index. Write out the following in full, as in Exercises 8, 9, and 10.

(a) $[8 + 3a_{ij}]$, $i = 1, 2, 3$; $j = 1, 2$

(b) $[b_{ij}]$, $j = 1, 2, 3$; $i = 1, 2$

(c) $[i^2 + 2i + 1]$, $i = 1, 2, 3, 4$; $j = 1, 2, 3$

(d) $[j]$, $i = 1, 2, 3$; $j = 1, 2, 3$

12. The square matrix $S_n = [(-1)^{i+j}]$ $(i, j = 1, \ldots, n)$ is called the **sign matrix** for an $n \times n$-matrix. Write out the matrices S_2, S_3, and S_4 in full.

13. The symbol δ is the lower case Greek letter delta. In mathematical and scientific literature, one often encounters the symbol δ_{ij}, called the **Kronecker delta.** It is defined by the relations

$$\delta_{ij} = \begin{cases} 0 & \text{if } i \neq j \\ 1 & \text{if } i = j \end{cases}.$$

Write out the matrix

$$[\delta_{ij}] \quad (i, j = 1, 2, 3)$$

in full. What is this matrix called? How is it related to the matrix

$$[\delta_{ij}] \quad (i = 1, 2, 3; j = 1, 2, 3, 4, 5)?$$

14. What is the main diagonal of the following matrices?

$$A = \begin{bmatrix} 2 & 5 & 6 \\ 0 & -1 & 7 \\ 5 & 6 & 7 \end{bmatrix}, \quad B = \begin{bmatrix} 4 & 6 & 8 & 7 \\ 5 & 7 & 9 & 1 \\ 1 & 4 & 5 & -6 \\ -2 & 5 & -4 & 7 \end{bmatrix}, \quad C = \begin{bmatrix} 1 & -1 \\ -1 & 1 \end{bmatrix},$$

$$D = \begin{bmatrix} 1 & -1 & 5 \\ 2 & -2 & 6 \\ 3 & -3 & 7 \end{bmatrix}, \quad E = [52], \quad F = \begin{bmatrix} a_{11} & a_{12} & a_{13} \\ a_{21} & a_{22} & a_{23} \\ a_{31} & a_{32} & a_{33} \end{bmatrix}.$$

Note that the concept of main diagonal makes sense only for **square** matrices.

15. Referring to the square matrices in Exercise 14:

(a) Find the sum of the main diagonal entries of A.

(b) Find the sum of the main diagonal entries of B.

(c) Construct a new matrix from C by subtracting 1 from each main diagonal entry.

(d) Construct a new matrix from D by subtracting 7 from each main diagonal entry.

(e) First subtract x from each diagonal entry of A. Then determine x so that the sum of the main diagonal entries is 0.

16. The matrix

$$H_n = \left[\frac{1}{i+j-1}\right], \qquad i,j = 1, 2, \ldots, n,$$

arises in applications and is called the **Hilbert matrix** of order n. It is also used as a test matrix for certain numerical algorithms involving square matrices. For $n = 4$ we have

$$H_4 = \begin{bmatrix} 1 & 1/2 & 1/3 & 1/4 \\ 1/2 & 1/3 & 1/4 & 1/5 \\ 1/3 & 1/4 & 1/5 & 1/6 \\ 1/4 & 1/5 & 1/6 & 1/7 \end{bmatrix}.$$

(a) What is the element in the second row and third column? In the fourth row and third column?

(b) Specify the rows and columns that contain the entry 1/5.

(c) Show that the (i, j)th element is equal to the (j, i)th element.

(d) What are the main diagonal elements of H_4?

(e) Show that H_3 is obtained from H_4 by deleting the last row and column of H_4.

(f) Write out the matrix H_5 in full.

17. *Without* writing out the full matrix, write the main diagonal of:

(a) $[2i + 3j]$, $i, j = 1, 2, 3$

(b) $[\sin i\pi + \cos j\pi]$, $i, j = 1, 2, 3, 4, 5$

(c) $[\delta_{ij} + ij]$, $i, j = 1, 2, 3, 4$

(d) $[k(-1)^i + (-1)^j]$, $i, j = 1, 2, 3, 4, 5$

(e) $[\sin (i\pi \cos j\pi]$, $i, j = 1, 2, 3, \ldots, 17$

18. The identity matrix is an example of a **diagonal matrix**. It has *nonzero* entries only on the main diagonal. Other examples of diagonal matrices are

$$U = \begin{bmatrix} 1 & 0 & 0 \\ 0 & 2 & 0 \\ 0 & 0 & 3 \end{bmatrix}, \quad V = \begin{bmatrix} \pi & 0 \\ 0 & \sqrt{10} \end{bmatrix}, \quad W = [14\tfrac{1}{2}],$$

$$X = \begin{bmatrix} -1 & 0 & 0 \\ 0 & 1 & 0 \\ 0 & 0 & -1 \end{bmatrix}, \quad Y = \begin{bmatrix} 2 & 0 & 0 & 0 \\ 0 & 3 & 0 & 0 \\ 0 & 0 & 5 & 0 \\ 0 & 0 & 0 & 0 \end{bmatrix}.$$

A shorthand abbreviation for a diagonal matrix $A = [a_{ij}]$ is

$$A = [a_{ij}] = \text{diag } \{a_{11}, a_{22}, \ldots, a_{nn}\}.$$

For example, matrix U above is written in this notation as

$$U = \text{diag } \{1, 2, 3\}.$$

Write the other examples above in this shorthand notation.

19. Consider the matrix $M = [U \ \ V]$. Write M in full for each of the following cases:

(a) $U = \begin{bmatrix} 1 & 4 \\ 2 & 5 \\ 3 & 6 \end{bmatrix}$, $V = \begin{bmatrix} 1 & 1 & 1 \\ 1 & 0 & 1 \\ 0 & 0 & 1 \end{bmatrix}$

(b) $U = \begin{bmatrix} 1 & 4 & 0 \\ 2 & 5 & 1 \\ 3 & 6 & 0 \end{bmatrix}$, $V = \begin{bmatrix} 1 & 1 \\ 0 & 1 \\ 0 & 1 \end{bmatrix}$

(c) $U = \begin{bmatrix} 1 \\ 2 \\ 3 \end{bmatrix}$, $V = \begin{bmatrix} 4 & 0 & 1 & 1 \\ 5 & 1 & 0 & 1 \\ 6 & 0 & 0 & 1 \end{bmatrix}$

20. Consider the block matrix

$$M = \begin{bmatrix} A & B \\ C & D \end{bmatrix}.$$

Here A, B, C, and D are submatrices of M. Write M in full in each of the following cases.

(a) $A = \begin{bmatrix} 1 & -1 \\ 2 & 3 \end{bmatrix}$, $B = \begin{bmatrix} 1 & 0 \\ -1 & 4 \end{bmatrix}$, $C = \begin{bmatrix} 0 & 0 \\ 0 & 0 \end{bmatrix}$, $D = \begin{bmatrix} 1 & 0 \\ 0 & 1 \end{bmatrix}$

(b) $A = \begin{bmatrix} 1 & -1 \\ 2 & 3 \end{bmatrix}$, $B = \begin{bmatrix} 1 \\ -1 \end{bmatrix}$, $C = [0 \quad 1]$, $D = [5]$

(c) $A = \begin{bmatrix} 1 \\ 2 \end{bmatrix}$, $B = \begin{bmatrix} -1 & 1 & 0 \\ 3 & -1 & 4 \end{bmatrix}$, $C = \begin{bmatrix} 0 \\ 0 \end{bmatrix}$, $D = \begin{bmatrix} 0 & 1 & 0 \\ 0 & 0 & 1 \end{bmatrix}$

(d) $A = [1]$, $B = [1 - 1]$, $C = \begin{bmatrix} 2 \\ 0 \\ 0 \end{bmatrix}$, $D = \begin{bmatrix} 3 & -1 \\ 0 & 1 \\ 0 & 0 \end{bmatrix}$

Why does A have the same number of rows as B and the same number of columns as C?

1.2   ADDITION, SUBTRACTION, AND SCALAR MULTIPLICATION

In this section we describe some of the basic operations on matrices. We shall see that matrices of the *same size* can be added or subtracted, while multiplying a matrix by a scalar is always valid. The operations of addition and subtraction are undefined for matrices of different sizes.

The simplest of these operations is *scalar multiplication*, which is defined as follows.

To multiply a matrix A by a scalar (number) s, we simply multiply each entry of A by s. The resulting matrix is designated by sA or As, whichever is convenient.
Briefly,

$$s[a_{ij}] = [sa_{ij}], \qquad [a_{ij}]s = [a_{ij}s] = [sa_{ij}].$$

For example, when $s = 6$ and

$$A = \begin{bmatrix} 2 & -1 & 3 \\ 4 & 0 & 6 \end{bmatrix},$$

the matrix $6A$ (and hence also $A6$) is

$$6A = \begin{bmatrix} 6(2) & 6(-1) & 6(3) \\ 6(4) & 6(0) & 6(6) \end{bmatrix} = \begin{bmatrix} 12 & -6 & 18 \\ 24 & 0 & 36 \end{bmatrix}.$$

Similarly,

$$(-6)A = \begin{bmatrix} -12 & 6 & -18 \\ -24 & 0 & -36 \end{bmatrix}, \qquad \frac{1}{2}A = \begin{bmatrix} 1 & -1/2 & 3/2 \\ 2 & 0 & 3 \end{bmatrix}.$$

Of course,

$$(0)A = \begin{bmatrix} 0 & 0 & 0 \\ 0 & 0 & 0 \end{bmatrix} = 0, \qquad (-1)A = \begin{bmatrix} -2 & 1 & -3 \\ -4 & 0 & -6 \end{bmatrix}.$$

For the general matrix $A = [a_{ij}]$ we have

$$(0)A = [0a_{ij}] = 0, \qquad 1A = [1a_{ij}] = A, \qquad (-1)A = [-a_{ij}].$$

The matrix $-A = (-1)A$ is called the **negative** of A. For arbitrary scalars r and s, we have the rule

$$r(sA) = (rs)A = s(rA).$$

This follows because

$$sA = [sa_{ij}], \qquad r(sA) = [rsa_{ij}] = (rs)A = s(rA).$$

For example, when $r = 2$, $s = 3$, and

$$A = \begin{bmatrix} 2 & -1 & 3 \\ 4 & 0 & 6 \end{bmatrix},$$

we have

$$3A = \begin{bmatrix} 6 & -3 & 9 \\ 12 & 0 & 18 \end{bmatrix}, \qquad 2(3A) = \begin{bmatrix} 12 & -6 & 18 \\ 24 & 0 & 36 \end{bmatrix} = 6A.$$

We proceed to define addition and subtraction of matrices. We begin with an example. In order to add two matrices of the same size such as

$$A = \begin{bmatrix} 2 & -1 & 3 \\ 4 & 0 & 6 \end{bmatrix}, \qquad B = \begin{bmatrix} 7 & 5 & -4 \\ 9 & -8 & 0 \end{bmatrix},$$

we add corresponding elements and obtain the matrix

$$A + B = \begin{bmatrix} 2+7 & -1+5 & 3+(-4) \\ 4+9 & 0+(-8) & 6+0 \end{bmatrix} = \begin{bmatrix} 9 & 4 & -1 \\ 13 & -8 & 6 \end{bmatrix}.$$

To obtain the difference $A - B$, we subtract corresponding elements in the given order as

follows:

$$A - B = \begin{bmatrix} 2-7 & -1-5 & 3-(-4) \\ 4-9 & 0-(-8) & 6-0 \end{bmatrix} = \begin{bmatrix} -5 & -6 & 7 \\ -5 & 8 & 6 \end{bmatrix}.$$

Observe that

$$-B = \begin{bmatrix} -7 & -5 & 4 \\ -9 & 8 & 0 \end{bmatrix}, \qquad A + (-B) = \begin{bmatrix} -5 & -6 & 7 \\ -5 & 8 & 6 \end{bmatrix}.$$

Consequently,

$$A - B = A + (-B) = A + (-1)B.$$

This example illustrates the following definition of addition and subtraction of matrices of the same size.

Let $A = [a_{ij}]$ and $B = [b_{ij}]$ be $m \times n$-matrices. The **sum** $A + B$ is defined to be the $m \times n$-matrix

$$A + B = [a_{ij} + b_{ij}]$$

and is obtained by adding corresponding elements a_{ij} and b_{ij} of A and B. Similarly, the **difference** $A - B$ is the $m \times n$-matrix given by the formula

$$A - B = [a_{ij} - b_{ij}].$$

As noted in our example, we have

$$A - B = A + (-B) = A + (-1)B.$$

Since the commutative law holds for addition of real numbers,

$$a_{ij} + b_{ij} = b_{ij} + a_{ij},$$

we have the *commutative law for matrix addition*:

$$A + B = B + A.$$

If $C = [c_{ij}]$ is a third $m \times n$-matrix, the associative law for the addition of real numbers,

$$(a_{ij} + b_{ij}) + c_{ij} = a_{ij} + (b_{ij} + c_{ij}) = a_{ij} + b_{ij} + c_{ij}$$

gives us the *associative law of matrix addition*:

$$(A + B) + C = A + (B + C) = A + B + C.$$

Just as we can solve the scalar equation

$$15 + x = 27$$

for x, one can solve the *matrix equation*

$$A + X = B$$

for the matrix X. Analogously, we obtain the unique solution

$$X = B - A.$$

We summarize the *basic properties of addition and scalar multiplication* of $m \times n$-matrices $A, B, C, \ldots$. Here r and s are arbitrary scalars (numbers). The properties of the difference $A - B = A + (-1)B$ follow from these.

1. $A + B = B + A$ (addition is commutative).
2. $(A + B) + C = A + (B + C)$ (addition is associative).
3. $A + 0 = 0 + A = A$, where 0 is the $m \times n$ zero matrix.
4. The equation $A + X = B$ has a unique solution, namely, $X = B - A$.
5. $sA = As$, $r(sA) = (rs)A$, $1A = A$, $0A = 0$, $(-1)A = -A$.
6. $(r + s)A = rA + sA$, $s(A + B) = sA + sB$.

We have already established properties 1, 2, and 5. The proofs of the others will be outlined in the Exercises.

We can verify the first relation in property 6 for 2×2-matrices by making the following computations. Let $A = \begin{bmatrix} a & b \\ c & d \end{bmatrix}$. Then

$$rA + sA = r\begin{bmatrix} a & b \\ c & d \end{bmatrix} + s\begin{bmatrix} a & b \\ c & d \end{bmatrix} = \begin{bmatrix} ra + sa & rb + sb \\ rc + sc & rd + sd \end{bmatrix}$$

$$= (r + s)\begin{bmatrix} a & b \\ c & d \end{bmatrix} = (r + s)A.$$

Continuing, if A, B, and C are matrices of the same size, and if the matrix X can be expressed in the form

$$X = aA + bB + cC,$$

where a b, and c are scalars, X is called a **linear combination** of the matrices A, B, and C. From the identity

$$\begin{bmatrix} a & b \\ c & d \end{bmatrix} = a\begin{bmatrix} 1 & 0 \\ 0 & 0 \end{bmatrix} + b\begin{bmatrix} 0 & 1 \\ 0 & 0 \end{bmatrix} + c\begin{bmatrix} 0 & 0 \\ 1 & 0 \end{bmatrix} + d\begin{bmatrix} 0 & 0 \\ 0 & 1 \end{bmatrix},$$

we conclude that every 2×2-matrix is a linear combination of the matrices

$$\begin{bmatrix} 1 & 0 \\ 0 & 0 \end{bmatrix}, \quad \begin{bmatrix} 0 & 1 \\ 0 & 0 \end{bmatrix}, \quad \begin{bmatrix} 0 & 0 \\ 1 & 0 \end{bmatrix}, \quad \begin{bmatrix} 0 & 0 \\ 0 & 1 \end{bmatrix}.$$

In particular,

$$\begin{bmatrix} 1 & -2 \\ 3 & 7 \end{bmatrix} = 1\begin{bmatrix} 1 & 0 \\ 0 & 0 \end{bmatrix} - 2\begin{bmatrix} 0 & 1 \\ 0 & 0 \end{bmatrix} + 3\begin{bmatrix} 0 & 0 \\ 1 & 0 \end{bmatrix} + 7\begin{bmatrix} 0 & 0 \\ 0 & 1 \end{bmatrix}.$$

Similarly, the relation

$$\begin{bmatrix} x \\ y \\ z \end{bmatrix} = x\begin{bmatrix} 1 \\ 0 \\ 0 \end{bmatrix} + y\begin{bmatrix} 0 \\ 1 \\ 0 \end{bmatrix} + z\begin{bmatrix} 0 \\ 0 \\ 1 \end{bmatrix}$$

tells us that every three-dimensional column vector is a linear combination of the coordinate column vectors

$$\begin{bmatrix} 1 \\ 0 \\ 0 \end{bmatrix}, \quad \begin{bmatrix} 0 \\ 1 \\ 0 \end{bmatrix}, \quad \begin{bmatrix} 0 \\ 0 \\ 1 \end{bmatrix}.$$

■ **EXAMPLE 1**

To show that the matrix

$$\begin{bmatrix} 2 & -2 \\ 3 & 6 \end{bmatrix}$$

is a linear combination of the matrices

$$\begin{bmatrix} 1 & -1 \\ 0 & 0 \end{bmatrix}, \quad \begin{bmatrix} 0 & 0 \\ 1 & 2 \end{bmatrix},$$

we form the equation

$$\begin{bmatrix} 2 & -2 \\ 3 & 6 \end{bmatrix} = a\begin{bmatrix} 1 & -1 \\ 0 & 0 \end{bmatrix} + b\begin{bmatrix} 0 & 0 \\ 1 & 2 \end{bmatrix} = \begin{bmatrix} a & -a \\ b & 2b \end{bmatrix}.$$

From this equation we conclude that $a = 2$ and $b = 3$, so that

$$\begin{bmatrix} 2 & -2 \\ 3 & 6 \end{bmatrix} = 2\begin{bmatrix} 1 & -1 \\ 0 & 0 \end{bmatrix} + 3\begin{bmatrix} 0 & 0 \\ 1 & 2 \end{bmatrix}.$$

■ **EXAMPLE 2**

In this example we will use the foregoing concepts in a slightly different way. We look for a number k so that the matrix

$$X = \begin{bmatrix} 3 & -2 & 3 \\ -2 & k & -2 \\ 3 & -2 & 3 \end{bmatrix}$$

can be written as a linear combination of the matrices

$$A = \begin{bmatrix} 1 & 0 & 1 \\ 0 & 2 & 0 \\ 1 & 0 & 1 \end{bmatrix}, \quad B = \begin{bmatrix} 0 & 1 & 0 \\ 1 & -1 & 1 \\ 0 & 1 & 0 \end{bmatrix}.$$

We work backwards. If X can be written as a linear combination of A and B, there are numbers a and b such that $X = a A + b B$, that is,

$$\begin{bmatrix} 3 & -2 & 3 \\ -2 & k & -2 \\ 3 & -2 & 3 \end{bmatrix} = a\begin{bmatrix} 1 & 0 & 1 \\ 0 & 2 & 0 \\ 1 & 0 & 1 \end{bmatrix} + b\begin{bmatrix} 0 & 1 & 0 \\ 1 & -1 & 1 \\ 0 & 1 & 0 \end{bmatrix}.$$

Combining the matrices on the right side of this equation, we have

$$\begin{bmatrix} 3 & -2 & 3 \\ -2 & k & -2 \\ 3 & -2 & 3 \end{bmatrix} = \begin{bmatrix} a & b & a \\ b & 2a - b & b \\ a & b & a \end{bmatrix}.$$

We observe that this equation is satisfied when

$$a = 3, \quad b = -2, \quad \text{and} \quad k = 2a - b = 6 + 2 = 8.$$

So, when $k = 8$, X is the linear combination $X = 3A - 2B$ of the matrices A and B.

The reader should be aware that given a collection of matrices $\{A_1, A_2, \ldots, A_n\}$ and given a matrix B, it is *not* always possible to express B as a linear combination of the A_i's. We illustrate this in a final example.

■ **EXAMPLE 3** ────────────────────────────────────

We will show that the matrix

$$X = \begin{bmatrix} 1 & 2 \\ 3 & 4 \end{bmatrix}$$

cannot be written as a linear combination of the matrices

$$A = \begin{bmatrix} 1 & 0 \\ 0 & 1 \end{bmatrix}, \qquad B = \begin{bmatrix} 0 & -1 \\ 1 & 0 \end{bmatrix}.$$

For if it could, there would be numbers a and b such that $X = aA + bB$, that is,

$$\begin{bmatrix} 1 & 2 \\ 3 & 4 \end{bmatrix} = a \begin{bmatrix} 1 & 0 \\ 0 & 1 \end{bmatrix} + b \begin{bmatrix} 0 & -1 \\ 1 & 0 \end{bmatrix}.$$

This would imply that

$$\begin{bmatrix} 1 & 2 \\ 3 & 4 \end{bmatrix} = \begin{bmatrix} a & -b \\ b & a \end{bmatrix},$$

an obvious contradiction. (Why?)

Exercises

1. Determine which of the following matrices can be added and subtracted. Find their sum and difference.

$$A = \begin{bmatrix} 2 & 0 \\ -1 & 1 \end{bmatrix}, \quad B = [0 \;\; 0 \;\; 0], \quad C = \begin{bmatrix} 5 & -2 \\ 2 & 3 \end{bmatrix},$$

$$D = \begin{bmatrix} 1 & 0 \\ 0 & 1 \\ 1 & 0 \end{bmatrix}, \quad E = [5 \;\; 6 \;\; 7], \quad F = \begin{bmatrix} 0 & 2 \\ 2 & 0 \\ 0 & -3 \end{bmatrix}.$$

2. Referring to Exercise 1, write out the following matrices in full: $5A$, $4B$, $-3C$, $7D$, $2E$, and $-6F$.

3. Referring to Exercise 1, compute the following linear combinations: $5A - 3C$, $5A + 3C$, $4B + 2E$, and $7D - 6F$.

4. Find the sum and difference, where possible, of the following:

$$G = \begin{bmatrix} \pi & 16 & 42 \\ 5 & 2 & 3 \\ 1/2 & 1 & 3/2 \end{bmatrix}, \quad H = \begin{bmatrix} 1 & -1 & 1 \\ -1 & 1 & -1 \end{bmatrix}, \quad J = [13], \quad K = \begin{bmatrix} \cos^2\theta & -\sin^2\theta \\ -\sin^2\theta & \cos^2\theta \end{bmatrix},$$

$$L = \begin{bmatrix} -1 & 1 & -1 \\ 1 & -1 & 1 \end{bmatrix}, \quad M = [3\sqrt{42}], \quad N = \begin{bmatrix} \sin^2\theta & -\cos^2\theta \\ -\cos^2\theta & \sin^2\theta \end{bmatrix}.$$

5. Referring to Exercise 4, write out the following matrices in full:

   (a) $\frac{2}{3}G$ (b) $-7H$ (c) $^3\sqrt{42}\,J$ (d) $\frac{1}{\cos^2\theta}K$ (cos $\theta \neq 0$)

   (e) $\sqrt{10}\,L$ (f) $0M$ (g) $\frac{1}{\sin^2\theta}N$ (sin $\theta \neq 0$)

6. Given the matrices

$$A = \begin{bmatrix} 3 & -1 \\ 5 & 2 \end{bmatrix}, \quad B = \begin{bmatrix} 1 & 3 & 7 \\ 0 & 4 & 6 \\ 0 & 0 & 5 \end{bmatrix}, \quad C = [4].$$

   (a) Construct the matrices $A - 3I, B - 4I$, and $C - 7I$, where I is the appropriate identity matrix.

   (b) What is the size of I in each case?

7. Referring to Exercise 1, compute the following:

   (a) $(7A - 6C) - (2A - 3C)$ (b) $3A + (2C + 7A) - (5A - 1C)$

   (c) $6B - 6E - 3(2B - 4E) + [3B - (2E + 3B)]$

   (d) $10D + 3F - 2[6D + 2F - (5D + 3F)]$

8. Solve the following linear equations for X, using the matrices from Exercises 1 and 4:

   (a) $A + X = C$ (b) $C + X = A$ (c) $-X = A$

   (d) $D + X = F$ (e) $5D + 3X = 7F$ (f) $13B - 27E + 16X = 33B$

   (g) $4H - 17L + \sqrt{\pi}\,X = 3(H + L)$ (h) $9B - (6E + 3B) + 2X = 5E - 2B$

9. Using the associative, commutative, and scalar product distributive laws, solve the following general matrix equations for X (assume that all matrices are of the same size and are general matrices, and that a and b are arbitrary scalars):

   (a) $(a + b)A + C - aA = X$ (b) $aA + (a + b)C + bX = A + C$ ($b \neq 0$)

   (c) $(A + C) + X + (C - A) = 0$ (d) $D + X - aF + bX = -(a + b)D$

   (e) $5X - (D + F)a = F - aD$

10. Referring to Exercise 1, is

$$\begin{bmatrix} -1 & 2 \\ -4 & 1 \end{bmatrix}$$

a linear combination $aA + cC$ of A and C? If so, what are the values of a and c?

11. Is the matrix

$$X = \begin{bmatrix} 0 & 3 \\ 2 & -5 \end{bmatrix}$$

a linear combination of the matrices

$$A = \begin{bmatrix} 0 & 1 \\ 0 & 0 \end{bmatrix}, \quad B = \begin{bmatrix} 0 & 0 \\ 1 & 0 \end{bmatrix}, \quad C = \begin{bmatrix} 0 & 0 \\ 0 & 1 \end{bmatrix}?$$

Is X a linear combination of the matrices

$$E = \begin{bmatrix} 0 & 1 \\ 1 & 0 \end{bmatrix}, \quad F = \begin{bmatrix} 0 & -1 \\ 1 & 0 \end{bmatrix}, \quad G = \begin{bmatrix} 0 & 1 \\ 1 & 1 \end{bmatrix}?$$

12. Let

$$I = \begin{bmatrix} 1 & 0 \\ 0 & 1 \end{bmatrix} \quad \text{and} \quad J = \begin{bmatrix} 0 & -1 \\ 1 & 0 \end{bmatrix}.$$

Set $X = \begin{bmatrix} 3 & -2 \\ 2 & 3 \end{bmatrix}$. Show that X is a linear combination of I and J. Verify that

$$\begin{bmatrix} a & -b \\ b & a \end{bmatrix} = aI + bJ$$

for all values of a and b.

13. Let $A = [a_{ij}]$ and $B = [b_{ij}]$ be $m \times n$-matrices.

  (a) Use the relations $a_{ij} + 0 = a_{ij} = 0 + a_{ij}$ to show that
  $A + 0 = A = 0 + A$, where 0 is the $m \times n$-zero matrix.
  (b) Show that $A - A = 0$ because $a_{ij} - a_{ij} = 0$.
  (c) Show that $X = B - A$ solves $A + X = B$.
  (d) For arbitrary scalars r and s, use the relations

$$ra_{ij} + sa_{ij} = (r + s)a_{ij}, \qquad sa_{ij} + sb_{ij} = s(a_{ij} + b_{ij})$$

to show that

$$rA + sA = (r + s)A, \qquad sA + sB = s(A + B).$$

14. Consider the block matrices

$$M = [A_1 \quad A_2 \quad A_3], \qquad N = [B_1 \quad B_2 \quad B_3]$$

where for each i, A_i and B_i are of the same size. Why is the formula

$$M + N = [A_1 + B_1 \quad A_2 + B_2 \quad C_1 + C_2]$$

valid? Verify this formula for the case in which

$$A_1 = \begin{bmatrix} 1 & 1 \\ -1 & 1 \end{bmatrix}, \quad A_2 = \begin{bmatrix} 1 & 0 & 1 \\ 0 & 1 & 1 \end{bmatrix}, \quad A_3 = \begin{bmatrix} 2 \\ -3 \end{bmatrix},$$

$$B_1 = \begin{bmatrix} 2 & -1 \\ 1 & 3 \end{bmatrix}, \quad B_2 = \begin{bmatrix} 0 & 1 & -1 \\ 1 & 0 & 1 \end{bmatrix}, \quad B_3 = \begin{bmatrix} 5 \\ 6 \end{bmatrix}.$$

15. Let M and N be the block matrices

$$M = \begin{bmatrix} A_1 & A_2 \\ A_3 & A_4 \end{bmatrix}, \quad N = \begin{bmatrix} B_1 & B_2 \\ B_3 & B_4 \end{bmatrix}.$$

What are the conditions on the sizes of the blocks A_1, A_2, A_3, A_4, B_1, B_2, B_3, B_4 for the formula

$$M + N = \begin{bmatrix} A_1 + B_1 & A_2 + B_2 \\ A_3 + B_3 & A_4 + B_4 \end{bmatrix}$$

to hold? Illustrate the result by a minimal case for the case when A_1 is 2×2, B_2 is 2×1, and A_3 is 3×2.

1.3   PRODUCTS OF MATRICES

In the preceding section we showed how to add two matrices of the same size and how to multiply a matrix by a scalar. We now turn to the problem of describing when and how the product of two matrices can be formed. To this end we begin with the notion of a row–column product RC of a row vector R and a column vector C.

When

$$R = [a \quad b], \quad C = \begin{bmatrix} x \\ y \end{bmatrix},$$

we define the **row–column product** by the formula

$$RC = [a \quad b]\begin{bmatrix} x \\ y \end{bmatrix} = ax + by.$$

In particular,

$$[3 \quad 4]\begin{bmatrix} x \\ y \end{bmatrix} = 3x + 4y, \quad [3 \quad 4]\begin{bmatrix} 2 \\ -5 \end{bmatrix} = 3(2) + 4(-5) = -14.$$

Similarly,

$$[x \quad y]\begin{bmatrix} 3 \\ 4 \end{bmatrix} = 3x + 4y, \quad [2 \quad -5]\begin{bmatrix} 3 \\ 4 \end{bmatrix} = 6 - 20 = -14.$$

Note that

$$[a_1 \quad a_2]\begin{bmatrix} b_1 \\ b_2 \end{bmatrix} = a_1b_1 + a_2b_2 = [b_1 \quad b_2]\begin{bmatrix} a_1 \\ a_2 \end{bmatrix},$$

so that "flipping things" around in this manner does not change the row–column product. When the row and column vectors

$$R = [a_1 \quad a_2 \quad a_3], \qquad C = \begin{bmatrix} b_1 \\ b_2 \\ b_3 \end{bmatrix}$$

have three entries we have the analogous ***row–column formula***

$$RC = [a_1 \quad a_2 \quad a_3]\begin{bmatrix} b_1 \\ b_2 \\ b_3 \end{bmatrix} = a_1b_1 + a_2b_2 + a_3b_3.$$

So

$$[5 \quad -1 \quad 3]\begin{bmatrix} 4 \\ 2 \\ 2 \end{bmatrix} = 5(4) + (-1)2 + (3)2 = 24,$$

$$[1 \quad 1 \quad 2]\begin{bmatrix} 1 \\ 1 \\ -1 \end{bmatrix} = 1 + 1 - 2 = 0,$$

$$[a \quad b \quad c]\begin{bmatrix} x \\ y \\ z \end{bmatrix} = ax + by + cz = [x \quad y \quad z]\begin{bmatrix} a \\ b \\ c \end{bmatrix}.$$

In the general case in which the row and column vectors

$$R = [a_1 \quad a_2 \quad \cdots \quad a_q], \qquad C = \begin{bmatrix} b_1 \\ b_2 \\ \vdots \\ b_q \end{bmatrix}$$

have the same number of entries, we define the row–column product RC by the formula

$$RC = [a_1 \quad a_2 \quad \cdots \quad a_q]\begin{bmatrix} b_1 \\ b_2 \\ \vdots \\ b_q \end{bmatrix} = a_1b_1 + a_2b_2 + \cdots + a_qb_q.$$

Note that this definition of the row–column product RC requires that R and C have the same number of entries. We have the following important relation:

$$[a_1 \quad a_2 \quad \cdots \quad a_q] \begin{bmatrix} b_1 \\ b_2 \\ \vdots \\ b_q \end{bmatrix} = [b_1 \quad b_2 \quad \cdots \quad b_q] \begin{bmatrix} a_1 \\ a_2 \\ \vdots \\ a_q \end{bmatrix}.$$

■ **EXAMPLE 1** _____

John took a course in linear algebra. In this course he took three tests and a final exam. His record in the course can be described by the row vector

$$\begin{array}{ccccc} & 1 & 2 & 3 & F \\ \text{John} & [72 & 84 & 66 & 90]. \end{array}$$

The instructor counted each test as 1/6 of the course grade and the final exam as 1/2 of the grade. John's course grade was, accordingly,

$$72(1/6) + 84(1/6) + 66(1/6) + 90(1/2) = 82.$$

Because this was a course in linear algebra, we write John's course grade as a row–column product:

$$\text{John} \quad [72 \quad 84 \quad 66 \quad 90] \begin{bmatrix} 1/6 \\ 1/6 \\ 1/6 \\ 1/2 \end{bmatrix} = 12 + 14 + 11 + 45 = 82.$$

Mary also took this course. Her grades were

$$\begin{array}{ccccc} & 1 & 2 & 3 & F \\ \text{Mary} & [78 & 72 & 72 & 90]. \end{array}$$

These computations can be combined as follows:

$$\begin{array}{c} \text{John} \\ \text{Mary} \end{array} \begin{bmatrix} 72 & 84 & 66 & 90 \\ 78 & 72 & 72 & 90 \end{bmatrix} \begin{bmatrix} 1/6 \\ 1/6 \\ 1/6 \\ 1/2 \end{bmatrix} = \begin{bmatrix} 82 \\ 82 \end{bmatrix} \begin{array}{l} \text{John's grade} \\ \text{Mary's grade.} \end{array}$$

■ **EXAMPLE 2** _____

We all use the concept of a row–column product in our daily life, except that we usually do not represent this concept in terms of rows and columns. For example, when we purchase 2 lb of apples at .35 a pound,

3 lb of oranges at .20 a pound, and 5 lb of bananas at .30 a pound, the total cost is given by

total cost $= 2(.35) + 3(.20) + 5(.30) = .70 + .60 + 1.50 = 2.80.$

This can be written in the form

$$[2 \quad 3 \quad 5] \begin{bmatrix} .35 \\ .20 \\ .30 \end{bmatrix}.$$

Here the row vector represents the items purchased and the column vector represents the cost per item. Of course, in this instance, one would not normally use the row–column representation of the total cost.

We now wish to define matrix multiplication of a matrix A with a matrix B. First, in our definition of matrix multiplication we will require that the matrices be a certain size. If we wish to form the product AB of a matrix A and a matrix B (note the order), we will require that the

number of columns of A = number of rows of B.

We begin with an example to illustrate how to form a matrix product. Let

$$A = \begin{bmatrix} 2 & 3 & 0 \\ 1 & 0 & 2 \end{bmatrix}, \quad B = \begin{bmatrix} -2 & 4 & 0 & 1 \\ 0 & 3 & 2 & 0 \\ 1 & 0 & 0 & 0 \end{bmatrix}.$$

We see that A is a 2×3-matrix and B is a 3×4-matrix. Since the (number of columns of A) = (number of rows of B), the product will turn out to make sense. The product AB will become a 2×4-matrix,

$$AB = \begin{bmatrix} * & * & * & * \\ * & * & * & * \end{bmatrix},$$

where the $*$ entries are to be determined. We now define how to find the product AB. To do so, we consider A to be a column of rows

$$A = \begin{bmatrix} R_1 \\ R_2 \end{bmatrix} \quad \text{where} \quad \begin{matrix} R_1 = [2 \quad 3 \quad 0] \\ R_2 = [1 \quad 0 \quad 2]. \end{matrix}$$

We consider B to be a row of columns

$$B = [C_1 \quad C_2 \quad C_3 \quad C_4],$$

where

$$C_1 = \begin{bmatrix} -2 \\ 0 \\ 1 \end{bmatrix}, \quad C_2 = \begin{bmatrix} 4 \\ 3 \\ 0 \end{bmatrix}, \quad C_3 = \begin{bmatrix} 0 \\ 2 \\ 0 \end{bmatrix}, \quad C_4 = \begin{bmatrix} 1 \\ 0 \\ 0 \end{bmatrix}.$$

To form the product, we need to determine each entry of AB. We define the $(1, 1)$-entry of AB to be the row column product $R_1 C_1$ of the first row R_1 of A and the first column C_1 of B. Hence

$$R_1 C_1 = [2 \quad 3 \quad 0] \begin{bmatrix} -2 \\ 0 \\ 1 \end{bmatrix} = 2(-2) + 3(0) + 0(1) = -4$$

is the $(1, 1)$-entry of AB. Thus

$$AB = \begin{bmatrix} -4 & * & * & * \\ * & * & * & * \end{bmatrix}.$$

Similarly, the $(1, 2)$-entry is the row column product

$$R_1 C_2 = [2 \quad 3 \quad 0] \begin{bmatrix} 4 \\ 3 \\ 0 \end{bmatrix} = 8 + 9 + 0 = 17$$

of the first row R_1 of A and the second column C_2 of B. So far we have

$$AB = \begin{bmatrix} -4 & 17 & * & * \\ * & * & * & * \end{bmatrix}.$$

In general, the (i, j)-entry of AB is the row column product $R_i C_j$ of the ith row R_i of A and the jth column C_j of B. Thus the $(2, 1)$-entry $R_2 C_1$ and the $(2, 2)$-entry $R_2 C_2$ of AB are

$$R_2 C_1 = 1(-2) + 0(0) + 2(1) = 0, \qquad R_2 C_2 = 1(4) + 0(3) + 2(0) = 4.$$

We now have four entries of the product

$$AB = \begin{bmatrix} -4 & 17 & * & * \\ 0 & 4 & * & * \end{bmatrix}$$

and the final entries are given by

$$(1, 3)\text{-entry} = R_1 C_3 = 2(0) + 3(2) + 0(0) = 6$$
$$(1, 4)\text{-entry} = R_1 C_4 = 2(1) + 3(0) + 0(0) = 2$$
$$(2, 3)\text{-entry} = R_2 C_3 = 1(0) + 0(2) + 2(0) = 0$$
$$(2, 4)\text{-entry} = R_2 C_4 = 1(1) + 0(0) + 2(0) = 1,$$

so that

$$AB = \begin{bmatrix} -4 & 17 & 6 & 2 \\ 0 & 4 & 0 & 1 \end{bmatrix}.$$

The following picture is helpful in checking whether or not the product AB makes sense:

$$A \qquad\qquad B$$

$$m \times q \qquad q \times n$$

$$\uparrow \text{equal} \uparrow$$

$$\llcorner\!\_\ m \times n\ \_\!\lrcorner$$

$$AB$$

It also shows how to determine the size $m \times n$ of the product AB. Now that we have seen how to do this for a specific example, we are ready to define the matrix product more generally, as follows:

The product AB of an $m \times q$-matrix A by a $q \times n$-matrix B is the $m \times n$-matrix whose (i, k)th entry is the row–column product $R_i C_k$ of the ith row R_i of A and the kth column C_k of B. Briefly,

$$AB = [R_i C_k], \qquad i = 1, \ldots, m; \quad k = 1, \ldots, n$$

As another example, consider the case in which

$$A = \begin{bmatrix} 2 & 3 & 0 & -1 \\ 1 & 5 & -6 & 4 \end{bmatrix}, \qquad B = \begin{bmatrix} 1 & 0 & 7 \\ -1 & 4 & -2 \\ 3 & -5 & 1 \\ 2 & 6 & 3 \end{bmatrix}$$

Here A is a 2×4-matrix and B is a 4×3-matrix. Each row of A and each column of B has four entries, so that our row–column product rule can be applied. To do so we view A to be a column

$$A = \begin{bmatrix} R_1 \\ R_2 \end{bmatrix}$$

of two rows

$$R_1 = [2 \quad 3 \quad 0 \quad -1], \qquad R_2 = [1 \quad 5 \quad -6 \quad 4]$$

and B to be a row

$$B = [C_1 \quad C_2 \quad C_3]$$

of three columns

$$C_1 = \begin{bmatrix} 1 \\ -1 \\ 3 \\ 2 \end{bmatrix}, \qquad C_2 = \begin{bmatrix} 0 \\ 4 \\ -5 \\ 6 \end{bmatrix}, \qquad C_3 = \begin{bmatrix} 7 \\ -2 \\ 1 \\ 3 \end{bmatrix}.$$

According to our definition, the product AB is given by the formula

$$AB = \begin{bmatrix} R_1 \\ R_2 \end{bmatrix} [C_1 \quad C_2 \quad C_3] = \begin{bmatrix} R_1C_1 & R_1C_2 & R_1C_3 \\ R_2C_1 & R_2C_2 & R_2C_3 \end{bmatrix}.$$

Observe that AB is a 2×3-matrix. The $(1, 1)$-entry of AB is

$$R_1C_1 = [2 \quad 3 \quad 0 \quad -1] \begin{bmatrix} 1 \\ -1 \\ 3 \\ 2 \end{bmatrix} = 2 - 3 + 0 - 2 = -3.$$

The $(1, 2)$-entry of AB is

$$R_1C_2 = [2 \quad 3 \quad 0 \quad -1] \begin{bmatrix} 0 \\ 4 \\ -5 \\ 6 \end{bmatrix} = 0 + 12 + 0 - 6 = 6$$

and the $(2, 3)$-entry of AB is

$$R_2C_3 = [1 \quad 5 \quad -6 \quad 4] \begin{bmatrix} 7 \\ -2 \\ 1 \\ 3 \end{bmatrix} = 7 - 10 - 6 + 12 = 3.$$

The remaining row–column products are

$$R_1C_3 = 5, \qquad R_2C_1 = -14, \qquad R_2C_2 = 74,$$

so that

$$AB = \begin{bmatrix} -3 & 6 & 5 \\ -14 & 74 & 3 \end{bmatrix}.$$

Actually, in applying our row–column product rule we do not exhibit the R_i's and C_k's explicitly but compute the row–column products mentally as follows:

$$\begin{bmatrix} 2 & 3 & 0 & -1 \\ 1 & 5 & -6 & 4 \end{bmatrix} \begin{bmatrix} 1 & 0 & 7 \\ -1 & 4 & -2 \\ 3 & -5 & 1 \\ 2 & 6 & 3 \end{bmatrix}$$

$$= \begin{bmatrix} 2 - 3 + 0 - 2 & 0 + 12 + 0 - 6 & 14 - 6 + 0 - 3 \\ 1 - 5 - 18 + 8 & 0 + 20 + 30 + 24 & 7 - 10 - 6 + 12 \end{bmatrix}.$$

Observe that in this case, the product BA cannot be formed because the number of columns of B is not equal to the number of rows of A.

In the example where

$$A = \begin{bmatrix} 2 & 3 \\ -1 & 5 \end{bmatrix}, \quad B = \begin{bmatrix} 8 & 1 & -6 \\ 0 & -4 & 7 \end{bmatrix}$$

we obtain the product AB by the row–column product rule as follows:

$$\begin{bmatrix} 2 & 3 \\ -1 & 5 \end{bmatrix}\begin{bmatrix} 8 & 1 & -6 \\ 0 & -4 & 7 \end{bmatrix} = \begin{bmatrix} 16+0 & 2-12 & -12+21 \\ -8+0 & -1-20 & 6+35 \end{bmatrix}$$

$$= \begin{bmatrix} 16 & -10 & 9 \\ -8 & -21 & 41 \end{bmatrix}.$$

Again we cannot form the product BA. Why?

As further examples, we have the matrix products

$$\begin{bmatrix} 2 & -1 & 3 \\ 0 & 1 & 5 \\ 6 & 4 & -1 \\ -1 & 2 & 1 \end{bmatrix}\begin{bmatrix} 1 & 2 \\ 2 & 2 \\ 3 & -1 \end{bmatrix} = \begin{bmatrix} 9 & -1 \\ 17 & -3 \\ 11 & 21 \\ 6 & 1 \end{bmatrix},$$

$$\begin{bmatrix} 1 & 1 & 1 & 1 \\ -1 & 1 & -1 & 1 \\ -1 & -1 & 1 & 1 \\ 1 & -1 & -1 & 1 \end{bmatrix}\begin{bmatrix} 1 & -1 & -1 & 1 \\ 1 & 1 & -1 & -1 \\ 1 & -1 & 1 & -1 \\ 1 & 1 & 1 & 1 \end{bmatrix} = \begin{bmatrix} 4 & 0 & 0 & 0 \\ 0 & 4 & 0 & 0 \\ 0 & 0 & 4 & 0 \\ 0 & 0 & 0 & 4 \end{bmatrix}.$$

In the special case in which

$$A = \begin{bmatrix} 1 \\ -2 \\ 1 \end{bmatrix}, \quad B = \begin{bmatrix} 2 & 0 & -3 & 4 \end{bmatrix},$$

each row of A and each column of B has only one entry. Our row–column product rule is therefore applicable. We have

$$\begin{bmatrix} 1 \\ -2 \\ 1 \end{bmatrix}\begin{bmatrix} 2 & 0 & -3 & 4 \end{bmatrix} = \begin{bmatrix} 2 & 0 & -3 & 4 \\ -4 & 0 & 6 & -8 \\ 2 & 0 & -3 & 4 \end{bmatrix}.$$

This illustrates the **column–row product**

$$\begin{bmatrix} a_1 \\ a_2 \\ \vdots \\ a_m \end{bmatrix}\begin{bmatrix} b_1 & b_2 & \cdots & b_n \end{bmatrix} = \begin{bmatrix} a_1b_1 & a_1b_2 & \cdots & a_1b_n \\ a_2b_1 & a_2b_2 & \cdots & a_2b_n \\ \vdots & \vdots & \vdots & \vdots \\ a_mb_1 & a_mb_2 & \cdots & a_mb_n \end{bmatrix}.$$

It is instructive to carry out this column–row product for small values of m and n, such as $m = 2, n = 3$ and $m = 4, n = 3$. It should be noted that in the column–row product, the column and row need not have the same number of entries.

We shall use the standard convention that when we write the expression AB it is understood that the number of columns of A is equal to the number of rows of B. Similarly, when we write $A + B$ or $A - B$, the matrices A and B are of the same size.

Just as with multiplication of real numbers, we shall see that the associative and distributive operational laws apply to matrix multiplication. To illustrate these facts, consider the matrices

$$A = \begin{bmatrix} 3 & 4 \\ 5 & -1 \end{bmatrix}, \quad B = \begin{bmatrix} 1 & 2 \\ 1 & -1 \end{bmatrix}, \quad C = \begin{bmatrix} 2 & 4 & 1 \\ 5 & -2 & -1 \end{bmatrix}$$

and their products

$$AB = \begin{bmatrix} 7 & 2 \\ 4 & 11 \end{bmatrix}, \quad BC = \begin{bmatrix} 12 & 0 & -1 \\ -3 & 6 & 2 \end{bmatrix}, \quad AC = \begin{bmatrix} 26 & 4 & -1 \\ 5 & 22 & 6 \end{bmatrix}.$$

We illustrate the associative law $(AB)C = A(BC)$ by observing that the computation

$$(AB)C = \begin{bmatrix} 7 & 2 \\ 4 & 11 \end{bmatrix} \begin{bmatrix} 2 & 4 & 1 \\ 5 & -2 & -1 \end{bmatrix} = \begin{bmatrix} 24 & 24 & 5 \\ 63 & -6 & -7 \end{bmatrix}$$

yields the same result as the computation

$$A(BC) = \begin{bmatrix} 3 & 4 \\ 5 & -1 \end{bmatrix} \begin{bmatrix} 12 & 0 & -1 \\ -3 & 6 & 2 \end{bmatrix} = \begin{bmatrix} 24 & 24 & 5 \\ 63 & -6 & -7 \end{bmatrix}.$$

Similarly, the distributive law $(A + B)C = AC + BC$ is illustrated by the fact that the following two computations yield the same result:

$$(A + B)C = \begin{bmatrix} 4 & 6 \\ 6 & -2 \end{bmatrix} \begin{bmatrix} 2 & 4 & 1 \\ 5 & -2 & -1 \end{bmatrix} = \begin{bmatrix} 38 & 4 & -2 \\ 2 & 28 & 8 \end{bmatrix}$$

$$AC + BC = \begin{bmatrix} 26 & 4 & -1 \\ 5 & 22 & 6 \end{bmatrix} + \begin{bmatrix} 12 & 0 & -1 \\ -3 & 6 & 2 \end{bmatrix} = \begin{bmatrix} 38 & 4 & -2 \\ 2 & 28 & 8 \end{bmatrix}.$$

We now ask the question: Does the operation of matrix multiplication follow the commutative law; that is, does

$$AB = BA?$$

To investigate this, we consider an example. Let

$$A = \begin{bmatrix} 3 & 4 \\ 5 & -1 \end{bmatrix}, \quad B = \begin{bmatrix} 1 & 2 \\ 1 & -1 \end{bmatrix}.$$

An easy computation yields

$$AB = \begin{bmatrix} 7 & 2 \\ 4 & 11 \end{bmatrix}, \qquad BA = \begin{bmatrix} 13 & 2 \\ -2 & 5 \end{bmatrix}.$$

It follows that $AB \neq BA$. We see that matrix multiplication does not, in general, obey the commutative law. This represents a major difference from the real number system, where the commutative law $ab = ba$ always holds.

There are special instances when matrix multiplication is commutative. For example, if

$$A = \begin{bmatrix} 3 & 4 \\ 5 & -1 \end{bmatrix}, \qquad C = \begin{bmatrix} 2 & 4 \\ 5 & -2 \end{bmatrix}, \qquad \text{then} \qquad AC = \begin{bmatrix} 26 & 4 \\ 5 & 22 \end{bmatrix} = CA.$$

It is of interest to note that we have chosen $C = A - I$. As a consequence, the relation $AC = CA$ can be verified as follows:

$$AC = A(A - I) = AA - A = (A - I)A = CA.$$

Another interesting property of matrices that differs from the real numbers is the following: Recall that if a and b are real numbers, and if the product $ab = 0$, it must follow that $a = 0$ or $b = 0$. This is no longer true for matrices. As an example, consider

$$A = \begin{bmatrix} 2 & -4 \\ -1 & 2 \end{bmatrix}, \qquad B = \begin{bmatrix} 2 & 2 \\ 1 & 1 \end{bmatrix}.$$

We have

$$AB = \begin{bmatrix} 2 & -4 \\ -1 & 2 \end{bmatrix} \begin{bmatrix} 2 & 2 \\ 1 & 1 \end{bmatrix} = \begin{bmatrix} 0 & 0 \\ 0 & 0 \end{bmatrix} = 0,$$

but $A \neq 0$ and $B \neq 0$. So we have matrices A and B such that $AB = 0$, but $A \neq 0$ and $B \neq 0$.

In Section 1.1 we mentioned how the identity matrix I played the same role for matrices as the number 1 does for the real number system. With real numbers,

$$1a = a \qquad \text{for all real } a$$

The same is true for matrices, with respect to I:

$$IA = A \qquad \text{for all } m \times n\text{-matrices } A$$

when I is $m \times m$. Similarly, $AI = A$ when I is $n \times n$. This is illustrated by the following multiplications:

$$\begin{bmatrix} 1 & 0 \\ 0 & 1 \end{bmatrix} \begin{bmatrix} a & b & c \\ d & e & f \end{bmatrix} = \begin{bmatrix} a & b & c \\ d & e & f \end{bmatrix} = \begin{bmatrix} a & b & c \\ d & e & f \end{bmatrix} \begin{bmatrix} 1 & 0 & 0 \\ 0 & 1 & 0 \\ 0 & 0 & 1 \end{bmatrix}.$$

As a further observation, note that

$$S = \begin{bmatrix} s & 0 \\ 0 & s \end{bmatrix} = s \begin{bmatrix} 1 & 0 \\ 0 & 1 \end{bmatrix} = sI.$$

A matrix S of the form $S = sI$ is called a **scalar matrix** because forming the product SA is equivalent to multiplying the matrix A by the scalar s, as is seen from the relations

$$SA = (sI)A = sIA = sA.$$

We now summarize the foregoing properties of matrix product operations. Here r and s are arbitrary scalars.

---

1. $sAB = (sA)B = A(sB) = ABs$.
2. $(AB)C = A(BC)$ (multiplication is associative). Hence we set $ABC = (AB)C = A(BC)$.
3. $(A + B)C = AC + BC$, $D(A + B) = DA + DB$. These are distributive laws. By (1) we have

    $$(rA + sB)C = rAC + sBC, \qquad D(rA + sB) = rDA + sDB.$$

4. If either $A = 0$ or $B = 0$, then $AB = 0$. However, there are nonzero matrices A and B having $AB = 0$.
5. If A is $m \times n$, then $IA = A$ when the identity I is m-dimensional and $AI = A$ when I is n-dimensional.
6. When A and B are square matrices of the same size, the products AB and BA can be formed but they need not be equal. When $AB = BA$, we say that A and B *commute* with respect to multiplication.
7. The integral powers $A^0, A^1, A^2, A^3, \ldots$ of a square matrix A are defined as follows:

    $$A^0 = I, \qquad A^1 = A, \qquad A^2 = AA, \qquad A^3 = AAA, \ldots.$$

We have the usual law of exponents

$$A^r A^s = A^{r+s}, \qquad (A^r)^s = A^{rs} \quad (r \geq 0, s \geq 0).$$

To establish the associative and distributive laws in the general case it will be convenient to use the *summation* symbol

$$\sum_{j=1}^{q},$$

which is read "sum from $j = 1$ to q of." For example, in each of the following equations the expression on the left is a short hand symbol for the sum on the right.

$$\sum_{j=1}^{4} a_j = a_1 + a_2 + a_3 + a_4,$$

$$\sum_{j=1}^{q} a_j = a_1 + a_2 + \cdots + a_q,$$

$$\sum_{i=1}^{3} a_i b_i = a_1 b_1 + a_2 b_2 + a_3 b_3,$$

$$\sum_{k=1}^{n} a_k b_k = a_1 b_1 + a_2 b_2 + \cdots + a_n b_n,$$

$$\sum_{j=1}^{q} a_{ij} b_{jk} = a_{i1} b_{1k} + a_{i2} b_{2k} + \cdots + a_{iq} b_{qk}.$$

Observe that the right-hand sum in the last equation is also the row–column product

$$[a_{i1} \quad a_{i2} \quad \cdots \quad a_{iq}] \begin{bmatrix} b_{1k} \\ b_{2k} \\ \vdots \\ b_{qk} \end{bmatrix} = R_i C_k$$

of the ith row R_i of the $m \times q$ matrix $A = [a_{ij}]$ and the kth column C_k of the $q \times n$-matrix $B = [b_{jk}]$. It follows that the matrix product $AB = [R_i C_k]$ is given by the compact (summation) formula

$$AB = \left[\sum_{j=1}^{q} a_{ij} b_{jk} \right] \qquad (i = 1, \ldots, m; \quad k = 1, \ldots, n).$$

Similarly, if $C = [c_{kh}]$ is an $n \times r$-matrix, we have

$$BC = \left[\sum_{k=1}^{n} b_{jk} c_{kh} \right].$$

Using the summation formula again, we find that

$$(AB)C = \left[\sum_{k=1}^{n} \left\{ \sum_{j=1}^{q} a_{ij} b_{jk} \right\} c_{kh} \right], \qquad A(BC) = \left[\sum_{j=1}^{q} a_{ij} \left\{ \sum_{k=1}^{n} b_{jk} c_{kh} \right\} \right].$$

These two matrices are equal because by the rules of elementary algebra,

$$\sum_{k=1}^{n} \left\{ \sum_{j=1}^{q} a_{ij} b_{jk} \right\} c_{kh} = \sum_{k=1}^{n} \left\{ \sum_{j=1}^{q} a_{ij} b_{jk} c_{kh} \right\} \qquad \{c_{kh} \text{ is independent of the sum on } j\}$$

$$= \sum_{j=1}^{q} \left\{ \sum_{k=1}^{n} a_{ij} b_{jk} c_{kh} \right\} \qquad \text{in a finite sum, the order of summation can be reversed}$$

$$= \sum_{j=1}^{q} a_{ij} \left\{ \sum_{k=1}^{n} b_{jk} c_{kh} \right\} \qquad \{a_{ij} \text{ is independent of the sum on } k\}$$

It is instructive to write out these summations for small values of q and n. For example, when $q = 2$ and $n = 3$ this sum is given by

$$a_{i1}b_{11}c_{1k} + a_{i1}b_{12}c_{2k} + a_{i1}b_{13}c_{3k} + a_{i2}b_{21}c_{1k} + a_{i2}b_{22}c_{2k} + a_{i2}b_{23}c_{3k}.$$

Similarly, the distributive law $(A + B)C = AC + BC$ for the matrices $A = [a_{ij}]$, $B = [b_{ij}]$, and $C = [c_{jk}]$ follows from the identity

$$\sum_{j=1}^{q} (a_{ij} + b_{ij})c_{jk} = \sum_{j=1}^{q} a_{ij}c_{jk} + \sum_{j=1}^{q} b_{ij}c_{jk}.$$

The distributive law $D(A + B) = DA + DB$ can be established in the same manner.

Exercises

1. Given the matrices

$$A = \begin{bmatrix} 1 & 1 \\ -1 & 1 \end{bmatrix}, \qquad B = \begin{bmatrix} 1 & 0 & 1 \\ 0 & 1 & 0 \end{bmatrix}, \qquad C = \begin{bmatrix} 3 & -1 & 0 \\ 2 & 0 & 1 \\ 1 & 1 & 0 \end{bmatrix},$$

$$X = \begin{bmatrix} 1 \\ -2 \\ 1 \end{bmatrix}, \qquad Y = \begin{bmatrix} 1 & 1 \\ 2 & -1 \\ 3 & 0 \end{bmatrix}, \qquad Z = \begin{bmatrix} 1 & -1 & 1 & 2 \end{bmatrix},$$

form all possible matrix products of two matrices.

2. Given the matrices

$$D = \begin{bmatrix} 5 & 4 \\ 3 & 2 \\ 1 & 0 \end{bmatrix}, \qquad E = \begin{bmatrix} 1 & -1 & 1 & -1 & 1 \end{bmatrix}, \qquad F = \begin{bmatrix} 1 & 1 \\ 1 & 1 \end{bmatrix},$$

$$U = \begin{bmatrix} 9 & 9 \\ 0 & 1 \\ 8 & 6 \\ 0 & 0 \\ 1 & 2 \end{bmatrix}, \qquad V = \begin{bmatrix} 8 \\ 1 \end{bmatrix}, \qquad W = \begin{bmatrix} 2 & 5 \end{bmatrix},$$

form all possible matrix products of two matrices.

3. For $I = \begin{bmatrix} 1 & 0 \\ 0 & 1 \end{bmatrix}$ and $A = \begin{bmatrix} 1 & 1 \\ -1 & 1 \end{bmatrix}$ compute the following matrices:

$$IA, \quad AI, \quad A^2 = AA, \quad A^3 = AAA, \quad \text{and} \quad B = 2I - 4A + 3A^2 - A^3.$$

4. Verify that $(AB)C = A(BC)$ for each case in which

$$A = \begin{bmatrix} 1 & 1 \\ -1 & 1 \end{bmatrix}, \qquad B = \begin{bmatrix} 1 & 0 & 1 \\ 0 & 1 & 0 \end{bmatrix}, \qquad C = \begin{bmatrix} 1 & 2 & -1 & 1 \\ -1 & 0 & 2 & 1 \\ 1 & 3 & 1 & 0 \end{bmatrix}.$$

5. Verify that $A(B + C) = AB + AC$ when

$$A = \begin{bmatrix} 1 & 1 \\ -1 & 1 \end{bmatrix}, \quad B = \begin{bmatrix} 1 & 0 & 1 \\ 0 & 1 & 0 \end{bmatrix}, \quad C = \begin{bmatrix} 0 & 1 & 0 \\ 2 & -1 & 3 \end{bmatrix}.$$

6. With A, B, and C as in Exercise 5, compute AB, AC, and $2B - 3C$. Then compute $2AB - 3AC$ by computing $A(2B - 3C)$.

7. Given that $A = \begin{bmatrix} 1 & 1 \\ 0 & 1 \end{bmatrix}$ and $I = \begin{bmatrix} 1 & 0 \\ 0 & 1 \end{bmatrix}$, show that

$$(A - I)^2 = A^2 - 2A + I = 0.$$

Hence show that

$$A(2I - A) = I, \quad A^2 = 2A - I, \quad A^3 = 3A - 2I, \quad A^4 = 4A - 3I.$$

8. Given that $A = \begin{bmatrix} 1 & 0 & 3 \\ -1 & 2 & 0 \\ 1 & 0 & 4 \end{bmatrix}$ and $I = \begin{bmatrix} 1 & 0 & 0 \\ 0 & 1 & 0 \\ 0 & 0 & 1 \end{bmatrix}$, show that $2I - 11A + 7A^2 - A^3 = 0$.

9. For each of the pairs of matrices,

(a) $A = \begin{bmatrix} 1 & 1 \\ -1 & 1 \end{bmatrix} \quad B = \begin{bmatrix} 3 & 4 \\ 5 & 6 \end{bmatrix}$

(b) $A = \begin{bmatrix} 2 & 3 & 4 \\ 4 & 3 & 2 \end{bmatrix}, \quad B = \begin{bmatrix} 1 & 2 \\ 1 & 0 \\ 4 & 4 \end{bmatrix}$

(c) $A = \begin{bmatrix} 2 & 4 & 0 \\ 5 & 0 & 3 \\ 0 & 1 & 0 \end{bmatrix}, \quad B = \begin{bmatrix} 5 & 50 \\ 1 & 60 \\ 1 & 86 \end{bmatrix}$

(d) $A = \begin{bmatrix} 2 & 3 \\ 0 & 1 \end{bmatrix}, \quad B = \begin{bmatrix} 8 & 1 \\ 1 & 2 \end{bmatrix}$

compute the products AB *and* BA when possible. Compare the results to see that $AB \neq BA$; that is, in general, matrix multiplication is not commutative. (This does not mean that matrices *will never* commute, as demonstrated in the following example:

(e) $A = \begin{bmatrix} 2 & 3 \\ 0 & 1 \end{bmatrix}, \quad B = \begin{bmatrix} 1 & 3 \\ 0 & 0 \end{bmatrix}$

where $AB = BA$.)

10. It was demonstrated in the text that if $A = 0$ or $B = 0$, then $AB = 0$. Using the following pairs of matrices,

(a) $A = \begin{bmatrix} 1 & 1 \\ 1 & 1 \end{bmatrix}, \quad B = \begin{bmatrix} 1 \\ -1 \end{bmatrix},$ **(b)** $A = \begin{bmatrix} 3 & 6 \\ 1 & 2 \end{bmatrix}, \quad B = \begin{bmatrix} -2 \\ 1 \end{bmatrix},$

(c) $A = \begin{bmatrix} 5 & 6 \end{bmatrix}, \quad B = \begin{bmatrix} -6 \\ 5 \end{bmatrix},$ **(d)** $A = \begin{bmatrix} 1 & 0 \\ 1 & 0 \end{bmatrix}, \quad B = \begin{bmatrix} 0 & 0 \\ 1 & 1 \end{bmatrix},$

compute the product *AB* in each set to demonstrate that the *converse* to the foregoing statement is *false;* that is, if *AB* = 0, then neither *A* nor *B* *must equal* 0. (This again demonstrates how different the matrix number system is from the real number system.)

11. Referring to Exercise 10, if *AB* = 0, can one conclude that *BA* = 0? Either prove or give a counterexample. (Assume that the product *BA* makes sense.) *Hint:* Consider part (d) of Exercise 10.

12. Given arbitrary square matrices *A*, *B*, *C*, and *X*, all of the same size, and arbitrary *a* and *b*, solve the following expressions for *X* using the associative, distributive, and general *noncommutativity* of matrix multiplication.

(a) $(B + C - bI)A = bX + CA + AB - bA$

(b) $X + aA - [C + B(aI - C)] = I + a(A - B) - C + BC$

(c) $B(I + C) + X = (A + I)B + BC$

13. Given the following vectors:

(i) $X = \begin{bmatrix} 1 \\ 0 \\ 1 \\ 2 \end{bmatrix}$, $Y = \begin{bmatrix} 3 & 4 & 5 & 6 \end{bmatrix}$ (ii) $X = \begin{bmatrix} 2 \\ 3 \\ 5 \\ 8 \end{bmatrix}$, $Y = \begin{bmatrix} 1 & 2 \end{bmatrix}$

(iii) $X = \begin{bmatrix} 0 \\ 1 \end{bmatrix}$, $Y = \begin{bmatrix} 0 & 2 & 6 & 3 & 5 \end{bmatrix}$

(a) Compute the *column–row* product *XY*.

(b) Compute, when possible, the *row–column* product *YX*.

14. Form the products *CD* and *DC* when

$$C = \begin{bmatrix} 6 & -2 & 0 \\ 2 & 0 & 1 \\ 1 & 1 & 0 \end{bmatrix}, \quad D = \begin{bmatrix} 2 & 0 & 0 \\ 0 & 4 & 0 \\ 0 & 0 & 6 \end{bmatrix}.$$

Note the special nature of *D*.

15.   A matrix *D* of the form

$$D = \begin{bmatrix} d_1 & 0 & 0 \\ 0 & d_2 & 0 \\ 0 & 0 & d_3 \end{bmatrix}$$

is called a ***diagonal*** matrix. Form the products *AD* and *DB* when

$$A = \begin{bmatrix} a_{11} & a_{12} & a_{13} \\ a_{21} & a_{22} & a_{23} \end{bmatrix}, \quad B = \begin{bmatrix} b_{11} & b_{12} & b_{13} & b_{14} \\ b_{21} & b_{22} & b_{23} & b_{24} \\ b_{31} & b_{32} & b_{33} & b_{34} \end{bmatrix}.$$

Observe how the main diagonal elements d_1, d_2, and d_3 appear in these products.

16. Matrices such as

$$U = \begin{bmatrix} d_1 & a & c \\ 0 & d_2 & b \\ 0 & 0 & d_3 \end{bmatrix}, \qquad V = \begin{bmatrix} e_1 & f & h \\ 0 & e_2 & g \\ 0 & 0 & e_3 \end{bmatrix},$$

whose entries below the main diagonal are zero, are called **upper triangular matrices.** Verify that the product $W = UV$ is also upper triangular. Verify that the main diagonal elements of W are d_1e_1, d_2e_2, and d_3e_3.

17. Show that the system of equations

$$2x + 3y = h$$
$$3x + 5y = k$$

is equivalent to the matrix equation

$$\begin{bmatrix} 2 & 3 \\ 3 & 5 \end{bmatrix} \begin{bmatrix} x \\ y \end{bmatrix} = \begin{bmatrix} h \\ k \end{bmatrix},$$

which is of the form

$$AX = H,$$

where

$$A = \begin{bmatrix} 2 & 3 \\ 3 & 5 \end{bmatrix}, \qquad X = \begin{bmatrix} x \\ y \end{bmatrix}, \qquad H = \begin{bmatrix} h \\ k \end{bmatrix}.$$

Solving the original system for x and y, we find that

$$x = 5h - 3k$$
$$y = -3h + 2k.$$

This solution can be written in the form

$$\begin{bmatrix} x \\ y \end{bmatrix} = \begin{bmatrix} 5 & -3 \\ -3 & 2 \end{bmatrix} \begin{bmatrix} h \\ k \end{bmatrix},$$

so that $X = BH$, where

$$B = \begin{bmatrix} 5 & -3 \\ -3 & 2 \end{bmatrix}.$$

Find x and y when $h = 5$, $k = 8$; also when $h = 2$, $k = 1$. Verify that $AB = I$.

18. Show that the system of linear equations

$$a_{11}x_1 + a_{12}x_2 + a_{13}x_3 = h_1$$
$$a_{21}x_1 + a_{22}x_2 + a_{23}x_3 = h_2$$
$$a_{31}x_1 + a_{32}x_2 + a_{33}x_3 = h_3$$

is equivalent to the matrix equation

$$AX = H,$$

where

$$
A = \begin{bmatrix} a_{11} & a_{12} & a_{13} \\ a_{21} & a_{22} & a_{23} \\ a_{31} & a_{32} & a_{33} \end{bmatrix}, \qquad X = \begin{bmatrix} x_1 \\ x_2 \\ x_3 \end{bmatrix}, \qquad H = \begin{bmatrix} h_1 \\ h_2 \\ h_3 \end{bmatrix}.
$$

Find A and H for the system

$$
x_1 + 3x_2 - 2x_3 = 2
$$
$$
x_2 + 5x_3 = 6
$$
$$
x_3 = 1,
$$

whose solution is $x_1 = 1$, $x_2 = 1$, and $x_3 = 1$. Show also that in this case, $X = BH$, where

$$
B = \begin{bmatrix} 1 & -3 & 17 \\ 0 & 1 & -5 \\ 0 & 0 & 1 \end{bmatrix}.
$$

Also verify that $AB = BA = I$, the identity.

19. Let $C_1, C_2, \ldots, C_n$ be the columns of an $m \times n$-matrix A. Let X be the column vector whose entries are $x_1, x_2, \ldots, x_n$. Show that

$$
AX = [C_1 \quad C_2 \quad \cdots \quad C_n] \begin{bmatrix} x_1 \\ x_2 \\ \vdots \\ x_n \end{bmatrix} = x_1 C_1 + x_2 C_2 + \cdots + x_n C_n.
$$

The vector AX is therefore a linear combination of the columns $C_1, C_2, \ldots, C_n$ of A. Similarly, if Y is a row vector, then YA is a linear combination of the rows $R_1, R_2, \ldots, R_m$ of A. This will be studied further in Section 1.5.

20.

$$
AB = \begin{bmatrix} 1 & 1 & 1 & 1 \\ -1 & 1 & -1 & 1 \\ -1 & -1 & 1 & 1 \end{bmatrix} \begin{bmatrix} 1 & -2 & -3 & 4 \\ 1 & 2 & -3 & -5 \\ 1 & -2 & 3 & -6 \\ 1 & 2 & 3 & 7 \end{bmatrix}
$$
$$
= \begin{bmatrix} 4 & 0 & 0 & 0 \\ 0 & 8 & 0 & 4 \\ 0 & 0 & 12 & 2 \end{bmatrix} = C.
$$

Delete the first row of A to obtain matrix D on the left below. The second matrix below is obtained from B by deleting the third column. Their product, DE, shown on the right, is obtained from C by deleting row 1 and column 3.

$$
DE = \begin{bmatrix} -1 & 1 & -1 & 1 \\ -1 & -1 & 1 & 1 \end{bmatrix} \begin{bmatrix} 1 & -2 & 4 \\ 1 & 2 & -5 \\ 1 & -2 & -6 \\ 1 & 2 & 7 \end{bmatrix} = \begin{bmatrix} 0 & 8 & 4 \\ 0 & 0 & 2 \end{bmatrix} = F.
$$

This illustrates the following result. Consider the product $AB = C$. Let D be a submatrix of A by deleting rows. Let E be a submatrix of B obtained by deleting columns. Show that the product $DE = F$ is a submatrix of the product $AB = C$. Show that every submatrix of C can be obtained in this manner.

21. Show that the product

$$\begin{bmatrix} a & b & d \\ c & a & b \\ e & c & a \end{bmatrix}\begin{bmatrix} 0 & 0 & 1 \\ 0 & 1 & 0 \\ 1 & 0 & 0 \end{bmatrix}$$

is a symmetric matrix. Generalize.

22. The product CR of a column C and a row R is called a column–row product. Verify that if

$$C = \begin{bmatrix} 2 \\ -1 \end{bmatrix}, \qquad R = [3 \quad -4 \quad 5],$$

then

$$CR = \begin{bmatrix} 2 \\ -1 \end{bmatrix}[3 \quad -4 \quad 5] = \begin{bmatrix} 6 & -8 & 10 \\ -3 & 4 & -5 \end{bmatrix}.$$

(a) Verify that each row of CR is a multiple of R.

(b) Verify that each column of CR is a multiple of C.

(c) Verify that parts (a) and (b) hold when

$$C = \begin{bmatrix} 1 \\ 0 \\ 1 \end{bmatrix}, \qquad R = [1 \quad -1 \quad 0 \quad 1].$$

Why do (a) and (b) hold no matter how column C and row R are chosen?

23. Write the matrices

$$A = \begin{bmatrix} 1 & 1 \\ 0 & 1 \\ 1 & 1 \end{bmatrix}, \qquad B = \begin{bmatrix} 2 & -4 \\ 4 & 5 \end{bmatrix}$$

in the form $A = [C_1 \quad C_2]$, $B = \begin{bmatrix} R_1 \\ R_2 \end{bmatrix}$, where C_1 and C_2 are columns of A and R_1 and R_2 are rows of B. Verify that product AB is given by the sum

$$AB = C_1R_1 + C_2R_2$$

of column–row products. This can be written as row–column product

$$AB = [C_1 \quad C_2]\begin{bmatrix} R_1 \\ R_2 \end{bmatrix} = C_1R_1 + C_2R_2$$

of a row of columns by a column of rows.

24. Let A and B be two matrices

$$A = [C_1 \quad C_2 \quad \cdots \quad C_m], \quad B = \begin{bmatrix} R_1 \\ R_2 \\ \vdots \\ R_m \end{bmatrix}$$

with A expressed as a row of m columns and B expressed as rows. Verify that the product AB is given by row–column product

$$AB = [C_1 \quad C_2 \quad \cdots \quad C_m] \begin{bmatrix} R_1 \\ R_2 \\ \vdots \\ R_m \end{bmatrix} = C_1R_1 + C_2R_2 + \cdots + C_mR_m$$

and so is the sum of m column–row products C_jR_j first in a numerical case and then in the general. *Hint:* In the general case with $A = [a_{ij}]$ and $B = [b_{jk}]$, interpret the formula

$$AB = \left[\sum_{j=1}^{m} a_{ij}b_{jk} \right] = \sum_{j=1}^{m} [a_{ij}b_{jk}],$$

observing that for each j, $[a_{ij}b_{jk}]$ is the column–row product

$$[a_{ij}][b_{jk}] = \begin{bmatrix} a_{1j} \\ \vdots \\ a_{pj} \end{bmatrix} [b_{j1} \quad \cdots \quad b_{jq}] = C_jR_j.$$

The formula

$$AB = C_1R_1 + C_2R_2 + \cdots + C_mR_m$$

can be used as a definition of the product AB of two matrices, $A = [a_{ij}]$ and $B = [b_{jk}]$.

A Topic for Self-Study

CR MATRICES

A *CR* matrix is a matrix that can be expressible as the product *CR* of a column vector C and a row vector R. In Exercise 24 for this section it was pointed out that the product AB is expressible as the sum

$$AB = C_1R_1 + C_2R_2 + \cdots + C_mR_m$$

of *CR* matrices. Here C_j is the jth column of A and R_j is the jth row of B. This is a basic application of *CR* matrices. Let us look at some of the properties of *CR*-matrices. The matrix $G = 0$ is expressible as a product $G = CR$ in which either $C = 0$ or $R = 0$, or $C = 0$ and $R = 0$. Some of the *CR* matrices appearing in the formula for the matrix product AB given above could be a zero *CR* matrix. For the moment, let us look at nonzero *CR* matrices. These matrices have another name. They are called ***matrices of rank 1*** according to the definition of rank of a matrix given later in the book.

Consider the matrix

$$A = \begin{bmatrix} 1 & 2 & -3 & -1 \\ 2 & 4 & -6 & -2 \\ 3 & 6 & -9 & -3 \end{bmatrix}.$$

This matrix has the special property that each column is a multiple of the first column. The second column is 2 times the first column, and so on. If we denote the first column by C, we can express this fact by writing

$$A = [C \quad 2C \quad -3C \quad -C].$$

If we factor out C, pulling it to the left, we have

$$A = C[1 \quad 2 \quad -3 \quad -1] = CR \quad \text{with} \quad R = [1 \quad 2 \quad -3 \quad -1].$$

Notice that R is the first row of A. Notice further that every row of A is a multiple of R. Let's try again using another column as our column C; for example, let C be column 3. We then have

$$A = [(-1/3)C \quad (-2/3)C \quad C \quad (1/3)C].$$

Factoring out C again, we find that

$$A = C[-1/3 \ -2/3 \ 1 \ 1/3] = CR \quad \text{with} \quad R = [-1/3 \ -2/3 \ 1 \ 1/3].$$

Again, the rows of A are multiples of R but in this case R is not a row of A. Now let us use rows instead of columns of A. Let R be the second row of A. We then have

$$A = \begin{bmatrix} (1/2)R \\ R \\ (3/2)R \end{bmatrix} = \begin{bmatrix} 1/2 \\ 1 \\ 3/2 \end{bmatrix} R = CR \quad \text{with} \quad C = \begin{bmatrix} 1/2 \\ 1 \\ 3/2 \end{bmatrix}.$$

Now the columns of A are multiples of C. We express this fact by writing

$$A = [2C \quad 4C \quad -6C \quad -2C] = C[2 \quad 4 \quad -6 \quad -2].$$

The new row is the second row again. Remember that columns are factored to the left while rows are factored to the right.

By the arguments just made for the special matrix A, we obtain the following result.

Let A be a nonzero matrix whose columns are multiples of a single column vector C. When we express this fact by writing each column as a multiple of C, we can factor out C to the left and obtain A expressed as the product $A = CR$, where R has the property that every row of A is a multiple of R. The roles of columns and rows can be interchanged, except that rows are factored to the right. We call a matrix of this type a **CR matrix.**

It will be seen later that CR matrices play a significant role as building blocks of matrices having special properties.

To describe a further application of CR matrices, consider a matrix A having three columns, C_1, C_2, and C_3. Let B be a matrix having three rows, R_1, R_2, and R_3. Finally, let D be a diagonal matrix whose main diagonal entries are d_1, d_2, and d_3. The columns of AD are d_1C_1, d_2C_2, and d_3C_3. Then by the result given above, we have

$$ADB = (AD)B = (d_1C_1)R_1 + (d_2C_2)R_2 + (d_3C_3)R_3$$
$$= d_1C_1R_1 + d_2C_2R_2 + d_3C_3R_3.$$

In the same manner we can establish the following result.

Let A be a matrix having m columns $C_1, C_2, \ldots, C_m$. Let B be a matrix having m rows $R_1, R_2, \ldots, R_m$. Let D be a diagonal matrix with main diagonal entries $d_1, d_2, \ldots, d_m$. Then

$$ADB = d_1C_1R_1 + d_2C_2R_2 + \cdots + d_mC_mR_m$$

is a sum of CR matrices.

This result is useful in a subject called **spectral theory.**

1.4   THE TRANSPOSE OF A MATRIX

Let A be an $m \times n$-matrix. By the *transpose* of A is meant the $n \times m$-matrix whose (j, i)th entry is the (i, j)th entry of A. We denote the transpose of A by A^T. The symbol A^T should not be confused with the power A^r of A.

The transpose of the matrix

$$A = \begin{bmatrix} a_{11} & a_{12} & a_{13} & a_{14} \\ a_{21} & a_{22} & a_{23} & a_{24} \\ a_{31} & a_{32} & a_{33} & a_{34} \end{bmatrix} \quad \text{is} \quad A^T = \begin{bmatrix} a_{11} & a_{21} & a_{31} \\ a_{12} & a_{22} & a_{32} \\ a_{13} & a_{23} & a_{33} \\ a_{14} & a_{24} & a_{34} \end{bmatrix}.$$

Observe that the entries in the first row of A are the entries of the first column of A^T. The entries in the second row of A are the entries of the second column of A^T. Similarly, the third row of A and the third column of A^T have the same entries. In other words, the rows of A become the columns of A^T.

> The transpose A^T of A is therefore obtained from A by interchanging rows and columns.

Here are some matrices and their transposes.

$$B = \begin{bmatrix} a & b \\ c & d \end{bmatrix}, \quad B^T = \begin{bmatrix} a & c \\ b & d \end{bmatrix},$$

$$C = \begin{bmatrix} 1 & 0 & 1 & 1 \\ -1 & 0 & -1 & 1 \\ 2 & 3 & -4 & 5 \\ 4 & 2 & -3 & 1 \end{bmatrix}, \quad C^T = \begin{bmatrix} 1 & -1 & 2 & 4 \\ 0 & 0 & 3 & 2 \\ 1 & -1 & -4 & -3 \\ 1 & 1 & 5 & 1 \end{bmatrix},$$

$$D = \begin{bmatrix} 1 & -1 & 1 \end{bmatrix}, \quad D^T = \begin{bmatrix} 1 \\ -1 \\ 1 \end{bmatrix}, \quad E = \begin{bmatrix} 2 \\ 3 \end{bmatrix}, \quad E^T = \begin{bmatrix} 2 & 3 \end{bmatrix}.$$

Observe that in each case the transpose of the transpose of a matrix is the matrix itself. Accordingly, we have the following property of transposition. For every matrix A

> $$(A^T)^T = A, \quad \text{that is,} \quad A^{TT} = A, \quad \text{where} \quad A^{TT} = (A^T)^T.$$

It is easily verified that we also have the relation

> $$(sA)^T = sA^T \quad \text{for every scalar } s.$$

In addition,

> $$(A + B)^T = A^T + B^T,$$
> that is, the transpose of a sum is equal to the sum of the transposes.

For example, when

$$A = \begin{bmatrix} a & b & c \\ d & e & f \end{bmatrix}, \quad B = \begin{bmatrix} u & v & w \\ x & y & z \end{bmatrix},$$

$$A + B = \begin{bmatrix} a + u & b + v & c + w \\ d + x & e + y & f + z \end{bmatrix},$$

we have

$$A^T = \begin{bmatrix} a & d \\ b & e \\ c & f \end{bmatrix}, \qquad B^T = \begin{bmatrix} u & x \\ v & y \\ w & z \end{bmatrix}, \qquad A^T + B^T = \begin{bmatrix} a + u & d + x \\ b + v & e + y \\ c + w & f + z \end{bmatrix}.$$

Clearly, $(A + B)^T = A^T + B^T$ in this case. The proof for the general case can be made in the same manner.

Another important property of transposition is the relation

$$(AB)^T = B^T A^T,$$

which states that the transpose of a product of matrices is equal to the product of their transposes in reverse order.

For a 2×2-matrix A and a 2×3-matrix B we have

$$A = \begin{bmatrix} a & b \\ c & d \end{bmatrix}, \qquad B = \begin{bmatrix} u & v & w \\ x & y & z \end{bmatrix},$$

$$AB = \begin{bmatrix} au + bx & av + by & aw + bz \\ cu + dx & cv + dy & cw + dz \end{bmatrix}, \qquad B^T = \begin{bmatrix} u & x \\ v & y \\ w & z \end{bmatrix}, \qquad A^T = \begin{bmatrix} a & c \\ b & d \end{bmatrix},$$

$$B^T A^T = \begin{bmatrix} au + bx & cu + dx \\ av + by & cv + dy \\ aw + bz & cw + dz \end{bmatrix}.$$

By inspection we see that $(AB)^T = B^T A^T$ in this case.

To establish this result in general we observe first that the transposes of the row and column vectors

$$R = [a_1 \quad a_2 \quad \cdots \quad a_n], \qquad C = \begin{bmatrix} b_1 \\ b_2 \\ \vdots \\ b_n \end{bmatrix}$$

are (in reverse order) the row and column vectors

$$C^T = [b_1 \quad b_2 \quad \cdots \quad b_n], \qquad R^T = \begin{bmatrix} a_1 \\ a_2 \\ \vdots \\ a_n \end{bmatrix}.$$

Forming row–column products, we find, as before, that

$$RC = C^T R^T = a_1 b_1 + a_2 b_2 + \cdots + a_n b_n.$$

The relation $RC = C^T R^T$ is the key idea in the remainder of the proof. Consider now the product $P = AB$ of two matrices A and B. The (j, i)th entry of its transpose $P^T = (AB)^T$ is the (i, j)th entry of $P = AB$, which in turn is the row–column product

$$R_i C_j = C_j^T R_i^T$$

of the ith row R_i of A and the jth column C_j of B. Because C_j^T is the jth row of B^T and R_i^T is the ith column of A^T, the element $C_j^T R_i^T$ is also the (j, i)th element of the product $B^T A^T$. It follows that the matrices $P^T = (AB)^T$ and $B^T A^T$ have the same entries and so are equal, as was to be proved.

A matrix A is said to be **symmetric** when $A^T = A$. A symmetric matrix is necessarily square. A matrix

$$A = [a_{ij}] \qquad (i, j = 1, 2, \ldots, n)$$

is symmetric if and only if the relations

$$a_{ij} = a_{ji}$$

hold. Accordingly, the matrix

$$\begin{bmatrix} 1 & 0 & 5 \\ x & 2 & -4 \\ z & y & 3 \end{bmatrix}$$

is symmetric if and only if $x = 0$, $y = -4$, and $z = 5$. All of the following matrices are symmetric.

$$A = \begin{bmatrix} 1 & 1 \\ 1 & 2 \end{bmatrix}, \qquad B = \begin{bmatrix} 1 & -2 \\ -2 & 2 \end{bmatrix}, \qquad 2A - 3B = \begin{bmatrix} -1 & 8 \\ 8 & -2 \end{bmatrix}.$$

If A and B are symmetric matrices of the same size, every linear combination $rA + sB$ of A and B is also symmetric. This follows because

$$(rA + sB)^T = rA^T + sB^T = rA + sB.$$

However, the product AB need not be symmetric. For if A and B are the symmetric 2×2-matrices given above, the product

$$AB = \begin{bmatrix} 1 & 1 \\ 1 & 2 \end{bmatrix} \begin{bmatrix} 1 & -2 \\ -2 & 2 \end{bmatrix} = \begin{bmatrix} -1 & 0 \\ -3 & 2 \end{bmatrix}$$

is not symmetric because its $(1, 2)$-entry 0 is not the same as its $(2, 1)$-entry -3. In general, for arbitrary symmetric matrices A and B of the same size, we have

$$(AB)^T = B^T A^T = BA.$$

This matrix equals AB if and only if AB is symmetric. It follows that

> The product AB of symmetric matrices A and B is symmetric if and only if $AB = BA$, that is, if and only if A and B commute.

Consequently, if A is symmetric, so are its powers A^2, A^3, A^4, ... and polynomials in A, such as

$$p(A) = 3I - 5A + 7A^2 + 11A^3 - A^4.$$

Although hitherto we have usually restricted scalars and entries of matrices to be real numbers, the results given remain valid when scalars and entries are permitted to be complex numbers. In the complex case we have two additional operations on matrices: the **conjugate** $\overline{A}$ of a matrix A and the **conjugate transpose** $A^* = \overline{A}^T$ of A. Here $\overline{A}^T = (\overline{A})^T$. Recall that the conjugate $\overline{z}$ of a complex number $z = x + iy$ is given by the formula $\overline{z} = x - iy$. Here x and y are real numbers and $i^2 = -1$. By the conjugate $\overline{A}$ of a matrix is meant the matrix obtained from A by replacing each entry by its conjugate. For example, the conjugate of

$$A = \begin{bmatrix} 1 + i & i & 3 - 4i \\ -i & 4 & -5 \end{bmatrix}$$

is the matrix

$$\overline{A} = \begin{bmatrix} 1 - i & -i & 3 + 4i \\ i & 4 & -5 \end{bmatrix}.$$

It is clear that the conjugate $\overline{\overline{A}}$ of the conjugate $\overline{A}$ of A is A itself. We have the relations

$$\overline{\overline{A}} = A, \quad \overline{sA} = \overline{s}\,\overline{A}, \quad \overline{A + B} = \overline{A} + \overline{B}, \quad \overline{AB} = \overline{A}\,\overline{B}, \quad (\overline{A^T}) = \overline{A}^T.$$

Clearly, $\overline{A} = A$ if and only if the entries of A are real.

Often we combine conjugation with transposition to obtain the new matrix A^* given by

$$A^* = \overline{A}^T = (\overline{A})^T.$$

For the complex matrix A given above we have

$$A^* = \begin{bmatrix} 1 - i & i \\ -i & 4 \\ 3 + 4i & -5 \end{bmatrix}$$

The matrix A^* is called the **conjugate transpose** of the matrix A. It is also called the **adjoint** of A but we shall not use this name here, because it could

be confused with a concept that arises in the study of determinants. When
A is real, that is, when the entries of A are real, the conjugate transpose $A*$
of A becomes the transpose A^T of A. In the theory of complex matrices the
conjugate transpose of matrices plays the same role as does the transpose
of a matrix for real matrices.

   The basic properties of the conjugate transpose are

$$A** = A, \quad (sA)* = \bar{s}A*, \quad (A + B)* = A* + B*, \quad (AB)* = B*A*.$$

Here $A** = (A*)*$. These reduce to properties of the transpose when all
numbers are real. If $A* = A$, then A is said to be a **Hermitian matrix**. When
A is a real matrix, it is Hermitian if and only if it is symmetric. Hermitian
matrices for complex matrices play the same role as do symmetric matrices
for real matrices.

Exercises

1. Find the transpose of each of the following matrices:

$$U = \begin{bmatrix} 3 & 2 \\ 5 & 4 \end{bmatrix}, \qquad V = \begin{bmatrix} 1 & 5 & 3 & 6 \\ 2 & 3 & 5 & 8 \end{bmatrix}, \qquad W = \begin{bmatrix} \pi \\ \sqrt{10} \\ 3.14 \\ 22/7 \end{bmatrix},$$

$$X = \begin{bmatrix} 6 & 28 & 496 \\ 0 & 1 & 0 \end{bmatrix}, \qquad Y = \begin{bmatrix} x_1 & x_2 & x_3 & x_4 \\ y_1 & y_2 & y_3 & y_4 \\ z_1 & z_2 & z_3 & z_4 \end{bmatrix}, \qquad Z = [14].$$

2. Given the matrices

$$A = \begin{bmatrix} 1 & 1 \\ -1 & 2 \end{bmatrix}, \qquad B = \begin{bmatrix} 0 & 0 & 0 \\ 1 & 1 & 1 \end{bmatrix}, \qquad C = \begin{bmatrix} 1 & -1 & 1 \\ -1 & 2 & 3 \\ 1 & 3 & 4 \end{bmatrix}$$

   find A^T, B^T, C^T, $P = AB$, $Q = B^T A^T$, P^T, $V = BC$, $W = C^T B^T$, V^T, and W^T.

3. Which of the following matrices are symmetric?

$$L = \begin{bmatrix} 3 & -2 \\ -2 & 5 \end{bmatrix}, \qquad M = \begin{bmatrix} 5 & 6 & 7 \\ 5 & 6 & 7 \\ 5 & 6 & 7 \end{bmatrix}, \qquad N = \begin{bmatrix} 4 & 1 & 4 \\ 1 & 4 & 1 \\ 4 & 1 & 4 \end{bmatrix},$$

$$R = \begin{bmatrix} 1 & 2 \\ 3 & 1 \end{bmatrix}, \qquad S = \begin{bmatrix} 1 & 3 & 5 & 7 \\ 3 & 7 & 2 & -9 \\ 5 & 2 & -9 & 6 \\ 2 & -9 & 6 & 16 \end{bmatrix}, \qquad T = \begin{bmatrix} 5 & 6 & 7 \\ 6 & 5 & 6 \\ 7 & 6 & 5 \end{bmatrix}.$$

4. **(a)** Show that the products AR and RA of the matrices

$$A = \begin{bmatrix} 3 & 4 & 5 \\ 2 & 3 & 4 \\ 1 & 2 & 3 \end{bmatrix}, \qquad R = \begin{bmatrix} 0 & 0 & 1 \\ 0 & 1 & 0 \\ 1 & 0 & 0 \end{bmatrix}$$

are symmetric.

(b) With R as in part (a) and

$$B = \begin{bmatrix} c & d & e \\ b & c & d \\ a & b & c \end{bmatrix},$$

show that the products BR and RB are symmetric.

5. Using the matrices in Exercises 1, 2, and 3, compute the following expressions using the properties of the transpose:

(a) $(U + A)^T$ **(b)** $(B + X)^T$ **(c)** $(U + A + L)^T$ **(d)** $(C + N)^T$

(e) $(UL)^T$ **(f)** $(LU)^T$ **(g)** $(AU)^T$ **(h)** $(AUP)^T$

6. For each of the following pairs of symmetric matrices,

(i) $A = \begin{bmatrix} 3 & 5 \\ 5 & 2 \end{bmatrix}$, $B = \begin{bmatrix} 0 & 1 \\ 1 & 0 \end{bmatrix}$ **(ii)** $A = \begin{bmatrix} 1 & 2 \\ 2 & 5 \end{bmatrix}$, $B = \begin{bmatrix} 3 & 3 \\ 3 & 3 \end{bmatrix}$

(iii) $A = \begin{bmatrix} 1 & 2 & 5 \\ 2 & 0 & 1 \\ 5 & 1 & 2 \end{bmatrix}$, $B = \begin{bmatrix} 0 & 2 & 0 \\ 2 & 0 & 2 \\ 0 & 2 & 0 \end{bmatrix}$

verify that:

(a) $2A + 3B$ is symmetric (this illustrates the additive property of symmetric matrices).

(b) AB is not symmetric even though A and B are symmetric.

7. Consider the symmetric matrices

$$A = \begin{bmatrix} 3 & 2 \\ 2 & 1 \end{bmatrix}, \qquad B = \begin{bmatrix} 2 & 2 \\ 2 & 0 \end{bmatrix}.$$

(a) Compute AB and verify that it is symmetric.

(b) Compute BA and verify that it is symmetric. [*Note:* Observe here that, A and B commute (i.e. $AB = BA$). This did not occur in Exercise 5.]

8. Using the matrices from part (a) of Exercise 5, compute the following matrix polynomials:

(a) $p(A) = A^2 + 2A + 3I$ **(b)** $q(B) = B^2 + 5I$

(c) $r(B) = B^3 + 3B^2 + 3B + I$ **(d)** $s(A,B) = A^2 + 2AB + B^2$

Which of the polynomial expressions above are symmetric matrices?

9. Compute the conjugate and the conjugate transpose of the following complex matrices:

$$A = \begin{bmatrix} 5 & 3+i \\ 3-i & 6 \end{bmatrix}, \quad B = \begin{bmatrix} 1+2i \\ 1-2i \end{bmatrix}, \quad C = \begin{bmatrix} 3-i & 6 \\ 2 & 5-5i \\ 3i & i \end{bmatrix},$$

$$D = \begin{bmatrix} i & 2i & 3i \\ 1 & 2 & 3 \end{bmatrix}, \quad E = \begin{bmatrix} 1 & i & 1+2i \\ -i & 2 & 3-4i \\ 1-2i & 3+4i & 3 \end{bmatrix}.$$

Which of the matrices above are Hermitian (i.e., satisfy $H^* = H$)?

10. If a square matrix is Hermitian, can a number of the form $a + bi$, $b \neq 0$ be on the main diagonal? Why or why not?

11. Using the matrices of Exercise 9, show that a complex matrix A can be "broken up" into its real and imaginary parts, namely that

$$A = X + iY = (\text{Re } A) + i(\text{Im } A),$$

where X and Y are real matrices. *Hint*:

$$A = \begin{bmatrix} 5 & 3+i \\ 3-i & 6 \end{bmatrix} = \begin{bmatrix} 5 & 3 \\ 3 & 6 \end{bmatrix} + i \begin{bmatrix} 0 & 1 \\ -1 & 0 \end{bmatrix} = (\text{Re } A) + i(\text{Im } A)$$

12. **(a)** Verify the following properties for complex numbers of the form $z = x + iy$:

   (i) $z + \bar{z} = 2x$ (real)   **(ii)** $z - \bar{z} = 2iy$ (purely imaginary)

   (b) Consider the general complex 2×2-matrix

$$A = \begin{bmatrix} a_{11} + ib_{11} & a_{12} + ib_{12} \\ a_{21} + ib_{21} & a_{22} + ib_{22} \end{bmatrix},$$

   where a_{ij} and b_{ij} are reals. Prove analogous properties for the matrices A and $\bar{A}$, that is,

   (i) $A + \bar{A} = 2(\text{Re } A)$ **(ii)** $A - \bar{A} = 2i(\text{Im } A)$

13. Let A and B be square matrices of the same size. Suppose that A is symmetric. Show that the matrices $C = B^T A B$ and $D = B A B^T$ are symmetric. Verify this result numerically when

$$A = \begin{bmatrix} 1 & -1 \\ -1 & 2 \end{bmatrix}, \quad B = \begin{bmatrix} 3 & -4 \\ 5 & 7 \end{bmatrix}.$$

14. Consider the matrices

$$I = \begin{bmatrix} 1 & 0 & 0 \\ 0 & 1 & 0 \\ 0 & 0 & 1 \end{bmatrix}, \quad P = \begin{bmatrix} 1 & 0 & 0 \\ 0 & 0 & 1 \\ 0 & 1 & 0 \end{bmatrix}, \quad Q = \begin{bmatrix} 0 & 0 & 1 \\ 0 & 1 & 0 \\ 1 & 0 & 0 \end{bmatrix}.$$

   (a) How are the columns of P, Q, $R = PQ$, $S = QP$, and $T = QR$ related to the columns of I?

   (b) Which of these matrices are symmetric?

   (c) Let X be any one of the matrices above. Show that $X^T X = I = X X^T$.

(d) Construct a matrix having three rows. Show that the matrix XA can be obtained from A by a permutation of its rows.

(e) Construct a matrix B having three columns. Show that the matrix BX can be obtained from B by a permutation of its columns. (*Note:* For this reason a matrix X of this type is called a ***permutation matrix***.)

15. Given that

$$A = \begin{bmatrix} 6 & 2 \\ 2 & 6 \end{bmatrix}, \quad P = \begin{bmatrix} 1 & x \\ 0 & 1 \end{bmatrix},$$

determine x so that $D = P^TAP$ is a diagonal matrix. Recall that a diagonal matrix is a square matrix whose nonzero entries lie on its main diagonal.

16. Given that a is not zero and that

$$A = \begin{bmatrix} a & b \\ b & c \end{bmatrix}, \quad P = \begin{bmatrix} 1 & x \\ 0 & 1 \end{bmatrix},$$

determine x so that $D = P^TAP$ is a diagonal matrix. Find D for this x.

17. Given that

$$A = \begin{bmatrix} 0 & b \\ b & c \end{bmatrix}, \quad P = \begin{bmatrix} x & 1-x \\ 1 & -1 \end{bmatrix},$$

determine x so that $D = P^TAP$ is a diagonal matrix. Find D for this x. Consider first the case $b = 1$ and $c = -3$.

18. Given that

$$I = \begin{bmatrix} 1 & 0 \\ 0 & 1 \end{bmatrix}, \quad J = \begin{bmatrix} 0 & -1 \\ 1 & 0 \end{bmatrix},$$

show that $J^2 = -I$ and that $J^T = -J$. Set

$$E(\theta) = I \cos \theta + J \sin \theta.$$

Show that $E(\theta)E(\beta) = E(\theta + \beta)$, $E(\theta)^T = E(-\theta)$, $E(\theta)^2 = E(2\theta)$, $E(\theta)^n = E(n\theta)$, and $E(\theta)E(-\theta) = I$.

19. (a) Show that if Y is a real column vector, then $Y^TY = 0$ if and only if $Y = 0$. *Hint:* Use the relation

$$Y^TY = y_1^2 + y_2^2 + \cdots + y_m^2,$$

where $y_1, y_2, \ldots, y_m$ are the components of Y.

(b) Given a real matrix A and a column vector Z, show that $A^TAZ = 0$ if and only if $AZ = 0$. *Hint:* Set $Y = AZ$. If $A^TAZ = 0$, then $Y^TY = Z^TA^TAZ = 0$ so that $Y = AZ = 0$.

(c) Show more generally that when X is a real matrix, we have $A^TAX = 0$ if and only if $AX = 0$. *Hint:* Apply part (b) to each column Z of X.

(d) Show that $AA^TX = 0$ if and only if $A^TX = 0$. [Interchange the roles of A and A^T in part (c).]

(e) Show that $AA^TAX = 0$ if and only if $A^TAX = 0$ and hence only if $AX = 0$. [Use part (d) and then part (c).] Conclude further that $A^TAA^TZ = 0$ if and only if $A^TZ = 0$.

(f) Show that $(A^TA)^kX = 0$ if and only if $AX = 0$. *Hint:* Use induction on k. Suppose true when $k \leq m$. Set $Z = A(A^TA)^{m-1}X$. Then $(A^TA)^{m+1}X = A^TAA^TZ = 0$ if and only if $A^TZ = (A^TA)^mX = 0$ and hence if and only if $AX = 0$.

(g) Use the results given above with $X = I$, the identity, to show that if one of the relations

$$A = 0, \quad A^TA = 0, \quad AA^TA = 0, \quad (A^TA)^k = 0, \quad A(A^TA)^k = 0$$

holds, they all hold.

20. (a) In the complex case let $1 + i$ and $1 - i$ be the components of a column vector Y. Show that $Y^TY = 0$ even though Y is not zero. However, Y^*Y is not zero.

    (b) Show that the results derived in Exercise 19 hold in the complex case when conjugate transpose is used instead of transpose.

21. Let A and B be $n \times n$-matrices that commute (i.e., $AB = BA$). Show that:

    (a) A^T and B^T commute.

    (b) A^p and B^q commute for all integers p and q.

    (c) $(AB)^m = A^mB^m$ for every integer m.

    (d) $p(A) = 3A^2 - 5A + 6I$ and $q(B) = 2B + 3I$ commute.

    (e) Every polynomial $P(A)$ of A commutes with every polynomial $q(B)$ of B.

22. Let A be a real symmetric matrix (i.e., $A^T = A$).

    (a) Show that A^k is symmetric.

    (b) Show that $A^2X = 0$ if and only if $AX = 0$. *Hint:* Observe that $A^2 = A^TA$ and use part (c) of Exercise 19.

    (c) Show that for every integer k, $A^kX = 0$ if and only if $AX = 0$. Conclude that $A^k = 0$ if and only if $A = 0$.

23. Let A be a matrix that commutes with its transpose (i.e., $A^TA = AA^T$). Such a matrix is said to be **normal**.

    (a) Show that $A^k = 0$ if and only if $A = 0$. *Hint:* Observe that $S = A^TA = AA^T$ is symmetric. By part (c) of Exercise 20,

    $$S^k = (A^TA)^k = (A^T)^kA^k.$$

    If $A^k = 0$, then successively $S^k = 0$, $S = A^TA = 0$, and $A = 0$.

    (b) Show that $AX = 0$ if and only if $A^TX = 0$. *Hint:* Use part (c) of Exercise 19 to establish the equivalence of the relations $AX = 0$, $A^TAX = 0$, $AA^TX = 0$, and $A^TX = 0$.

24. In the complex case, a matrix A is said to be **normal** if it commutes with its conjugate transpose A^*. Show that if A is normal, then $A^k = 0$ if and only if $A = 0$.

1.5 AN INTERPRETATION OF MATRIX PRODUCTS

In elementary algebra we studied solutions of systems of linear equations. In the following examples we present these equations in two forms, in component form on the left and in matrix form on the right. In the first case we have

$$
\begin{array}{r}
3x + 2y + z = 11 \\
x \quad\ - z = 3; \\
2y + 3z = 3
\end{array}
\qquad
\begin{bmatrix} 3 & 2 & 1 \\ 1 & 0 & -1 \\ 0 & 2 & 3 \end{bmatrix}
\begin{bmatrix} x \\ y \\ z \end{bmatrix}
=
\begin{bmatrix} 11 \\ 3 \\ 3 \end{bmatrix}.
$$

In a second case we have

$$
\begin{array}{r}
3x + 2y + z = 6 \\
x \quad\ - z = 0 \\
2y + 3z = 5 \\
x + y + z = 3
\end{array};
\qquad
\begin{bmatrix} 3 & 2 & 1 \\ 1 & 0 & -1 \\ 0 & 2 & 3 \\ 1 & 1 & 1 \end{bmatrix}
\begin{bmatrix} x \\ y \\ z \end{bmatrix}
=
\begin{bmatrix} 6 \\ 0 \\ 5 \\ 3 \end{bmatrix}.
$$

The equations on the right are of the form $AX = H$, with A and H given and X to be determined. There is a second interpretation of these equations. In the first case

$$
\begin{array}{r}
3x + 2y + z = 11 \\
x \quad\ - z = 3; \\
2y + 3z = 3
\end{array}
\qquad
x\begin{bmatrix} 3 \\ 1 \\ 0 \end{bmatrix}
+ y\begin{bmatrix} 2 \\ 0 \\ 2 \end{bmatrix}
+ z\begin{bmatrix} 1 \\ -1 \\ 3 \end{bmatrix}
=
\begin{bmatrix} 11 \\ 3 \\ 3 \end{bmatrix}.
$$

In a second case we have

$$
\begin{array}{r}
3x + 2y + z = 6 \\
x \quad\ - z = 0 \\
2y + 3z = 5 \\
x + y + z = 3
\end{array};
\qquad
x\begin{bmatrix} 3 \\ 1 \\ 0 \\ 1 \end{bmatrix}
+ y\begin{bmatrix} 2 \\ 0 \\ 2 \\ 1 \end{bmatrix}
+ z\begin{bmatrix} 1 \\ -1 \\ 3 \\ 1 \end{bmatrix}
=
\begin{bmatrix} 6 \\ 0 \\ 5 \\ 3 \end{bmatrix}.
$$

The equations on the right are of the form

$$
xC_1 + yC_2 + zC_3 = H,
$$

where we are given three column vectors, C_1, C_2, and C_3, and seek to express a vector H as a linear combination of the vectors C_1, C_2, and C_3. We see that a solution x, y, and z of

$$
AX = A\begin{bmatrix} x \\ y \\ z \end{bmatrix} = H
$$

yields the same numbers x, y, and z, which allows one to express H as a linear combination

$$
xC_1 + yC_2 + zC_3 = H
$$

of columns C_1, C_2, and C_3 of A. In a more general situation, with

$$A = \begin{bmatrix} a_1 & a_2 & a_3 & a_4 \\ b_1 & b_2 & b_3 & b_4 \\ c_1 & c_2 & c_3 & c_4 \end{bmatrix}, \qquad X = \begin{bmatrix} x_1 \\ x_2 \\ x_3 \\ x_4 \end{bmatrix}, \qquad H = \begin{bmatrix} h_1 \\ h_2 \\ h_3 \end{bmatrix},$$

we normally interpret the equation $AX = H$ to be a matrix version of the system of linear equations

$$a_1 x_1 + a_2 x_2 + a_3 x_3 + a_4 x_4 = h_1$$
$$b_1 x_1 + b_2 x_2 + b_3 x_3 + b_4 x_4 = h_2$$
$$c_1 x_1 + c_2 x_2 + c_3 x_3 + c_4 x_4 = h_3.$$

However, as noted above, there is a second interpretation of equal importance. To obtain this second interpretation, we rewrite these equations in the form

$$x_1 \begin{bmatrix} a_1 \\ b_1 \\ c_1 \end{bmatrix} + x_2 \begin{bmatrix} a_2 \\ b_2 \\ c_2 \end{bmatrix} + x_3 \begin{bmatrix} a_3 \\ b_3 \\ c_3 \end{bmatrix} + x_4 \begin{bmatrix} a_4 \\ b_4 \\ c_4 \end{bmatrix} = \begin{bmatrix} h_1 \\ h_2 \\ h_3 \end{bmatrix} = H.$$

This equation states that H is a linear combination

$$H = x_1 C_1 + x_2 C_2 + x_3 C_3 + x_4 C_4$$

of the columns

$$C_1 = \begin{bmatrix} a_1 \\ b_1 \\ c_1 \end{bmatrix}, \qquad C_2 = \begin{bmatrix} a_2 \\ b_2 \\ c_2 \end{bmatrix}, \qquad C_3 = \begin{bmatrix} a_3 \\ b_3 \\ c_3 \end{bmatrix}, \qquad C_4 = \begin{bmatrix} a_4 \\ b_4 \\ c_4 \end{bmatrix}$$

of A, provided that x_1, x_2, x_3, and x_4 are solutions of the system of linear equations above and hence also a solution to $AX = H$.

■ **EXAMPLE 1** _____

Consider the first system of equations above:

$$3x + 2y + z = 11$$
$$x \qquad - z = 3$$
$$2y + 3z = 3.$$

The reader should verify that the numbers $x = 2$, $y = 3$, and $z = -1$ solve the foregoing system. This is equivalent to the fact that the vector

$$X = \begin{bmatrix} 2 \\ 3 \\ -1 \end{bmatrix} \qquad \text{solves} \qquad AX = \begin{bmatrix} 3 & 2 & 1 \\ 1 & 0 & -1 \\ 0 & 2 & 3 \end{bmatrix} \begin{bmatrix} x \\ y \\ z \end{bmatrix} = \begin{bmatrix} 11 \\ 3 \\ 3 \end{bmatrix} = H.$$

From our second interpretation of this system, we see that the right-hand vector H can be written as a linear combination

$$H = \begin{bmatrix} 11 \\ 3 \\ 3 \end{bmatrix} = 2 \begin{bmatrix} 3 \\ 1 \\ 0 \end{bmatrix} + 3 \begin{bmatrix} 2 \\ 0 \\ 2 \end{bmatrix} - 1 \begin{bmatrix} 1 \\ -1 \\ 3 \end{bmatrix}$$

of the columns of A. Observe that the numbers $2, 3$, and -1 are the same x, y, and z that solve $AX = H$.

In the general case, when A is $m \times n$, the equation $H = AX$, if true, states that the column vector H is a linear combination

$$H = x_1 C_1 + x_2 C_2 + \cdots + x_n C_n$$

of the columns $C_1, C_2, \ldots, C_n$ of A. With this in mind we proceed to define some new terminology:

A **column vector H** is said to be **linearly dependent** on a set of column vectors $C_1, C_2, \ldots, C_n$ if H *can* be written as a **linear combination** of $C_1, C_2, \ldots, C_n$. That is, one can find numbers $x_1, \ldots, x_n$ such that

$$H = x_1 C_1 + x_2 C_2 + \cdots + x_n C_n.$$

We also say that the vectors $H, C_1, C_2, \ldots, C_n$ form a **linearly dependent set of vectors**.

■ **EXAMPLE 1 (continued)** ─────────────────────────────

In the example above we saw that the vectors

$$H = \begin{bmatrix} 11 \\ 3 \\ 3 \end{bmatrix}, \quad C_1 = \begin{bmatrix} 3 \\ 1 \\ 0 \end{bmatrix}, \quad C_2 = \begin{bmatrix} 2 \\ 0 \\ 2 \end{bmatrix}, \quad C_3 = \begin{bmatrix} 1 \\ -1 \\ 3 \end{bmatrix}$$

satisfy the relation

$$H = 2C_1 + 3C_2 + (-1)C_3.$$

This tells us that H is linearly dependent on the vectors C_1, C_2, and C_3. Relabeling H as C_4, so that $C_4 = H$, we obtain the relation

$$2C_1 + 3C_2 + (-1)C_3 + (-1)C_4 = 0.$$

Observe that C_1, C_2, C_3, and C_4 are the columns of the matrix

$$\begin{bmatrix} 3 & 2 & 1 & 11 \\ 1 & 0 & -1 & 3 \\ 0 & 2 & 3 & 3 \end{bmatrix}.$$

This leads us to the following equivalent definition of linear dependence of n vectors $C_1, C_2, \ldots, C_n$.

> The column vectors $C_1, C_2, \ldots, C_n$ are said to be ***linearly dependent*** if one can find scalars $x_1, x_2, \ldots, x_n$ *not all* 0 such that
>
> $$x_1 C_1 + x_2 C_2 + \cdots + x_n C_n = 0.$$

Equivalently, the column vectors $C_1, C_2, \ldots, C_n$ are linearly dependent if and only if one of them can be written as a linear combination of the remaining ones.

We see that in the example above, at least one of the scalars was nonzero. (In fact, here, all of them were nonzero, although this is more than the definition requires.) It follows that this equivalent definition of linear dependence is satisfied.

We note that the equation

$$x_1 C_1 + x_2 C_2 + \cdots + x_n C_n = 0$$

always has the solution

$$x_1 = x_2 = \cdots = x_n = 0.$$

However, to satisfy the definition of linear dependence, one must find $x_1, \ldots, x_n$ *not all* 0 satisfying the equation above.

■ **EXAMPLE 2** _____

Consider the column vectors

$$C_1 = \begin{bmatrix} 2 \\ 1 \\ 0 \end{bmatrix}, \quad C_2 = \begin{bmatrix} 0 \\ 5 \\ 1 \end{bmatrix}, \quad C_3 = \begin{bmatrix} 0 \\ 0 \\ 2 \end{bmatrix}$$

and form the equation

$$x_1 C_1 + x_2 C_2 + x_3 C_3 = x_1 \begin{bmatrix} 2 \\ 1 \\ 0 \end{bmatrix} + x_2 \begin{bmatrix} 0 \\ 5 \\ 1 \end{bmatrix} + x_3 \begin{bmatrix} 0 \\ 0 \\ 2 \end{bmatrix} = \begin{bmatrix} 0 \\ 0 \\ 0 \end{bmatrix} = 0.$$

We see that

$$2x_1 = 0, \qquad x_1 + 5x_2 = 0, \qquad x_2 + 2x_3 = 0.$$

It follows that the *only* solution is

$$x_1 = x_2 = x_3 = 0.$$

Since we could not find x_1, x_2, and x_3 *not all* 0, we conclude that the set C_1, C_2, and C_3 are not linearly dependent. We say that the vectors form a *linearly independent set.* This motivates the following definition of linear independence of n vectors $C_1, C_2, \ldots, C_n$:

The vectors $C_1, C_2, \ldots, C_n$ are said to be *linearly independent* if the only solution to the equation

$$x_1 C_1 + x_2 C_2 + \cdots + x_n C_n = 0$$

is $x_1 = x_2 = \cdots = x_n = 0$.

We note that $C_1, C_2, \ldots, C_n$ are linearly independent if and only if no one of them can be written as a linear combination of the remaining ones. This is clearly illustrated in the preceding example, where we saw that the only solution of

$$x_1 \begin{bmatrix} 2 \\ 1 \\ 0 \end{bmatrix} + x_2 \begin{bmatrix} 0 \\ 5 \\ 1 \end{bmatrix} + x_3 \begin{bmatrix} 0 \\ 0 \\ 2 \end{bmatrix} = \begin{bmatrix} 0 \\ 0 \\ 0 \end{bmatrix}$$

was $x_1 = x_2 = x_3 = 0$.

The only way one could express one of the vectors C_1, C_2, and C_3 above as a linear combination of the remaining ones would require division by 0, an impossibility.

■ **EXAMPLE 3** _____

Consider the matrix

$$A = \begin{bmatrix} 1 & 0 & -3 \\ 2 & 4 & 0 \\ 1 & -1 & 0 \end{bmatrix}.$$

The columns of A are

$$C_1 = \begin{bmatrix} 1 \\ 2 \\ 1 \end{bmatrix}, \qquad C_2 = \begin{bmatrix} 0 \\ 4 \\ -1 \end{bmatrix}, \qquad C_3 = \begin{bmatrix} -3 \\ 0 \\ 0 \end{bmatrix}.$$

Forming the equation

$$x_1 C_1 + x_2 C_2 + x_3 C_3 = x_1 \begin{bmatrix} 1 \\ 2 \\ 1 \end{bmatrix} + x_2 \begin{bmatrix} 0 \\ 4 \\ -1 \end{bmatrix} + x_3 \begin{bmatrix} -3 \\ 0 \\ 0 \end{bmatrix} = \begin{bmatrix} 0 \\ 0 \\ 0 \end{bmatrix} = 0,$$

we see that

$$x_1 - 3x_3 = 0, \qquad 2x_1 + 4x_2 = 0, \qquad x_1 - x_2 = 0.$$

Solving, once again the only solution is

$$x_1 = x_2 = x_3 = 0.$$

It follows that the columns C_1, C_2, and C_3 of A form a linearly independent set. From an earlier interpretation of matrix products, we can write

$$x_1C_1 + x_2C_2 + x_3C_3 = 0 \quad \text{as} \quad AX = \begin{bmatrix} 1 & 0 & -3 \\ 2 & 4 & 0 \\ 1 & -1 & 0 \end{bmatrix} \begin{bmatrix} x_1 \\ x_2 \\ x_3 \end{bmatrix} = \begin{bmatrix} 0 \\ 0 \\ 0 \end{bmatrix}.$$

Since C_1, C_2, and C_3 form a linearly independent set, we see that the only solution of $AX = 0$ in this case is $X = 0$. This condition that $AX = 0$ must imply that $X = 0$ will be of fundamental importance in Chapter 2.

Summarizing the conditions for linear independence and linear dependence of the columns of a matrix A, we have:

> The columns of a matrix A are *linearly independent* if and only if $X = 0$ is the only solution of $AX = 0$.
>
> The columns of A are *linearly dependent* if and only if $AX = 0$ has a nonzero solution X.

So far we have interpreted the product $H = AX$ only for the case in which X is a column vector. In this case we noted that the product H is a linear combination of the columns of A. What special interpretations can be given to the product $H = AX$ when X is a general matrix having k columns? To give an answer to this question, consider the case in which X has three columns X_1, X_2, and X_3. Then X is the matrix

$$X = [X_1 \quad X_2 \quad X_3].$$

Moreover,

$$H = AX = A[X_1 \quad X_2 \quad X_3] = [AX_1 \quad AX_2 \quad AX_3].$$

The columns of H are accordingly

$$H_1 = AX_1, \qquad H_2 = AX_2, \qquad H_3 = AX_3,$$

each of which is a linear combination of the columns of A. This result also holds when X has k columns. Accordingly, we have the following interpretations of the product $P = AX$ of two matrices A and X.

> Each column of the product $H = AX$ of two matrices A and X is a linear combination of the columns of A.
>
> Each row of $H = AX$ is a linear combination of the rows of X.

The last statement in the box follows from the first by looking at the product $H^T = X^T A^T$. Each column of H^T is a linear combination of the columns of X^T. Since the columns of the transpose of a matrix are the rows of the matrix itself, it follows that each row of H is a linear combination of the rows of X.

Remark. The concepts of linear independence and linear dependence are not limited to column vectors and row vectors. It is applicable to any system that behaves like vectors in the sense that the operations of addition and scalar multiplication are valid subject to the usual rules. For example, real-valued functions $f(t)$ on an interval have this property. In particular, the functions $\sin t$ and $\cos t$ are linearly independent on the real axis. For if the relation

$$a \sin t + b \cos t = 0$$

holds for all t, then, setting $t = 0$, this equation becomes

$$a(0) + b(1) = 0.$$

Hence $b = 0$. Similarly, $a = 0$, signifying that $\sin t$ and $\cos t$ are linearly independent. These and related concepts are discussed more fully in Chapter 5.

Exercises

1. In the manner shown at the beginning of this section, express the following systems of linear equations in the matrix form $AX = H$. Next express H as a linear combination of the columns of A. In each case obtain the solution of the linear equations.

   (a) $x + 2y = -2$
   $x + 3y = -6$

   (b) $x \quad + \quad z = 7$
   $y + z = 8$
   $x \quad + 2z = 12$

   (c) $x + y + z = 4$
   $2x - y + z = 8$
   $3x + 2y - 3z = -5$

   (d) $x + 5y + 6z = -12$
   $x + 2y + 3z = -6$
   $3y + 3z = -6$
   $z = -1$

   (e) $2x_1 + 3x_2 + 5x_3 = 8$
   $x_1 + 6x_2 - 4x_3 = 13$

   (f) $x_1 - 2x_2 + 3x_3 = -10$
   $x_1 + 2x_2 \quad = 3$
   $3x_1 + 2x_2 \quad = 1$
   $5x_1 + 2x_2 \quad = 9$

   (g) $13x_1 - 2x_2 + 11x_3 - 5x_4 = 10$

2. Verify that the following sets of vectors are linearly dependent. In each case do so by expressing one vector as a linear combination of the remaining vectors.

   (a) $\begin{bmatrix} 1 \\ -1 \end{bmatrix}, \begin{bmatrix} 1 \\ 1 \end{bmatrix}, \begin{bmatrix} 2 \\ 0 \end{bmatrix}$

   (b) $\begin{bmatrix} 1 \\ 2 \\ -1 \end{bmatrix}, \begin{bmatrix} 0 \\ 1 \\ 0 \end{bmatrix}, \begin{bmatrix} 1 \\ 4 \\ -1 \end{bmatrix}$

(c) $\begin{bmatrix} 2 \\ 5 \\ 5 \end{bmatrix}$, $\begin{bmatrix} 1 \\ 0 \\ 0 \end{bmatrix}$, $\begin{bmatrix} 0 \\ 1 \\ 0 \end{bmatrix}$, $\begin{bmatrix} 0 \\ 1 \\ 0 \end{bmatrix}$
    (d) $\begin{bmatrix} 2 \\ 4 \\ -1 \\ 0 \end{bmatrix}$, $\begin{bmatrix} 0 \\ 3 \\ 1 \\ 2 \end{bmatrix}$, $\begin{bmatrix} 4 \\ 8 \\ -2 \\ 0 \end{bmatrix}$

3. Continuing with Exercise 2, in each case construct the matrix A whose columns are the given vectors in the order displayed. In each case find a nonzero solution X of the equation $AX = 0$.

4. Determine whether or not the columns of the following matrices are linearly independent.

$$A = \begin{bmatrix} 1 & 1 \\ -1 & 1 \end{bmatrix}, \qquad B = \begin{bmatrix} 2 & -3 \\ -2 & 3 \end{bmatrix}, \qquad C = \begin{bmatrix} 1 & 1 & 1 \\ -2 & 1 & 1 \end{bmatrix},$$

$$D = \begin{bmatrix} 1 & 2 & 3 \\ 0 & 1 & 2 \\ 0 & 0 & 1 \end{bmatrix}, \qquad E = \begin{bmatrix} 3 & 2 & 1 & 10 \\ 2 & 1 & 0 & 6 \\ 1 & 0 & 0 & 2 \end{bmatrix}.$$

5. Determine whether or not the rows of the matrices displayed in Exercise 4 are linearly independent.

6. Show that the functions 1, t, and t^2 are linearly independent. That is, show that $a + bt + ct^2 = 0$ for all values of t only if $a = b = c = 0$.

7. (a) Show that the rows of a matrix A are linearly dependent if and only if the columns of its transpose A^T are linearly dependent.

   (b) Show that the rows of A are linearly independent if and only if the columns of A^T are linearly independent.

   (c) Show that the rows of A are linearly dependent if and only if the equation $YA = 0$ has a nonzero solution Y.

   (d) Show that if $Y = 0$ is the only solution of $YA = 0$, the rows of A are linearly independent.

8. (a) Show that the columns of the matrix

$$A = \begin{bmatrix} 1 \\ 2 \end{bmatrix} \begin{bmatrix} 3 & 4 \end{bmatrix}$$

   are linearly dependent. Are its rows linearly dependent?

   (b) Show that the rows of the matrix

$$B = \begin{bmatrix} 1 \\ 0 \\ 1 \end{bmatrix} \begin{bmatrix} 3 & 4 & -5 \end{bmatrix}$$

   are linearly dependent. Are its columns linearly dependent?

9. Consider the 2×2-matrix

$$A = \begin{bmatrix} a & b \\ c & d \end{bmatrix}.$$

(a) Show that the columns of A are linearly dependent if and only if $ad - bc = 0$.

(b) Show that the rows of A are linearly dependent if and only if $ad - bc = 0$.

(c) Show that the columns of A are linearly dependent if and only if its rows are linearly dependent.

(d) Conclude that the columns of our 2×2-matrix A are linearly independent if and only if its rows are linearly dependent.

We shall see later in the book that for square matrices, its columns are linearly independent if and only if its rows are linearly independent.

10. Use the result in part (a) of Exercise 8 to determine whether or not the columns of the following matrices are linearly dependent:

(a) $\begin{bmatrix} 2 & 4 \\ -1 & -2 \end{bmatrix}$,   (b) $\begin{bmatrix} 5 & 1 \\ 7 & 3 \end{bmatrix}$,   (c) $\begin{bmatrix} 3 & 6 \\ 4 & 8 \end{bmatrix}$.

11. Show that if a matrix has a column of zeros, its columns are linearly dependent. Conclude by using transposes that if a matrix has a row of zeros, its rows are linearly dependent.

12. (a) Show that if a matrix has two identical columns, its columns are linearly dependent. Use this result to conclude that the columns of the matrix C in Exercise 4 are linearly dependent.

(b) Show that if one column of a matrix is a multiple of another, its columns are linearly dependent.

13. Consider the product AB of two matrices A and B.

(a) Show that if the columns of B are linearly dependent, the columns of AB are linearly dependent.

(b) Show that if the rows of A are linearly dependent, the rows of AB are linearly dependent.

14. (a) Let B be a matrix obtained from a matrix A by rearranging its columns. Show that the columns of B are linearly independent if and only if the columns of A are linearly independent.

(b) Let B be a matrix obtained from a matrix A by rearranging its rows. Show that the rows of B are linearly independent if and only if the rows of A are linearly independent.

15. Let B be a matrix obtained from a matrix A by deleting some of its rows. Verify that

(a) $BX = 0$ whenever $AX = 0$.

(b) If the columns of A are linearly dependent, the columns of B are also linearly dependent.

(c) If the columns of B are linearly independent, the columns of A are linearly independent.

(d) If B was obtained from A by deleting zero rows, the columns of A are linearly independent if and only if the columns of B are linearly independent.

1.6 VECTORS AND THEIR GEOMETRIC INTERPRETATION

Column vectors and row vectors have an important property in common; namely, they are matrix representations of n-tuples of numbers. The column and row vectors

$$\begin{bmatrix} x \\ y \end{bmatrix}, \qquad [x \quad y]$$

are matrix representations of the pair of numbers (x, y). The column and row vectors

$$\begin{bmatrix} x \\ y \\ z \end{bmatrix}, \qquad [x \quad y \quad z]$$

are matrix representations of the triple of numbers (x, y, z). The matrix representations of the n-tuple $(x_1, x_2, \ldots, x_n)$ are the column and row vectors

$$\begin{bmatrix} x_1 \\ x_2 \\ \vdots \\ x_n \end{bmatrix}, \qquad [x_1 \quad x_2 \quad \cdots \quad x_n].$$

An n-tuple of numbers $(x_1, x_2, \ldots, x_n)$ will be called an ***n-dimensional vector*** *or* an ***n-dimensional point.*** For example, in the familiar xy-plane, the number pair (x, y) represents a point whose coordinates are x and y. Alternatively, it represents an arrow (a vector) whose tip is the point (x, y) and whose tail is at the origin. A similar interpretation holds in the three-dimensional case and in the n-dimensional case. Vectors and points will be denoted by lowercase boldface letters, such as $\mathbf{x}, \mathbf{y}, \mathbf{z}, \mathbf{u}, \mathbf{v}, \mathbf{w}, \ldots$. Thus for the case $n = 3$,

$$\mathbf{x} = (x_1, x_2, x_3), \qquad \mathbf{y} = (y_1, y_2, y_3)$$

are three-dimensional vectors. The rules for addition and scalar multiplication of vectors are the same as for row and column vectors. That is,

$$\mathbf{x} + \mathbf{y} = (x_1 + y_1, x_2 + y_2, x_3 + y_3), \qquad s\mathbf{x} = (sx_1, sx_2, sx_3),$$

with similar formulas in the n-dimensional case.

It will be convenient to identify a vector $\mathbf{x}$ with its column vector representation. This means that the matrix representation of the vector $\mathbf{x} = (x_1, x_2, x_3)$ is the column vector

$$\mathbf{x} = \begin{bmatrix} x_1 \\ x_2 \\ x_3 \end{bmatrix}$$

and that $\mathbf{x}^T$ is the corresponding row vector

$$\mathbf{x}^T = [x_1 \quad x_2 \quad x_3].$$

This convention enables us to carry out matrix operations on vectors. It also enables us to interpret a column vector as a point in a Euclidean space.

We now turn to geometrical interpretations of n-tuples of numbers as points and vectors in a Euclidean space. In the two-dimensional case, our points and vectors are denoted by $\mathbf{x} = (x, y)$ as well as by $\mathbf{x} = (x_1, x_2)$. The set of all two-dimensional points is called a **two-dimensional space** or simply a **2-space**. It is customary to denote this space by $\mathbb{R}^2$. The length of a vector $\mathbf{x} = (x, y)$ in $\mathbb{R}^2$ is

$$\|\mathbf{x}\| = [x^2 + y^2]^{1/2}.$$

It is the distance from the point $\mathbf{x} = (x, y)$ to the origin. Observe that in terms of matrices, we have

$$\mathbf{x}^T\mathbf{x} = [x \quad y] \begin{bmatrix} x \\ y \end{bmatrix} = x^2 + y^2.$$

It follows that the length $\|\mathbf{x}\|$ of $\mathbf{x}$ is given by the formula.

$$\|\mathbf{x}\| = [\mathbf{x}^T\mathbf{x}]^{1/2}.$$

This formula holds in general, whatever the dimension. In the three-dimensional case with $\mathbf{x} = (x_1, x_2, x_3)$, the length $\|\mathbf{x}\|$ of $\mathbf{x}$ is

$$\|\mathbf{x}\| = [\mathbf{x}^T\mathbf{x}]^{1/2} = [x_1^2 + x_2^2 + x_3^2]^{1/2}.$$

The set of all three-dimensional points $\mathbf{x} = (x_1, x_2, x_3)$ will be denoted by $\mathbb{R}^3$. It is a **3-space**. Similarly, we denote by $\mathbb{R}^n$, the n-dimensional space of points or vectors $\mathbf{x} = (x_1, x_2, \ldots, x_n)$. It is an **n-space**. The length $\|\mathbf{x}\|$ of a vector $\mathbf{x}$ is given by the formula

$$\|\mathbf{x}\| = [\mathbf{x}^T\mathbf{x}]^{1/2} = [x_1^2 + x_2^2 + \cdots + x_n^2]^{1/2}.$$

Here, as before, $\|\mathbf{x}\|$ is the distance of point $\mathbf{x}$ to the origin. We have the following property:

$$\|\mathbf{x}\| > 0 \text{ unless } \mathbf{x} = \mathbf{0}, \qquad \|\mathbf{0}\| = 0.$$

Incidentally, when $n = 1$, our space $\mathbb{R}^1$ is the set of all real numbers. We usually denote this space by $\mathbb{R}$ instead of $\mathbb{R}^1$.

A vector $\mathbf{u}$ is called a **unit vector** if its length $\|\mathbf{u}\| = 1$. Every nonzero vector $\mathbf{x}$ has associated with it a unique unit vector $\mathbf{u}$ defined by either of the equivalent relations

$$\mathbf{x} = \|\mathbf{x}\|\mathbf{u}, \qquad \mathbf{u} = \frac{\mathbf{x}}{\|\mathbf{x}\|}.$$

Observe that to get **u**, all we have done is to divide **x** by its length $\|\mathbf{x}\|$. We refer to **u** as the *direction* of **x**.

■ **EXAMPLE 1** _____

Returning to the two-dimensional case, we note that with $\mathbf{x} = (x, y)$, the equation

$$\|\mathbf{x}\| = r,$$

when written in coordinate form, is equivalent to the familiar equation

$$x^2 + y^2 = r^2$$

of a circle of radius r with the origin as its center. Similarly, in the three-dimensional case, with $\mathbf{x} = (x, y, z)$, the equation

$$\|\mathbf{x}\| = r$$

is equivalent to the equation

$$x^2 + y^2 + z^2 = r^2$$

of a sphere of radius r with the origin as its center. In general, in the n-dimensional case, the set of points **x** satisfying the equation $\|\mathbf{x}\| = r$ forms an $(n - 1)$-sphere about the origin. It is comprised of all points **x** at a distance r from the origin.

We now turn to a geometrical interpretation of scalar multiplication. To this end, let $\mathbf{v} = (v_1, v_2)$ be a two-dimensional vector emanating from the origin O as shown in Figure 1. The vector **v** lies on a line through the origin. When $t > 0$, the vector $\mathbf{x} = t\mathbf{v} = (tv_1, tv_2)$ also lies on this line and is t times as long as **v** and has the same direction. When $t < 0$, the vector $\mathbf{x} = t\mathbf{v}$ is oppositely directed and is $|t|$ times as long as **v**. Considering t as a variable parameter, the equation $\mathbf{x} = t\mathbf{v}$, or more conventionally, $\mathbf{x} = \mathbf{v}t$, is a parametric equation of a straight line through the origin O in the direction of **v**. If **v** is interpreted to be a velocity vector and t is time, then $\mathbf{x} = \mathbf{v}t$ is

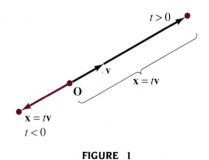

FIGURE I

the equation of motion of a particle moving at a constant velocity **v** starting at the origin O at the time $t = 0$. For purposes of illustration, we chose **v** to be a two-dimensional vector. However, the remarks made are equally valid when **v** is a three-dimensional vector or an n-dimensional vector. In all cases, with t as a parameter, the equation $\mathbf{x} = \mathbf{v}t$ represents a straight line through the origin in the direction of **v**, except when $\mathbf{v} = \mathbf{0}$. We have the fundamental relations

$$\|t\mathbf{v}\| = |t|\,\|\mathbf{v}\|, \qquad \|-\mathbf{v}\| = \|\mathbf{v}\|.$$

We next give a geometric interpretation of vector addition. Again we begin with the two-dimensional case. Consider two vectors $\mathbf{x} = (a, b)$ and $\mathbf{y} = (c, d)$. Algebraically, their sum is $\mathbf{x} + \mathbf{y} = (a + c, b + d)$. To add them geometrically, we refer to Figure 2. To form the sum $\mathbf{x} + \mathbf{y}$ of two vectors **x** and **y**, we construct the parallelogram shown in the figure with **x** and **y** as adjacent sides emanating from the origin O. Then the diagonal of the parallelogram emanating from the origin is the sum $\mathbf{x} + \mathbf{y}$ of **x** and **y**. Alternatively, to add the vector **y** to the vector **x**, move the vector **y** so that

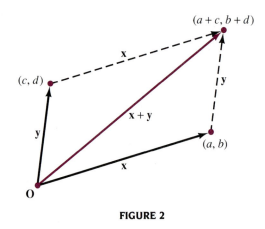

FIGURE 2

its tail is at the tip of **x**. Then the tip of **y** becomes the tip of the vector $\mathbf{x} + \mathbf{y}$. Similarly, to add **x** to **y**, we move **x** so that its tail is at the tip of **y**. Then the tip of **x** is at the tip of $\mathbf{y} + \mathbf{x} = \mathbf{x} + \mathbf{y}$. Of course, in this motion we preserve the length and direction of the vector moved. In Figure 2 we have chosen the case in which **x** and **y** are not multiples of each other, that is, the case in which **x** and **y** are linearly independent. The construction still holds in the degenerate case in which they are linearly dependent. This geometric construction of the sum of two vectors is valid for vectors of all dimensions. Only the components are different.

As an application, if we wish to obtain the parametric equation of a line **L** through a point $\mathbf{x}_0$ in the direction of **v**, we need only replace the vector **x** in the equation

$$\mathbf{x} = \mathbf{v}t$$

by $\mathbf{x} - \mathbf{x}_0$. This gives us the equation

$$\mathbf{x} - \mathbf{x}_0 = \mathbf{v}t, \qquad \text{or equivalently,} \qquad \mathbf{x} = \mathbf{x}_0 + \mathbf{v}t$$

for the line **L**. We have the situation shown in Figure 3. This formula holds in any dimension.

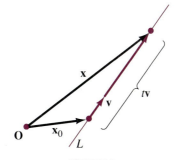

FIGURE 3

■ **EXAMPLE 2** _____

In the two-dimensional case with $\mathbf{x} = (x, y)$, $\mathbf{x}_0 = (2, 3)$, and $\mathbf{v} = (4, 5)$ our equation of line **L** through $\mathbf{x}_0$ in the direction of **v** becomes

$$\mathbf{x} - \mathbf{x}_0 = (x - 2, y - 3) = t(4, 5) \quad \text{or} \quad (x, y) = (2, 3) + t(4, 5).$$

Hence

$$(x, y) = (2 + 4t, 3 + 5t).$$

In component form this becomes

$$x = 2 + 4t, \qquad y = 3 + 5t.$$

To make a geometric construction of the difference $\mathbf{z} = \mathbf{y} - \mathbf{x}$ of two vectors **x** and **y**, we form the triangle shown in Figure 4. Looking at this triangle we see that the vector **y** is the sum $\mathbf{y} = \mathbf{x} + \mathbf{z}$ and hence that $\mathbf{z} = \mathbf{y} - \mathbf{x}$, as desired. Observe that we can view the vector $\mathbf{z} = \mathbf{y} - \mathbf{x}$ to be a vector (or arrow) having the point **x** as its tail and the point **y** as its tip. Referring to Figure 2, we see that the vector $\mathbf{y} - \mathbf{x}$ is the other diagonal of the parallelogram.

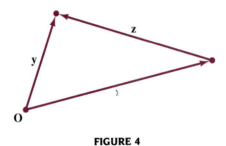

FIGURE 4

In Figure 5 we have the same triangle as in Figure 4 but with different labels for its sides. Here the letters a, b, and c denote the lengths of the corresponding sides. Accordingly, we have

$$a = \|\mathbf{x}\|, \qquad b = \|\mathbf{y}\|, \qquad c = \|\mathbf{z}\| = \|\mathbf{y} - \mathbf{x}\|.$$

Denote the angle at the vertex O by the Greek lowercase letter θ (theta). By the law of cosines we have

$$c^2 = a^2 + b^2 - 2ab \cos \theta.$$

Also because $a^2 = \|\mathbf{x}\|^2 = \mathbf{x}^T\mathbf{x}$ and $b^2 = \|\mathbf{y}\|^2 = \mathbf{y}^T\mathbf{y}$, we have

$$c^2 = (\mathbf{y} - \mathbf{x})^T(\mathbf{y} - \mathbf{x}) = \mathbf{y}^T\mathbf{y} - 2\mathbf{x}^T\mathbf{y} + \mathbf{x}^T\mathbf{x} = b^2 + a^2 - 2\mathbf{x}^T\mathbf{y}.$$

Here we have used the relation $\mathbf{y}^T\mathbf{x} = \mathbf{x}^T\mathbf{y}$. Comparing these two formulas for c^2, we see that

$$\mathbf{x}^T\mathbf{y} = \|\mathbf{x}\| \, \|\mathbf{y}\| \cos \theta,$$

where θ is the angle between the vectors $\mathbf{x}$ and $\mathbf{y}$. The value of θ is therefore determined by the formula

$$\cos \theta = \frac{\mathbf{x}^T\mathbf{y}}{\|\mathbf{x}\| \, \|\mathbf{y}\|}.$$

When $\mathbf{x} = (12, 5, 0)$ and $\mathbf{y} = (0, 3, 4)$, we have $\|\mathbf{x}\| = 13$, $\|\mathbf{y}\| = 5$, $\mathbf{x}^T\mathbf{y} =$

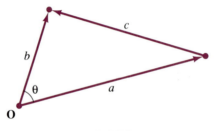

FIGURE 5

15, and cos θ = 15/65 = 3/13. Since $-1 \leq \cos \theta \leq 1$, the formula for cos θ gives us the inequality

$$-\|\mathbf{x}\| \, \|\mathbf{y}\| \leq \mathbf{x}^T\mathbf{y} \leq \|\mathbf{x}\| \, \|\mathbf{y}\|.$$

This inequality is known as the **Schwarz inequality**. As a consequence of this inequality, we have the further inequality

$$\|\mathbf{x} - \mathbf{y}\|^2 = [\|\mathbf{x}\| + \|\mathbf{y}\|]^2 - 2[\|\mathbf{x}\| \, \|\mathbf{y}\| + \mathbf{x}^T\mathbf{y}] \leq [\|\mathbf{x}\| + \|\mathbf{y}\|]^2.$$

Taking square roots, we obtain the first of the **triangle inequalities,**

$$\|\mathbf{x} - \mathbf{y}\| \leq \|\mathbf{x}\| + \|\mathbf{y}\|, \qquad \|\mathbf{x} + \mathbf{y}\| \leq \|\mathbf{x}\| + \|\mathbf{y}\|.$$

The second follows from the first by replacing $\mathbf{y}$ by $-\mathbf{y}$. Geometrically, the triangle inequality states that the length of one side of a triangle cannot exceed the sum of the lengths of the other two sides.

The row–column product $\mathbf{x}^T\mathbf{y}$ is called the **inner product** of the vectors $\mathbf{x}$ and $\mathbf{y}$. When $\mathbf{x} = (x_1, x_2, x_3)$ and $\mathbf{y} = (y_1, y_2, y_3)$, their inner product is given by the formula

$$\mathbf{x}^T\mathbf{y} = [x_1 \quad x_2 \quad x_3] \begin{bmatrix} y_1 \\ y_2 \\ y_3 \end{bmatrix} = x_1y_1 + x_2y_2 + x_3y_3.$$

A similar formula holds in the n-dimensional case. The vectors $\mathbf{x}$ and $\mathbf{y}$ are orthogonal (perpendicular) to each other when the angle θ between them is 90° or equivalently when cos θ = 0. Thus *two nonzero vectors* $\mathbf{x}$ *and* $\mathbf{y}$ *are orthogonal if and only if they satisfy the relation* $\mathbf{x}^T\mathbf{y} = 0$. When $\mathbf{x} = \mathbf{0}$, we have $\mathbf{x}^T\mathbf{y} = 0$ for every vector $\mathbf{y}$. Accordingly, we say that a zero vector is orthogonal to every vector $\mathbf{y}$, even though it does not have a perpendicularity interpretation. The vectors $\mathbf{x} = (1, -1, 1)$ and $\mathbf{y} = (1, 2, 1)$ are orthogonal because $\mathbf{x}^T\mathbf{y} = 1 - 2 + 1 = 0$. Similarly, the vectors $\mathbf{x} = (1, 2)$ and $\mathbf{y} = (2, -1)$ are orthogonal.

Our space $\mathbb{R}^n$ has the important property that if $\mathbf{x}$ and $\mathbf{y}$ are in $\mathbb{R}^n$, so also is every linear combination $\mathbf{z} = a\mathbf{x} + b\mathbf{y}$ of these vectors. Because of this property we call $\mathbb{R}^n$ a linear space. Linear spaces will be studied in Chapter 5. In our space $\mathbb{R}^n$ we have a number of geometric configurations, such as lines, 2-planes, 3-planes, k-planes, and so on. We shall describe these configurations for the case $n = 3$.

In analytic geometry we learn that in xyz-space, namely $\mathbb{R}^3$, the equation

$$ax + by + cz = h$$

represents a plane (a 2-plane) **P** having the vector $\mathbf{n} = (a, b, c)$ as a normal. Here it is understood that $\mathbf{n}$ is nonzero. It is perpendicular to the plane. For two distinct values h_1 and h_2 of h, the planes

$$ax + by + cz = h_1, \qquad ax + by + cz = h_2$$

have the same normal and so are parallel planes. They have no points in common. If (x_0, y_0, z_0) is a point in the plane **P**,

$$ax + by + cz = h,$$

then

$$ax_0 + by_0 + cz_0 = h.$$

Subtracting, we obtain the equation

$$a(x - x_0) + b(y - y_0) + c(z - z_0) = 0$$

of a plane through the point $\mathbf{x}_0 = (x_0, y_0, z_0)$ with $\mathbf{n} = (a, b, c)$ as its normal. In vector form with $\mathbf{x} = (x, y, z)$ the two equations for our plane **P** become

$$\mathbf{n}^T \mathbf{x} = h \qquad \text{and} \qquad \mathbf{n}^T(\mathbf{x} - \mathbf{x}_0) = 0.$$

The last equation states that **x** is in **P** if and only if the vector $\mathbf{v} = \mathbf{x} - \mathbf{x}_0$ is orthogonal to **n**. Let $\mathbf{x}_1$ be a point in **P** distinct from $\mathbf{x}_0$. The vector $\mathbf{v}_1 = \mathbf{x}_1 - \mathbf{x}_0$ and hence also $t_1 \mathbf{v}_1$ is orthogonal to **n** for every scalar t_1. It follows that for each choice of t_1, the point

$$\mathbf{x} = \mathbf{x}_0 + t_1 \mathbf{v}_1 = \mathbf{x}_0 + t_1(\mathbf{x}_1 - \mathbf{x}_0)$$

is in **P**. These points form a line through $\mathbf{x}_0$ and $\mathbf{x}_1$. Let $\mathbf{x}_2$ be a point in **P** not on this line. The vector $\mathbf{v}_2 = \mathbf{x}_2 - \mathbf{x}_0$ and its multiple $t_2 \mathbf{v}_2$ are orthogonal to **n**. It follows that for all choices of t_1 and t_2, the point

$$\mathbf{x} = \mathbf{x}_0 + t_1 \mathbf{v}_1 + t_2 \mathbf{v}_2 = \mathbf{x}_0 + t_1(\mathbf{x}_1 - \mathbf{x}_0) + t_2(\mathbf{x}_2 - \mathbf{x}_0)$$

is such that the vector $\mathbf{v} = \mathbf{x} - \mathbf{x}_0$ is orthogonal to **n**. The point **x** is therefore in **P**. Moreover, every point **x** in **P** is expressible in this form. This equation with t_1 and t_2 as parameters is a parametric equation of our 2-plane **P**.

■ **EXAMPLE 3** _____

In a numerical case in which $\mathbf{n} = (3, 2, 5)$ and $\mathbf{x}_0 = (-1, 7, 4)$, the equation of our plane **P** is

$$3(x + 1) + 2(y - 7) + 5(z - 4) = 0 \qquad or \qquad 3x + 2y + 5z = 31.$$

Because the vectors $\mathbf{v}_1 = (2, -3, 0)$ and $\mathbf{v}_2 = (0, -5, 2)$ are orthogonal to **n**, the points

$$\mathbf{x}_1 = \mathbf{x}_0 + \mathbf{v}_1 = (1, 4, 4), \qquad \mathbf{x}_2 = \mathbf{x}_0 + \mathbf{v}_2 = (-1, 2, 6)$$

are in **P**. The parametric equation

$$\mathbf{x} = \mathbf{x}_0 + t_1 \mathbf{v}_1 + t_2 \mathbf{v}_2$$

of **P** when put in component form becomes

$$x = -1 + 2t_1, \qquad y = 7 - 3t_1 - 5t_2, \qquad z = 4 + 2t_2.$$

Consider next two planes $\mathbf{P}_1$ and $\mathbf{P}_2$ defined by the equations

$$\mathbf{n}_1^T\mathbf{x} = h_1, \qquad \mathbf{n}_2^T\mathbf{x} = h_2$$

and having linearly independent normals $\mathbf{n}_1 = (a_1, b_1, c_1)$ and $\mathbf{n}_2 = (a_2, b_2, c_2)$. Their intersection is a line perpendicular to $\mathbf{n}_1$ and $\mathbf{n}_2$. In terms of xyz-coordinates, our line is the set of points $\mathbf{x} = (x, y, z)$ that satisfy the pair of equations

$$a_1x + b_1y + c_1z = h_1$$

$$a_2x + b_2y + c_2z = h_2.$$

These equations can be written as a single matrix equation

$$A\mathbf{x} = \mathbf{h},$$

where

$$A = \begin{bmatrix} a_1 & b_1 & c_1 \\ a_2 & b_2 & c_2 \end{bmatrix}, \qquad \mathbf{h} = \begin{bmatrix} h_1 \\ h_2 \end{bmatrix}.$$

The equation

$$A\mathbf{x} = \mathbf{h}$$

therefore represents a straight line when the rows of the 2×3-matrix A are linearly independent. When a point $\mathbf{x}_0$ on this line is known, our equation can be written in the alternative form

$$A(\mathbf{x} - \mathbf{x}_0) = 0.$$

■ **EXAMPLE 4**

When

$$A = \begin{bmatrix} 3 & 2 & 5 \\ 1 & 1 & 2 \end{bmatrix}, \qquad \mathbf{h} = \begin{bmatrix} 31 \\ 14 \end{bmatrix}$$

the equation $A\mathbf{x} = \mathbf{h}$ gives us the line whose parametric equation is

$$\mathbf{x} = \mathbf{x}_0 + t\mathbf{v},$$

where $\mathbf{x}_0 = (-1, 7, 4)$ and $\mathbf{v} = (1, 1, -1)$. This result can be verified by substitution.

Generalizing to $\mathbb{R}^n$, let $\mathbf{n}$ be a fixed nonzero vector in $\mathbb{R}^n$ and let h a fixed number. The set $\mathbf{H}$ of points $\mathbf{x}$ satisfying the equation

$$\mathbf{n}^T\mathbf{x} = h$$

is called a *hyperplane* or an $(n - 1)$-*plane*. The vector $\mathbf{n}$ is called its *normal*.

If $\mathbf{x}_0$ is a point in $\mathbf{H}$, we have $\mathbf{n}^T\mathbf{x}_0 = h$, so that our equation for $\mathbf{H}$ can be written in the form

$$\mathbf{n}^T(\mathbf{x} - \mathbf{x}_0) = 0.$$

When we have k hyperplanes

$$\mathbf{n}_1^T\mathbf{x} = h_1, \qquad \mathbf{n}_2^T\mathbf{x} = h_2, \ldots, \mathbf{n}_k^T\mathbf{x} = h_k$$

whose normals $\mathbf{n}_1, \mathbf{n}_2, \ldots, \mathbf{n}_k$ are linearly independent, their intersection is a $(n - k)$-plane. These normals are the rows of a $k \times n$-matrix A. Consequently, our $(n - k)$-plane is determined by the matrix equation

$$A\mathbf{x} = \mathbf{h}$$

with $\mathbf{h} = (h_1, h_2, \ldots, h_k)$. Later we shall see that this equation has a solution of the form

$$\mathbf{x} = \mathbf{x}_0 + t_1\mathbf{v}_1 + t_2\mathbf{v}_2 + \cdots + t_{n-k}\mathbf{v}_{n-k},$$

where $\mathbf{v}_1, \mathbf{v}_2, \ldots, \mathbf{v}_{n-k}$ are linearly independent vectors orthogonal to the normals $\mathbf{n}_1, \mathbf{n}_2, \ldots, \mathbf{n}_k$. This equation is a parametric equation of our $(n - k)$-plane.

Exercises

1. Find the lengths of the vectors

$$\mathbf{x} = (3, -4), \qquad \mathbf{y} = (-12, 5), \qquad \mathbf{z} = (-1, 1),$$

$$\mathbf{u} = (3, -4, 12), \qquad \mathbf{v} = (0, 2, 1), \qquad \mathbf{w} = (1, 1, 1, 1).$$

2. Find the unit vector associated with each of the vectors given in Exercise 1.

3. In each of the following cases draw the parallelogram determined by the vectors $\mathbf{x}$ and $\mathbf{y}$.

   (a) $\mathbf{x} = (1, 5)$, $\mathbf{y} = (4, 2)$ (b) $\mathbf{x} = (1, 5)$, $\mathbf{y} = (4, -2)$

   In each case construct the diagonals $\mathbf{x} + \mathbf{y}$ and $\mathbf{y} - \mathbf{x}$. Compare the lengths of these diagonals. Show that the diagonal $\mathbf{y} - \mathbf{x}$ is longer than the diagonal $\mathbf{x} + \mathbf{y}$ if and only if the angle θ between $\mathbf{x}$ and $\mathbf{y}$ exceeds 90°.

4. Find the value of the inner products between the following pairs of vectors.

   (a) $\mathbf{x} = (1, 1)$, $\mathbf{y} = (1, 2)$ (b) $\mathbf{x} = (4, 3)$, $\mathbf{y} = (3, -4)$

   (c) $\mathbf{x} = (1, 1, 1)$, $\mathbf{y} = (1, -2, 3)$ (d) $\mathbf{x} = (1, 1, 1, 1)$, $\mathbf{y} = (1, -1, 1, -1)$

   In each case find the cosine of the angle between $\mathbf{x}$ and $\mathbf{y}$.

5. In each of the following cases, determine the number k so that the following vectors $\mathbf{u}$ and $\mathbf{v}$ are orthogonal.

   (a) $\mathbf{u} = (1, k)$, $\mathbf{v} = (6, 3)$ (b) $\mathbf{u} = (1, 0, 1)$, $\mathbf{v} = (k, 4, -9)$

   (c) $\mathbf{u} = (5, 2, 0)$, $\mathbf{v} = (2, -5, k)$

6. With $\mathbf{v} = (2, 3)$ and $\mathbf{x} = (x, y)$, draw the line having $\mathbf{x} = \mathbf{v}t$ as its parametric equation. Set $\mathbf{x}_0 = (3, 1)$. Plot the path defined by the equation $\mathbf{x} = \mathbf{x}_0 + \mathbf{v}t$. Show that this path is a straight line parallel to the line $\mathbf{x} = \mathbf{v}t$.

7. Use the expansions

$$\|\mathbf{x} + \mathbf{y}\|^2 = \|\mathbf{x}\|^2 + 2\mathbf{x}^T\mathbf{y} + \|\mathbf{y}\|^2, \qquad \|\mathbf{x} - \mathbf{y}\|^2 = \|\mathbf{x}\|^2 - 2\mathbf{x}^T\mathbf{y} + \|\mathbf{y}\|^2$$

to establish the identity

$$\|\mathbf{x} + \mathbf{y}\|^2 + \|\mathbf{x} - \mathbf{y}\|^2 = 2\|\mathbf{x}\|^2 + 2\|\mathbf{y}\|^2$$

for the vectors $\mathbf{x}$ and $\mathbf{y}$. Geometrically, looking at the parallelogram determined by $\mathbf{x}$ and $\mathbf{y}$, show that this equation states that the sum of the squares of the lengths of the diagonals of a parallelogram is equal to twice the sum of the squares of the lengths of its sides. This identity is known as the **parallelogram law**.

8. Use the expansions given in Exercise 7 to establish the second identity

$$\mathbf{x}^T\mathbf{y} = (1/4)[\|\mathbf{x} + \mathbf{y}\|^2 - \|\mathbf{x} - \mathbf{y}\|^2]$$

for the vectors $\mathbf{x}$ and $\mathbf{y}$. Show that $\mathbf{x}$ and $\mathbf{y}$ are orthogonal if and only if $\|\mathbf{x} + \mathbf{y}\| = \|\mathbf{x} - \mathbf{y}\|$. Conclude that the parallelogram determined by $\mathbf{x}$ and $\mathbf{y}$ is a rectangle if and only if its diagonals are of equal length. This fact is used by builders to construct a rectangular foundation for a building.

9. Two linearly independent vectors $\mathbf{x}$ and $\mathbf{y}$ determine a parallelogram as shown in the following figure.

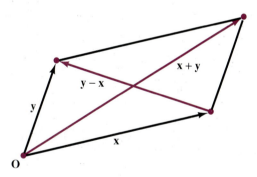

(a) Show that the diagonals $\mathbf{x} + \mathbf{y}$ and $\mathbf{y} - \mathbf{x}$ bisect each other.

(b) Show that the diagonals are orthogonal if and only if its sides are of equal length.

(c) Show that its diagonals are of equal length if and only if the parallelogram is a rectangle.

10. Let $\mathbf{x}$ and $\mathbf{y}$ be nonzero vectors.

(a) Determine a number c such that the vector $\mathbf{h} = \mathbf{y} - c\mathbf{x}$ is orthogonal to $\mathbf{x}$. Show that c is given by the formula

$$c = \frac{\mathbf{x}^T\mathbf{y}}{\mathbf{x}^T\mathbf{x}} = \frac{\mathbf{x}^T\mathbf{y}}{\|\mathbf{x}\|^2}.$$

(b) Determine **h** for the case **x** = (12, 5) and **y** = (3, 4).

(c) Draw the parallelogram determined by **x** and **y**. Show that when we choose **x** to be the base of this parallelogram, **h** represents the altitude of the parallelogram. Note that if we place the tip of **h** at the tip of **y**, the tail of **h** lies on the vector **x** at the tip of the vector **p** = *c***x**. The vector **p** is called the *orthogonal projection* of **y** on **x**. Find the vector **p** for the two vectors given above.

(d) Compute the product $\|\mathbf{x}\| \, \|\mathbf{h}\|$. This product is the area of the parallelogram. Why?

(e) Use the formula for **h** to show that

$$(\text{area})^2 = \|\mathbf{x}\|^2 \, \|\mathbf{h}\|^2 = \|\mathbf{x}\|^2 \, \|\mathbf{y}\|^2 - (\mathbf{x}^T\mathbf{y})^2.$$

Conclude that the area of the parallelogram determined by **x** and **y** can be computed without finding the altitude vector **h**.

11. In three-dimensional vector analysis it is customary to use the notation

$$\mathbf{i} = (1, 0, 0), \qquad \mathbf{j} = (0, 1, 0), \qquad \mathbf{k} = (0, 0, 1)$$

for the unit coordinate vectors. We choose our coordinate system so that they form a right-handed system. A vector **x** = (*x*, *y*, *z*) is expressible in the form

$$\mathbf{x} = x\mathbf{i} + y\mathbf{j} + z\mathbf{k}.$$

If

$$\mathbf{u} = u\mathbf{i} + v\mathbf{j} + w\mathbf{k}$$

is a second vector, the scalar product

$$\mathbf{x}^T\mathbf{u} = xu + yv + zw$$

is often denoted by the symbol **x·u** and is called the *dot product.* Thus

$$\mathbf{x}\cdot\mathbf{u} = xu + yv + zw = \|\mathbf{x}\| \, \|\mathbf{u}\| \cos\theta,$$

where θ is the angle between **x** and **u**. We have

$$\mathbf{i}\cdot\mathbf{i} = \mathbf{j}\cdot\mathbf{j} = \mathbf{k}\cdot\mathbf{k} = 1, \quad \mathbf{i}\cdot\mathbf{j} = \mathbf{j}\cdot\mathbf{i} = 0, \quad \mathbf{i}\cdot\mathbf{k} = \mathbf{k}\cdot\mathbf{i} = 0, \quad \mathbf{j}\cdot\mathbf{k} = \mathbf{k}\cdot\mathbf{j} = 0.$$

There is another product of **x** and **u** called the *cross product,* **x** × **u**. For the vectors **i**, **j**, and **k** the cross product is defined by the relations

$$\mathbf{i} \times \mathbf{i} = 0, \quad \mathbf{j} \times \mathbf{j} = 0, \quad \mathbf{k} \times \mathbf{k} = 0,$$

$$\mathbf{i} \times \mathbf{j} = \mathbf{k} = -\mathbf{j} \times \mathbf{i}, \quad \mathbf{j} \times \mathbf{k} = \mathbf{i} = -\mathbf{k} \times \mathbf{j}, \quad \mathbf{k} \times \mathbf{i} = \mathbf{j} = -\mathbf{i} \times \mathbf{k}.$$

The cross product **x** × **u** is then obtained by using the distribution law. This gives us

$$\mathbf{x} \times \mathbf{u} = (yw - zv)\mathbf{i} + (zv - xw)\mathbf{j} + (xv - yv)\mathbf{k}.$$

When **x** and **u** are parallel, we have **x** × **u** = 0. Otherwise, the vectors **x**, **u**, and **x** × **u** form a right-handed system. The magnitude of **x** × **u** is given in the formula

$$\|\mathbf{x} \times \mathbf{u}\| = \|\mathbf{x}\| \, \|\mathbf{u}\| \sin\theta,$$

where again θ is the angle between **x** and **u**. This magnitude is the area of the parallelogram having **x** and **u** as its sides.

As an example, the cross product of the vectors

$$\mathbf{x} = 2\mathbf{i} - \mathbf{j} + \mathbf{k}, \qquad \mathbf{u} = \mathbf{j} + 2\mathbf{k}$$

is

$$\mathbf{z} = \mathbf{x} \times \mathbf{u} = (-2 - 1)\mathbf{i} + (2 - 4)\mathbf{j} + 2\mathbf{k} = -3\mathbf{i} - 4\mathbf{j} + 2\mathbf{k}.$$

(a) Verify that $\mathbf{x} \cdot \mathbf{z} = 0$, $\mathbf{u} \cdot \mathbf{z} = 0$.

(b) Find the cross product of the two vectors

$$\mathbf{x} = 3\mathbf{i} + \mathbf{j} - \mathbf{k} \qquad \text{and} \qquad \mathbf{u} = 2\mathbf{i} - \mathbf{j} + 4\mathbf{k}.$$

2

Inverses and Solutions of Linear Matrix Equations

2.1 INVERTIBLE MATRICES

In the preceding chapter we found that we can always multiply a matrix by a scalar, we can add and subtract matrices of the same size, and we can multiply two matrices of appropriate sizes. In addition, we introduced the concept of the transpose of a matrix and looked at how we can interpret a system of linear equations in terms of matrices. In particular, we showed how the system

$$a_{11}x_1 + a_{12}x_2 + a_{13}x_3 = h_1$$

$$a_{21}x_1 + a_{22}x_2 + a_{23}x_3 = h_2$$

$$a_{31}x_1 + a_{32}x_2 + a_{33}x_3 = h_3$$

can be written as a single linear matrix equation $AX = H$,

$$\begin{bmatrix} a_{11} & a_{12} & a_{13} \\ a_{21} & a_{22} & a_{23} \\ a_{31} & a_{32} & a_{33} \end{bmatrix} \begin{bmatrix} x_1 \\ x_2 \\ x_3 \end{bmatrix} = \begin{bmatrix} h_1 \\ h_2 \\ h_3 \end{bmatrix}.$$

We would now like to explore how to solve the matrix equation $AX = H$, where A and H are known matrices. The question arises: Can an analog of division for ordinary numbers be defined for matrices under suitable conditions? To pursue this idea further, let us recall a definition of division of numbers that admits a generalization to matrices.

Given the scalar equation

$$ax = h$$

we solve for x by multiplying both sides by $1/a$ to obtain the solution

$$x = \frac{h}{a} = \frac{1}{a}h = a^{-1}h.$$

The number a^{-1} is called the **inverse** or **reciprocal** of a. Observe that a^{-1} has the property that

$$aa^{-1} = a^{-1}a = 1.$$

Equivalently, $x = a^{-1}$ satisfies the equations

$$ax = xa = 1.$$

It is important to note that this equation $ax = h$ involving ordinary numbers can be solved only when $a \neq 0$, since division by zero is not allowed. This condition that $a \neq 0$ can be written in a more abstract equivalent form, by saying: The real number equation $ax = h$ has a unique solution if and only if the special equation

$$ax = 0$$

has only *one* solution: $x = 0$.

This form will be useful when we generalize to matrices. The proof of this statement is easily argued by saying that if $a = 0$, any real number x_0 will satisfy $ax = 0$, since

$$ax_0 = 0x_0 \equiv 0 \qquad \text{for any real number } x_0.$$

Conversely, if $a \neq 0$, the only solution to $ax = 0$ must be $x = 0$, since if the product of any two reals is 0, one of the multiplicands must be 0.

With the foregoing notions in mind, we can generalize to matrices in the following way: Let A be a given *square* matrix and consider the matrix equation

$$AX = H$$

for a given matrix H. We seek the analog of a reciprocal for the matrix A. Such a notion will make sense only when the condition of *nonsingularity* of A is satisfied:

A square matrix A is said to be **nonsingular** if the *only* solution of $AX = 0$ is $X = 0$. If there is an $X \neq 0$ such that $AX = 0$, then A is said to be **singular**.

We have the following result:

Let A be a square matrix. If A is nonsingular, the equation

$$AX = I$$

has a unique (one and only one) solution $X = A^{-1}$, called the **inverse** of A. The matrix A^{-1} satisfies the relations

$$AA^{-1} = A^{-1}A = I.$$

This result will be established in Section 2.3. (It should be observed that the foregoing criteria for matrices is the exact analog for the conditions that allow the division of real numbers.)

If A is *singular*, then A^{-1} does not exist, for if A is singular, there is an $X \neq 0$ such that $AX = 0$. If A^{-1} does exist, then

$$X = A^{-1}(AX) = A^{-1}0 = 0$$

(i.e., $X = 0$, a contradiction).

Let A be a square matrix. We say that A is **invertible** if there is a matrix A^{-1} such that

$$AA^{-1} = A^{-1}A = I.$$

It follows that A is nonsingular if and only if A is invertible. As a consequence, the solution X of the equation

$$AX = H$$

is given by the formula

$$X = A^{-1}H.$$

This can be seen by the substitution

$$AX = A(A^{-1}H) = (AA^{-1})H = IH = H.$$

Notice that the order in which this is written is very important, since matrix multiplication is, in general, noncommutative.

It should be noted that when A is a nonsingular matrix, the equation $AX = H$ can have at most one solution, and since we have found a solution, it is unique. To show this, let us suppose, to the contrary, that there are two solutions, say X and Y. Then $AX = H$ and $AY = H$. It follows that $AX = AY$, so that $A(X - Y) = 0$. Since A is nonsingular, we have that $X - Y = 0$, and hence $X = Y$, as was to be shown.

■ **EXAMPLE 1** _____

Consider the matrix equation $AX = H$, given by

$$\begin{bmatrix} 8 & 3 \\ 5 & 2 \end{bmatrix} \begin{bmatrix} x_1 \\ x_2 \end{bmatrix} = \begin{bmatrix} 2 \\ -1 \end{bmatrix}.$$

We first determine whether this equation has a solution by checking that

$$A = \begin{bmatrix} 8 & 3 \\ 5 & 2 \end{bmatrix}$$

is nonsingular. To do so, consider

$$AX = \begin{bmatrix} 8 & 3 \\ 5 & 2 \end{bmatrix} \begin{bmatrix} x_1 \\ x_2 \end{bmatrix} = \begin{bmatrix} 0 \\ 0 \end{bmatrix} = 0.$$

We have

$$8x_1 + 3x_2 = 0$$

$$5x_1 + 2x_2 = 0.$$

Substituting $x_2 = -\frac{8}{3}x_1$ from the first equation into the second equation we obtain

$$5x_1 + 2\left(-\frac{8}{3}x_1\right) = 0.$$

This gives $x_1 = 0$, which in turn gives $x_2 = 0$ as the only solution.

Since A is nonsingular, it follows that A^{-1} exists. Later in this chapter we show how to find the inverse of any nonsingular matrix. It turns out that here

$$A^{-1} = \begin{bmatrix} 2 & -3 \\ -5 & 8 \end{bmatrix}.$$

The reader should verify that

$$AA^{-1} = \begin{bmatrix} 8 & 3 \\ 5 & 2 \end{bmatrix} \begin{bmatrix} 2 & -3 \\ -5 & 8 \end{bmatrix} = \begin{bmatrix} 1 & 0 \\ 0 & 1 \end{bmatrix} = I$$

and

$$A^{-1}A = \begin{bmatrix} 2 & -3 \\ -5 & 8 \end{bmatrix} \begin{bmatrix} 8 & 3 \\ 5 & 2 \end{bmatrix} = \begin{bmatrix} 1 & 0 \\ 0 & 1 \end{bmatrix} = I.$$

As a consequence, the solution of $AX = H$ is

$$X = A^{-1}H = \begin{bmatrix} 2 & -3 \\ -5 & 8 \end{bmatrix} \begin{bmatrix} 2 \\ -1 \end{bmatrix} = \begin{bmatrix} 7 \\ -18 \end{bmatrix}.$$

So $x_1 = 7$ and $x_2 = -18$ is the unique solution.

■ **EXAMPLE 2**

We return to our partial fractions example in Section 1.0. We had the system of linear equations

$$\begin{array}{rcl} a + b & = & 2 \\ a \quad\quad + c & = & -4 \\ a \quad\quad\quad & = & 3, \end{array}$$

so our coefficient matrix is

$$A = \begin{bmatrix} 1 & 1 & 0 \\ 1 & 0 & 1 \\ 1 & 0 & 0 \end{bmatrix}.$$

If

$$\begin{bmatrix} 1 & 1 & 0 \\ 1 & 0 & 1 \\ 1 & 0 & 0 \end{bmatrix} \begin{bmatrix} x_1 \\ x_2 \\ x_3 \end{bmatrix} = \begin{bmatrix} 0 \\ 0 \\ 0 \end{bmatrix},$$

then

$$x_1 + x_2 = 0$$

$$x_1 + x_3 = 0$$

$$x_1 \quad\quad = 0.$$

Therefore, $x_1 = 0$, $x_2 = 0$, and $x_3 = 0$. So A^{-1} exists.

Here it turns out that

$$A^{-1} = \begin{bmatrix} 0 & 0 & 1 \\ 1 & 0 & -1 \\ 0 & 1 & -1 \end{bmatrix}.$$

Again we can verify that

$$\begin{bmatrix} 0 & 0 & 1 \\ 1 & 0 & -1 \\ 0 & 1 & -1 \end{bmatrix} \begin{bmatrix} 1 & 1 & 0 \\ 1 & 0 & 1 \\ 1 & 0 & 0 \end{bmatrix} = \begin{bmatrix} 1 & 1 & 0 \\ 1 & 0 & 1 \\ 1 & 0 & 0 \end{bmatrix} \begin{bmatrix} 0 & 0 & 1 \\ 1 & 0 & -1 \\ 0 & 1 & -1 \end{bmatrix}$$

$$= \begin{bmatrix} 1 & 0 & 0 \\ 0 & 1 & 0 \\ 0 & 0 & 1 \end{bmatrix}.$$

Therefore, the solution of

$$\begin{bmatrix} 1 & 1 & 0 \\ 1 & 0 & 1 \\ 1 & 0 & 0 \end{bmatrix} \begin{bmatrix} a \\ b \\ c \end{bmatrix} = \begin{bmatrix} 2 \\ -4 \\ 3 \end{bmatrix}$$

is

$$\begin{bmatrix} a \\ b \\ c \end{bmatrix} = A^{-1} \begin{bmatrix} 2 \\ -4 \\ 3 \end{bmatrix} = \begin{bmatrix} 0 & 0 & 1 \\ 1 & 0 & -1 \\ 0 & 1 & -1 \end{bmatrix} \begin{bmatrix} 2 \\ -4 \\ 3 \end{bmatrix} = \begin{bmatrix} 3 \\ -1 \\ -7 \end{bmatrix},$$

so

$$\frac{2x^2 - 4x + 3}{x^3 + x^2 + x} = \frac{3}{x} + \frac{-x - 7}{x^2 + x + 1}.$$

■ **EXAMPLE 3**

Consider the equation

$$\begin{bmatrix} 1 & 3 \\ -2 & -6 \end{bmatrix} \begin{bmatrix} x_1 \\ x_2 \end{bmatrix} = \begin{bmatrix} 2 \\ 3 \end{bmatrix}.$$

The matrix

$$A = \begin{bmatrix} 1 & 3 \\ -2 & -6 \end{bmatrix}$$

is singular as seen by the following computation:

$$AX = \begin{bmatrix} 1 & 3 \\ -2 & -6 \end{bmatrix} \begin{bmatrix} x_1 \\ x_2 \end{bmatrix} = \begin{bmatrix} 0 \\ 0 \end{bmatrix} = 0$$

gives

$$x_1 + 3x_2 = 0$$
$$-2x_1 - 6x_2 = 0.$$

This has infinitely many nonzero solutions. One such solution is

$$x_1 = -3, \qquad x_2 = 1.$$

The singularity of A guarantees that A^{-1} does *not* exist. The reader should check that the equation

$$AX = \begin{bmatrix} 1 & 3 \\ -2 & -6 \end{bmatrix} \begin{bmatrix} x_1 \\ x_2 \end{bmatrix} = \begin{bmatrix} 2 \\ 3 \end{bmatrix} = H$$

does *not* have a solution.

However, the equation

$$AX = \begin{bmatrix} 1 & 3 \\ -2 & -6 \end{bmatrix} \begin{bmatrix} x_1 \\ x_2 \end{bmatrix} = \begin{bmatrix} 4 \\ -8 \end{bmatrix} = H$$

has a solution: $x_1 = 1$, $x_2 = 1$. Also, $x_1 = -3 - 3c$, $x_2 = 1 + c$ is a solution for all choices of the number c. This illustrates the fact that when A is singular, the equation $AX = H$ has solutions X for some choices of H but not for *all* choices of H. Moreover, if it does have a solution, it has many solutions.

In the next sections we give a general method for finding inverses, but for now we turn to some algebraic properties of the inverse.

If A is invertible, then A is the inverse of its inverse A^{-1}. Accordingly, we have

$$(A^{-1})^{-1} = A.$$

Now consider the matrices

$$A = \begin{bmatrix} 8 & 3 \\ 5 & 2 \end{bmatrix}, \qquad X = \begin{bmatrix} 2 & -3 \\ -5 & 8 \end{bmatrix},$$

$$A^T = \begin{bmatrix} 8 & 5 \\ 3 & 2 \end{bmatrix}, \qquad X^T = \begin{bmatrix} 2 & -5 \\ -3 & 8 \end{bmatrix}.$$

Performing the multiplications

$$AX = \begin{bmatrix} 8 & 3 \\ 5 & 2 \end{bmatrix} \begin{bmatrix} 2 & -3 \\ -5 & 8 \end{bmatrix} = \begin{bmatrix} 1 & 0 \\ 0 & 1 \end{bmatrix} = \begin{bmatrix} 2 & -3 \\ -5 & 8 \end{bmatrix} \begin{bmatrix} 8 & 3 \\ 5 & 2 \end{bmatrix} = XA,$$

we see that $X = A^{-1}$ and that $A = X^{-1}$. A similar computation will show that X^T is the inverse of A^T and that A^T is the inverse of X^T. These computations motivate the following result:

> If A is invertible, so is A^T. Moreover, the transpose of A^{-1} is the inverse of A^T. That is,
>
> $$(A^{-1})^T = (A^T)^{-1}.$$

This follows because, by transposition, the equations $AA^{-1} = A^{-1}A = I$ become $(A^{-1})^T A^T = A^T (A^{-1})^T = I$, which imply that the transpose of A^{-1} is the inverse of A^T. Consider a second 2×2-matrix B and its inverse

$$B = \begin{bmatrix} 1 & 0 \\ -2 & 1 \end{bmatrix}, \qquad B^{-1} = \begin{bmatrix} 1 & 0 \\ 2 & 1 \end{bmatrix}.$$

With A chosen as above, we find that

$$AB = \begin{bmatrix} 8 & 3 \\ 5 & 2 \end{bmatrix} \begin{bmatrix} 1 & 0 \\ -2 & 1 \end{bmatrix} = \begin{bmatrix} 2 & 3 \\ 1 & 2 \end{bmatrix},$$

$$B^{-1}A^{-1} = \begin{bmatrix} 1 & 0 \\ 2 & 1 \end{bmatrix} \begin{bmatrix} 2 & -3 \\ -5 & 8 \end{bmatrix} = \begin{bmatrix} 2 & -3 \\ -1 & 2 \end{bmatrix}.$$

It is easily seen, by mental computations, that $B^{-1}A^{-1}$ is the inverse of AB. Again these computations motivate the following result:

> If A and B are invertible, so is AB. Moreover, $B^{-1}A^{-1}$ is the inverse of AB. That is, the inverse of a product is the product of their inverses in *reverse order.* Briefly,
>
> $$(AB)^{-1} = B^{-1}A^{-1}.$$

This follows from the observation that

$$(AB)(B^{-1}A^{-1}) = A(BB^{-1})A^{-1} = AIA^{-1} = I.$$

Similarly, $B^{-1}A^{-1}AB = I$.

In constructing the 2×2-matrices used above, we made use of the fact that the inverse of the matrix

$$A = \begin{bmatrix} a & b \\ c & d \end{bmatrix}$$

is given by the formula

$$A^{-1} = \begin{bmatrix} d/\Delta & -b/\Delta \\ -c/\Delta & a/\Delta \end{bmatrix}, \qquad \begin{array}{l} \Delta = ad - bc \quad (\Delta \text{ is the Greek capital} \\ \text{letter delta)} \end{array}$$

when the number $\Delta = ad - bc$ is not zero. The number $\Delta = ad - bc$ is called the **determinant** of A. This result can be verified by performing the multiplication

$$\begin{bmatrix} a & b \\ c & d \end{bmatrix} \begin{bmatrix} d/\Delta & -b/\Delta \\ -c/\Delta & a/\Delta \end{bmatrix} = \begin{bmatrix} \dfrac{ad - bc}{\Delta} & \dfrac{-ab + ab}{\Delta} \\ \dfrac{cd - dc}{\Delta} & \dfrac{-bc + ad}{\Delta} \end{bmatrix} = \begin{bmatrix} 1 & 0 \\ 0 & 1 \end{bmatrix}$$

and a similar multiplication with the order of the matrices reversed. When we study determinants in Chapter 3, we give a formula for the inverse of a general $n \times n$-matrix A. However, this formula is not useful for computational purposes except when $n = 2$ and $n = 3$. A practical procedure for finding inverses is presented in Section 2.3.

To see what happens when $\Delta = ad - bc = 0$, while a, b, c, and d are not all zero, we introduce the nonzero matrix

$$X = \begin{bmatrix} d & -b \\ -c & a \end{bmatrix}.$$

We have

$$AX = \begin{bmatrix} a & b \\ c & d \end{bmatrix} \begin{bmatrix} d & -b \\ -c & a \end{bmatrix} = \begin{bmatrix} ad - bc & 0 \\ 0 & ad - bc \end{bmatrix} = \begin{bmatrix} 0 & 0 \\ 0 & 0 \end{bmatrix},$$

so that $AX = 0$ has a solution X that is not zero. This signifies that A is singular and so is not invertible. It will be seen that the determinant plays a key role in the idea of invertibility.

Finally, we ask what happens to the inverse of A when A is multiplied by a nonzero scalar s? To see what happens, consider the matrices

$$A = \begin{bmatrix} 8 & 3 \\ 5 & 2 \end{bmatrix}, \qquad A^{-1} = \begin{bmatrix} 2 & -3 \\ -5 & 8 \end{bmatrix},$$

$$B = 2A = \begin{bmatrix} 16 & 6 \\ 10 & 4 \end{bmatrix}, \qquad B^{-1} = \begin{bmatrix} 1 & -3/2 \\ -5/2 & 4 \end{bmatrix}.$$

Clearly, $B^{-1} = (1/2)A^{-1}$. This illustrates the following result, which is easily verified.

If A is invertible and s is not zero, then $B = sA$ is invertible and

$$B^{-1} = \left(\frac{1}{s}\right) A^{-1}.$$

Exercises

1. Verify that the following pairs of matrices are inverses of each other.

(a) $A_1 = \begin{bmatrix} 13 & 5 \\ 5 & 2 \end{bmatrix}$,　$A_2 = \begin{bmatrix} 2 & -5 \\ -5 & 13 \end{bmatrix}$

(b) $B_1 = \begin{bmatrix} 5 & 2 \\ 1 & -7 \end{bmatrix}$,　$B_2 = \begin{bmatrix} 7/37 & 2/37 \\ 1/37 & -5/37 \end{bmatrix}$

(c) $C_1 = \begin{bmatrix} 3 & -1 & 1 \\ -2 & 1 & 1 \\ 1 & -1 & -2 \end{bmatrix}$,　$C_2 = \begin{bmatrix} -1 & -3 & -2 \\ -3 & -7 & -5 \\ 1 & 2 & 1 \end{bmatrix}$

(d) $D_1 = \begin{bmatrix} 3 & 0 & 0 \\ -2 & 1 & 0 \\ 1 & 0 & 1 \end{bmatrix}$,　$D_2 = \begin{bmatrix} 1/3 & 0 & 0 \\ 2/3 & 1 & 0 \\ -1/3 & 0 & 1 \end{bmatrix}$

(e) $E_1 = \begin{bmatrix} 3 & 0 & 0 \\ 0 & 5 & 0 \\ 0 & 0 & -2 \end{bmatrix}$,　$E_2 = \begin{bmatrix} 1/3 & 0 & 0 \\ 0 & 1/5 & 0 \\ 0 & 0 & -1/2 \end{bmatrix}$

(f) $F_1 = \begin{bmatrix} 0 & 1 & 0 \\ 0 & 0 & 1 \\ 1 & 0 & 0 \end{bmatrix}$,　$F_2 = \begin{bmatrix} 0 & 0 & 1 \\ 1 & 0 & 0 \\ 0 & 1 & 0 \end{bmatrix} = F_1^T$

2. Referring to Exercise 1, compute the inverses of the following matrices:

(a) $5A_1$　　(b) A_1B_1　　(c) $2C_1(E_1^T)$　　(d) $D_2C_1D_1$　　(e) A_1^2

Hint: Recall the properties of inverses of products!

3. By the use of the definition of nonsingularity, show that the matrices

$$A_1 = \begin{bmatrix} 13 & 5 \\ 5 & 2 \end{bmatrix},　B_1 = \begin{bmatrix} 5 & 2 \\ 1 & -7 \end{bmatrix}$$

are, in fact, nonsingular.

4. Referring to Exercise 1, solve the following matrix equations by using inverses.

(a) $\begin{bmatrix} 13 & 5 \\ 5 & 2 \end{bmatrix}\begin{bmatrix} x_1 \\ x_2 \end{bmatrix} = \begin{bmatrix} 3 \\ -4 \end{bmatrix}$　　(b) $\begin{bmatrix} 5 & 2 \\ 1 & -7 \end{bmatrix}\begin{bmatrix} x_1 \\ x_2 \end{bmatrix} = \begin{bmatrix} 0 \\ 1 \end{bmatrix}$

(c) $\begin{bmatrix} 3 & -1 & 1 \\ -2 & 1 & 1 \\ 1 & -1 & -2 \end{bmatrix}\begin{bmatrix} x_1 \\ x_2 \\ x_3 \end{bmatrix} = \begin{bmatrix} 3 \\ 6 \\ 9 \end{bmatrix}$　　(d) $\begin{bmatrix} 2 & -5 \\ -5 & 13 \end{bmatrix}\begin{bmatrix} x_1 \\ x_2 \end{bmatrix} = \begin{bmatrix} -2 \\ 3 \end{bmatrix}$

5. In each of the following cases, determine the scalar b so that B is the inverse of A.

(a) $A = \begin{bmatrix} 1 & 2 \\ 3 & 4 \end{bmatrix}$,　$B = b\begin{bmatrix} 4 & -2 \\ -3 & 1 \end{bmatrix}$

(b) $A = \begin{bmatrix} 3 & 0 & 0 \\ 0 & 1 & 0 \\ 0 & 0 & 1 \end{bmatrix}$, $\quad B = b\begin{bmatrix} 1 & 0 & 0 \\ 0 & 3 & 0 \\ 0 & 0 & 3 \end{bmatrix}$

(c) $A = \begin{bmatrix} 1 & 1 & 0 \\ 0 & 2 & 0 \\ 0 & 3 & 1 \end{bmatrix}$, $\quad B = \begin{bmatrix} 1 & -b & 0 \\ 0 & b & 0 \\ 0 & -3b & 1 \end{bmatrix}$

(d) $A = \begin{bmatrix} 1 & 0 & 4 \\ 0 & 1 & 0 \\ 0 & 0 & 1 \end{bmatrix}$, $\quad B = \begin{bmatrix} 1 & 0 & b \\ 0 & 1 & 0 \\ 0 & 0 & 1 \end{bmatrix}$

6. Show that the following matrices are singular.

$$A = \begin{bmatrix} 1 & 1 \\ 1 & 1 \end{bmatrix}, \qquad B = \begin{bmatrix} 2 & 3 \\ -4 & -6 \end{bmatrix},$$

$$C = \begin{bmatrix} 1 & 1 & 1 \\ 0 & 0 & 0 \\ 0 & 0 & 1 \end{bmatrix}, \qquad D = \begin{bmatrix} 1 & 0 & 3 & 1 \\ 0 & 1 & 4 & 1 \\ 0 & 0 & 0 & 1 \\ 0 & 0 & 0 & 2 \end{bmatrix}.$$

7. Which of the following matrices are nonsingular and which are singular?

$$A = \begin{bmatrix} 1 & 3 \\ 1 & 2 \end{bmatrix}, \qquad B = \begin{bmatrix} -1 & -2 \\ -1 & -2 \end{bmatrix}, \qquad C = \begin{bmatrix} 0 & -3 \\ 0 & 4 \end{bmatrix},$$

$$D = \begin{bmatrix} 1 & 1 \\ 1 & 0 \end{bmatrix}, \qquad E = \begin{bmatrix} 3 & 2 & 0 \\ 1 & 1 & 2 \\ 4 & 3 & 2 \end{bmatrix}.$$

8. By the formula for the inverse of a 2×2 matrix given in the text, determine the inverses of the following matrices, where possible. When not possible, explain why not.

$$A = \begin{bmatrix} 1 & 2 \\ 2 & 5 \end{bmatrix}, \qquad B = \begin{bmatrix} 3 & -2 \\ 7 & 4 \end{bmatrix}, \qquad C = \begin{bmatrix} 1 & 1 \\ 1 & 1 \end{bmatrix},$$

$$D = \begin{bmatrix} -3 & 0 \\ 2 & 6 \end{bmatrix}, \qquad E = \begin{bmatrix} 18 & 54 \\ 2 & 6 \end{bmatrix}.$$

9. (a) Let A be an invertible matrix. Show that A^{-2} is the inverse of A^2, where

$$A^{-2} = (A^{-1})^2.$$

(b) What is the inverse of A^3? A^5?

10. (a) Consider the matrices

$$A = \begin{bmatrix} 1 & 1 & 0 \\ 0 & 1 & 1 \\ 0 & 0 & 1 \end{bmatrix}, \qquad E = \begin{bmatrix} 0 & 1 & 0 \\ 0 & 0 & 1 \\ 0 & 0 & 0 \end{bmatrix}.$$

Show that $E^3 = 0$.

(b) Show that $A = I + E$ and that

$$A^{-1} = I - E + E^2 = \begin{bmatrix} 1 & -1 & 1 \\ 0 & 1 & -1 \\ 0 & 0 & 1 \end{bmatrix}.$$

Notice that $AA^{-1} = (I + E)(I - E + E^2) = I + E^3 = I$.

11. Referring to Exercise 10 as an example, E is called a **nilpotent matrix** because $E^3 = 0$. In general, a matrix A is **nilpotent** if $A^n = 0$ for some integer n. Prove that a nilpotent matrix is always singular. *Hint:* If $n = 1$, then $A = 0$, so A is singular. Otherwise, select the smallest integer n such that $A^n = 0$. Then $X = A^{n-1}$ is a nonzero solution to $AX = 0$. Conclude that A is singular.

12. Show that if a matrix A satisfies the polynomial equation

$$x^3 + 3x^2 - 5x + 1 = 0,$$

then A is invertible. That is, if

$$A^3 + 3A^2 - 5A + I = 0,$$

then A is invertible. (Find the expression for its inverse.) *Hint:* Write the equation in the form $A(-A^2 - 3A + 5) = I$.

13. Determine whether the following matrix equations have no solutions, or infinitely many solutions. If solutions exist, find them.

(a) $\begin{bmatrix} 1 & 1 \\ 1 & 1 \end{bmatrix} \begin{bmatrix} x_1 \\ x_2 \end{bmatrix} = \begin{bmatrix} 1 \\ 2 \end{bmatrix}$ **(b)** $\begin{bmatrix} 1 & 1 \\ 1 & 1 \end{bmatrix} \begin{bmatrix} x_1 \\ x_2 \end{bmatrix} = \begin{bmatrix} -3 \\ -3 \end{bmatrix}$

(c) $\begin{bmatrix} -4 & 2 \\ -6 & 3 \end{bmatrix} \begin{bmatrix} x_1 \\ x_2 \end{bmatrix} = \begin{bmatrix} 6 \\ 9 \end{bmatrix}$ **(d)** $\begin{bmatrix} -4 & 2 \\ -6 & 3 \end{bmatrix} \begin{bmatrix} x_1 \\ x_2 \end{bmatrix} = \begin{bmatrix} -1 \\ 1 \end{bmatrix}.$

Notice how the change in the column vector on the right affects the type of solution category.

14. Determine whether the following matrix equations have no solution, a unique solution, or infinitely many solutions.

(a) $\begin{bmatrix} 1 & 2 \\ 4 & 3 \end{bmatrix} \begin{bmatrix} x_1 \\ x_2 \end{bmatrix} = \begin{bmatrix} 5 \\ -2 \end{bmatrix}$ **(b)** $\begin{bmatrix} 3 & 0 \\ 1 & 0 \end{bmatrix} \begin{bmatrix} x_1 \\ x_2 \end{bmatrix} = \begin{bmatrix} 0 \\ 5 \end{bmatrix}$

(c) $\begin{bmatrix} 4 & -1 \\ 8 & -2 \end{bmatrix} \begin{bmatrix} x_1 \\ x_2 \end{bmatrix} = \begin{bmatrix} 10 \\ 5 \end{bmatrix}$

15. Show that if an equation $AX = H$ has two distinct solutions (so that A is singular), it has infinitely many solutions. *Hint:* Since A is singular, there exists $X_0 \neq 0$ such that $AX_0 = 0$. Look at the expression $Y + kX_0$, where $AY = H$ and k is any real number.

16. If A is nonsquare, why does there not exist a matrix X such that

$$AX = I, \qquad XA = I$$

even when the two *I*'s are of different sizes? *Hint:* If *A* is *m* × *n* with *m* > *n*, there is a nonzero *Y* such that *YA* = 0, contrary to the relation

$$0 = YAX = YI = Y.$$

Hence *AX* = *I* cannot hold.

17. Let *A* be an *m* × *n*-matrix. It can be shown that there exists a unique *n* × *m*-matrix *X* such that the square matrices

$$E = AX, \quad F = XA$$

are symmetric and satisfy the relations

$$EA = A, \quad FX = X.$$

The matrix *X* is called the *pseudoinverse* of *A*. Verify that

$$AF = A, \quad XE = X, \quad E^2 = E, \quad F^2 = F.$$

Verify that if *A* is a nonsingular square matrix, then *E* = *F* = *I* and *X* = *A*⁻¹.

18. Show that the inverse of the matrix

$$E = \begin{bmatrix} \cos\theta & \sin\theta \\ -\sin\theta & \cos\theta \end{bmatrix}$$

is given by *E*⁻¹ = *E*ᵀ.

19. Suppose that *a* and *b* are not both zero. Find the inverse of

$$A = \begin{bmatrix} a & b \\ -b & a \end{bmatrix}.$$

Show that there is a number *c* such that *A*⁻¹ = *cA*ᵀ.

(a) Find a formula for *c*.

(b) What are *c* and *A*⁻¹ for the case in which *a* = 4, *b* = 3? For the case *a* = 12, *b* = −5?

20. Let *A* be a matrix with the property that *A*⁻¹ = *cA*ᵀ. Show that the matrix

$$B = \begin{bmatrix} A & A \\ -A & A \end{bmatrix}$$

is of the form *B*⁻¹ = *dB*ᵀ.

(a) Find a formula for *d*.

(b) Find *B*⁻¹ for the case

$$A = \begin{bmatrix} 4 & 3 \\ -3 & 4 \end{bmatrix}.$$

(c) Find the inverse for the case in which *A* is the 1 × 1-matrix *A* = [1].

(d) Find a formula for the inverse of the matrix

$$C = \begin{bmatrix} B & B \\ -B & B \end{bmatrix}.$$

21. Starting with a nonzero 1×1-matrix A, construct successively the matrices

$$B = \begin{bmatrix} A & A \\ -A & A \end{bmatrix}, \qquad C = \begin{bmatrix} B & B \\ -B & B \end{bmatrix}, \qquad D = \begin{bmatrix} C & C \\ -C & C \end{bmatrix}.$$

(a) What is the size of B, C, and D?

(b) Use the results given in Exercise 20 to construct the inverses of B, C, and D.

(c) Recall that two column vectors X and Y are orthogonal if $X^T Y = 0$. Show that the columns of B, of C, and of D are mutually orthogonal.

(d) Show that the columns of B, of C, and of D are linearly independent.

22. Let M be a square matrix. Verify the following statements.

(a) M is invertible if and only if M^T is invertible.

(b) M is nonsingular if and only if M^T is nonsingular.

(c) M is nonsingular if and only if its columns are linearly independent.

(d) M is singular if and only if its columns are linearly dependent.

(e) M is nonsingular if and only if its rows are linearly independent.

(f) M is singular if and only if its rows are linearly dependent.

(g) The columns of M are linearly independent if and only if its rows are linearly independent.

(h) The columns of M are linearly dependent if and only if its rows are linearly dependent.

23. Consider a square matrix of the form

$$M = \begin{bmatrix} A & 0 \\ P & B \end{bmatrix},$$

where A is $p \times p$ and B is $q \times q$. Verify the following statements.

(a) If A is singular, so is M.

(b) If B is singular, so is M.

(c) If M is nonsingular, so are A and B.

(d) If A and B are invertible, so is M. The formula for M^{-1} is given by the formula

$$M^{-1} = \begin{bmatrix} A^{-1} & 0 \\ -B^{-1}PA^{-1} & B^{-1} \end{bmatrix}.$$

24. Consider the following matrices and their inverse.

$$Q = \begin{bmatrix} 2 & 0 \\ 5 & 3 \end{bmatrix}, \qquad Q^{-1} = \begin{bmatrix} 1/2 & 0 \\ -5/6 & 1/3 \end{bmatrix},$$

$$S = \begin{bmatrix} 2 & 0 & 0 \\ 5 & 3 & 0 \\ 5 & 9 & 4 \end{bmatrix}, \qquad S^{-1} = \begin{bmatrix} 1/2 & 0 & 0 \\ -5/6 & 1/3 & 0 \\ 5/4 & -3/4 & 1/4 \end{bmatrix}$$

(a) Referring to Exercise 23, show that if we set $A = [2]$, $B = [3]$, $P = [5]$, we find that $M = Q$. Verify that Q^{-1} can be obtained by the formula for M^{-1} given in Exercise 23.

(b) Referring to Exercise 23, set $A = Q$, $B = [4]$, $P = [5 \quad 9]$. We find that $M = S$. Verify that S^{-1} can be obtained by the formula for M^{-1} given in Exercise 23.

25. Consider the matrices

$$H = \begin{bmatrix} 3 & 2 \\ 4 & 3 \end{bmatrix}, \quad K = \begin{bmatrix} 3 & 2 & 0 \\ 4 & 3 & 0 \\ 2 & 1 & 1 \end{bmatrix}, \quad L = \begin{bmatrix} 3 & 2 & 0 & 0 \\ 4 & 3 & 0 & 0 \\ 2 & 1 & 1 & 0 \\ 1 & -1 & 2 & 1 \end{bmatrix}.$$

(a) Verify that their inverses are

$$H^{-1} = \begin{bmatrix} 3 & -2 \\ -4 & 3 \end{bmatrix}, \quad K^{-1} = \begin{bmatrix} 3 & -2 & 0 \\ -4 & 3 & 0 \\ -2 & 1 & 1 \end{bmatrix},$$

$$L^{-1} = \begin{bmatrix} 3 & -2 & 0 & 0 \\ -4 & 3 & 0 & 0 \\ -2 & 1 & 1 & 0 \\ -3 & -1 & -2 & 1 \end{bmatrix}.$$

(b) Obtain the inverse of the 2×2-matrix H by the formula given in the text.

(c) Observe that $M = K$ is of the form given in Exercise 23 with $A = H$, $B = [1]$, $P = [-2 \quad 1]$. Use the formula for the inverse M^{-1} given there to obtain K^{-1}.

(d) Observe that $M = L$ is of the form given in Exercise 23 with $A = K$, $B = [1]$, $P = [1 \quad -1 \quad 2]$. Use the formula for M^{-1} given there to obtain L^{-1}.

(e) Observe that $M = L$ is also of the form given in Exercise 23 with $A = H$ and with

$$B = \begin{bmatrix} 1 & 0 \\ 2 & 1 \end{bmatrix}, \quad P = \begin{bmatrix} 2 & 1 \\ 1 & -1 \end{bmatrix}, \quad B^{-1} = \begin{bmatrix} 1 & 0 \\ -2 & 1 \end{bmatrix}.$$

Use the formula for M^{-1} to obtain L^{-1}.

26. Consider the general matrices

$$H = \begin{bmatrix} a & b \\ c & d \end{bmatrix}, \quad K = \begin{bmatrix} a & b & 0 \\ c & d & 0 \\ e & f & 1 \end{bmatrix},$$

$$L = \begin{bmatrix} a & b & 0 & 0 \\ c & d & 0 & 0 \\ e & f & 1 & 0 \\ g & h & k & 1 \end{bmatrix}.$$

(a) As noted in the text, H is singular if and only if $\Delta = ad - bc = 0$. Conclude, by Exercise 23, that K and L are singular if and only if $\Delta = 0$.

(b) Suppose that Δ is not zero. Then the inverses of H, K, and L are given by the formulas

$$H^{-1} = \frac{1}{\Delta}\begin{bmatrix} d & -b \\ -c & a \end{bmatrix}, \quad K^{-1} = \frac{1}{\Delta}\begin{bmatrix} d & -b & 0 \\ -c & a & 0 \\ p & q & \Delta \end{bmatrix},$$

$$L^{-1} = \frac{1}{\Delta}\begin{bmatrix} d & -b & 0 & 0 \\ -c & a & 0 & 0 \\ p & q & \Delta & 0 \\ r & s & -k\Delta & \Delta \end{bmatrix},$$

where

$$p = -ed + fc, \qquad q = eb - fa, \qquad r = -gd + hc - kp,$$
$$s = gb - ha - kq.$$

27. Use the formulas given in Exercise 26 to obtain the inverses of the matrices H, K, and L described in Exercise 25.

28. Referring to Exercise 26, construct the matrices H, K, and L and their inverses for the cases in which:

(a) $a = d = 1, \quad b = c = 0, \quad e = f = g = h = 1, \quad k = 0$

(b) $a = 5, \quad b = -1, \quad c = 2, \quad d = 3, \quad e = h = 1, \quad f = g = 0, \quad k = 5$

2.2 ELEMENTARY MATRICES: SOME EASILY INVERTIBLE MATRICES

The easiest matrix to invert is the identity I. It is its own inverse. By simple modifications of I we obtain other square matrices that are easy to invert. We shall consider primarily three types of matrices, called *elementary matrices*. As we shall see, elementary matrices are the building blocks of invertible matrices.

Elementary Matrices of the First Kind

Matrices obtained by rearranging (permuting) any *two* of the rows or columns of the identity I are called *elementary matrices of the first kind*. These are also called *elementary permutation matrices* because, as we shall see, they can be used to permute any two rows or columns of a matrix. We begin by considering the matrices

$$P_{12} = \begin{bmatrix} 0 & 1 & 0 \\ 1 & 0 & 0 \\ 0 & 0 & 1 \end{bmatrix}, \quad P_{13} = \begin{bmatrix} 0 & 0 & 1 \\ 0 & 1 & 0 \\ 1 & 0 & 0 \end{bmatrix}, \quad P_{23} = \begin{bmatrix} 1 & 0 & 0 \\ 0 & 0 & 1 \\ 0 & 1 & 0 \end{bmatrix}.$$

These are the 3×3 *elementary matrices of the first kind.* Observe that the matrix P_{12} is obtained from the identity

$$I = \begin{bmatrix} 1 & 0 & 0 \\ 0 & 1 & 0 \\ 0 & 0 & 1 \end{bmatrix}$$

by interchanging the first and second rows (or columns) of I. Similarly, P_{13} is obtained by interchanging rows 1 and 3 of I (or columns 1 and 3 of I), and P_{23} is found by interchanging rows 2 and 3 (or columns 2 and 3) of I. These matrices are symmetric and are their own inverses. That is,

$$P_{ij}^{T} = P_{ij}, \qquad P_{ij}^{-1} = P_{ij}.$$

Observe further that if A is a three-rowed matrix, such as

$$A = \begin{bmatrix} a & b & c & d \\ p & q & r & s \\ w & x & y & z \end{bmatrix},$$

then the matrix

$$P_{13}A = \begin{bmatrix} 0 & 0 & 1 \\ 0 & 1 & 0 \\ 1 & 0 & 0 \end{bmatrix} \begin{bmatrix} a & b & c & d \\ p & q & r & s \\ w & x & y & z \end{bmatrix} = \begin{bmatrix} w & x & y & z \\ p & q & r & s \\ a & b & c & d \end{bmatrix}$$

is obtained from A by interchanging rows 1 and 3. A similar computation will show that forming the product $P_{12}A$ is equivalent to interchanging rows 1 and 2 in A. Also, $P_{23}A$ can be obtained from A by interchanging rows 2 and 3.

Similarly, we also have the property that the product BP_{ij} is obtained from B by interchanging columns i and j of B (notice that the order of the product is reversed for column operations). For example, if

$$B = \begin{bmatrix} e & l & r \\ f & m & s \\ g & n & t \\ h & p & u \end{bmatrix},$$

then

$$
BP_{13} = \begin{bmatrix} e & l & r \\ f & m & s \\ g & n & t \\ h & p & u \end{bmatrix} \begin{bmatrix} 0 & 0 & 1 \\ 0 & 1 & 0 \\ 1 & 0 & 0 \end{bmatrix} = \begin{bmatrix} r & l & e \\ s & m & f \\ t & n & g \\ u & p & h \end{bmatrix}.
$$

In the general case, we have:

The $n \times n$-matrix P_{ij} obtained from the $n \times n$ identity matrix I by interchanging rows i and j (or columns i and j) of I is called an *elementary permutation matrix* or an *elementary matrix of the first kind*. Here, P_{ij} is symmetric and is its own inverse.

The product $P_{ij}A$ is obtained from A by interchanging *rows* i and j of A. Similarly, the product BP_{ij} is obtained from B by interchanging *columns* i and j of B.

The operation of interchanging two rows of A is called an *elementary row operation of the first kind*. Interchanging two columns of A is an *elementary column operation of the first kind*.

Diagonal Matrices

Before we consider elementary matrices of the second kind, we shall examine more closely the concept of a *diagonal matrix*. Diagonal matrices are square matrices of the form

$$
\begin{bmatrix} a & 0 \\ 0 & b \end{bmatrix}, \quad \begin{bmatrix} a & 0 & 0 \\ 0 & b & 0 \\ 0 & 0 & c \end{bmatrix}, \quad \begin{bmatrix} a & 0 & 0 & 0 \\ 0 & b & 0 & 0 \\ 0 & 0 & c & 0 \\ 0 & 0 & 0 & d \end{bmatrix}, \quad \cdots
$$

They are called *diagonal matrices* because their only nonzero entries, if any, appear on the main diagonal. A diagonal matrix D is invertible if and only if no main diagonal entry of D is zero. When its main diagonal entries are all different from zero, the inverse of a diagonal matrix D is obtained by replacing each diagonal entry d by its reciprocal, $1/d$. For example, when the numbers a, b, c, and d appearing in the diagonal matrices displayed above are not zero, the inverses of these matrices are

$$
\begin{bmatrix} 1/a & 0 \\ 0 & 1/b \end{bmatrix}, \quad \begin{bmatrix} 1/a & 0 & 0 \\ 0 & 1/b & 0 \\ 0 & 0 & 1/c \end{bmatrix}, \quad \begin{bmatrix} 1/a & 0 & 0 & 0 \\ 0 & 1/b & 0 & 0 \\ 0 & 0 & 1/c & 0 \\ 0 & 0 & 0 & 1/d \end{bmatrix}.
$$

It is easily seen that the diagonal matrices

$$\begin{bmatrix} 3 & 0 & 0 \\ 0 & -5 & 0 \\ 0 & 0 & 7/4 \end{bmatrix}, \quad \begin{bmatrix} 1/3 & 0 & 0 \\ 0 & -1/5 & 0 \\ 0 & 0 & 4/7 \end{bmatrix}$$

are inverses of each other. However, the diagonal matrices

$$\begin{bmatrix} 0 & 0 & 0 \\ 0 & -5 & 0 \\ 0 & 0 & 7/4 \end{bmatrix}, \quad \begin{bmatrix} 1/3 & 0 & 0 \\ 0 & -1/5 & 0 \\ 0 & 0 & 0 \end{bmatrix}, \quad \begin{bmatrix} 0 & 0 & 0 \\ 0 & -4 & 0 \\ 0 & 0 & 0 \end{bmatrix}$$

are not invertible, because each matrix has at least one zero on the main diagonal.

To see the effect of muliplying a matrix A on the left by a diagonal matrix D, consider the matrices

$$D = \begin{bmatrix} r & 0 & 0 \\ 0 & s & 0 \\ 0 & 0 & t \end{bmatrix}, \quad A = \begin{bmatrix} a & b & c & d \\ e & f & g & h \\ w & x & y & z \end{bmatrix}.$$

By performing the multiplication

$$DA = \begin{bmatrix} r & 0 & 0 \\ 0 & s & 0 \\ 0 & 0 & t \end{bmatrix} \begin{bmatrix} a & b & c & d \\ e & f & g & h \\ w & x & y & z \end{bmatrix} = \begin{bmatrix} ra & rb & rc & rd \\ se & sf & sg & sh \\ tw & tx & ty & tz \end{bmatrix}$$

it is seen that the product DA is obtained from A by multiplying the first row of A by the first main diagonal entry of D, the second row of A by the second main diagonal entry of D, and the third row of A by the third main diagonal entry of D. If

$$B = \begin{bmatrix} c & l & u \\ d & m & v \\ e & n & w \\ f & p & x \end{bmatrix},$$

then looking at the product

$$BD = \begin{bmatrix} c & l & u \\ d & m & v \\ e & n & w \\ f & p & x \end{bmatrix} \begin{bmatrix} r & 0 & 0 \\ 0 & s & 0 \\ 0 & 0 & t \end{bmatrix} = \begin{bmatrix} rc & sl & tu \\ rd & sm & tv \\ re & sn & tw \\ rf & sp & tx \end{bmatrix},$$

we see that multiplying a matrix B on the right by a diagonal matrix D is equivalent to multiplying each column of B by the corresponding main diagonal entry of D. We are now ready to proceed with elementary matrices of the second kind.

Elementary Matrices of the Second Kind

Consider the special diagonal matrices

$$S_1 = \begin{bmatrix} r & 0 & 0 \\ 0 & 1 & 0 \\ 0 & 0 & 1 \end{bmatrix}, \quad S_2 = \begin{bmatrix} 1 & 0 & 0 \\ 0 & s & 0 \\ 0 & 0 & 1 \end{bmatrix}, \quad S_3 = \begin{bmatrix} 1 & 0 & 0 \\ 0 & 1 & 0 \\ 0 & 0 & t \end{bmatrix},$$

where r, s, and t are not zero. These are the 3×3 *elementary matrices of the second kind.* These nonsingular diagonal matrices are obtained from the identity matrix I by multiplying a *single* row (or column) of I by a *nonzero* number.

As an example, S_1 here is obtained from I by multiplying the first row (or column) of I by the number r. Since these matrices are diagonal, their inverses are

$$S_1^{-1} = \begin{bmatrix} 1/r & 0 & 0 \\ 0 & 1 & 0 \\ 0 & 0 & 1 \end{bmatrix}, \quad S_2^{-1} = \begin{bmatrix} 1 & 0 & 0 \\ 0 & 1/s & 0 \\ 0 & 0 & 1 \end{bmatrix},$$

$$S_3^{-1} = \begin{bmatrix} 1 & 0 & 0 \\ 0 & 1 & 0 \\ 0 & 0 & 1/t \end{bmatrix}.$$

Clearly, these inverses are also elementary matrices of the second kind.

By looking at relations of the form

$$S_2 A = \begin{bmatrix} 1 & 0 & 0 \\ 0 & s & 0 \\ 0 & 0 & 1 \end{bmatrix} \begin{bmatrix} a & b & c & d \\ e & f & g & h \\ w & x & y & z \end{bmatrix}$$

$$= \begin{bmatrix} a & b & c & d \\ se & sf & sg & sh \\ w & x & y & z \end{bmatrix},$$

we see that the product $S_i A$ can be obtained from A by multiplying the scalar s throughout a *single row,* the ith row of A. The operation of multiplying a row of a matrix A by a nonzero scalar is called an *elementary row operation of the second kind.* A similar result may be obtained for column operations, by multiplying on the right by S_i, instead of on the left. For example, if

$$B = \begin{bmatrix} c & l & u \\ d & m & v \\ e & n & w \\ f & p & x \end{bmatrix},$$

then

$$BS_3 = \begin{bmatrix} c & l & u \\ d & m & v \\ e & n & w \\ f & p & x \end{bmatrix} \begin{bmatrix} 1 & 0 & 0 \\ 0 & 1 & 0 \\ 0 & 0 & t \end{bmatrix} = \begin{bmatrix} c & l & tu \\ d & m & tv \\ e & n & tw \\ f & p & tx \end{bmatrix}.$$

The basic properties of an elementary matrix of the second kind can be summarized as follows:

An *elementary matrix of the second kind* S_i is obtained from the identity matrix I by multiplying the ith row of I by the nonzero scalar s.

The inverse of S_i is obtained from S_i by replacing s by $1/s$.

The product $S_i A$ is obtained from A by multiplying the ith row of A by the nonzero scalar s. Similarly, the product BS_i is obtained from B by multiplying the ith column of B by the nonzero scalar s.

Since division of a row by a nonzero scalar s is the same as multiplying the row by $1/s$, it follows that division of a row by s is also an elementary row operation of the second kind, which corresponds to an appropriate elementary matrix of the second kind.

Elementary Matrices of the Third Kind

We now consider matrices of a third type, called *elementary matrices of the third kind*. For the case $n = 3$ they are of the form

$$T_{12} = \begin{bmatrix} 1 & 0 & 0 \\ a & 1 & 0 \\ 0 & 0 & 1 \end{bmatrix}, \qquad T_{13} = \begin{bmatrix} 1 & 0 & 0 \\ 0 & 1 & 0 \\ b & 0 & 1 \end{bmatrix},$$

$$T_{21} = \begin{bmatrix} 1 & c & 0 \\ 0 & 1 & 0 \\ 0 & 0 & 1 \end{bmatrix}, \qquad T_{23} = \begin{bmatrix} 1 & 0 & 0 \\ 0 & 1 & 0 \\ 0 & d & 1 \end{bmatrix},$$

$$T_{31} = \begin{bmatrix} 1 & 0 & e \\ 0 & 1 & 0 \\ 0 & 0 & 1 \end{bmatrix}, \qquad T_{32} = \begin{bmatrix} 1 & 0 & 0 \\ 0 & 1 & f \\ 0 & 0 & 1 \end{bmatrix}.$$

Observe that each matrix T_{ij} is obtained from the identity I by adding a multiple of its ith row to its jth row. Here it is understood that the indices i and j are distinct. The operation of adding a multiple of one row of a matrix to another row is called an **elementary row operation of the third kind.** Looking at the products

$$T_{31}A = \begin{bmatrix} 1 & 0 & e \\ 0 & 1 & 0 \\ 0 & 0 & 1 \end{bmatrix} \begin{bmatrix} a & b & c & d \\ p & q & r & s \\ w & x & y & z \end{bmatrix}$$

$$= \begin{bmatrix} a + ew & b + ex & c + ey & d + ez \\ p & q & r & s \\ w & x & y & z \end{bmatrix},$$

$$T_{32}A = \begin{bmatrix} 1 & 0 & 0 \\ 0 & 1 & f \\ 0 & 0 & 1 \end{bmatrix} \begin{bmatrix} a & b & c & d \\ p & q & r & s \\ w & x & y & z \end{bmatrix}$$

$$= \begin{bmatrix} a & b & c & d \\ p + fw & q + fx & r + fy & s + fz \\ w & x & y & z \end{bmatrix},$$

we see that performing an elementary row operation of the third kind on a matrix A is equivalent to multiplying A on the left by a suitably chosen elementary matrix of the third kind.

The inverse of an elementary matrix of the third kind is obtained by replacing the nonzero off diagonal entry in T_{ij} by the negative of that entry. For example, in the matrix

$$T_{13} = \begin{bmatrix} 1 & 0 & 0 \\ 0 & 1 & 0 \\ b & 0 & 1 \end{bmatrix},$$

we have, replacing b by $-b$,

$$T_{13}^{-1} = \begin{bmatrix} 1 & 0 & 0 \\ 0 & 1 & 0 \\ -b & 0 & 1 \end{bmatrix}.$$

This is verified by the computation

$$T_{13}T_{13}^{-1} = \begin{bmatrix} 1 & 0 & 0 \\ 0 & 1 & 0 \\ b & 0 & 1 \end{bmatrix} \begin{bmatrix} 1 & 0 & 0 \\ 0 & 1 & 0 \\ -b & 0 & 1 \end{bmatrix} = \begin{bmatrix} 1 & 0 & 0 \\ 0 & 1 & 0 \\ 0 & 0 & 1 \end{bmatrix} = I.$$

In the two-dimensional case, we have

$$T_{12} = \begin{bmatrix} 1 & 0 \\ s & 1 \end{bmatrix}, \qquad T_{21} = \begin{bmatrix} 1 & s \\ 0 & 1 \end{bmatrix},$$

$$T_{12}^{-1} = \begin{bmatrix} 1 & 0 \\ -s & 1 \end{bmatrix}, \qquad T_{21}^{-1} = \begin{bmatrix} 1 & -s \\ 0 & 1 \end{bmatrix}$$

as elementary matrices of the third kind.

We summarize the basic properties of elementary matrices of the third kind as follows:

An *elementary matrix of the third kind* T_{ij} is obtained from the identity matrix I by adding s times the ith row of I to the jth row of I.

The inverse of T_{ij} is obtained from T_{ij} by replacing s by $-s$.

The product $T_{ij}A$ is obtained from A by multiplying the ith row of A by s and adding to the jth row of A. Similarly, the product BT_{ij} is obtained from B by multiplying the jth column by s and adding to the ith column of B.

In the next section we will see how the elementary matrices can be used to find the inverse of any nonsingular matrix.

Just as the prime numbers are the building blocks for the integers in the sense that any integer

$$n = p_1 p_2 \cdots p_q$$

can be expressed as a product of primes, we shall see that any nonsingular, square matrix

$$A = E_1 E_2 \cdots E_q$$

can be expressed as a product of elementary matrices of the first, second, and third kinds. Accordingly, a nonsingular square matrix A is invertible and its inverse is given by the formula

$$A^{-1} = E_q^{-1} E_{q-1}^{-1} \cdots E_2^{-1} E_1^{-1}.$$

The proof of this result follows from the Gaussian elimination method for finding the inverse of a nonsingular matrix, which is the topic of the next section.

As a summary of this section, we include the following table:

Elementary matrix	Obtained from I by:	Effect on A (or B) by multiplication on left (right)
P_{ij}—first kind	Interchanging rows (columns) i and j of I	$P_{ij}A$—interchanges *rows* i and j of A BP_{ij}—interchanges *columns* i and j of B
S_i—second kind	Multiplying the ith row (column) of I by a nonzero scalar	S_iA—multiplies *row* i of A by a nonzero scalar BS_i—multiplies *column* i of B by a nonzero scalar
T_{ij}—third kind	Multiplying the ith row (jth column) by a scalar and adding it to the jth row (ith column) of I	$T_{ij}A$—multiplies *row* i of A by a scalar and adds it to row j BT_{ij}—multiplies *column* j of B by a scalar and adds it to column i

Exercises

1. Which of the following matrices are elementary?

$$A = \begin{bmatrix} 1 & 4 \\ 0 & 1 \end{bmatrix}, \qquad B = \begin{bmatrix} 0 & 1 \\ 1 & 0 \end{bmatrix}, \qquad C = \begin{bmatrix} 0 & 1 \\ 3 & 0 \end{bmatrix},$$

$$D = \begin{bmatrix} 1 & 0 & 0 \\ -6 & 1 & 0 \\ 0 & 0 & 0 \end{bmatrix}, \quad E = \begin{bmatrix} 1 & 0 & 0 \\ 0 & -4 & 0 \\ 0 & 0 & 1 \end{bmatrix}, \quad F = \begin{bmatrix} 0 & 1 & 0 \\ 1 & 0 & 0 \\ 0 & 0 & 1 \end{bmatrix},$$

$$G = \begin{bmatrix} 3 & 0 & 0 \\ 0 & 1 & 0 \\ -2 & 0 & 1 \end{bmatrix}, \quad H = \begin{bmatrix} 0 & 0 & 1 \\ 1 & 0 & 0 \\ 0 & 1 & 0 \end{bmatrix}, \quad I = \begin{bmatrix} 6 & 0 & 0 \\ 0 & 4 & 0 \\ 0 & 0 & 1 \end{bmatrix}.$$

2. For the matrices in Exercise 1 that are elementary, give their inverses.

3. For the matrices in Exercise 1 that are *not* elementary, write these matrices as a *product* of elementary matrices, and determine the inverses of these nonelementary matrices via the inverse product rule.

4. (a) Write the following matrices as a product of elementary matrices, as illustrated by matrix A below.

$$A = \begin{bmatrix} 4 & 0 \\ 2 & 1 \end{bmatrix} = \begin{bmatrix} 1 & 0 \\ 1/2 & 1 \end{bmatrix} \begin{bmatrix} 4 & 0 \\ 0 & 1 \end{bmatrix} \begin{bmatrix} 1 & 0 \\ 0 & 1 \end{bmatrix}, \qquad B = \begin{bmatrix} 1 & 2 \\ 0 & 3 \end{bmatrix},$$

$$C = \begin{bmatrix} 2 & 1 \\ 4 & 0 \end{bmatrix}, \qquad D = \begin{bmatrix} -2 & 2 \\ -6 & 7 \end{bmatrix}, \qquad E = \begin{bmatrix} 1 & 0 & 0 \\ 0 & 4 & 0 \\ 3 & 0 & 1 \end{bmatrix}.$$

(b) Find the inverses of the matrices in part (a) by using the inverse product rule $(AB)^{-1} = B^{-1}A^{-1}$ and your knowledge of the inverses of elementary matrices.

5. Referring to Exercise 4, use the inverses you have found to solve the following equations.

(a) $AX = \begin{bmatrix} 3 \\ 4 \end{bmatrix}$ (b) $BX = \begin{bmatrix} -2 \\ -2 \end{bmatrix}$ (c) $CX = \begin{bmatrix} 0 \\ 1 \end{bmatrix}$

(d) $DX = \begin{bmatrix} \pi \\ \sqrt{10} \end{bmatrix}$ (e) $EX = \begin{bmatrix} 2 \\ 0 \\ -1 \end{bmatrix}$

6. (a) If the matrices $A_1, A_2, \ldots, A_n$ are all nonsingular, show that the product $A_1A_2 \cdots A_n$ is again nonsingular. (Use the idea in the proof of two factors.)

(b) If the matrices $A_1, A_2, \ldots, A_n$ are all nonsingular except for one A_i that is singular, show that $A_1A_2 \cdots A_i \cdots A_n$ is singular.

(c) If the matrices $A_1, \ldots, A_n$ are singular, their product is singular.

7. Can a singular matrix be written as the product of elementary matrices?

8. Perform elementary row operations to transform the following matrices into matrices that have a 1 in the (1, 1)-entry position. (Give the resultant matrix and the operations used.)

(a) $\begin{bmatrix} 0 & 1 \\ 3 & 0 \end{bmatrix}$ (b) $\begin{bmatrix} 0 & 1 & 0 \\ 0 & 4 & 1 \\ 5 & 6 & 3 \end{bmatrix}$ (c) $\begin{bmatrix} 7 & -4 & 0 \\ 1 & 6 & 2 \\ 5 & -1 & 3 \end{bmatrix}$

9. Using *only* elementary row operations of the third type, transform the following matrices into matrices that have a 2 in the (1, 2)-entry position.

(a) $\begin{bmatrix} 5 & 10 \\ 6 & -4 \end{bmatrix}$ (b) $\begin{bmatrix} -1 & -7 \\ 3 & 4 \end{bmatrix}$ (c) $\begin{bmatrix} 1 & 0 & 0 \\ 6 & 3 & 0 \\ -4 & -2 & 1 \end{bmatrix}$

10. Show that in the following matrices, it is not possible to obtain a 1 in the (2, 2)-entry position *and* a 0 in the (2, 1)-entry position via row operations.

(a) $\begin{bmatrix} 1 & 2 \\ 1 & 2 \end{bmatrix}$ (b) $\begin{bmatrix} 4 & -2 \\ -2 & 1 \end{bmatrix}$ (c) $\begin{bmatrix} 5 & 7 & -3 \\ 0 & 0 & 0 \\ 15 & 21 & -9 \end{bmatrix}$

11. Matrices of the form

$$Q_1 = \begin{bmatrix} 0 & 0 & 1 \\ 1 & 0 & 0 \\ 0 & 1 & 0 \end{bmatrix}, \qquad Q_2 = \begin{bmatrix} 0 & 1 & 0 \\ 0 & 0 & 1 \\ 1 & 0 & 0 \end{bmatrix}$$

are called *permutation matrices* (which are *not* elementary matrices). They are formed from the identity by interchanging more than two rows (or columns) of *I*.

(a) Show that Q_1 and Q_2 can be written as a product of *elementary* permutation matrices.

(b) Show that $Q_1^{-1} = Q_1^T$ and $Q_2^{-1} = Q_2^T$.

(c) Using the fact that a permutation matrix Q can be written as a product $Q = P_1 P_2 P_3 \cdots P_n$, of elementary matrices P_i, show that the inverse of Q is given by

$$Q^{-1} = Q^T.$$

Hint: What is P_i^{-1}?

12. Show that there are $n!$ permutation matrices of size $n \times n$. How many of these are elementary permutation matrices (if you consider *I* as a permutation matrix)?

13. Find the inverses of the following matrices:

$$D = \begin{bmatrix} 2 & 0 & 0 \\ 0 & -3 & 0 \\ 0 & 0 & 5 \end{bmatrix}, \qquad X = \begin{bmatrix} 1 & 2 & 0 \\ 0 & 1 & 0 \\ 0 & 3 & 1 \end{bmatrix}, \qquad Y = \begin{bmatrix} 1 & 2 & 0 \\ 0 & -5 & 0 \\ 0 & 3 & 1 \end{bmatrix},$$

$$P = \begin{bmatrix} 0 & 0 & 1 \\ 1 & 0 & 0 \\ 0 & 1 & 0 \end{bmatrix}, \qquad Q = \begin{bmatrix} 1 & 0 & 7 \\ 0 & 1 & 0 \\ 0 & 0 & 1 \end{bmatrix}, \qquad R = \begin{bmatrix} 1 & 0 & 0 \\ 2 & -5 & 3 \\ 0 & 0 & 1 \end{bmatrix},$$

$$U = \begin{bmatrix} 1 & 2 & 0 \\ 0 & -3 & 0 \\ 0 & 0 & 1 \end{bmatrix}, \qquad V = \begin{bmatrix} 1 & 0 & 0 \\ 0 & 1 & 0 \\ 0 & 3 & 1 \end{bmatrix}, \qquad W = \begin{bmatrix} 0 & 0 & 2 \\ 1 & 0 & -1 \\ 0 & 1 & 1 \end{bmatrix}.$$

14. Observe that an elementary matrix T_{ij} of the third kind differs from the identity *I* only in an off-diagonal entry in its *i*th column. There is a larger class of easily invertible matrices comprised of matrices T_i which differ from *I* only in one or more of its off-diagonal entries in its *i*th column. For example, for the case $n = 3$, the matrices

$$T_1 = \begin{bmatrix} 1 & 0 & 0 \\ a & 1 & 0 \\ b & 0 & 1 \end{bmatrix}, \qquad T_2 = \begin{bmatrix} 1 & c & 0 \\ 0 & 1 & 0 \\ 0 & d & 1 \end{bmatrix}, \qquad T_3 = \begin{bmatrix} 1 & 0 & e \\ 0 & 1 & f \\ 0 & 0 & 1 \end{bmatrix}$$

are of this type. Verify that their inverses are

$$T_1^{-1} = \begin{bmatrix} 1 & 0 & 0 \\ -a & 1 & 0 \\ -b & 0 & 1 \end{bmatrix}, \qquad T_2^{-1} = \begin{bmatrix} 1 & -c & 0 \\ 0 & 1 & 0 \\ 0 & -d & 1 \end{bmatrix}, \qquad T_3^{-1} = \begin{bmatrix} 1 & 0 & -e \\ 0 & 1 & -f \\ 0 & 0 & 1 \end{bmatrix}.$$

Observe that the inverse of T_i is obtained by replacing its nonzero off-diagonal entries by their negatives.

15. The matrix

$$R = \begin{bmatrix} 0 & 0 & 1 \\ 0 & 1 & 0 \\ 1 & 0 & 0 \end{bmatrix}$$

is sometimes called the *reverse* **3 × 3-*matrix.***

(a) Show that R is symmetric and that $R^{-1} = R$.

(b) Show that the matrix $B = AR$ is obtained from A by reversing its columns.

(c) Show that $C = RA$ is obtained from A by reversing its rows. What is the 4×4-reverse matrix?

2.3   CONSTRUCTION OF INVERSES

In Section 2.1 we stated that a nonsingular square matrix is invertible. In this section we show how this result can be established. We actually construct the inverse of a 3×3-matrix, thereby illustrating the procedure in the $n \times n$ case.

In doing so, let us recall from Section 2.2 that multiplying a matrix A on the left by elementary matrices of the first, second, or third kinds resulted, respectively, in the following changes in the matrix A:

1. Interchanged two rows of A.

2. Multiplied a row of A by a scalar s.

3. Multiplied one row of A by a scalar and added it to another row of A.

One might expect that after a finite number q of elementary row operations on a nonsingular $n \times n$ matrix A, that A could be transformed into the identity matrix I. Pictorially,

$$A \xrightarrow[\text{row operations}]{\text{via elementary}} I$$

The q successive row operations are equivalent to q successive multiplications on the left by elementary matrices $E_1, E_2, \ldots, E_q$. We have

$$E_q(E_{q-1}(\cdots (E_2(E_1A)) \cdots)) = E_q E_{q-1} \cdots E_2 E_1 A = I$$

This is of the form $BA = I$, where

$$B = E_q E_{q-1} \cdots E_2 E_1 = E_q E_{q-1} \cdots E_2 E_1 I$$

so that $B = A^{-1}$. Observe that $B = A^{-1}$ is obtained by multiplying the identity matrix I on the left successively by the elementary matrices

$E_1, E_2, \ldots, E_q$. These are the same elementary matrices that reduce A via multiplication on the left to the identity matrix I.

With this in mind, we introduce the special **augmented matrix**

$$[A \mid I],$$

which is obtained by placing the identity matrix I (whose size is that of A) on the right of A. As we perform row operations on A, we perform the same operations on the identity matrix I. In terms of the corresponding elementary matrices $E, E_2, \ldots, E_q$, this will generate the following scheme, known as a *Gaussian elimination method:*

$$[A \mid I]$$

$$[E_1 A \mid E_1 I]$$

$$[E_2 E_1 A \mid E_2 E_1 I]$$

$$\vdots$$

$$[E_q E_{q-1} \cdots E_2 E_1 A \mid E_q E_{q-1} \cdots E_2 E_1 I].$$

This last step of the scheme gives us the matrix

$$[I \mid A^{-1}]$$

because, by our choice of $E_1, E_2, \ldots, E_q$, we have

$$E_q E_{q-1} \cdots E_2 E_1 A = I, \qquad E_q E_{q-1} \cdots E_2 E_1 I = A^{-1}.$$

■ **EXAMPLE 1** ──────────────────────────────────

Consider the matrix

$$A = \begin{bmatrix} 0 & -1 & 1 \\ 3 & 6 & 1 \\ 2 & 4 & -2 \end{bmatrix}.$$

The augmented matrix is

$$[A \mid I] = \begin{bmatrix} 0 & -1 & 1 & \mid & 1 & 0 & 0 \\ 3 & 6 & 1 & \mid & 0 & 1 & 0 \\ 2 & 4 & -2 & \mid & 0 & 0 & 1 \end{bmatrix}.$$

The matrix A is 3×3 and the augmented matrix is 3×6. Recall that in our elimination procedure we seek to transform A into the identity matrix I by suitably chosen row operations. We shall carry out this elimination procedure as follows: In the first step we use row operations to "fix up" column 1 of A, that is, make column 1 to be the first column of the identity I. Next, we fix up column 2 without disturbing column 1. That is, we use row operations to change column 2 into the second column of I. In the final step for this example, we fix up column 3 to be the third column of I. In doing so, we have transformed $[A \mid I]$ into $[I \mid A^{-1}]$. Pictorially, we have

$$[A \mid I] \longrightarrow \begin{bmatrix} 1 & * & * & | & * & * & * \\ 0 & * & * & | & * & * & * \\ 0 & * & * & | & * & * & * \end{bmatrix} \longrightarrow \begin{bmatrix} 1 & 0 & * & | & * & * & * \\ 0 & 1 & * & | & * & * & * \\ 0 & 0 & * & | & * & * & * \end{bmatrix}$$

$$\longrightarrow \begin{bmatrix} 1 & 0 & 0 & | & * & * & * \\ 0 & 1 & 0 & | & * & * & * \\ 0 & 0 & 1 & | & * & * & * \end{bmatrix} = [I \mid A^{-1}],$$

where the $*$'s are appropriate numbers. The reader is reminded that whatever row operation is performed on A is being performed on the entire augmented matrix $[A \mid I]$.

Each step is comprised of several substeps. In step 1 we first look at the $(1, 1)$-entry of $[A \mid I]$. We find that it is zero. Since we want it to be 1 and since we cannot divide by zero, we interchange rows 1 and 3. Thus

$$\begin{bmatrix} 0 & -1 & 1 & | & 1 & 0 & 0 \\ 3 & 6 & 1 & | & 0 & 1 & 0 \\ 2 & 4 & -2 & | & 0 & 0 & 1 \end{bmatrix} \xrightarrow[\substack{\text{interchange} \\ \text{rows 1 and 3}}]{P_{13}} \begin{bmatrix} 2 & 4 & -2 & | & 0 & 0 & 1 \\ 3 & 6 & 1 & | & 0 & 1 & 0 \\ 0 & -1 & 1 & | & 1 & 0 & 0 \end{bmatrix}.$$

The $(1, 1)$-entry is now 2. To make this entry 1, we divide row 1 by 2 or, equivalently, we multiply row 1 by 1/2:

$$\begin{bmatrix} 2 & 4 & -2 & | & 0 & 0 & 1 \\ 3 & 6 & 1 & | & 0 & 1 & 0 \\ 0 & -1 & 1 & | & 1 & 0 & 0 \end{bmatrix} \xrightarrow[\substack{\text{multiply} \\ \text{row 1 by 1/2}}]{S_1} \begin{bmatrix} 1 & 2 & -1 & | & 0 & 0 & 1/2 \\ 3 & 6 & 1 & | & 0 & 1 & 0 \\ 0 & -1 & 1 & | & 1 & 0 & 0 \end{bmatrix}.$$

The $(1, 1)$-entry is now 1. We wish to get 0's below this 1 in the first column. To do so, we use elementary row operations of the third kind. To change the $(2, 1)$-entry to 0, we multiply the first row by -3 and add it to the second row:

$$\begin{bmatrix} 1 & 2 & -1 & | & 0 & 0 & 1/2 \\ 3 & 6 & 1 & | & 0 & 1 & 0 \\ 0 & -1 & 1 & | & 1 & 0 & 0 \end{bmatrix} \xrightarrow[\substack{\text{add} -3\,(\text{row 1}) \\ \text{to row 2}}]{T_{12}} \begin{bmatrix} 1 & 2 & -1 & | & 0 & 0 & 1/2 \\ 0 & 0 & 4 & | & 0 & 1 & -3/2 \\ 0 & -1 & 1 & | & 1 & 0 & 0 \end{bmatrix}.$$

If we did not already have a 0 in the $(3, 1)$-entry of our augmented matrix, we would again use an elementary row operation of the third kind to obtain it.

The augmented matrix $[A \mid I]$ has been transformed into the form

$$\begin{bmatrix} 1 & * & * & | & * & * & * \\ 0 & * & * & | & * & * & * \\ 0 & * & * & | & * & * & * \end{bmatrix}.$$

We continue by fixing up column 2 so that it is the second column of *I*. We note that the (2, 2)-entry is zero. We seek to make it nonzero. This could be accomplished by interchanging rows 1 and 2, but this is not permissible since it alters column 1. We therefore interchange rows 2 and 3. Thus

$$\begin{bmatrix} 1 & 2 & -1 & | & 0 & 0 & 1/2 \\ 0 & 0 & 4 & | & 0 & 1 & -3/2 \\ 0 & -1 & 1 & | & 1 & 0 & 0 \end{bmatrix} \xrightarrow[\substack{\text{interchange} \\ \text{rows 2 and 3}}]{P_{23}} \begin{bmatrix} 1 & 2 & -1 & | & 0 & 0 & 1/2 \\ 0 & -1 & 1 & | & 1 & 0 & 0 \\ 0 & 0 & 4 & | & 0 & 1 & -3/2 \end{bmatrix}.$$

The (2, 2)-entry in the matrix on the right is -1. To make it 1, we multiply row 2 by -1. We have

$$\begin{bmatrix} 1 & 2 & -1 & | & 0 & 0 & 1/2 \\ 0 & -1 & 1 & | & 1 & 0 & 0 \\ 0 & 0 & 4 & | & 0 & 1 & -3/2 \end{bmatrix}$$

$$\xrightarrow[\substack{\text{multiply} \\ \text{row 2 by } -1}]{S_2} \begin{bmatrix} 1 & 2 & -1 & | & 0 & 0 & 1/2 \\ 0 & 1 & -1 & | & -1 & 0 & 0 \\ 0 & 0 & 4 & | & 0 & 1 & -3/2 \end{bmatrix}.$$

We now wish to get 0's in the remaining entries of column 2. Observe that the (3, 2)-entry is already 0. We make the (1, 2)-entry in the matrix on the right equal to 0 as follows:

$$\begin{bmatrix} 1 & 2 & -1 & | & 0 & 0 & 1/2 \\ 0 & 1 & -1 & | & -1 & 0 & 0 \\ 0 & 0 & 4 & | & 0 & 1 & -3/2 \end{bmatrix}$$

$$\xrightarrow[\substack{\text{add } -2 \, (\text{row 2}) \\ \text{to row 1}}]{T_{21}} \begin{bmatrix} 1 & 0 & 1 & | & 2 & 0 & 1/2 \\ 0 & 1 & -1 & | & -1 & 0 & 0 \\ 0 & 0 & 4 & | & 0 & 1 & -3/2 \end{bmatrix}.$$

The resulting matrix is now of the form

$$\begin{bmatrix} 1 & 0 & * & | & * & * & * \\ 0 & 1 & * & | & * & * & * \\ 0 & 0 & * & | & * & * & * \end{bmatrix},$$

whose first two columns are the first two columns of *I*.

Finally, we transform the third column into the third column of I as follows:

$$
\begin{bmatrix}
1 & 0 & 1 & | & 2 & 0 & 1/2 \\
0 & 1 & -1 & | & -1 & 0 & 0 \\
0 & 0 & 4 & | & 0 & 1 & -3/2
\end{bmatrix}
\xrightarrow[\substack{\text{multiply}\\ \text{row 3 by 1/4}}]{S_3}
\begin{bmatrix}
1 & 0 & 1 & | & 2 & 0 & 1/2 \\
0 & 1 & -1 & | & -1 & 0 & 0 \\
0 & 0 & 1 & | & 0 & 1/4 & -3/8
\end{bmatrix}
$$

$$
\xrightarrow[\substack{\text{add row 3}\\ \text{to row 2}}]{T_{32}}
\begin{bmatrix}
1 & 0 & 1 & | & 2 & 0 & 1/2 \\
0 & 1 & 0 & | & -1 & 1/4 & -3/8 \\
0 & 0 & 1 & | & 0 & 1/4 & -3/8
\end{bmatrix}
$$

$$
\xrightarrow[\substack{\text{add} - \text{(row 3)}\\ \text{to row 1}}]{T_{31}}
\begin{bmatrix}
1 & 0 & 0 & | & 2 & -1/4 & 7/8 \\
0 & 1 & 0 & | & -1 & 1/4 & -3/8 \\
0 & 0 & 1 & | & 0 & -1/4 & -3/8
\end{bmatrix}.
$$

The matrix on the right is the desired matrix $[I \mid A^{-1}]$, so that

$$
A^{-1} = \begin{bmatrix}
2 & -1/4 & 7/8 \\
-1 & 1/4 & -3/8 \\
0 & 1/4 & -3/8
\end{bmatrix}.
$$

It is easily verified, by computation, that $A^{-1}A = AA^{-1} = I$.

Since we have kept track of the elementary matrices that correspond to each elementary row operation used, we have

$$
T_{31}T_{32}S_3T_{21}S_2P_{23}T_{12}S_1P_{13}A = I.
$$

Equivalently,

$$
A^{-1} = T_{31}T_{32}S_3T_{21}S_2P_{23}T_{12}S_1P_{13}.
$$

We see from the above, and by virtue of the fact that the inverse of an elementary matrix is again an elementary matrix, that the nonsingular matrix A can be written as a product of elementary matrices:

$$
A = (A^{-1})^{-1} = (T_{31}T_{32}S_3T_{21}S_2P_{23}T_{12}S_1P_{13})^{-1}
$$
$$
= P_{13}^{-1}S_1^{-1}T_{12}^{-1}P_{23}^{-1}S_2^{-1}T_{21}^{-1}S_3^{-1}T_{32}^{-1}T_{31}^{-1}.
$$

For our 3×3-matrix A, the elementary matrices displayed in these equations play the role of the elementary matrices $E_1, E_2, \ldots, E_q$ described at the beginning of this section.

Keeping track of the elementary matrices that transform $[A \mid I]$ into $[I \mid A^{-1}]$ enabled us to write the nonsingular matrix A as a product of elementary matrices. The Gaussian elimination method allows us, in general, to make the following statement:

> Every nonsingular matrix A can be written as a product of elementary matrices.

When we are only interested in computing the inverse A^{-1} of A, as is usually the case, it is unnecessary to keep track or even consider the elementary matrices that are implicitly involved.

■ **EXAMPLE 2**

Consider the matrix

$$\begin{bmatrix} 1 & 2 & 1 \\ 2 & 4 & 3 \\ 3 & 6 & 4 \end{bmatrix}.$$

To find A^{-1}, we construct the augmented matrix

$$[A \mid I] = \begin{bmatrix} 1 & 2 & 1 & \mid & 1 & 0 & 0 \\ 2 & 4 & 3 & \mid & 0 & 1 & 0 \\ 3 & 6 & 4 & \mid & 0 & 0 & 1 \end{bmatrix}.$$

We compute A^{-1} as follows:

$$\begin{bmatrix} 1 & 2 & 1 & \mid & 1 & 0 & 0 \\ 2 & 4 & 3 & \mid & 0 & 1 & 0 \\ 3 & 6 & 4 & \mid & 0 & 0 & 1 \end{bmatrix} \xrightarrow[\text{to row 2}]{\text{add} -2\,(\text{row 1})} \begin{bmatrix} 1 & 2 & 1 & \mid & 1 & 0 & 0 \\ 0 & 0 & 1 & \mid & -2 & 1 & 0 \\ 3 & 6 & 4 & \mid & 0 & 0 & 1 \end{bmatrix}$$

$$\xrightarrow[\text{to row 3}]{\text{add} -3\,(\text{row 1})} \begin{bmatrix} 1 & 2 & 1 & \mid & 1 & 0 & 0 \\ 0 & 0 & 1 & \mid & -2 & 1 & 0 \\ 0 & 0 & 1 & \mid & -3 & 0 & 1 \end{bmatrix} \xrightarrow[\text{and the (3, 2)-entries are both zero}]{\text{we cannot proceed because the (2, 2)-}}$$

The reason this occurs is because the matrix A, which we started with, is singular and so has no inverse.

For a singular matrix, our Gaussian elimination method will not produce an inverse since an inverse does not exist. However, it will tell us that the matrix is singular, because the method will break down.

As shown above, the Gaussian elimination just employed allowed us to find the inverse A^{-1} of a nonsingular matrix A. If we wish to find the unique solution $X = A^{-1}H$ of the equation $AX = H$, we could multiply H on the left by A^{-1}, which was just found. However, a simple modification of the Gaussian method enables us to find the unique solution

$$X = A^{-1}H$$

directly. To do so we use the augmented matrix $[A \mid H]$ in place of the augmented matrix $[A \mid I]$. Proceeding as before with elementary matrices (row operations), we generate the scheme

$$[A \mid H]$$

$$[E_1A \mid E_1H]$$

$$\vdots$$

$$[E_qE_{q-1} \cdots E_2E_1A \mid E_qE_{q-1} \cdots E_2E_1H].$$

The last step of the scheme gives us the matrix

$$[I \mid A^{-1}H]$$

because, by our choice of $E_1, E_2, \ldots, E_q$, we have

$$E_qE_{q-1} \cdots E_2E_1A = I, \qquad E_qE_{q-1} \cdots E_2E_1 = A^{-1}.$$

Observe that this modified procedure, while producing the unique solution $X = A^{-1}H$ of $AX = H$, does not give A^{-1} explicitly.

As an example, we seek to solve the following system.

■ **EXAMPLE 3** _____

$$AX = \begin{bmatrix} 0 & -1 & 1 \\ 3 & 6 & 1 \\ 2 & 4 & -2 \end{bmatrix} \begin{bmatrix} x_1 \\ x_2 \\ x_3 \end{bmatrix} = \begin{bmatrix} 2 \\ -5 \\ -3 \end{bmatrix} = H,$$

where A is the matrix used in our first example. The new augmented matrix $[A \mid H]$ is

$$[A \mid H] = \begin{bmatrix} 0 & -1 & 1 & \mid & 2 \\ 3 & 6 & 1 & \mid & -5 \\ 2 & 4 & -2 & \mid & -3 \end{bmatrix}.$$

Carrying out our elimination method, we find that

$$
\begin{bmatrix} 0 & -1 & 1 & | & 2 \\ 3 & 6 & 1 & | & -5 \\ 2 & 4 & -2 & | & -3 \end{bmatrix} \xrightarrow[\substack{\text{interchange} \\ \text{rows 1 and 3}}]{} \begin{bmatrix} 2 & 4 & -2 & | & -3 \\ 3 & 6 & 1 & | & -5 \\ 0 & -1 & 1 & | & 2 \end{bmatrix} \xrightarrow[\substack{\text{multiply} \\ \text{row 1 by 1/2}}]{}
$$

$$
\begin{bmatrix} 1 & 2 & -1 & | & -3/2 \\ 3 & 6 & 1 & | & -5 \\ 0 & -1 & 1 & | & 2 \end{bmatrix} \xrightarrow[\substack{\text{add } -3 \text{ (row 1)} \\ \text{to row 2}}]{} \begin{bmatrix} 1 & 2 & -1 & | & -3/2 \\ 0 & 0 & 4 & | & -1/2 \\ 0 & -1 & 1 & | & 2 \end{bmatrix} \xrightarrow[\substack{\text{interchange} \\ \text{rows 2 and 3}}]{}
$$

$$
\begin{bmatrix} 1 & 2 & -1 & | & -3/2 \\ 0 & -1 & 1 & | & 2 \\ 0 & 0 & 4 & | & -1/2 \end{bmatrix} \xrightarrow[\substack{\text{multiply} \\ \text{row 2 by } -1}]{} \begin{bmatrix} 1 & 2 & -1 & | & -3/2 \\ 0 & 1 & -1 & | & -2 \\ 0 & 0 & 4 & | & -1/2 \end{bmatrix} \xrightarrow[\substack{\text{add } -2 \text{ (row 2)} \\ \text{to row 1}}]{}
$$

$$
\begin{bmatrix} 1 & 0 & 1 & | & 5/2 \\ 0 & 1 & -1 & | & -2 \\ 0 & 0 & 4 & | & -1/2 \end{bmatrix} \xrightarrow[\substack{\text{multiply} \\ \text{row 3 by 1/4}}]{} \begin{bmatrix} 1 & 0 & 1 & | & 5/2 \\ 0 & 1 & -1 & | & -2 \\ 0 & 0 & 1 & | & -1/8 \end{bmatrix} \xrightarrow[\substack{\text{add row 3} \\ \text{to row 2}}]{}
$$

$$
\begin{bmatrix} 1 & 0 & 1 & | & 5/2 \\ 0 & 1 & 0 & | & -17/8 \\ 0 & 0 & 1 & | & -1/8 \end{bmatrix} \xrightarrow[\substack{\text{add } - \text{ (row 3)} \\ \text{to row 1}}]{} \begin{bmatrix} 1 & 0 & 0 & | & 21/8 \\ 0 & 1 & 0 & | & -17/8 \\ 0 & 0 & 1 & | & -1/8 \end{bmatrix}.
$$

The last matrix is the matrix $[I \mid A^{-1}H]$, where $A^{-1}H$ is the solution of $AX = H$. The solution of our equation is therefore

$$x_1 = 21/8, \qquad x_2 = -17/8, \qquad x_3 = -1/8.$$

Note that the row operations used here are the same row operations that we used to find the inverse of A.

As another example, consider the following:

■ **EXAMPLE 4** _____

Electronic engineers and technicians, in designing and troubleshooting television sets, computers, and almost all electronic equipment, often need to determine the amount of current which flows through the various loops of a circuit. This is accomplished using **Kirchoff's current law**, which states that the current that flows into a particular junction must equal the current which flows out of that junction. Also useful in circuit theory is **Kirchoff's voltage law**, which states that the sum of the voltages around a closed loop of the circuit is equal to the total voltage of that loop. In a circuit that involves only resistances and voltage sources, the voltage V across each resistor is determined by Ohm's law and is given by $V = IR$, where I is the current measured in amperes and R is the resistance measured in ohms. As an illustration of the theory above, consider the following simple electrical circuit:

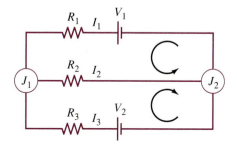

If we apply Kirchhoff's current law to either junction J_1 or junction J_2, we get

$$I_1 + I_3 = I_2$$

or equivalently,

$$I_1 - I_2 + I_3 = 0.$$

Using Kirchhoff's voltage law on each of the two loops yields for the upper loop

$$I_1R_1 + I_2R_2 = V_1$$

and for the lower loop

$$I_2R_2 + I_3R_3 = V_2.$$

If we know the values of the resistors R_1, R_2, and R_3 and the voltage sources, V_1 and V_2, we can determine the currents I_1, I_2, and I_3 from the last three equations above. For example, suppose that $R_1 = 8$ ohms, $R_2 = 6$ ohms, and $R_3 = 12$ ohms. If $V_1 = 5$ volts and $V_2 = 18$ volts, the three equations that determine the currents I_1, I_2, and I_3 are given by

$$I_1 - I_2 + I_3 = 0$$
$$8I_1 + 6I_2 \qquad = 5$$
$$6I_2 + 12I_3 = 18.$$

The augmented matrix associated with this system of equations is

$$\begin{bmatrix} 1 & -1 & 1 & | & 0 \\ 8 & 6 & 0 & | & 5 \\ 0 & 6 & 12 & | & 18 \end{bmatrix}.$$

We have

$$
\begin{bmatrix} 1 & -1 & 1 & | & 0 \\ 8 & 6 & 0 & | & 5 \\ 0 & 6 & 12 & | & 18 \end{bmatrix} \xrightarrow{T_{12}(-8)} \begin{bmatrix} 1 & -1 & 1 & | & 0 \\ 0 & 14 & -8 & | & 5 \\ 0 & 6 & 12 & | & 18 \end{bmatrix}
$$

$$
\xrightarrow{P_{23}} \begin{bmatrix} 1 & -1 & 1 & | & 0 \\ 0 & 6 & 12 & | & 18 \\ 0 & 14 & -8 & | & 5 \end{bmatrix} \xrightarrow{S_2(1/6)} \begin{bmatrix} 1 & -1 & 1 & | & 0 \\ 0 & 1 & 2 & | & 3 \\ 0 & 14 & -8 & | & 5 \end{bmatrix}
$$

$$
\xrightarrow{T_{21}(1)} \begin{bmatrix} 1 & 0 & 3 & | & 3 \\ 0 & 1 & 2 & | & 3 \\ 0 & 14 & -8 & | & 5 \end{bmatrix} \xrightarrow{T_{22}(-14)} \begin{bmatrix} 1 & 0 & 3 & | & 3 \\ 0 & 1 & 2 & | & 3 \\ 0 & 0 & -36 & | & -37 \end{bmatrix}
$$

$$
\xrightarrow{S_3(-1/36)} \begin{bmatrix} 1 & 0 & 3 & | & 3 \\ 0 & 1 & 2 & | & 3 \\ 0 & 0 & 1 & | & 37/36 \end{bmatrix} \xrightarrow{T_{32}(-2)} \begin{bmatrix} 1 & 0 & 3 & | & 3 \\ 0 & 1 & 0 & | & 34/36 \\ 0 & 0 & 1 & | & 37/36 \end{bmatrix}
$$

$$
\xrightarrow{T_{31}(-3)} \begin{bmatrix} 1 & 0 & 0 & | & -3/36 \\ 0 & 1 & 0 & | & 34/36 \\ 0 & 0 & 1 & | & 37/36 \end{bmatrix}.
$$

Therefore,

$$
I_1 = -3/36 \text{ amperes}, \qquad I_2 = 34/36 \text{ amperes}, \qquad I_3 = 37/36 \text{ amperes}.
$$

In the examples given above the amount of writing is lengthy. To cut down on the amount of writing in carrying out the elimination, we introduce a scheme based on the following rules:

1. Place each row of $[A \mid H]$ in a suitable-sized box, as indicated by the $\rightarrow$ below.
2. In carrying out a row operation, write the new row directly below the row that is changed. Record the order of the operation by a number on the left. Record the row operation used on the right.

Then at each stage the bottom row in each box gives us that matrix with which we are dealing. Using this scheme in Example 3,

$$
\begin{bmatrix} 0 & -1 & 1 \\ 3 & 6 & 1 \\ 2 & 4 & -2 \end{bmatrix} \begin{bmatrix} x_1 \\ x_2 \\ x_3 \end{bmatrix} = \begin{bmatrix} 2 \\ -5 \\ -3 \end{bmatrix},
$$

we have

→	0	−1	1	2	
1	2	4	−2	−3	Interchange rows 1 and 3
2	1	2	−1	−3/2	Multiply row 1 by 1/2
6	1	0	1	5/2	Add −2(row 2) to row 1
9	1	0	0	**21/8**	Add −(row 3) to row 1

→	3	6	1	−5	
3	0	0	4	−1/2	Add −3(row 1) to row 2
4	0	−1	1	2	Interchange rows 2 and 3
5	0	1	−1	−2	Multiply row 2 by −1
8	0	1	0	**−17/8**	Add row 3 to row 2

→	2	4	−2	−3	
1	0	−1	1	2	Interchange rows 1 and 3
4	0	0	4	−1/2	Interchange rows 2 and 3
7	0	0	1	**−1/8**	Multiply row 3 by 1/4

Reading the boldface numbers in the last column, we find that the solution to our equation $AX = H$ is

$$x_1 = 21/8, \qquad x_2 = -17/8, \qquad x_3 = -1/8.$$

Exercises

1. Find the inverses of the following matrices using Gaussian elimination.

(a) $\begin{bmatrix} 4 & 1 \\ -1 & 0 \end{bmatrix}$ (b) $\begin{bmatrix} 4 & 3 \\ 2 & 1 \end{bmatrix}$ (c) $\begin{bmatrix} 2 & -1 & 0 \\ -1 & 1 & 0 \\ 1 & -1 & 1 \end{bmatrix}$

(d) $\begin{bmatrix} 4 & 3 & 1 \\ 0 & 2 & 6 \\ 1 & 0 & 1 \end{bmatrix}$ (e) $\begin{bmatrix} 1 & 1 & 1 & 1 \\ -1 & 1 & -1 & 1 \\ 1 & -1 & 1 & 1 \\ 1 & 1 & -1 & 1 \end{bmatrix}$

2. Use the Gaussian elimination method to obtain the inverse of each of the following matrices, when the inverse exists.

$$A = \begin{bmatrix} 1 & 0 & 2 \\ 1 & 1 & -1 \\ 1 & 1 & 1 \end{bmatrix}, \qquad B = \begin{bmatrix} 2 & -3 & -1 \\ 0 & 1 & 3 \\ 1 & 0 & 2 \end{bmatrix}, \qquad C = \begin{bmatrix} 1 & 1 & 1 \\ 1 & 2 & 3 \\ 0 & 1 & 2 \end{bmatrix}.$$

3. **(a)** Write each of the matrices in Exercise 1 as a product of elementary matrices.

   (b) Write each of the inverses found in Exercise 1 as a product of elementary matrices.

4. Consider $AX = H$ for each matrix A in Exercise 1 where

$$H = \begin{bmatrix} 2 \\ 3 \end{bmatrix} \text{ in parts (a) and (b)}, \quad H = \begin{bmatrix} 1 \\ -1 \\ 2 \end{bmatrix} \text{ in (c) and (d)}, \quad \text{and } H = \begin{bmatrix} 1 \\ 0 \\ 0 \\ 1 \end{bmatrix} \text{ in (e)}.$$

   (a) Solve for X by using the inverse A^{-1} found in Exercise 1.

   (b) Solve for X by not computing A^{-1} explicitly.

5. Two columns C_1 and C_2 of a matrix A are orthogonal if $C_1^T C_2 = 0$.

   (a) Verify that the columns C_1, C_2, and C_3 of the matrix

$$A = \begin{bmatrix} 2 & 0 & -1 \\ 1 & 1 & 1 \\ 1 & -1 & 1 \end{bmatrix}$$

   are mutually orthogonal.

   (b) The square of the length of a column vector C is the number $C^T C$. Find the square of the lengths of the columns of A.

   (c) Form a new matrix B by dividing each column of A by the square of its length. Show that

$$B = \begin{bmatrix} 1/3 & 0 & -1/3 \\ 1/6 & 1/2 & 1/3 \\ 1/6 & -1/2 & 1/3 \end{bmatrix}.$$

   (d) Show that B^T is the inverse of A.

6. Continuing with Exercise 5, let X be the matrix obtained from A by dividing each column by its length.

   (a) Show that

$$X = \begin{bmatrix} 2/\sqrt{6} & 0 & -1/\sqrt{3} \\ 1/\sqrt{6} & 1/\sqrt{2} & 1/\sqrt{3} \\ 1/\sqrt{6} & -1/\sqrt{2} & 1/\sqrt{3} \end{bmatrix}.$$

   (b) Show that X^T is the inverse of X.

7. The matrices

$$P = \begin{bmatrix} 0 & 1 & 0 & 0 \\ 0 & 0 & 0 & 1 \\ 1 & 0 & 0 & 0 \\ 0 & 0 & 1 & 0 \end{bmatrix}, \quad P^T = \begin{bmatrix} 0 & 0 & 1 & 0 \\ 1 & 0 & 0 & 0 \\ 0 & 0 & 0 & 1 \\ 0 & 1 & 0 & 0 \end{bmatrix}$$

can be obtained by permuting the columns (or rows) of the identity matrix. They are called permutation matrices. P and P^T are inverses of each other. Apply the

Gauss elimination routine for finding the inverses of P and show thereby that P and P^T are products of elementary matrices of the first kind.

8. Choose nonzero numbers r, s, and t and construct the diagonal matrices

$$D = \begin{bmatrix} r & 0 & 0 \\ 0 & s & 0 \\ 0 & 0 & t \end{bmatrix}, \quad D^{-1} = \begin{bmatrix} 1/r & 0 & 0 \\ 0 & 1/s & 0 \\ 0 & 0 & 1/t \end{bmatrix}.$$

Apply the Gauss elimination to D to find its inverse D^{-1} and show thereby that D is the product of elementary matrices of the second kind.

9. The matrices

$$L = \begin{bmatrix} 1 & 0 & 0 \\ a & 1 & 0 \\ c & b & 1 \end{bmatrix}, \quad L^{-1} = \begin{bmatrix} 1 & 0 & 0 \\ -a & 1 & 0 \\ -c + ab & -b & 1 \end{bmatrix}$$

are lower triangular matrices. Obtain L^{-1} from L by the Gauss elimination method and show thereby that L is the product of elementary matrices of the third kind, each of which is also lower triangular.

10. Choose a nonzero number a. Construct the matrices

$$A = \begin{bmatrix} a & 0 & 0 \\ b & 1 & 0 \\ c & 0 & 1 \end{bmatrix}, \quad D = \begin{bmatrix} 1 & c & 0 \\ 0 & a & 0 \\ 0 & b & 1 \end{bmatrix}, \quad F = \begin{bmatrix} 1 & 0 & b \\ 0 & 1 & c \\ 0 & 0 & a \end{bmatrix}.$$

By the Gauss elimination or some other means, show that their inverses are

$$A^{-1} = \begin{bmatrix} 1/a & 0 & 0 \\ -b/a & 1 & 0 \\ -c/a & 0 & 1 \end{bmatrix}, \quad D^{-1} = \begin{bmatrix} 1 & -c/a & 0 \\ 0 & 1/a & 0 \\ 0 & -b/a & 1 \end{bmatrix},$$

$$F^{-1} = \begin{bmatrix} 1 & 0 & -b/a \\ 0 & 1 & -c/a \\ 0 & 0 & 1/a \end{bmatrix}.$$

11. Consider the matrices

$$A = \begin{bmatrix} 2 & 0 & 0 \\ 4 & 1 & 0 \\ 6 & 0 & 1 \end{bmatrix}, \quad B = \begin{bmatrix} 2 & 8 & 0 \\ 4 & 14 & 0 \\ 6 & 20 & 1 \end{bmatrix}, \quad D = \begin{bmatrix} 1 & 4 & 0 \\ 0 & -2 & 0 \\ 0 & -4 & 1 \end{bmatrix}.$$

The matrices A and D are of the type described in Exercise 10. Notice how A is related to B. Show that $D = A^{-1}B$ and $B = AD$.

12. Assume that the matrices

$$A = \begin{bmatrix} a_1 & 0 & 0 \\ a_2 & 1 & 0 \\ a_3 & 0 & 1 \end{bmatrix}, \quad B = \begin{bmatrix} a_1 & b_1 & 0 \\ a_2 & b_2 & 0 \\ a_3 & b_3 & 1 \end{bmatrix}, \quad C = \begin{bmatrix} a_1 & b_1 & c_1 \\ a_2 & b_2 & c_2 \\ a_3 & b_3 & c_3 \end{bmatrix}$$

are nonsingular or, equivalently, assume that the matrices

$$[a_1], \quad \begin{bmatrix} a_1 & b_1 \\ a_2 & b_2 \end{bmatrix}, \quad \begin{bmatrix} a_1 & b_1 & c_1 \\ a_2 & b_2 & c_2 \\ a_3 & b_3 & c_3 \end{bmatrix}$$

are nonsingular.

(a) Show that the matrices $D = A^{-1}B$ and $F = B^{-1}C$ are of the form

$$D = \begin{bmatrix} 1 & d_1 & 0 \\ 0 & d_2 & 0 \\ 0 & d_3 & 1 \end{bmatrix}, \quad F = \begin{bmatrix} 1 & 0 & f_1 \\ 0 & 1 & f_2 \\ 0 & 0 & f_3 \end{bmatrix}.$$

(b) Show that $B = AD$, $B^{-1} = D^{-1}A^{-1}$, $F = D^{-1}A^{-1}C$, $C = ADF$, and $C^{-1} = F^{-1}D^{-1}A^{-1}$.

   Remark. It follows that A^{-1} can be obtained by successively computing the following matrices:

$$A^{-1}, \quad D = A^{-1}B, \quad D^{-1}, \quad B^{-1} = D^{-1}A^{-1}, \quad F = B^{-1}C, \quad F^{-1}, \quad A^{-1} = F^{-1}B^{-1}.$$

Referring to Exercise 12, the inverse of

$$C = \begin{bmatrix} a_1 & b_1 & c_1 \\ a_2 & b_2 & c_2 \\ a_3 & b_3 & c_3 \end{bmatrix}$$

can be obtained by the following steps:

1. As in Exercise 12, the first column of C determines a matrix A whose inverse is the first of the matrices

$$A^{-1} = \begin{bmatrix} 1/a_1 & 0 & 0 \\ -a_2/a_1 & 1 & 0 \\ -a_3/a_1 & 0 & 1 \end{bmatrix}, \quad E = A^{-1}C = \begin{bmatrix} 1 & d_1 & e_1 \\ 0 & d_2 & e_2 \\ 0 & d_3 & e_3 \end{bmatrix}.$$

2. As in Exercise 12, the second column of E determines a matrix D whose inverse is the first of the matrices

$$D^{-1} = \begin{bmatrix} 1 & -d_1/d_2 & 0 \\ 0 & 1/d_2 & 0 \\ 0 & -d_3/d_2 & 1 \end{bmatrix}, \quad F = D^{-1}E = \begin{bmatrix} 1 & 0 & f_1 \\ 0 & 1 & f_2 \\ 0 & 0 & f_3 \end{bmatrix}.$$

3. The inverse of C is given by the following computations.

$$C^{-1} = F^{-1}D^{-1}A^{-1}.$$

   In making the matrix multiplications we see that the d's, e's, and f's are given by the formulas

$$d_1 = b_1, \qquad d_2 = b_2 - \frac{b_1 a_2}{a_1}, \qquad d_3 = b_3 - \frac{b_1 a_3}{a_1},$$

$$e_1 = \frac{c_1}{a_1}, \qquad e_3 = c_3 - \frac{c_1 a_2}{a_1}, \qquad e_1 = c_3 - \frac{c_1 a_3}{a_1},$$

$$f_1 = e_1 - \frac{e_2 d_1}{d_2}, \qquad f_2 = \frac{e_2}{d_2}, \qquad f_3 = e_3 - \frac{e_2 d_3}{d_2}.$$

These are computations that are made in the Gauss elimination routine when no pivoting is involved. This means that the procedure described in Exercise 12 is a variant of the Gauss elimination method without pivoting.

Consider the matrix

$$C = \begin{bmatrix} 2 & 8 & -4 \\ 4 & 14 & 0 \\ 6 & 20 & 2 \end{bmatrix}.$$

Find the inverse of C by the method described in Exercise 12. Proceed as follows.

(a) As in Exercise 12, the first column of C determines a matrix whose inverse is the first of the matrices

$$A^{-1} = \begin{bmatrix} 1/2 & 0 & 0 \\ -2 & 1 & 0 \\ -3 & 0 & 1 \end{bmatrix}, \quad E = A^{-1}C = \begin{bmatrix} 1 & 4 & -2 \\ 0 & -2 & 6 \\ 0 & -4 & 14 \end{bmatrix}.$$

(b) As in Exercise 12, the second column of E determines a matrix D whose inverse is the first of the matrices

$$D^{-1} = \begin{bmatrix} 1 & 2 & 0 \\ 0 & -1/2 & 0 \\ 0 & -2 & 1 \end{bmatrix}, \quad F = D^{-1}E = \begin{bmatrix} 1 & 0 & 14 \\ 0 & 1 & -4 \\ 0 & 0 & -2 \end{bmatrix}.$$

(c) The inverse of C is given by the product $C^{-1} = F^{-1}(D^{-1}A^{-1})$, so that

$$C^{-1} = \begin{bmatrix} 1 & 0 & 7 \\ 0 & 1 & -2 \\ 0 & 0 & -1/2 \end{bmatrix} \begin{bmatrix} -7/2 & 2 & 0 \\ 1 & -1/2 & 0 \\ 1 & -2 & 1 \end{bmatrix} = \begin{bmatrix} 8/2 & -12 & 7 \\ -1 & 7/2 & -2 \\ -1/2 & 1 & -1/2 \end{bmatrix}.$$

2.4   SOLUTIONS OF *AX* = *H* FOR A GENERAL MATRIX *A*

In the last section we used a Gaussian elimination method to find the inverse of a nonsingular matrix A. We then showed how to extend the method for solving $AX = H$ when A is nonsingular. The unique solution was given by $X = A^{-1}H$. In arriving at this solution, the technique used consisted of forming the augmented matrix $[A \mid H]$. By elementary row operations, we transformed $[A \mid H]$ into a matrix of the form $[I \mid A^{-1}H]$.

The technique used can be modified so that we can solve an arbitrary system of equations

$$AX = H,$$

where A need not be square, and if square, A could be singular. We shall see that here, there are systems that have more than one solution, in fact, infinitely many solutions. On the other hand, we shall also see that there are situations when $AX = H$ does not always have a solution X for arbitrary choices of H. However, even when our equation $AX = H$ is not solvable, our elimination procedure can be carried out and will inform us in the last step why the equation is not solvable.

Before discussing a general case, let us examine a simple case that one might encounter in a course in elementary algebra. Let us consider the system of equations

$$x - 2y + z = 1$$
$$2x - 4y - 3z = -8.$$

Multiplying the first equation by -2 and adding the result to the second equation, we obtain the equivalent equations

$$x - 2y + z = 1$$
$$-5z = -10.$$

Multiplying the last equation by $-1/5$, our equations become

$$x - 2y + z = 1$$
$$z = 2.$$

Subtracting the last equation from the first, we find that

$$x - 2y = -1$$
$$z = 2.$$

Setting y equal to any number b, we obtain the solution

$$x = 2b - 1, \qquad y = b, \qquad z = 2.$$

Observe that our solution contains an arbitrary constant b so that we have many solutions, one for each choice of b. This is a common phenomenon for a general system of linear equations.

When the equations we just solved are written in the matrix form

$$AX = H,$$

we have

$$A = \begin{bmatrix} 1 & -2 & 1 \\ 2 & -4 & -3 \end{bmatrix}, \qquad X = \begin{bmatrix} x \\ y \\ z \end{bmatrix}, \qquad H = \begin{bmatrix} 1 \\ -8 \end{bmatrix}.$$

The corresponding augmented matrix $[A \mid H]$ is

$$[A \mid H] = \begin{bmatrix} 1 & -2 & 1 & | & 1 \\ 2 & -4 & -3 & | & -8 \end{bmatrix}.$$

The operations we used above to solve our equations are equivalent to the following row operations on matrices. First, we add -2 times the first row to the second. This gives us the matrix

$$\begin{bmatrix} 1 & -2 & 1 & | & 1 \\ 0 & 0 & -5 & | & -10 \end{bmatrix}.$$

We now multiply the second row by $-1/5$. This yields the matrix

$$\begin{bmatrix} 1 & -2 & 1 & | & 1 \\ 0 & 0 & 1 & | & 2 \end{bmatrix}.$$

Adding -1 times the second row to the first row gives us the matrix

$$[F \mid K] = \begin{bmatrix} 1 & -2 & 0 & | & -1 \\ 0 & 0 & 1 & | & 2 \end{bmatrix},$$

where

$$F = \begin{bmatrix} 1 & -2 & 0 \\ 0 & 0 & 1 \end{bmatrix}, \qquad K = \begin{bmatrix} -1 \\ 2 \end{bmatrix}.$$

Our original equation $AX = H$ is equivalent to the equation

$$FX = K,$$

which is easily solved by writing it in the component form

$$x - 2y = -1$$
$$z = 2.$$

The solution is $x = 2b - 1$, $y = b$, $z = 2$, as noted above.

For the simple problem just described, there is very little to be gained by transforming our problem into a matrix problem, except to illustrate the procedure to be used in a more general situation. The procedure we used is as follows. Starting with the matrix equation

$$AX = H.$$

we made the following steps.

1. We formed the augmented matrix

$$[A \mid H].$$

2. By a Gaussian elimination method using only elementary row operations we transformed $[A \mid H]$ into a new augmented matrix $[F \mid K]$ called its ***reduced row echelon matrix***.

The new system of equations $FX = K$ has the same solutions as the original system of equations $AX = H$. The major advantage is that one can readily solve the new system $FX = K$ since it has a much easier form.

We proceed to define formally the reduced row echelon form by returning to the example above. We have seen that the augmented matrix

$$[A \mid H] = \begin{bmatrix} 1 & -2 & 1 & \mid & 1 \\ 2 & -4 & -3 & \mid & -8 \end{bmatrix}$$

was transformed by elementary row operations into

$$[F \mid K] = \begin{bmatrix} 1 & -2 & 0 & \mid & -1 \\ 0 & 0 & 1 & \mid & 2 \end{bmatrix}.$$

We notice that $[F \mid K]$ has the following properties:

> **1.** Each row (which does not consist entirely of zeros) has a first nonzero entry of 1, called the *leading entry* of that row.
>
> **2.** All rows consisting entirely of zeros have been moved to the bottom of the matrix. (In this example, there are none.)
>
> **3.** The leading entries of the rows, as one moves from top to bottom, are successively to the right. (In this example, the leading entry 1 in row 2 is to the right of the leading entry 1 in row 1.)
>
> **4.** All entries above and below a leading entry 1 are 0.

These four properties characterize what is known as the *reduced row echelon form* of a matrix. Whether or not a matrix is augmented is of no concern in verifying these properties. The following matrices are in reduced row echelon form:

$$\begin{bmatrix} 1 & 0 & \mid & 5 \\ 0 & 1 & \mid & 2 \end{bmatrix}, \quad \begin{bmatrix} 1 & -7 \\ 0 & 0 \\ 0 & 0 \end{bmatrix}, \quad \begin{bmatrix} 0 & 1 & 4 & 0 \\ 0 & 0 & 0 & 1 \\ 0 & 0 & 0 & 0 \end{bmatrix},$$

$$\begin{bmatrix} 0 & 0 & 1 & -2 & 0 & 0 & \mid & 1 \\ 0 & 0 & 0 & 0 & 1 & 0 & \mid & 2 \\ 0 & 0 & 0 & 0 & 0 & 1 & \mid & 3 \\ 0 & 0 & 0 & 0 & 0 & 0 & \mid & 0 \end{bmatrix}.$$

The following matrices are not in reduced row echelon form. The reader should verify that at least one of the criteria for a matrix to be in reduced row echelon form is not satisfied.

$$\begin{bmatrix} 1 & 0 & 0 \\ 0 & 2 & 1 \end{bmatrix}, \quad \begin{bmatrix} 0 & 1 & 2 & 0 & | & 0 \\ 0 & 0 & 0 & 1 & | & 0 \\ 0 & 0 & 0 & 0 & | & 0 \\ 0 & 0 & 0 & 0 & | & 1 \end{bmatrix},$$

$$\begin{bmatrix} 0 & 0 & 1 \\ 1 & 0 & 0 \end{bmatrix}, \quad \begin{bmatrix} 1 & 0 & 0 & 0 \\ 0 & 1 & 0 & 2 \\ 0 & -1 & 1 & 0 \end{bmatrix}.$$

We illustrate the general method by looking at another example.

■ **EXAMPLE 1** _____

Suppose that we are given the system of equations

$$2x + 2y + z + 7w = -6$$
$$x + y + 2z + 8w = 0$$
$$x + y + 2w = -4.$$

Here we have three equations in four unknowns. These equations can be written in the matrix form

$$AX = H,$$

where

$$A = \begin{bmatrix} 2 & 2 & 1 & 7 \\ 1 & 1 & 2 & 8 \\ 1 & 1 & 0 & 2 \end{bmatrix}, \quad X = \begin{bmatrix} x \\ y \\ z \\ w \end{bmatrix}, \quad H = \begin{bmatrix} -6 \\ 0 \\ -4 \end{bmatrix}.$$

The corresponding augmented matrix $[A \mid H]$ is

$$[A \mid H] = \begin{bmatrix} 2 & 2 & 1 & 7 & | & -6 \\ 1 & 1 & 2 & 8 & | & 0 \\ 1 & 1 & 0 & 2 & | & -4 \end{bmatrix}.$$

We proceed to transform this matrix into reduced row echelon form using only elementary row operations.

To begin, we seek a nonzero row whose leading entry is as far to the left as possible. In this example, any row can be used. For simplicity, to save division, we choose row 3 and interchange it with row 1 to obtain

$$\begin{bmatrix} 1 & 1 & 0 & 2 & | & -4 \\ 1 & 1 & 2 & 8 & | & 0 \\ 2 & 2 & 1 & 7 & | & -6 \end{bmatrix}.$$

We now must obtain zeros in the first column below the leading entry 1. Row operations of the third kind yield

$$\begin{bmatrix} 1 & 1 & 0 & 2 & | & -4 \\ 0 & 0 & 2 & 6 & | & 4 \\ 0 & 0 & 1 & 3 & | & 2 \end{bmatrix}.$$

We now seek, if possible, a leading entry in row 2. Here the first nonzero entry in row 2 lies in column 3. Multiplying the second row by 1/2 yields

$$\begin{bmatrix} 1 & 1 & 0 & 2 & | & -4 \\ 0 & 0 & 1 & 3 & | & 2 \\ 0 & 0 & 1 & 3 & | & 2 \end{bmatrix}.$$

The leading entry 1 in the second row is therefore in the (2, 3) position. We now wish to get zeros in the remaining entries of column 3. We have, using an elementary row operation of the third kind:

$$\begin{bmatrix} 1 & 1 & 0 & 2 & | & -4 \\ 0 & 0 & 1 & 3 & | & 2 \\ 0 & 0 & 0 & 0 & | & 0 \end{bmatrix}.$$

This matrix is in reduced row echelon form. We see that the original augmented matrix $[A \mid H]$ has been transformed by elementary row operations into the reduced row echelon matrix $[F \mid K]$. In terms of a system of equations, our matrix $[F \mid K]$ yields

$$\begin{bmatrix} 1 & 1 & 0 & 2 \\ 0 & 0 & 1 & 3 \\ 0 & 0 & 0 & 0 \end{bmatrix} \begin{bmatrix} x \\ y \\ z \\ w \end{bmatrix} = \begin{bmatrix} -4 \\ 2 \\ 0 \end{bmatrix},$$

which is equivalent to our original matrix equation $AX = H$.

In component form, we have

$$x + y \quad + 2w = -4$$
$$z + 3w = \quad 2.$$

Since w occurs in both equations, we set $w = a$, where a stands for any real number. The number a is called a **parameter**. Our system becomes

$$x + y = -2a - 4$$
$$z = -3a + 2$$
$$w = a.$$

The choice $w = a$ has determined the variable z, but as we see, has not completely determined x and y. We need to assign still another parameter to either x or y. Setting $y = b$ yields

$$x = -b - 2a - 4, \qquad y = b, \qquad z = -3a + 2, \qquad w = a$$

as the solution to our original system of equations. Since a and b are parameters that can take on any real values, there are an infinite number of solutions of the form

$$\begin{bmatrix} x \\ y \\ z \\ w \end{bmatrix} = \begin{bmatrix} -b - 2a - 4 \\ b \\ -3a + 2 \\ a \end{bmatrix}.$$

It is sometimes said that there are a "doubly infinite" amount of solutions since here we have "two" parameters of freedom.

■ **EXAMPLE 2** _____

There are many practical and very interesting applications in which the theory of **networks** comes into play. In all of these applications, some type of flow takes place, and the basic rule employed is Kirchoff's current law, which, as we have seen, states that the amount of current flowing into any junction of a circuit must equal the amount of current flowing out of the junction. Typical applications may include irrigation systems, traffic analysis, and electrical circuits. Consider the following:

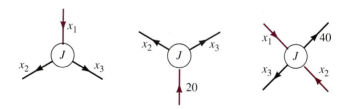

In each of the three junctions illustrated, we employ the fundamental principle which states that the total flow coming into the junction must equal the total flow going out of it. In the first junction above, we see that $x_1 = x_2 + x_3$. In the second junction, we have $20 = x_2 + x_3$, and in the third junction $x_1 + x_2 = 40 + x_3$. In all three cases, the fundamental principle yields a linear equation. If we wish to analyze a network that consists of many junctions, we can determine the flow in the network by solving a system of linear equations. Consider the following network:

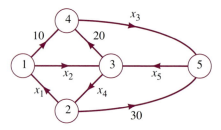

Each of the five junctions above yields a linear equation. We have

Junction 1: $\qquad x_1 - x_2 \qquad\qquad\qquad = \quad 10$

Junction 2: $\qquad x_1 \qquad\qquad - x_4 \qquad\quad = \ -30$

Junction 3: $\qquad\qquad x_2 \qquad - x_4 + x_5 = \quad 20$

Junction 4: $\qquad\qquad\qquad x_3 \qquad\qquad\quad = \quad 30$

Junction 5: $\qquad\qquad\qquad x_3 \qquad - x_5 = \ -30$

We can apply Gaussian elimination on this system of equations by considering the augmented matrix

$$\left[\begin{array}{ccccc|c} 1 & -1 & 0 & 0 & 0 & 10 \\ 1 & 0 & 0 & -1 & 0 & -30 \\ 0 & 1 & 0 & -1 & 1 & 20 \\ 0 & 0 & 1 & 0 & 0 & 30 \\ 0 & 0 & 1 & 0 & -1 & -30 \end{array}\right].$$

The reader can easily verify that the matrix above transforms to the augmented matrix in reduced row echelon form

$$\left[\begin{array}{ccccc|c} 1 & 0 & 0 & -1 & 0 & -30 \\ 0 & 1 & 0 & -1 & 0 & -40 \\ 0 & 0 & 1 & 0 & 0 & 30 \\ 0 & 0 & 0 & 0 & 1 & 60 \\ 0 & 0 & 0 & 0 & 0 & 0 \end{array}\right].$$

This yields

$$x_1 - x_4 = -30, \quad x_2 - x_4 = -40, \quad x_3 = 30, \quad x_5 = 60.$$

In the first two equations above, we let $x_4 = k$. Then

$$x_1 = k - 30, \quad x_2 = k - 40, \quad x_3 = 30, \quad x_4 = k, \quad x_5 = 60,$$

is the solution to our problem. There are an infinite number of solutions. If this network represents water flow in a drip irrigation system (under water rationing) one might be advised to restrict the water flow through, say, branch x_4 by assigning it a particular value. When this is done, one and only one solution to the problem is obtained. This same type of restriction would allow one to control the flow of traffic through a neighborhood or the flow of current through an electrical circuit, and so on.

■ **EXAMPLE 3** _____

Consider the case in which the matrix equation

$$AX = H$$

is the equation

$$\begin{bmatrix} -1 & 1 & -2 & 0 & -2 & 0 \\ -2 & 2 & -4 & 4 & -6 & -2 \\ 2 & -2 & 4 & -2 & 5 & 1 \\ 2 & -2 & 4 & 4 & 2 & 0 \end{bmatrix} \begin{bmatrix} x_1 \\ x_2 \\ x_3 \\ x_4 \\ x_5 \\ x_6 \end{bmatrix} = \begin{bmatrix} -3 \\ -8 \\ 7 \\ 6 \end{bmatrix}.$$

In this case the augmented matrix $[A \mid H]$ is the matrix

$$[A \mid H] = \begin{bmatrix} -1 & 1 & -2 & 0 & -2 & 0 & | & -3 \\ -2 & 2 & -4 & 4 & -6 & -2 & | & -8 \\ 2 & -2 & 4 & -2 & 5 & 1 & | & 7 \\ 2 & -2 & 4 & 4 & 2 & 0 & | & 6 \end{bmatrix}.$$

By the use of the Gaussian elimination method described above we transform $[A \mid H]$ into its row echelon matrix

$$[F \mid K] = \begin{bmatrix} 1 & -1 & 2 & 0 & 2 & 0 & | & 3 \\ 0 & 0 & 0 & 1 & -1/2 & 0 & | & 0 \\ 0 & 0 & 0 & 0 & 0 & 1 & | & 1 \\ 0 & 0 & 0 & 0 & 0 & 0 & | & 0 \end{bmatrix}.$$

Note that F is the row echelon matrix for A. It has the same number of nonzero rows as $[F \mid K]$. This is essential for our equation $AX = H$ to be solvable as we shall see in Proposition 3.

As in previous examples, the equation $AX = H$ is equivalent to the equation $FX = K$. Writing $FX = K$ in component form, we obtain the equations

$$x_1 - x_2 + 2x_3 \quad + 2x_5 \quad = 3$$
$$x_4 - \tfrac{1}{2}x_5 \quad = 0$$
$$x_6 = 1.$$

Observe that the number of equations in this set is equal to the number $r = 3$ of nonzero rows in the row echelon matrix F of A. To solve these equations we work backwards. We first note that $x_6 = 1$. Since x_5 is common to the first two equations, we assign a parameter to x_5, say $x_5 = a$. Our system becomes

$$x_1 - x_2 + 2x_3 = 3 - 2a, \qquad x_4 = \tfrac{1}{2}a, \qquad x_5 = a, \qquad x_6 = 1.$$

By the choice $x_5 = a$, we have now also determined x_4. From the first equation we see that the variables x_1, x_2, and x_3 are still to be determined. We introduce the new parameters

$$x_3 = b, \qquad x_2 = c.$$

This yields all solutions

$$x_1 = 3 - 2a - 2b + c, \quad x_2 = c, \quad x_3 = b, \quad x_4 = \tfrac{1}{2}a, \quad x_5 = a, \quad x_6 = 1.$$

We see that in this example, once again there are an infinite number of solutions, which we might call "triply infinite" in the spirit of the terminology used in Example 2.

This example illustrates the following result:

> The number q of basic parameters in the solution of $AX = H$ obtained by using its reduced row echelon form $FX = K$ is
>
> $q =$ (number of variables) $-$ (number of nonzero equations)
>
> or equivalently, $q = n - r$, where n is the number of columns of A and r is the number of nonzero rows in the row echelon matrix F for A.

Remark. The number r has special significance. It will give the **rank** of the matrix A, according to a definition that we will encounter later in the book.

Returning to our example, we write the solution in vector form. We have

$$X = \begin{bmatrix} 3 \\ 0 \\ 0 \\ 0 \\ 0 \\ 1 \end{bmatrix} + a \begin{bmatrix} -2 \\ 0 \\ 0 \\ 1/2 \\ 1 \\ 0 \end{bmatrix} + b \begin{bmatrix} -2 \\ 0 \\ 1 \\ 0 \\ 0 \\ 0 \end{bmatrix} + c \begin{bmatrix} 1 \\ 1 \\ 0 \\ 0 \\ 0 \\ 0 \end{bmatrix}.$$

This is of the form

$$X = Y + aZ_1 + bZ_2 + cZ_3,$$

where

$$Y = \begin{bmatrix} 3 \\ 0 \\ 0 \\ 0 \\ 0 \\ 1 \end{bmatrix}, \quad Z_1 = \begin{bmatrix} -2 \\ 0 \\ 0 \\ 1/2 \\ 1 \\ 0 \end{bmatrix}, \quad Z_2 = \begin{bmatrix} -2 \\ 0 \\ 1 \\ 0 \\ 0 \\ 0 \end{bmatrix}, \quad Z_3 = \begin{bmatrix} 1 \\ 1 \\ 0 \\ 0 \\ 0 \\ 0 \end{bmatrix}.$$

Setting $a = b = c = 0$, we see that $X = Y$ is a solution of $AX = H$. The remaining part

$$Z = X - Y = aZ_1 + bZ_2 + cZ_3$$

has the property that

$$AZ = AX - AY = H - H = 0.$$

Therefore,

$$Z = aZ_1 + bZ_2 + cZ_3$$

is a solution of $AX = 0$ for all choices of a, b, c. Setting $a = 1, b = c = 0$, we have $Z = Z_1$, so that $AZ_1 = 0$. Similarly, we see that $AZ_2 = 0$ by choosing $a = c = 0, b = 1$. Finally, by selecting $a = b = 0, c = 1$, we find that $AZ_3 = 0$. We can also verify this result by direct computation. For example, we see that $AZ_1 = 0$ by the computation

$$AZ_1 = \begin{bmatrix} -1 & 1 & -2 & 0 & -2 & 0 \\ -2 & 2 & -4 & 4 & -6 & -2 \\ 2 & -2 & 4 & -2 & 5 & 1 \\ 2 & -2 & 4 & 4 & 2 & 0 \end{bmatrix} \begin{bmatrix} -2 \\ 0 \\ 0 \\ 1/2 \\ 1 \\ 0 \end{bmatrix} = \begin{bmatrix} 0 \\ 0 \\ 0 \\ 0 \end{bmatrix}.$$

Similarly, by computation, it is seen that $AZ_2 = 0$ and $AZ_3 = 0$. It is easily seen that the vectors $Z_1, Z_2,$ and Z_3 are linearly independent.

This illustrates the following general result.

PROPOSITION 1. Let r be the number of nonzero rows in the row echelon matrix F of an $m \times n$-matrix A. If $r < n$, there exist $q = n - r$ solutions $Z_1, Z_2, \ldots, Z_q$ of $AX = 0$ having the following two properties:

  (**a**) They are linearly independent.

  (**b**) Every solution Z of $AX = 0$ is a linear combination

$$Z = a_1 Z_1 + a_2 Z_2 + \cdots + a_q Z_q$$

  of these solutions.

The number q is called the ***nullity*** of A. If $r = n$, then $q = 0$; that is, $X = 0$ is the only solution of $AX = 0$.

This result is discussed in detail in Chapter 6. There it is shown that if $r < n$ and $Z_1, Z_2, \ldots, Z_q$ are q solutions having properties (a) and (b), then $q = n - r$. Thus we can associate with the term ***nullity*** the adjective *largest,* as follows:

The nullity of A is the largest integer q for which there exist q linearly independent solutions of $AX = 0$. If there are no nonzero solutions of $AX = 0$, then $q = 0$. We have the formula $q = n - r$, where r is the number of rows of the reduced row echelon matrix F for A.

In our example, $r = 3$, $n = 6$, and $q = 6 - 3 = 3$. Our example also illustrates the following general result, which is studied more fully in Chapter 6.

PROPOSITION 2. Suppose that Y is a solution of $AX = H$. If $Z_1, Z_2, \ldots, Z_q$ are each solutions of $AX = 0$, then

$$X = Y + a_1 Z_1 + a_2 Z_2 + \cdots a_q Z_q, \qquad a_i \text{ any constants}$$

is a solution of $AX = H$. If $Z_1, Z_2, \ldots, Z_q$ have properties (a) and (b) of Proposition 1, every solution X of $AX = H$ can be obtained in this manner. If $X = 0$ is the only solution of $AX = 0$, then Y is the only solution of $AX = H$.

Thus every solution X of $AX = H$ is of the form

$$
\begin{array}{ccc}
\textbf{\textit{particular solution}} & & \textbf{\textit{linear combination}} \\
Y & & a_1Z_1 + a_2Z_2 + \cdots + a_qZ_q \\
X \quad = \quad \textbf{\textit{of}} & + & \text{of solutions } Z_i \text{ of} \\
AX = H & & AX = 0
\end{array}
$$

provided that we have "enough solutions Z_i" of $AX = 0$.

There are cases for which the equation $AX = H$ has no solution. A test for solvability of $AX = H$ is given in the following

---

PROPOSITION 3. The equation $AX = H$ is solvable if and only if the row echelon matrices F and $[F \mid K]$ of A and $[A \mid H]$, respectively, have the same number of nonzero rows.

This follows from the Gaussian elimination procedure described above.

We continue by considering an example of $AX = H$, which has no solution. We have already encountered such an example in Section 2.1 when A was square and singular. We now look at a case when A is a nonsquare matrix.

■ **EXAMPLE 4** _____

Let

$$
AX = \begin{bmatrix} 1 & 3 & 0 & -2 \\ 0 & 0 & 1 & 0 \\ 3 & 9 & -1 & -6 \end{bmatrix} \begin{bmatrix} x \\ y \\ z \\ w \end{bmatrix} = \begin{bmatrix} 5 \\ 3 \\ 13 \end{bmatrix} = H.
$$

We attempt to transform the augmented matrix

$$
[A \mid H] = \begin{bmatrix} 1 & 3 & 0 & -2 & \mid & 5 \\ 0 & 0 & 1 & 0 & \mid & 3 \\ 3 & 9 & -1 & -6 & \mid & 13 \end{bmatrix}
$$

into reduced row echelon form. Proceeding yields

$$
[A \mid H] = \begin{bmatrix} 1 & 3 & 0 & -2 & \mid & 5 \\ 0 & 0 & 1 & 0 & \mid & 3 \\ 3 & 9 & -1 & -6 & \mid & 13 \end{bmatrix} \longrightarrow \begin{bmatrix} 1 & 3 & 0 & -2 & \mid & 5 \\ 0 & 0 & 1 & 0 & \mid & 3 \\ 0 & 0 & -1 & 0 & \mid & -2 \end{bmatrix}
$$

$$
\longrightarrow \begin{bmatrix} 1 & 3 & 0 & -2 & \mid & 5 \\ 0 & 0 & 1 & 0 & \mid & 3 \\ 0 & 0 & 0 & 0 & \mid & 1 \end{bmatrix} \longrightarrow \begin{bmatrix} 1 & 3 & 0 & -2 & \mid & 0 \\ 0 & 0 & 1 & 0 & \mid & 0 \\ 0 & 0 & 0 & 0 & \mid & 1 \end{bmatrix} = [F \mid K].
$$

We notice that the number of nonzero rows of the matrix F is 2, while the number of nonzero rows of the augmented matrix $[F \mid K]$ is 3. We conclude from Proposition 3 that the equation $AX = H$ does not have a solution.

In component form, we have

$$x + 3y \qquad - 2w = 0$$

$$z \qquad = 0$$

$$0x + 0y + 0z + 0w = 1.$$

It is obvious that the third equation can never be satisfied. Therefore, in this example, the system $AX = H$ does not have a solution. In general, the equation $AX = H$ will have no solution if in transforming the augmented matrix $[A \mid H]$ into reduced row echelon form $[F \mid K]$, the system of equations $FX = K$ yields an inconsistency.

We conclude this section by considering the chemical equation we had in Chapter 1.

■ **EXAMPLE 5** _____

If

$$x\mathrm{H_2SO_3} + y\mathrm{HBrO_3} \quad \longrightarrow \quad z\mathrm{H_2SO_4} + w\mathrm{Br_2} + u\mathrm{H_2O}$$

is balanced, we have

$$2x + y - 2z \qquad - 2u = 0$$

$$x \qquad - z \qquad = 0$$

$$3x + 3y - 4z \qquad - u = 0$$

$$y \qquad - 2w \qquad = 0.$$

The coefficient matrix is given by

$$\begin{bmatrix} 2 & 1 & -2 & 0 & -2 \\ 1 & 0 & -1 & 0 & 0 \\ 3 & 3 & -4 & 0 & -1 \\ 0 & 1 & 0 & -2 & 0 \end{bmatrix}.$$

One can easily verify that the reduced row echelon form of this matrix is

$$F = \begin{bmatrix} 1 & 0 & 0 & 0 & -5 \\ 0 & 1 & 0 & 0 & -2 \\ 0 & 0 & 1 & 0 & -5 \\ 0 & 0 & 0 & 1 & -1 \end{bmatrix}.$$

We see from the matrix F that

$$x \quad\quad - 5u = 0$$

$$y \quad\quad - 2u = 0$$

$$z \quad - 5u = 0$$

$$w - \quad u = 0.$$

It follows that

$$x = 5u, \quad y = 2u, \quad z = 5u, \quad w = u.$$

Letting $u = 1$, we see that $x = z = 5$, $y = 2$, and $w = 1$. The original chemical equation can now be balanced. We have

$$5H_2SO_3 + 2HBrO_3 \quad\longrightarrow\quad 5H_2SO_4 + Br_2 + H_2O.$$

Exercises

1. Determine whether or not the following matrices are in reduced row echelon form.

   (a) $\begin{bmatrix} 1 & 0 & 3 \\ 0 & 1 & 2 \end{bmatrix}$

   (b) $\begin{bmatrix} 2 & 0 & | & 6 \\ 0 & 1 & | & 5 \end{bmatrix}$

   (c) $\begin{bmatrix} 0 & 1 & 0 & | & 0 \\ 0 & 0 & 0 & | & 0 \\ 0 & 0 & 0 & | & 0 \end{bmatrix}$

   (d) $\begin{bmatrix} 1 & 2 & 0 & | & 5 \\ 0 & 1 & 0 & | & 4 \\ 0 & 0 & 0 & | & 0 \end{bmatrix}$

   (e) $\begin{bmatrix} 1 & 0 \\ 0 & 1 \end{bmatrix}$

   (f) $\begin{bmatrix} 1 & 0 & 0 & 5 \\ 0 & 1 & 0 & 7 \\ 0 & 0 & 0 & 0 \\ 0 & 0 & 1 & 4 \end{bmatrix}$

2. Represent the following systems of linear equation in terms of matrices, and solve using Gaussian elimination.

   (a) $x - y - \quad z = 2$
   $\quad\quad 2y - 10z = 1$

   (b) $2x \quad + 2z - 4w = -12$
   $\quad\quad y + 2z - \quad w = \quad 3$
   $\quad\quad 2y + 5z - 4w = \quad 9$

   (c) $3x + 6y - 12z = 9$
   $\quad 2x + 4y - \quad 8z = 6$

   (d) $-2x - 4z \quad\quad = -6$
   $\quad\quad x + 2z + 3w = -3$
   $\quad\quad 3x + 6z + \quad w = \quad 9$

3. For the matrices that are not in reduced row echelon form in Exercise 1, put them in reduced row echelon form.

4. Find the general solution of the following systems of equations:

   (a) $x_1 + 2x_2 \quad + x_4 = \quad 4$
   $\quad\quad x_3 - 2x_4 = -1$

   (b) $x_1 \quad + x_3 \quad + x_5 = 7$
   $\quad\quad x_2 + 2x_3 \quad - x_5 = 1$
   $\quad\quad\quad\quad x_4 - 2x_5 = 2$

5. Show that the nonzero rows of a matrix F in reduced row echelon form are linearly independent.

6. Prove that every equation $AX = H$ has one of the following:

   (i) 0 solutions,   (ii) 1 solution,   (iii) Infinitely many solutions.

   Hint: One only needs to show that two solutions imply infinitely many. If X_0 and X_1 $(X_0 \neq X_1)$ are solutions of $AX = H$, consider $X_0 + t(X_1 - X_0)$ for t any real number.

7. Show that if a matrix A has more columns than rows, there is a nonzero column vector X such that $AX = 0$. Conclude that in this event, the columns of A are linearly dependent. *Hint:* Transform A into its reduced row echelon form.

8. Show that if A has more rows than columns, there is a nonzero row vector Y such that $YA = 0$. Conclude that in this event, the rows of A are linearly dependent. *Hint:* Apply the result given in Exercise 7 to A^T.

9. Let A be an $m \times n$-matrix whose columns are linearly independent. Show that the reduced row echelon matrix F associated with A is of the form

$$F = \begin{bmatrix} I \\ 0 \end{bmatrix}$$

   when $m > n$ and that $F = I$ when $m = n$. Here I is the $m \times m$-identity matrix.

10. Let F be the reduced row echelon matrix for a matrix A. Show that there is a nonsingular matrix B such that $F = BA$. *Hint:* B is the product of elementary matrices.

The following topics are geometric applications of $AX = H$ and are intended for further study and verification by the reader.

11. In analytic geometry, we learned that the equation

$$3x + 4y - 2z = 1$$

   is the equation of a plane **P**, a 2-plane, in the three-dimensional space of points $X = (x, y, z)$.

   (a) This equation is of the form $AX = H$, where

$$A = [3 \quad 4 \quad -2], \qquad H = [1] = 1.$$

   (b) The vector A is orthogonal (perpendicular) to the plane **P** and is called a normal of **P**.

   (c) The number H determines the position of **P**. If we change H, we move **P** parallel to itself. For example, if we select $H = 5$, we obtain the plane

$$3x + 4y - 2z = 5,$$

   which is parallel to **P**. If we choose $H = 0$, we obtain

$$3x + 4y - 2z = 0.$$

   This plane is also parallel to **P**. It passes through the origin $X = 0$.

(d) The point $Y = (5, -3, 1)$ lies in **P**. This follows because $AY = 1$.

(e) The solution $Z = (2, -1, 1)$ of $AZ = 0$ has two interpretations. It is a point in the plane $AX = 0$. On the other hand, it can be interpreted to be a direction vector in **P**. In particular, if Y is in **P**, then, for every number t, the point

$$X = Y + tZ$$

is a point in **P**. If we consider t to be variable, these points determine a line **L** in **P** through Y in the direction Z. This equation is a parametric equation for **L**. In component form, with $Y = (5, -3, 1)$, this equation becomes the parametric equations

$$x = 5 + 2t, \qquad y = -3 - t, \qquad z = 1 + t$$

for the line **L**.

(f) The vector $Z_2 = (-2, 2, 1)$, as well as the vector $Z_1 = (2, -1, 1)$, is a direction vector in **P**. They satisfy $AZ = 0$. The vectors Z_1 and Z_2 are linearly independent. According to Proposition 1, every direction vector Z in **P** is expressible in the form

$$Z = sZ_1 + tZ_2,$$

where s and t are suitably chosen numbers (parameters). According to Proposition 2, with Y in **P**, every point X in **P** is expressible in the form

$$X = Y + sZ_1 + tZ_2.$$

An equation of this type is called a ***parametric equation*** for our 2-plane **P**. In component form it becomes

$$x = 5 + 2s - 2t, \qquad y = -3 - s + 2t, \qquad z = 1 + s + t.$$

Parametric equations for **P** are not unique. There is one for each choice of Y, Z_1, and Z_2.

(g) The equation $AX = H$ is called a ***nonparametric equation*** for **P**. It is not unique. This follows because $2AX = 2H$; that is,

$$6x + 8y - 4z = 2$$

is also a nonparametric equation for **P**. So also are $5AX = 5H$ and $cAX = cH$ for every nonzero number c.

12. As in Exercise 11, discuss the case in which **P** is a 2-plane defined by the equation

$$2x - 3y + 5z = 4.$$

In particular, find a parametric equation for **P**.

13. When we have two equationas

$$3x + 4y - 2z = 1$$
$$2x + 3y - z = 0,$$

we have two planes. They intersect in a line **L**. These two equations are therefore equations defining **L**.

(a) These equations can be written in matrix form

$$AX = H$$

with

$$A = \begin{bmatrix} 3 & 4 & -2 \\ 2 & 3 & -1 \end{bmatrix}, \quad H = \begin{bmatrix} 1 \\ 0 \end{bmatrix}, \quad X = \begin{bmatrix} x \\ y \\ z \end{bmatrix}.$$

(b) The row vectors of A are normals of the line **L**. They are linearly independent. Every normal of **L** is a linear combination of the rows of A. The matrix A can be viewed to be a "matrix normal" of **L**.

(c) The vector H determines the position of the line **L**. It is a "position vector" for **L**. If we replace H by a new vector, say $H_1 = (1, 0)$, the new line $AX = H_1$ is parallel to **L**. A second vector, say $H_2 = (0, 1)$, gives us a second line $AX = H_2$ parallel to **L**. In fact, it can be shown that every line parallel to **L** is determined by an equation of the form

$$AX = h_1 H_1 + h_2 H_2.$$

In component form this becomes the pair of equations

$$3x + 4y - 2z = h_1, \quad 2x + 3y - z = h_2.$$

(d) The point $Y = (5, -3, 1)$ lies on **L**.

(e) The solution $Z = (2, -1, 1)$ of $AZ = 0$ has two interpretations. It is a point on the line $AX = 0$ parallel to **L**. On the other hand, it is a direction vector in **L**. For a fixed point Y in **L** and for each number t, the point

$$X = Y + tZ$$

is in **L**. In fact, with t as a parameter, this equation is a parametric equation for **L**. In component form, with $Y = (5, -3, 1)$, this equation becomes

$$x = 5 + 2t, \quad y = -3 - t, \quad z = 1 + t.$$

(f) With Z as a row vector, the equation

$$ZX = 2x - y + z = 0$$

is the plane orthogonal to **L** which passes through the origin.

(g) For each nonsingular 2×2-matrix P, the equation

$$PAX = PH$$

is an alternative equation for the line **L**. If we select

$$P = \begin{bmatrix} 3 & -4 \\ -2 & 3 \end{bmatrix}$$

and set

$$F = PA = \begin{bmatrix} 1 & 0 & -2 \\ 0 & 1 & 1 \end{bmatrix}, \quad K = PH = \begin{bmatrix} 3 \\ -2 \end{bmatrix},$$

we obtain the equation $FX = K$ for **L**, in which F is in reduced row echelon form.

14. Consider the line described by $AX = H$ for the case in which

$$A = \begin{bmatrix} 2 & 10 & 5 \\ 3 & 15 & 8 \end{bmatrix}, \quad H = \begin{bmatrix} 17 \\ 26 \end{bmatrix}, \quad X = \begin{bmatrix} x \\ y \\ z \end{bmatrix}.$$

   (a) Find two planes that intersect in this line.

   (b) Find a point Y on this line.

   (c) Find a direction vector Z for this line.

   (d) Find a parametric equation of this line.

15. Consider the equation $AX = H$, where

$$A = \begin{bmatrix} 3 & 4 & -2 \\ 2 & 3 & -1 \\ 1 & 1 & -1 \end{bmatrix}, \quad H = \begin{bmatrix} 1 \\ 0 \\ 1 \end{bmatrix}, \quad X = \begin{bmatrix} x \\ y \\ z \end{bmatrix}.$$

   (a) Observe that the first row is the sum of the other two rows. That is,

$$VH = 0,$$

   where V is the row vector $V = \begin{bmatrix} 1 & -1 & -1 \end{bmatrix}$. However, the first two rows are linearly independent. This means that if $WA = 0$, then W is a multiple of V.

   (b) The relation $VH = 0$ tells us that the equation $AX = H$ is solvable. One solution is $Y = (5, -3, 1)$.

   (c) In component form, the equation $AX = H$ is

$$3x + 4y - 2z = 1$$
$$2x + 3y - z = 0$$
$$x - y - z = 1.$$

   These are the equations of three planes. The first two intersect in a line **L**. Since the last equation is the difference of the first two, the plane defined by the last equation also contains **L**. Consequently, we have three planes that intersect in a line. Because the first two planes are the planes given in Exercise 13, the line **L** coincides with the line described in Exercise 13.

   (d) Again, the vector H determines the position of the line $AX = H$. However, it is not arbitrary. It must be orthogonal to the row vector V described in part (b).

   (e) The line **L** is also defined by the equation $PAX = PH$ for every nonsingular matrix P. When we select

$$P = \begin{bmatrix} 3 & -4 & 0 \\ -2 & 3 & 0 \\ 1 & -1 & -1 \end{bmatrix}$$

and set $F = PA$, $K = PH$, we obtain the equation

$$FX = K,$$

where

$$F = \begin{bmatrix} 1 & 0 & -2 \\ 0 & 1 & 1 \\ 0 & 0 & 0 \end{bmatrix}, \qquad H = \begin{bmatrix} 1 \\ 0 \\ 0 \end{bmatrix}, \qquad X = \begin{bmatrix} x \\ y \\ z \end{bmatrix}.$$

The matrix F is in reduced row echelon form.

A Topic for Self-Study

NEWTON'S METHOD FOR INVERTING MATRICES

In the early days of high-speed computers, some of the computers did not have a built-in division algorithm. The programmer had to write his or her own division routine. One method is to use the standard long-division routine. However, there is another routine that is faster. It consists of finding the reciprocal a^{-1} of a positive number a by a Newton's method. To divide b by a, we multiply b by a^{-1}. Newton's method for inverting a number can be extended so as to obtain a method for inverting a nonsingular matrix.

Newton's method for inverting a number a proceeds as follows. Select a number x_1 so that

$$e_1 = 1 - ax_1$$

is less than 1 in absolute value. For example, we can select x_1 to be the number obtained in the first step of long division. Then perform the following iteration:

$$x_2 = x_1 + x_1e_1, \qquad e_2 = 1 - ax_2 \qquad \text{or} \qquad e_2 = e_1^2$$

$$x_3 = x_2 + x_2e_2, \qquad e_3 = 1 - ax_3 \qquad \text{or} \qquad e_3 = e_2^2$$

$$\vdots \qquad\qquad\qquad \vdots \qquad\qquad\qquad\qquad \vdots$$

$$x_{k+1} = x_k + x_ke_k, \qquad e_{k+1} = 1 - ax_{k+1} \qquad \text{or} \qquad e_{k+1} = e_k^2$$

$$\vdots \qquad\qquad\qquad \vdots \qquad\qquad\qquad\qquad \vdots$$

Terminate when e_k is so small that x_k is a satisfactory estimate of a^{-1}.

Observe that we have two ways for computing e_k. Either can be used. They yield the same number. For if $e_k = 1 - ax_k$, then

$$e_{k+1} = 1 - ax_{k+1} = 1 - a(x_k + x_ke_k) = e_k - ax_ke_k$$

$$= (1 - ax_k)e_k = (1 - ax_k)e_k = e_k^2.$$

To illustrate this algorithm, let us invert $a = 3$. Choose $x_1 = .3$. Then $e_1 = 1 - .9 = .1 < 1$. Then our algorithm yields the estimates

$$x_1 = .3, \qquad\qquad e_1 = .1$$
$$x_2 = .33, \qquad\qquad e_2 = .01$$
$$x_3 = .3333, \qquad\qquad e_3 = .0001$$
$$x_4 = .33333333, \qquad\qquad e_4 = .00000001$$
$$x_5 = .3333333333333333, \qquad e_5 = .0000000000000001$$
$$\vdots \qquad\qquad\qquad \vdots$$

As remarked above, the algorithm just described can be extended to obtain a procedure for finding the inverse of a nonsingular $n \times n$-matrix A. To carry out this procedure, we select an initial estimate X_1 of A^{-1} such that the absolute value e_1 of the largest entry in the error matrix

$$E_1 = I - AX_1$$

is less than $1/n$. Thus $e = ne_1 < 1$. We then perform the iteration, for $k = 1, 2, 3, \ldots$,

$$X_{k+1} = X_k + X_k E_k, \qquad E_{k+1} = I - AX_{k+1} = E_k^2.$$

Again we have two ways for computing E_{k+1}. Their equality follows because if $E_k = I - AX_k$, then

$$E_{k+1} = I - AX_{k+1} = I - A(X_k + X_k E_k) = E_k - AX_k E_k = (I - AX_k)E_k$$
$$= (1 - AX_k)E_k = E_k^2.$$

We terminate when E_k is so small that X_k can be taken to be a suitable estimate of the inverse of A.

To see that E_k approaches the zero matrix as k increases indefinitely, let e_k denote the maximum absolute values of the entries in E_k. We shall establish the inequality

$$ne_k \le e^{2^{k-1}},$$

where $e = ne_1 < 1$. This inequality holds when $k = 1$. Suppose that it holds for a given integer k. Then since $E_{k+1} = E_k^2$, we have

$$e_{k+1} \le ne_k^2 \quad \text{so that} \quad ne_{k+1} \le (ne_k)^2 \le (e^{2^{k-1}})^2 = e^{2^k}.$$

Because $e < 1$, e_k tends to zero quadratically and E_k tends to the zero matrix, as was to be proved.

The difficulty encountered in applying Newton's algorithm is in the selection of the initial estimate X_1. For this reason the principal use for this procedure is to improve an estimate X_1 of A^{-1} obtained by some other procedure, such as Gaussian elimination. In computing the inverse on a computer, round-off errors occur. Accordingly, the estimate X_1 of the inverse obtained may not be a satisfactory estimate of the inverse. By applying Newton's method *with high-precision arithmetic* we can obtain improved estimates of A^{-1}.

To see how this algorithm works, consider the 2×2 matrices

$$A = \begin{bmatrix} 2 & -1 \\ 1 & 2 \end{bmatrix}, \qquad A^{-1} = \begin{bmatrix} .4 & .2 \\ -.2 & .4 \end{bmatrix}, \qquad X_1 = \begin{bmatrix} .5 & .1 \\ -.3 & .3 \end{bmatrix}.$$

We take X_1 as an initial estimate of A^{-1}. Then

$$E_1 = I - AX_1 = \begin{bmatrix} -.3 & .1 \\ .1 & .3 \end{bmatrix}, \qquad E_2 = E_1^2 = \begin{bmatrix} .1 & 0 \\ 0 & .1 \end{bmatrix},$$

$$E_3 = E_2^2 = \begin{bmatrix} .01 & 0 \\ 0 & .01 \end{bmatrix}, \qquad E_4 = E_3^2 = \begin{bmatrix} .0001 & 0 \\ 0 & .0001 \end{bmatrix}, \qquad \text{etc.}$$

The corresponding estimates $X_{k+1} = X_k + X_k E_k$ of A^{-1} are

$$X_2 = \begin{bmatrix} .36 & .18 \\ -.18 & .36 \end{bmatrix}, \qquad X_3 = \begin{bmatrix} .396 & .198 \\ -.198 & .396 \end{bmatrix},$$

$$X_4 = \begin{bmatrix} .3996 & .1998 \\ -.1998 & .3998 \end{bmatrix}.$$

Computing further we see that

$$X_5 = \begin{bmatrix} .3999996 & .1999998 \\ -.1999998 & .3999996 \end{bmatrix}.$$

This simple example illustrates the effectiveness of Newton's method for finding inverses.

3

Determinants and Their Properties

In elementary algebra we learned that the determinant of the 2×2-matrix

$$A = \begin{bmatrix} a & b \\ c & d \end{bmatrix}$$

is given by the formula

$$\det A = \begin{vmatrix} a & b \\ c & d \end{vmatrix} = ad - bc.$$

We see that the determinant of A is a number and that this number is easily remembered by use of the diagram

The determinant of A is the product of the entries a and d on the "$+$" arrow minus the product of the entries b and c on the "$-$" arrow. For example,

$$\begin{vmatrix} 7 & -1 \\ 2 & 3 \end{vmatrix} = 7(3) - 2(-1) = 21 + 2 = 23.$$

Observe that we also have

$$\begin{vmatrix} 7 & 2 \\ -1 & 3 \end{vmatrix} = 7(3) - 2(-1) = 21 + 2 = 23.$$

In general it is clear that the 2×2-matrices

$$A = \begin{bmatrix} a & b \\ c & d \end{bmatrix}, \qquad A^T = \begin{bmatrix} a & c \\ b & d \end{bmatrix}$$

have the same determinant $ad - bc$. The relation

$$\det A = \det A^T$$

is a fundamental property of determinants.

We proceed to define the determinant of a 3×3-matrix

$$A = \begin{bmatrix} a_1 & b_1 & c_1 \\ a_2 & b_2 & c_2 \\ a_3 & b_3 & c_3 \end{bmatrix}.$$

The determinant

$$\det A = \begin{vmatrix} a_1 & b_1 & c_1 \\ a_2 & b_2 & c_2 \\ a_3 & b_3 & c_3 \end{vmatrix}$$

of A is given by the formula

$$\det A = a_1b_2c_3 + b_1c_2a_3 + c_1a_2b_3 - a_3b_2c_1 - b_3c_2a_1 - c_3a_2b_1.$$

This formula is easily remembered by use of the arrow diagrams

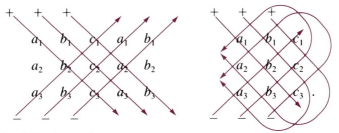

We have the following rule:

$\det A$ = sum of products of the entries on "$+$" arrows $-$ sum of products of the entries on "$-$" arrows.

For example, using the second diagram, we find that

$$\begin{vmatrix} 5 & -1 & 3 \\ 1 & 2 & 0 \\ 0 & 1 & 1 \end{vmatrix} = \begin{array}{l} 5(2)(1) + (-1)(0)(0) + 3(1)(1) - 0(2)(3) \\ - 1(0)(5) - 1(1)(-1) = 14 \end{array}$$

Using the formula on the transpose

$$A^T = \begin{bmatrix} a_1 & a_2 & a_3 \\ b_1 & b_2 & b_3 \\ c_1 & c_2 & c_3 \end{bmatrix},$$

we find that $\det A^T$ is the same as $\det A$ so that

$$\det A^T = \det A$$

also holds for 3×3-matrices.

Unfortunately, the arrow diagrams, which are so useful for two- or three-dimensional determinants, become cumbersome in higher-dimensional determinants. However, there is another scheme which generalizes easily to higher-dimensional determinants. To describe this scheme, we begin by writing the formula for

$$\det A = \begin{vmatrix} a_1 & b_1 & c_1 \\ a_2 & b_2 & c_2 \\ a_3 & b_3 & c_3 \end{vmatrix}$$

in six different ways. Recall that

$$\det A = a_1b_2c_3 + b_1c_2a_3 + c_1a_2b_3 - a_3b_2c_1 - b_3c_2a_1 - c_3a_2b_1.$$

One way is to rearrange the terms as follows:

$$\det A = a_1(b_2c_3 - b_3c_2) - a_2(b_1c_3 - b_3c_1) + a_3(b_1c_2 - b_2c_1).$$

The quantities in parentheses are 2×2-determinants. Hence this formula for det A can be rewritten as the first of the following six expansions for det A. The others can be obtained in the same manner.

$$\det A = a_1 \begin{vmatrix} b_2 & c_2 \\ b_3 & c_3 \end{vmatrix} - a_2 \begin{vmatrix} b_1 & c_1 \\ b_3 & c_3 \end{vmatrix} + a_3 \begin{vmatrix} b_1 & c_1 \\ b_2 & c_2 \end{vmatrix}$$

$$\det A = -b_1 \begin{vmatrix} a_2 & c_2 \\ a_3 & c_3 \end{vmatrix} + b_2 \begin{vmatrix} a_1 & c_1 \\ a_3 & c_3 \end{vmatrix} - b_3 \begin{vmatrix} a_1 & c_1 \\ a_2 & c_2 \end{vmatrix}$$

$$\det A = c_1 \begin{vmatrix} a_2 & b_2 \\ a_3 & b_3 \end{vmatrix} - c_2 \begin{vmatrix} a_1 & b_1 \\ a_3 & b_3 \end{vmatrix} + c_3 \begin{vmatrix} a_1 & b_1 \\ a_2 & b_2 \end{vmatrix}$$

$$\det A = a_1 \begin{vmatrix} b_2 & c_2 \\ b_3 & c_3 \end{vmatrix} - b_1 \begin{vmatrix} a_2 & c_2 \\ a_3 & c_3 \end{vmatrix} + c_1 \begin{vmatrix} a_2 & b_2 \\ a_3 & b_3 \end{vmatrix}$$

$$\det A = -a_2 \begin{vmatrix} b_1 & c_1 \\ b_3 & c_3 \end{vmatrix} + b_2 \begin{vmatrix} a_1 & c_1 \\ a_3 & c_3 \end{vmatrix} - c_2 \begin{vmatrix} a_1 & b_1 \\ a_3 & b_3 \end{vmatrix}$$

$$\det A = a_3 \begin{vmatrix} b_1 & c_1 \\ b_2 & c_2 \end{vmatrix} - b_3 \begin{vmatrix} a_1 & c_1 \\ a_2 & c_2 \end{vmatrix} + c_3 \begin{vmatrix} a_1 & b_1 \\ a_2 & b_2 \end{vmatrix}$$

The first three of these formulas are called *column expansions* of det A and the last three are *row expansions* for det A. In each case the 2×2-determinant in the formula is called a *minor* in det A of the entry, by which it is multiplied. The minor of an entry is obtained by deleting the row and column containing the entry. For example, the minor of a_1 in det A is obtained by deleting the first row and the first column. Thus

$$\text{minor of } a_1 \text{ in } \begin{vmatrix} a_1 & b_1 & c_1 \\ a_2 & b_2 & c_2 \\ a_3 & b_3 & c_3 \end{vmatrix} = \begin{vmatrix} b_2 & c_2 \\ b_3 & c_3 \end{vmatrix}.$$

Similarly,

$$\text{minor of } c_2 \text{ in } \begin{vmatrix} a_1 & b_1 & c_1 \\ a_2 & b_2 & c_2 \\ a_3 & b_3 & c_3 \end{vmatrix} = \begin{vmatrix} a_1 & b_1 \\ a_3 & b_3 \end{vmatrix}.$$

Each minor of an entry has associated with it a sign determined by the position of the entry. The sign is chosen according to the following chart:

$$\begin{vmatrix} + & - & + \\ - & + & - \\ + & - & + \end{vmatrix}.$$

Looking at this chart we obtain the rule

sign to be attached to the minor of the (*i*, *j*)-entry = sign of $(-1)^{i+j}$.

Note that when $i + j$ is even, the sign attached to the minor is $+$ and when $i + j$ is odd, the sign attached to the minor is $-$. To attach its sign to a minor we introduce the concept of the *cofactor* of an entry as follows:

cofactor of the (*i*, *j*)-entry is $(-1)^{i+j}$ times its minor.

For example,

$$\text{for our } 3 \times 3\text{-matrix } A = \begin{bmatrix} a_1 & b_1 & c_1 \\ a_2 & b_2 & c_2 \\ a_3 & b_3 & c_3 \end{bmatrix},$$

$$\text{cofactor of } a_1 = (-1)^{1+1} \begin{vmatrix} b_2 & c_2 \\ b_3 & c_3 \end{vmatrix} = \begin{vmatrix} b_2 & c_2 \\ b_3 & c_3 \end{vmatrix},$$

$$\text{cofactor of } c_2 = (-1)^{2+3} \begin{vmatrix} a_1 & b_1 \\ a_3 & b_3 \end{vmatrix} = - \begin{vmatrix} a_1 & b_1 \\ a_3 & b_3 \end{vmatrix}.$$

With this concept in mind we can describe the expansion of det A by a column or row as follows:

Choose a column (or row) of A. The expansion of det A about this column (or row) is given by the formula

det A = sum of the products of the entries of the column (or row) by their cofactors.

■ **EXAMPLE 1** _____

In a numerical case using the first column expansion we have

$$\begin{vmatrix} 5 & -1 & 3 \\ 1 & 2 & 0 \\ 0 & 1 & 1 \end{vmatrix} = 5 \begin{vmatrix} 2 & 0 \\ 1 & 1 \end{vmatrix} - 1 \begin{vmatrix} -1 & 3 \\ 1 & 1 \end{vmatrix} + 0 \begin{vmatrix} -1 & 3 \\ 2 & 0 \end{vmatrix}.$$

Evaluating the 2×2 determinants we see that the result is 14. Using the second row expansion we see that

$$\begin{vmatrix} 5 & -1 & 3 \\ 1 & 2 & 0 \\ 0 & 1 & 1 \end{vmatrix} = -1 \begin{vmatrix} -1 & 3 \\ 1 & 1 \end{vmatrix} + 2 \begin{vmatrix} 5 & 3 \\ 0 & 1 \end{vmatrix} - 0 \begin{vmatrix} 5 & -1 \\ 0 & 1 \end{vmatrix}.$$

Again the result is 14.

We now turn to the definition of the determinant of an $n \times n$-matrix. Our motivation for its definition will follow along the same lines as the way we defined the determinant of a 3×3-matrix in terms of determinants of 2×2-matrices. We proceed as follows:

$$A = [a_{ij}] = \begin{bmatrix} a_{11} & a_{12} & \cdots & a_{1n} \\ a_{21} & a_{22} & \cdots & a_{2n} \\ \vdots & \vdots & & \vdots \\ a_{n1} & a_{n2} & \cdots & a_{nn} \end{bmatrix}.$$

We denote the determinant of A by various symbols such as

$$\det A, \qquad |A|, \qquad \begin{vmatrix} a_{11} & a_{12} & \cdots & a_{1n} \\ a_{21} & a_{22} & \cdots & a_{2n} \\ \vdots & \vdots & & \vdots \\ a_{n1} & a_{n2} & \cdots & a_{nn} \end{vmatrix}, \qquad \det [a_{ij}].$$

When $n = 1$ we set $\det A = a_{11}$. Henceforth assume that $n > 1$. Assume further that the $(n - 1)$-dimensional determinants have been defined. Denote by $A(i|j)$ the $(n - 1) \times (n - 1)$-matrix obtained by deleting the ith row and jth column of A. Then we define

minor of a_{ij} to be $\det A(i|j)$

and

cofactor of a_{ij} to be $(-1)^{i+j} \det A(i|j)$.

For example, when $n = 5, i = 3, j = 4$,

$$A(3|4) = \begin{bmatrix} a_{11} & a_{12} & a_{13} & a_{14} & a_{15} \\ a_{21} & a_{22} & a_{23} & a_{24} & a_{25} \\ a_{31} & a_{32} & a_{33} & a_{34} & a_{35} \\ a_{41} & a_{42} & a_{43} & a_{44} & a_{45} \\ a_{51} & a_{52} & a_{53} & a_{54} & a_{55} \end{bmatrix} = \begin{bmatrix} a_{11} & a_{12} & a_{13} & a_{15} \\ a_{21} & a_{22} & a_{23} & a_{25} \\ a_{41} & a_{42} & a_{43} & a_{45} \\ a_{51} & a_{52} & a_{53} & a_{55} \end{bmatrix},$$

$$\textit{minor of } a_{34} = \det A(3|4) = \begin{vmatrix} a_{11} & a_{12} & a_{13} & a_{15} \\ a_{21} & a_{22} & a_{23} & a_{25} \\ a_{41} & a_{42} & a_{43} & a_{45} \\ a_{51} & a_{52} & a_{53} & a_{55} \end{vmatrix},$$

$$\textit{cofactor of } a_{34} = (-1)^{3+4} \det A(3|4) = - \begin{vmatrix} a_{11} & a_{12} & a_{13} & a_{15} \\ a_{21} & a_{22} & a_{23} & a_{25} \\ a_{41} & a_{42} & a_{43} & a_{45} \\ a_{51} & a_{52} & a_{53} & a_{55} \end{vmatrix}.$$

We have the product of a_{ij} by its cofactor $= (-1)^{i+j} a_{ij} \det A(i|j)$. It can be shown (see Exercises 5 and 6) that the sums

$$\sum_{i=1}^{n} (-1)^{i+j} a_{ij} \det A(i|j), \qquad \sum_{j=1}^{n} (-1)^{i+j} a_{ij} \det A(i|j)$$

have the same value no matter how j is chosen in the first sum and i is chosen in the second sum. The first sum is the jth column expansion and the second sum is the ith row expansion of our determinant. We can take any one of these sums to be the definition of det A in the n-dimensional case. It will be convenient to select the first column expansion,

$$\det A = \sum_{i=1}^{n} (-1)^{i+1} a_{i1} \det A(i|1),$$

to be our definition of det A. Later we shall show that the other expansions yield the determinant of A, so that

$$\det A = \sum_{i=1}^{n} (-1)^{i+j} a_{ij} \det A(i|j) = \sum_{j=1}^{n} (-1)^{i+j} a_{ij} \det A(i|j).$$

In the four-dimensional case,

$$A = [a_{ij}] = \begin{bmatrix} a_{11} & a_{12} & a_{13} & a_{14} \\ a_{21} & a_{22} & a_{23} & a_{24} \\ a_{31} & a_{32} & a_{33} & a_{34} \\ a_{41} & a_{42} & a_{43} & a_{44} \end{bmatrix}.$$

The cofactors of the entries a_{11}, a_{21}, a_{31}, and a_{41} of column 1 of A are, respectively,

$$c_{11} = (-1)^{1+1} \begin{vmatrix} a_{11} & a_{12} & a_{13} & a_{14} \\ a_{21} & a_{22} & a_{23} & a_{24} \\ a_{31} & a_{32} & a_{33} & a_{34} \\ a_{41} & a_{42} & a_{43} & a_{44} \end{vmatrix} = \begin{vmatrix} a_{22} & a_{23} & a_{24} \\ a_{32} & a_{33} & a_{34} \\ a_{42} & a_{43} & a_{44} \end{vmatrix},$$

$$c_{21} = (-1)^{2+1} \begin{vmatrix} a_{11} & a_{12} & a_{13} & a_{14} \\ a_{21} & a_{22} & a_{23} & a_{24} \\ a_{31} & a_{32} & a_{33} & a_{34} \\ a_{41} & a_{42} & a_{43} & a_{44} \end{vmatrix} = - \begin{vmatrix} a_{12} & a_{13} & a_{14} \\ a_{32} & a_{33} & a_{34} \\ a_{42} & a_{43} & a_{44} \end{vmatrix},$$

$$c_{31} = (-1)^{3+1} \begin{vmatrix} a_{11} & a_{12} & a_{13} & a_{14} \\ a_{21} & a_{22} & a_{23} & a_{24} \\ a_{31} & a_{32} & a_{33} & a_{34} \\ a_{41} & a_{42} & a_{43} & a_{44} \end{vmatrix} = \begin{vmatrix} a_{12} & a_{13} & a_{14} \\ a_{22} & a_{23} & a_{24} \\ a_{42} & a_{43} & a_{44} \end{vmatrix},$$

$$c_{41} = (-1)^{4+1} \begin{vmatrix} a_{11} & a_{12} & a_{13} & a_{14} \\ a_{21} & a_{22} & a_{23} & a_{24} \\ a_{31} & a_{32} & a_{33} & a_{34} \\ a_{41} & a_{42} & a_{43} & a_{44} \end{vmatrix} = - \begin{vmatrix} a_{12} & a_{13} & a_{14} \\ a_{22} & a_{23} & a_{24} \\ a_{32} & a_{33} & a_{34} \end{vmatrix}.$$

The determinant

$$\det A = a_{11}c_{11} + a_{21}c_{21} + a_{31}c_{31} + a_{41}c_{41}$$

of A is therefore given by the formula

$$\det A = a_{11} \begin{vmatrix} a_{22} & a_{23} & a_{24} \\ a_{32} & a_{33} & a_{34} \\ a_{42} & a_{43} & a_{44} \end{vmatrix} - a_{21} \begin{vmatrix} a_{12} & a_{13} & a_{14} \\ a_{32} & a_{33} & a_{34} \\ a_{42} & a_{43} & a_{44} \end{vmatrix}$$

$$+ a_{31} \begin{vmatrix} a_{12} & a_{13} & a_{14} \\ a_{22} & a_{23} & a_{24} \\ a_{42} & a_{43} & a_{44} \end{vmatrix} - a_{41} \begin{vmatrix} a_{12} & a_{13} & a_{14} \\ a_{22} & a_{23} & a_{24} \\ a_{32} & a_{33} & a_{34} \end{vmatrix}.$$

This first column expansion can be written down directly from the matrix by the use of the concept of cofactors. We so do in the following example.

■ **EXAMPLE 2**

$$\begin{vmatrix} 1 & 2 & -1 & 1 \\ 2 & 5 & 0 & 2 \\ -1 & 0 & 6 & 0 \\ 1 & 2 & 0 & 3 \end{vmatrix} = 1 \begin{vmatrix} 5 & 0 & 2 \\ 0 & 6 & 0 \\ 2 & 0 & 3 \end{vmatrix} - 2 \begin{vmatrix} 2 & -1 & 1 \\ 0 & 6 & 0 \\ 2 & 0 & 3 \end{vmatrix}$$

$$+ (-1) \begin{vmatrix} 2 & -1 & 1 \\ 5 & 0 & 2 \\ 2 & 0 & 3 \end{vmatrix} - 1 \begin{vmatrix} 2 & -1 & 1 \\ 5 & 0 & 2 \\ 0 & 6 & 0 \end{vmatrix}$$

$$= (90 - 24) - 2(36 - 12) - (11) - (-6)(-1) = 1.$$

It is simpler to use the third-row expansion. When this is done and zero entries are taken into account, we have

$$\begin{vmatrix} 1 & 2 & -1 & 1 \\ 2 & 5 & 0 & 2 \\ -1 & 0 & 6 & 0 \\ 1 & 2 & 0 & 3 \end{vmatrix} = -1 \begin{vmatrix} 2 & -1 & 1 \\ 5 & 0 & 2 \\ 2 & 0 & 3 \end{vmatrix}$$

$$+ 6 \begin{vmatrix} 1 & 2 & 1 \\ 2 & 5 & 2 \\ 1 & 2 & 3 \end{vmatrix} = -11 + 12 = 1.$$

■ **EXAMPLE 3**

Again by the first column expansion, we see that

$$\begin{vmatrix} 1 & 0 & 0 & 0 \\ 0 & 1 & 0 & 0 \\ 0 & 0 & 1 & 0 \\ 0 & 0 & 0 & 1 \end{vmatrix} = \begin{vmatrix} 1 & 0 & 0 \\ 0 & 1 & 0 \\ 0 & 0 & 1 \end{vmatrix} = \begin{vmatrix} 1 & 0 \\ 0 & 1 \end{vmatrix} = 1.$$

This illustrates the fact that det $I = 1$ for every identity matrix I, whatever the dimension. Similarly, by first-column expansions, we find that

$$\begin{vmatrix} a & 0 & 0 & 0 \\ 0 & b & 0 & 0 \\ 0 & 0 & c & 0 \\ 0 & 0 & 0 & d \end{vmatrix} = a \begin{vmatrix} b & 0 & 0 \\ 0 & c & 0 \\ 0 & 0 & d \end{vmatrix} = ab \begin{vmatrix} c & 0 \\ 0 & d \end{vmatrix} = abcd.$$

In this manner we see that the *determinant of a diagonal matrix is the product of its main diagonal entries*. In particular, the *determinant of an elementary matrix of the second kind*, such as

$$\begin{vmatrix} s & 0 & 0 \\ 0 & 1 & 0 \\ 0 & 0 & 1 \end{vmatrix}, \quad \begin{vmatrix} 1 & 0 & 0 \\ 0 & s & 0 \\ 0 & 0 & 1 \end{vmatrix}, \quad \begin{vmatrix} 1 & 0 & 0 \\ 0 & 1 & 0 \\ 0 & 0 & s \end{vmatrix}$$

is s. In view of the expansion of the type

$$\begin{vmatrix} 1 & 0 & 0 & 0 \\ 0 & 1 & 0 & 0 \\ 0 & s & 1 & 0 \\ 0 & 0 & 0 & 1 \end{vmatrix} = \begin{vmatrix} 1 & 0 & 0 \\ s & 1 & 0 \\ 0 & 0 & 1 \end{vmatrix} = \begin{vmatrix} 1 & 0 \\ 0 & 1 \end{vmatrix} = 1,$$

we conclude that the *determinant of an elementary matrix of the third kind is* **1**. Finally, the reader should verify that the *determinant of an elementary matrix of the first kind*, such as

$$\begin{vmatrix} 0 & 1 & 0 \\ 1 & 0 & 0 \\ 0 & 0 & 1 \end{vmatrix}, \quad \begin{vmatrix} 0 & 0 & 1 \\ 0 & 1 & 0 \\ 1 & 0 & 0 \end{vmatrix}, \quad \begin{vmatrix} 1 & 0 & 0 \\ 0 & 0 & 1 \\ 0 & 1 & 0 \end{vmatrix}$$

is **−1**. Some of the results above will be examined in greater detail in the next section.

Remarks on the Definition of Determinants

There are many ways to define a determinant. We have chosen an inductive definition based on row and column expansions. We do so because it immediately gives us a standard method for evaluating a determinant. However, there is an alternative definition of a determinant based on the concept of even and odd permutations of n objects. We essentially used this permutation method for the two-and three-dimensional cases. For the case $n = 4$, we have the matrix

$$A = \begin{bmatrix} a_1 & b_1 & c_1 & d_1 \\ a_2 & b_2 & c_2 & d_2 \\ a_3 & b_3 & c_3 & d_3 \\ a_4 & b_4 & c_4 & d_4 \end{bmatrix}.$$

The even and odd permutations of $(1, 2, 3, 4)$ are

$+(1, 2, 3, 4),$	$-(2, 1, 3, 4),$	$+(3, 1, 2, 4),$	$-(4, 1, 2, 3)$
$-(1, 2, 4, 3),$	$+(2, 1, 4, 3),$	$-(3, 1, 4, 2),$	$+(4, 1, 3, 2)$
$+(1, 3, 4, 2),$	$-(2, 3, 4, 1),$	$+(3, 2, 4, 1),$	$-(4, 2, 3, 1)$
$-(1, 3, 2, 4),$	$+(2, 3, 1, 4),$	$-(3, 2, 1, 4),$	$+(4, 2, 1, 3)$
$+(1, 4, 2, 3),$	$-(2, 4, 1, 3),$	$+(3, 4, 1, 2),$	$-(4, 3, 1, 2)$
$-(1, 4, 3, 2),$	$+(2, 4, 3, 1),$	$-(3, 4, 2, 1),$	$+(4, 3, 2, 1).$

We use $+$ to designate the even permutation and $-$ to designate the odd permutations. Each 4-tuple is obtained from each of its neighbors by the interchange of two of its elements. Using the 4-tuples as subscripts on a, b, c, and d and attaching their signs, we obtain

$$a_1 b_2 c_3 d_4 - a_2 b_1 c_3 d_4 + a_3 b_1 c_2 d_4 - a_4 b_1 c_2 d_3$$
$$- a_1 b_2 c_4 d_3 + a_2 b_1 c_4 d_3 - a_3 b_1 c_4 d_2 + a_4 b_1 c_3 d_2$$
$$+ a_1 b_3 c_4 d_2 - a_2 b_3 c_4 d_1 + a_3 b_2 c_4 d_1 - a_4 b_2 c_3 d_1$$
$$- a_1 b_3 c_2 d_4 + a_2 b_3 c_1 d_4 - a_3 b_2 c_1 d_4 + a_4 b_2 c_1 d_3$$
$$+ a_1 b_4 c_2 d_3 - a_2 b_4 c_1 d_3 + a_3 b_4 c_1 d_2 - a_4 b_3 c_1 d_2$$
$$- a_1 b_4 c_3 d_2 + a_2 b_4 c_3 d_1 - a_3 b_4 c_2 d_1 + a_4 b_3 c_2 d_1.$$

Performing the indicated operations of addition and subtraction, we obtain det A. Observe that the sum of the first column is the product $a_1 C_1$ of a_1 by its cofactor C_1. The sum of the second column is the product $a_2 C_2$ by its cofactor C_2. Similarly, the sum of the third column is $a_3 C_3$ and the sum of the last column is $a_4 C_4$, where C_3 and C_4 are the cofactors of a_3 and a_4, respectively. This gives us the first column expansion

$$\det A = a_1 C_1 + a_2 C_2 + a_3 C_3 + a_4 C_4$$

used in our original definition of det A.

Exercises

1. Evaluate the 2×2-determinants

$$\begin{vmatrix} 2 & -5 \\ 1 & 3 \end{vmatrix}, \quad \begin{vmatrix} 2 & -5 \\ 3 & 9 \end{vmatrix}, \quad \begin{vmatrix} 1 & -1 \\ -1 & 1 \end{vmatrix}, \quad \begin{vmatrix} a & b \\ -b & a \end{vmatrix},$$

$$\begin{vmatrix} \cos\theta & \sin\theta \\ -\sin\theta & \cos\theta \end{vmatrix}, \quad \begin{vmatrix} 3/5 & 4/5 \\ -4/5 & 3/5 \end{vmatrix}, \quad \begin{vmatrix} 1 & 0 \\ 2 & 0 \end{vmatrix}.$$

2. Evaluate the 3 × 3-determinants

$$\begin{vmatrix} 1 & 2 & 3 \\ 2 & 3 & 0 \\ 3 & 0 & 0 \end{vmatrix}, \quad \begin{vmatrix} 3 & 2 & 1 \\ 0 & 3 & 2 \\ 0 & 0 & 3 \end{vmatrix}, \quad \begin{vmatrix} 5 & -2 & 0 \\ 2 & 1 & 0 \\ 7 & -5 & 3 \end{vmatrix}.$$

3. Find the determinants of the transposes of the matrices associated with the determinants in Exercise 2. Observe that in these cases det A = det A^T. As we shall see later, this is true in all cases.

4. Evaluate the 4 × 4-determinants

$$\begin{vmatrix} 5 & -2 & 0 & 0 \\ 2 & 1 & 0 & 0 \\ 9 & 8 & 5 & -2 \\ 6 & -6 & 2 & 1 \end{vmatrix}, \quad \begin{vmatrix} 1 & 1 & 2 & 2 \\ 2 & 2 & 4 & 4 \\ 1 & -2 & 3 & -4 \\ 1 & 0 & 1 & 0 \end{vmatrix}, \quad \begin{vmatrix} 1 & 1 & -1 & 0 \\ 0 & 2 & 0 & 3 \\ 2 & 0 & -2 & 0 \\ 0 & 1 & 0 & 1 \end{vmatrix},$$

$$\begin{vmatrix} 0 & 0 & 5 & 2 \\ 0 & 0 & -2 & 1 \\ 2 & 1 & 0 & 0 \\ -1 & 2 & 0 & 0 \end{vmatrix}, \quad \begin{vmatrix} 2 & 1 & 0 & 0 \\ 1 & 2 & 1 & 0 \\ 0 & 1 & 2 & 1 \\ 0 & 0 & 1 & 2 \end{vmatrix}, \quad \begin{vmatrix} 1 & 0 & 0 & 0 \\ 1 & 2 & 1 & 0 \\ 0 & 1 & 2 & 1 \\ 0 & 0 & 1 & 2 \end{vmatrix}.$$

5. Evaluate the 5 × 5-determinants

$$\begin{vmatrix} 5 & -2 & 0 & 0 & 0 \\ 2 & 1 & 0 & 0 & 0 \\ 0 & 0 & 2 & -1 & -3 \\ 0 & 0 & 1 & 3 & 4 \\ 0 & 0 & 0 & 0 & 1 \end{vmatrix}, \quad \begin{vmatrix} 2 & 1 & 0 & 0 & 0 \\ 1 & 2 & 1 & 0 & 0 \\ 0 & 1 & 2 & 1 & 0 \\ 0 & 0 & 1 & 2 & 1 \\ 0 & 0 & 0 & 1 & 2 \end{vmatrix}.$$

6. Verify that

$$\begin{vmatrix} a & e & g \\ 0 & b & f \\ 0 & 0 & c \end{vmatrix} = abc, \quad \begin{vmatrix} a & 0 & 0 \\ e & b & 0 \\ g & f & c \end{vmatrix} = abc.$$

Conclude that the determinant of an upper triangular matrix is the product of its main diagonal entries. The same is true for a lower triangular matrix.

7. In the next section we shall see that interchanging two columns in a determinant changes its sign. In the following matrices

$$\begin{vmatrix} 1 & 1 & 1 & 1 \\ -1 & 1 & -1 & 1 \\ -1 & -1 & 1 & 1 \\ 1 & -1 & -1 & 1 \end{vmatrix}, \quad \begin{vmatrix} 1 & 1 & 1 & 1 \\ -1 & -1 & 1 & 1 \\ 1 & -1 & -1 & 1 \\ -1 & 1 & -1 & 1 \end{vmatrix}, \quad \begin{vmatrix} 1 & 1 & 1 & 1 \\ 1 & -1 & 1 & -1 \\ 1 & -1 & -1 & 1 \\ 1 & 1 & -1 & -1 \end{vmatrix},$$

the determinant of the first matrix is 16. The second matrix is obtained from the first by moving column 3 to the column 1 position while preserving the order of the other columns. This can be done by $2 = 3 - 1$ adjacent column interchanges. Verify that the determinant of the second matrix is $(-1)^{3-1}16 = 16$. The third matrix is obtained from the first by moving the fourth column to the front. This involves $3 = 4 - 1$ adjacent column interchanges. Its determinant is $(-1)^{4-1}16 = -16$. Conclude that if a matrix B is obtained from a matrix A by moving its jth column to the first position while preserving the order of the remaining columns, B can be obtained from A by $j - 1$ adjacent column interchanges so that $\det B = (-1)^{j-1}\det A$.

8. Let A and B be square matrices, not necessarily of the same size. It can be shown that if M and N are matrices of appropriate sizes,

$$\det \begin{bmatrix} A & M \\ 0 & B \end{bmatrix} = \det \begin{bmatrix} A & 0 \\ N & B \end{bmatrix} = \det A \det B.$$

Use this result to evaluate the following determinants:

$$\begin{bmatrix} 5 & -2 & 3 \\ 2 & 1 & 4 \\ 0 & 0 & 6 \end{bmatrix}, \quad \begin{vmatrix} 5 & -2 & 0 & 0 \\ 2 & 1 & 0 & 0 \\ 3 & 4 & 8 & 5 \\ 2 & 6 & 3 & 2 \end{vmatrix}, \quad \begin{vmatrix} 5 & -2 & 7 & 9 \\ 2 & 1 & 0 & 4 \\ 0 & 0 & 2 & 5 \\ 0 & 0 & 2 & 3 \end{vmatrix}.$$

9. Show, by Exercise 8, that if A, B, and C are square matrices, then

$$\det \begin{bmatrix} A & 0 & 0 \\ P & B & 0 \\ Q & R & C \end{bmatrix} = \det A \det \begin{bmatrix} B & 0 \\ R & C \end{bmatrix} = \det A \det B \det C.$$

Apply this result to show that

$$\begin{vmatrix} 5 & -2 & 0 & 0 & 0 & 0 \\ 2 & 1 & 0 & 0 & 0 & 0 \\ 6 & 7 & 2 & 3 & 0 & 0 \\ 0 & 1 & 1 & 1 & 0 & 0 \\ -7 & 8 & 9 & 0 & 4 & -1 \\ 5 & 6 & 7 & 8 & 1 & 2 \end{vmatrix} = (9)(-1)(9) = -81.$$

10. Let $B = [b_{ik}]$ be the matrix obtained from $A = [a_{ik}]$ by moving its jth column into the first column position while preserving the order of the other columns. Show that $b_{i1} = a_{ij}$ and $B\,(i|1) = A\,(i|j)$. As noted in Exercise 7, $\det B = (-1)^{j-1}\det A$. We also have

$$\det B = \sum_{i=1}^{n}(-1)^{i+1}b_{i1}\det B\,(i|1) = \sum_{i=1}^{n}(-1)^{i+1}a_{ij}\det A\,(i|j),$$

so that

$$\det A = (-1)^{j-1}\det B = \sum_{i=1}^{n}(-1)^{i+j}A(i|j).$$

Conclude that any column expansion can be used to compute $\det A$.

11. Set $B = A^T$. In the next section we show that det B = det A. Observe that $b_{ji} = a_{ij}$ and $B(j|i) = A(i|j)^T$. Conclude that det $B(j|i)$ = det $A(i|j)$ and that

$$\det B = \sum_{j=1}^{n}(-1)^{j+i}b_{ji}\det B(j|i) = \sum_{j=1}^{n}(-1)^{i+j}a_{ij}\det A(i|j).$$

Since det A = det B we have the ith row expansion

$$\det A = \sum_{j=1}^{n}(-1)^{i+j}a_{ij}\det A(i|j)$$

for det A. Observe that the row expansions for det A are column expansions for det A^T. Similarly, row expansions for det A are column expansions for det A^T.

3.2   PROPERTIES OF DETERMINANTS

In this section we study some of the properties of determinants that will be useful in simplifying the calculations when one evaluates a determinant. In establishing these properties, we will use the first-column expansion definition of determinants.

As we have seen in Chapter 2, the elementary matrices were, in a sense, the building blocks of nonsingular matrices. We shall, as a result, establish some important properties of these elementary matrices and their determinants as we proceed.

The first property of determinants that we establish can be motivated by considering the 2×2-case.

■ **EXAMPLE 1**

Suppose that $A = \begin{bmatrix} 1 & 2 \\ 3 & 4 \end{bmatrix}$, so that det $A = -2$. If we interchange *rows* 1 and 2 above, and call the new matrix

$$B = \begin{bmatrix} 3 & 4 \\ 1 & 2 \end{bmatrix},$$

we see that det $B = 2 = -$det A.

Similarly, if we interchange *columns* 1 and 2 of A,

$$\det\begin{bmatrix} 2 & 1 \\ 4 & 3 \end{bmatrix} = 2 = -\det\begin{bmatrix} 1 & 2 \\ 3 & 4 \end{bmatrix}.$$

In general, we have the following:

PROPERTY 1. If B is obtained from A by an elementary row (or column) operation of the first kind, that is, by interchanging two rows (or columns) of A, then

$$\det B = -\det A.$$

The proof of Property 1 is given at the end of the section.

From Property 1 we can calculate the determinant of an elementary matrix of the first kind. Recall that an elementary matrix of the first kind P_{ij} is constructed from the identity matrix I by performing an elementary row operation of the first kind on this matrix I. Pictorially,

$$I \xrightarrow[\text{of first kind}]{\substack{\text{elementary} \\ \text{row operation}}} P_{ij}.$$

Therefore, by Property 1,

$$\det P_{ij} = -\det I.$$

But $\det I = 1$, so

$$\det P_{ij} = -1$$

for any elementary matrix P_{ij} of the first kind.

Recall that if A is any matrix, a row operation of the first kind on A is equivalent to multiplying A on the *left* by an elementary matrix of the first kind, to form a matrix B, where

$$B = P_{ij}A.$$

By Property 1,

$$\det B = \det (P_{ij}A) = -\det A.$$

But, by the argument above,

$$\det P_{ij} = -1.$$

Therefore,

$$\det (P_{ij}A) = \det P_{ij} \det A,$$

where P_{ij} is an elementary matrix of the first kind.

In summary,

If P_{ij} is an elementary matrix of the first kind, then

    1. det $P_{ij} = -1$.
    2. det $(P_{ij}A) = $ det P_{ij} det A.

for every matrix A.

We now wish to evaluate the determinant of an elementary matrix of the second kind. Once again, we will motivate this second property of determinants by looking at an example in the two-dimensional case.

■ **EXAMPLE 2** _____

Consider the matrix

$$A = \begin{bmatrix} 2 & 3 \\ 1 & 4 \end{bmatrix}, \qquad \text{whose determinant is 5.}$$

If we multiply the second row of A by 6, we get the matrix

$$B = \begin{bmatrix} 2 & 3 \\ 6 & 24 \end{bmatrix}, \qquad \text{whose determinant is 30.}$$

We see that

$$30 = \det B = \begin{vmatrix} 2 & 3 \\ 6 & 24 \end{vmatrix} = 6 \begin{vmatrix} 2 & 3 \\ 1 & 4 \end{vmatrix} = 6 \cdot 5 = 6 \det A.$$

This has the effect of allowing us to *factor* common factors from a row of a determinant. The same is true for the column of a determinant.

The general property is the following:

PROPERTY 2. If B is obtained from A by an elementary row (or column) operation of the second kind, that is, by multiplying the kth row (or column) of A by s, then

$$\det B = s \det A.$$

This property is quite useful in calculating determinants of matrices whose entries are "large numbers."

■ **EXAMPLE 3**

Consider

$$\begin{vmatrix} 16 & 20 \\ -7 & 35 \end{vmatrix} = 4 \begin{vmatrix} 4 & 5 \\ -7 & 35 \end{vmatrix} = 4(7) \begin{vmatrix} 4 & 5 \\ -1 & 5 \end{vmatrix} = 4(7)(5) \begin{vmatrix} 4 & 1 \\ -1 & 1 \end{vmatrix}.$$

The value of the determinant is, therefore, $140(5) = 700$. In the first step we factored 4 from the first row. In the second step we factored 7 from the second row, and in the third step, we factored 5 from the second column. Of course, in this two-dimensional case, it would be simpler to evaluate this determinant directly by the computation

$$16(35) - (-7)(20) = 560 + 140 = 700.$$

Factorizations of this type usually simplify computations of higher-dimensional determinants, as can be seen in the next example.

■ **EXAMPLE 4**

$$\begin{vmatrix} 22 & 22 & 22 \\ 105 & 210 & 105 \\ -19 & -19 & 38 \end{vmatrix} = 22 \begin{vmatrix} 1 & 1 & 1 \\ 105 & 210 & 105 \\ -19 & -19 & 38 \end{vmatrix}$$

$$= 22(105) \begin{vmatrix} 1 & 1 & 1 \\ 1 & 2 & 1 \\ -19 & -19 & 38 \end{vmatrix}$$

$$= 22(105)(19) \begin{vmatrix} 1 & 1 & 1 \\ 1 & 2 & 1 \\ -1 & -1 & 2 \end{vmatrix}.$$

The calculation of the latter determinant is certainly simpler than the original and one is burdened with at most three multiplications of large numbers in contrast to evaluating the original determinant.

The proof of Property 2 is given at the end of the section. Property 2 will enable us to calculate the determinant of an elementary matrix of the second kind. Recall that an elementary matrix of the second kind is constructed from the identity matrix I by multiplying a row (or column) of I by a nonzero scalar s [a row (or column) operation of the second kind]. Pictorially:

**elementary
row operation**

$$I \xrightarrow{\quad\text{of the second kind}\quad} S_i$$

identity   elementary
matrix   matrix of the
                            second kind

Therefore, by Property 2,

$$\det S_i = s \det I.$$

Again, since $\det I = 1$, we have

$$\det S_i = s,$$

where s is the nonzero scalar in the main diagonal of the elementary matrix of the second kind, S_i.

Recall once again that a row operation of the second kind on an arbitrary matrix A is equivalent to multiplying A on the left by an elementary matrix of the second kind S_i. We write

$$B = S_i A.$$

By Property 2 we have

$$\det B = \det (S_i A) = s \det A,$$

where s is the nonzero scalar on the main diagonal of S_i. But

$$\det S_i = s.$$

It follows that

$$\det (S_i A) = \det S_i \det A.$$

In summary, we have:

If S_i is an elementary matrix of the second kind, with a nonzero scalar s on its main diagonal, then

1. $\det S_i = s.$
2. $\det (S_i A) = \det S_i \det A.$

for any matrix A.

We have established that

$$\det (P_{ij} A) = \det P_{ij} \det A$$

$$\det (S_i A) = \det S_i \det A.$$

One might expect a similar property for elementary matrices of the third kind. To effect this, we will need some more properties of determinants.

The first is quite straightforward.

PROPERTY 3. If all the entries of a row (or column) of a matrix A are 0, then $\det A = 0$.

For suppose that the kth row of A is zero. Multiplying this row by 2 gives a second matrix B. Since $B = A$ we have det B = det A. By Property 2, we also have det B = 2 det A. This is possible only if det $A = 0$.

According to this result, we have

$$\begin{vmatrix} 2 & 0 \\ 3 & 0 \end{vmatrix} = 0, \qquad \begin{vmatrix} 7 & -5 & 9 \\ -1 & 3 & 6 \\ 0 & 0 & 0 \end{vmatrix} = 0.$$

To motivate the next property of determinants, consider

■ **EXAMPLE 5** ──────────────────────────────────────

$$\begin{vmatrix} 1 & 3 \\ 2 & 6 \end{vmatrix} = 0.$$

Observe that in this determinant, row 2 = 2(row 1).

■ **EXAMPLE 6** ──────────────────────────────────────

As a second example, consider

$$\begin{vmatrix} 1 & -1 & 1 \\ 7 & -1 & 6 \\ 3 & -3 & 3 \end{vmatrix}.$$

Expanding by minors of column 1, we see that

$$\begin{vmatrix} 1 & -1 & 1 \\ 7 & -1 & 6 \\ 3 & -3 & 3 \end{vmatrix} = 1(15) - 7(0) + 3(-5) = 0.$$

In this determinant, row 3 = 3(row 1).

■ **EXAMPLE 7** ──────────────────────────────────────

Finally, one can verify that

$$\begin{vmatrix} 1 & 1 & 2 \\ 7 & -1 & 14 \\ 5 & 8 & 10 \end{vmatrix} = 0.$$

In this last example, column 3 = 2(column 1).

In general, we have

> **PROPERTY 4.** If one row of a matrix A is a multiple of another row of A, then det $A = 0$. Similarly, if one column of a matrix A is a multiple of another column of A, then det $A = 0$.

We note that if, in particular, A has two identical rows (or columns) then det $A = 0$. We prove Property 4 as follows: Suppose that the ith row R_i is a multiple of the jth row R_j of A. If R_i or R_j is 0, then det $A = 0$, by Property 3. Suppose, therefore, that these rows are not zero. We have two cases:

Case I. $R_i = R_j$. If $R_i = R_j$, then interchange rows R_i and R_j to obtain a new matrix B. Since this will not alter A, we have that $A = B$. But by Property 1,

$$\det B = -\det A.$$

But det $B = $ det A since $A = B$, so

$$\det A = -\det A.$$

This is true only when det $A = 0$.

Case II. $R_i = sR_j$, $s \neq 0$. If $R_i = sR_j$, then multiply R_j by s to obtain a new matrix C whose ith and jth rows are identical. By case I,

$$\det C = 0.$$

But, by Property 2,

$$\det C = s \det A.$$

Since $s \neq 0$, it follows that det $A = 0$.

We now motivate another property of determinants with the following example.

■ **EXAMPLE 8** _____

Suppose that

$$A = \begin{bmatrix} 3 & -4 & 2 \\ 2 & -3 & -1 \\ 1 & 2 & -4 \end{bmatrix}, \quad \det A = 28 \quad \text{and} \quad R = \begin{bmatrix} 1 & 0 & 2 \end{bmatrix}.$$

If we construct a new matrix B by **adding R to the third row of A,** we see that

$$B = \begin{bmatrix} 3 & -4 & 2 \\ 2 & -3 & -1 \\ 2 & 2 & -2 \end{bmatrix}, \quad \det B = 36.$$

Similarly, if we construct a new matrix M by **replacing the third row of A by R,** we obtain

$$M = \begin{bmatrix} 3 & -4 & 2 \\ 2 & -3 & -1 \\ 1 & 0 & 2 \end{bmatrix}, \quad \det M = 8.$$

We have the relationship

$$\det B = \det A + \det M.$$

Observe that in this example, the equality holds when the third row of B is obtained by *adding only the third rows* of the matrices A and M. We have in general:

PROPERTY 5. Let $R = [r_1 \quad r_2 \quad \cdots \quad r_n]$ be a row vector. Let

(a) M be the matrix obtained by *replacing* the ith row of A by R.

(b) B be the matrix obtained by *adding* the vector R to the ith row of A.

Then

$$\det B = \det A + \det M.$$

The proof of this result is given in the Supplementary Proofs at the end of the section. In this proof we are permitted only to use the first column expansion for det A, because we used this expansion as the definition of det A. However, Property 5 will be more easily understood if we assume that we have already shown that row expansions can be used to obtain det A. We illustrate the situation for the 3×3 case, and when $i = 2$. The general proof is left for the Supplementary Proofs at the end of the section. Let

$$A = \begin{bmatrix} a_{11} & a_{12} & a_{13} \\ a_{21} & a_{22} & a_{23} \\ a_{31} & a_{32} & a_{33} \end{bmatrix}$$

and $R = [r_1 \quad r_2 \quad r_3]$. Construct

$$B = \begin{bmatrix} a_{11} & a_{12} & a_{13} \\ a_{21} + r_1 & a_{22} + r_2 & a_{23} + r_3 \\ a_{31} & a_{32} & a_{33} \end{bmatrix}, \qquad M = \begin{bmatrix} a_{11} & a_{12} & a_{13} \\ r_1 & r_2 & r_3 \\ a_{31} & a_{32} & a_{33} \end{bmatrix}.$$

We calculate the determinants of A, B, and M by cofactor expansions involving row 2, the row of interest. We have

$$\det M = -r_1(a_{12}a_{33} - a_{13}a_{32}) + r_2(a_{11}a_{33} - a_{13}a_{31})$$
$$- r_3(a_{11}a_{32} - a_{12}a_{31})$$

$$\det B = -(a_{21} + r_1)(a_{12}a_{33} - a_{13}a_{32}) + (a_{22} + r_2)(a_{11}a_{33} - a_{13}a_{31})$$
$$- (a_{23} + r_3)(a_{11}a_{32} - a_{12}a_{31}).$$

Distributing, we see that

$$\det B = [-a_{21}(a_{12}a_{33} - a_{13}a_{32}) + a_{22}(a_{11}a_{33} - a_{13}a_{31}) - a_{23}(a_{11}a_{32} - a_{12}a_{31})]$$
$$+ [-r_1(a_{12}a_{33} - a_{13}a_{32}) + r_2(a_{11}a_{33} - a_{13}a_{31}) - r_3(a_{11}a_{32} - a_{12}a_{31})]$$
$$= \det A + \det M.$$

Expressing this result in a more descriptive way, suppose, for example, that our matrices are 4×4's, i.e. $n = 4$, and that if we construct M and B by altering the third row of A so that $i = 3$, we have

$$\begin{vmatrix} R_1 \\ R_2 \\ R_3 + R \\ R_4 \end{vmatrix} = \begin{vmatrix} R_1 \\ R_2 \\ R_3 \\ R_4 \end{vmatrix} + \begin{vmatrix} R_1 \\ R_2 \\ R \\ R_4 \end{vmatrix}.$$

The reader should be cautioned that this property does not claim that the determinant of a sum is the sum of the determinants, which is, in fact, false. That is,

$$\det (A + B) \neq (\det A) + (\det B)$$

unless A and B have special properties.

The next property of determinants is recognizable as a property having to do with elementary matrices of the third kind since it involves adding a multiple of one row of a matrix to another row of a matrix.

■ **EXAMPLE 9** _____

If

$$A = \begin{bmatrix} 3 & -4 & 2 \\ 2 & -3 & -1 \\ 1 & 2 & -4 \end{bmatrix}$$

multiplying the third row of A by -2 and adding this to the second row of A yields

$$B = \begin{bmatrix} 3 & -4 & 2 \\ 0 & -7 & 7 \\ 1 & 2 & -4 \end{bmatrix}.$$

A simple calculation will show that $\det A = 28 = \det B$. Similarly, if one multiplies the third row of B by -3 and adds this to the first row of B, one has

$$C = \begin{bmatrix} 0 & -10 & 14 \\ 0 & -7 & 7 \\ 1 & 2 & -4 \end{bmatrix}, \qquad \text{whose determinant is also 28..}$$

These examples illustrate the general property that adding a multiple of one row to another row does not change the value of a determinant. We have:

PROPERTY 6   If B is the matrix obtained from A by a row (or column) operation of the third kind, that is, by adding a multiple of one row (column) of A to another row (column) of A, then

$$\det B = \det A.$$

From the examples above, we see that

$$A = \begin{bmatrix} 3 & -4 & 2 \\ 2 & -3 & -1 \\ 1 & 2 & -4 \end{bmatrix}$$

has the same determinant as

$$C = \begin{bmatrix} 0 & -10 & 14 \\ 0 & -7 & 7 \\ 1 & 2 & -4 \end{bmatrix}.$$

It is obvious that a first column expansion of C will yield the value of that determinant much quicker than any row or column expansion of A. Property 6 should therefore be seen as extremely useful when evaluating determinants. Repeated applications of Property 6 will yield many 0's in a row or column, which will aid in the calculation of the determinant. Since all of the properties can be interpreted for both row and column operations, one may use *both column and row operations* in "reducing" matrices so that their determinants are more easily calculated.

Using Property 4 and Property 5, we proceed with a proof of Property 6 in the special case when $n = 5$ and when we add a multiple of the second row to the fourth row. The proof in the general case follows in exactly the same way. Let R_1, R_2, R_3, R_4, and R_5 be the rows of A. Then

$$A = \begin{bmatrix} R_1 \\ R_2 \\ R_3 \\ R_4 \\ R_5 \end{bmatrix}, \qquad B = \begin{bmatrix} R_1 \\ R_2 \\ R_3 \\ R_4 + sR_2 \\ R_5 \end{bmatrix}, \qquad M = \begin{bmatrix} R_1 \\ R_2 \\ R_3 \\ sR_2 \\ R_5 \end{bmatrix},$$

where B is obtained from A by adding sR_2 to R_4 and M is obtained from A by replacing R_4 in A by sR_2. By Property 5,

$$\det B = \det A + \det M.$$

But since the fourth row of M is a multiple of its second row, by Property 4

$$\det M = 0.$$

Therefore,

$$\det B = \det A.$$

As a consequence of Property 6, we have the analogous result for elementary matrices of the third kind T_{ij}. Because T_{ij} is constructed from I by an elementary row operation of the third kind performed on I,

$$I \quad \xrightarrow[\text{of third kind}]{\text{row operation}} \quad T_{ij},$$

it follows from Property 6 that

$$\det T_{ij} = \det I = 1.$$

Again, recall that an elementary row operation of the third kind on a matrix A is equivalent to multiplying A on the left by an elementary matrix of the third kind T_{ij}. We write

$$B = T_{ij}A.$$

Therefore,

$$\det B = \det (T_{ij}A) = \det A.$$

It follows that

$$\det (T_{ij}A) = \det T_{ij} \det A.$$

In summary,

If T_{ij} is an elementary matrix of the third kind, then

   1. $\det T_{ij} = 1$.
   2. $\det (T_{ij}A) = \det T_{ij} \det A$ for every matrix A.

It should be observed that by repeated applications of Property 6, if B is a matrix obtained from A by adding to the ith row (or column) of A, a *linear combination* of the remaining rows (or columns) of A, then

$$\det B = \det A.$$

■ **EXAMPLE 10** _____

$$A = \begin{bmatrix} 3 & -4 & 2 \\ 2 & -3 & -1 \\ 1 & 2 & -4 \end{bmatrix} = \begin{bmatrix} R_1 \\ R_2 \\ R_3 \end{bmatrix},$$

whose determinant $\det A = 28$. Construct B so that

$$B = \begin{bmatrix} R_1 \\ (3R_1 - R_3) + R_2 \\ R_3 \end{bmatrix} = \begin{bmatrix} 3 & -4 & 2 \\ 10 & -17 & 9 \\ 1 & 2 & -4 \end{bmatrix}.$$

It is easy to check that $\det B = 28$, verifying that

$$\det A = \det B.$$

If it should happen that one row of A, say the ith row R_i, is a linear combination R of the other rows of A, we can add the negative $-R$ of this linear combination to R_1 to produce a matrix B whose ith row is zero and which has the same determinant as A. Since $\det B = 0$, we have $\det A = 0$ also. This gives the following corollary of Property 6.

If one row of A is a linear combination of the other rows, that is, if A is singular, then $\det A = 0$.

We have seen that

$$\det (EA) = \det E \det A$$

for any matrix A and any elementary matrix E. This tells us that if E_1, E_2, and E_3 are elementary matrices, then

$$\det (E_1E_2E_3A) = \det E_1 \det (E_2E_3A) = \det E_1 \det E_2 \det (E_3A)$$
$$= \det E_1 \det E_2 \det E_3 \det A.$$

Similarly,

$$\det (E_1E_2E_3) = \det E_1 \det E_2 \det E_3.$$

Consequently,

$$\det (E_1E_2E_3A) = \det (E_1E_2E_3) \det A.$$

In the same manner it is seen that if E_1, E_2, . . . , E_q are q elementary matrices, then

$$\det (E_1E_2 \cdots E_qA) = \det (E_1E_2 \cdots E_q) \det A.$$

But every nonsingular matrix B is expressible as a product

$$B = E_1E_2 \cdots E_q$$

of elementary matrices E_1, E_2, . . . , E_q. Hence

$$\det (BA) = \det (E_1E_2 \cdots E_qA) = \det (E_1E_2 \cdots E_q) \det A = \det B \det A.$$

We have accordingly established, in part, the following property of determinants for the case in which B is nonsingular.

PROPERTY 7. If A and B are $n \times n$-matrices, then

$$\det(BA) = \det B \det A = \det(AB).$$

That is, "determinant of a product = product of the determinants."

When B is singular, there is a nonzero row vector Y such that $YB = 0$ and hence such that $YBA = 0$. The matrix BA is therefore singular whenever B is singular. Moreover, by the corollary to Property 6, the determinants of BA and B are zero, so that

$$0 = \det(BA) = 0\ (\det A) = \det B \det A.$$

It follows that $\det(BA) = \det B \det A$ in all cases. Interchanging B and A, we see that $\det(AB) = \det A \det B$. Consequently,

$$\det(AB) = \det A \det B = \det B \det A = \det(BA),$$

as stated in Property 7.

As an immediate consequence of this result, we have the following

---

COROLLARY. If A is invertible, then

$$\det A^{-1} = \frac{1}{\det A}.$$

For, if A is invertible, then $A^{-1}A = I$, so that $\det(A^{-1}A) = \det I = 1$. By Property 7, with $B = A^{-1}$, we have $\det(A^{-1}A) = (\det A^{-1})(\det A) = 1$. The result follows.

From this corollary we see that a nonsingular matrix *must* have a nonzero determinant since the determinant of its inverse must exist. Similarly, we see that a singular matrix *cannot* have an inverse, for if the inverse did exist, its determinant would be undefined. We summarize this in the following property.

PROPERTY 8. Let A be an $n \times n$-matrix. If

$\det A \neq 0$	$\det A = 0$
Then	Then
1. A is *nonsingular*.	**1.** A is *singular*.
2. A is *invertible*.	**2.** A is *not invertible*.

These are often the main working criteria for nonsingularity and singularity of an $n \times n$-matrix A in computational exercises.

Finally, by virtue of Property 7, we can now prove a property that we have already seen was true in the case of two and three dimensions.

PROPERTY 9.

$$\det A^T = \det A.$$

The proof is given in the Supplementary Proofs at the end of the section.

Exercises

1. We seek to evaluate the first of the following determinants using row and column operations in order to simplify the computation:

$$d = \begin{vmatrix} 1 & 0 & 2 & 6 & 0 \\ 2 & -4 & 0 & 0 & 0 \\ -1 & 5 & 7 & 6 & -5 \\ 1 & 3 & -5 & 2 & 0 \\ 1 & -2 & 3 & -1 & 1 \end{vmatrix} = \begin{vmatrix} 1 & 2 & 2 & 6 & 0 \\ 2 & 0 & 0 & 0 & 0 \\ -1 & 3 & 7 & 6 & -5 \\ 1 & 5 & -5 & 2 & 0 \\ 1 & 0 & 3 & -1 & 1 \end{vmatrix}.$$

To do so, observe the three zeros in row 2 leads one to seek another 0 there and then expand the determinant by that row. We multiply the first column by 2 and add the result to the second column. We thereby obtain the second matrix displayed above. By Property 6 the value of the determinant is unchanged. Expanding by the second row, we find that the first determinant is

$$d = -2 \begin{vmatrix} 2 & 2 & 6 & 0 \\ 3 & 7 & 6 & -5 \\ 5 & -5 & 2 & 0 \\ 0 & 3 & -1 & 1 \end{vmatrix} = -2 \begin{vmatrix} 2 & 2 & 6 & 0 \\ 3 & 22 & 1 & 0 \\ 5 & -5 & 2 & 0 \\ 0 & 3 & -1 & 1 \end{vmatrix}.$$

Multiplying row 4 by 5 and adding the result to row 2, we obtain the next determinant, which has three zeros in its last column. Expanding by column 4 gives

$$d = -2(1) \begin{vmatrix} 2 & 2 & 6 \\ 3 & 22 & 1 \\ 5 & -5 & 2 \end{vmatrix} = -2(1)2 \begin{vmatrix} 1 & 1 & 3 \\ 3 & 22 & 1 \\ 5 & -5 & 2 \end{vmatrix}.$$

By Property 2 we can factor out 2 from row 1 to get the quantity on the right. In the last determinant we first subtract column 1 from column 2 and then multiply column 1 by 3 and subtract the result from column 3. We find that

$$d = -4 \begin{vmatrix} 1 & 0 & 0 \\ 3 & 19 & -8 \\ 5 & -10 & -13 \end{vmatrix} = -4 \begin{vmatrix} 19 & -8 \\ -10 & -13 \end{vmatrix} = -4[19(-13)-(-8)(-10)]$$

and hence that $d = 1308$. The reader should fill in all of the steps in the computation above.

2. Evaluate the determinants

$$
\begin{vmatrix} 1 & 0 & 1 & 0 \\ 0 & 1 & 0 & 1 \\ 0 & 0 & 2 & 1 \\ 0 & 0 & 1 & 2 \end{vmatrix},
\quad
\begin{vmatrix} 1 & 0 & 0 & 0 \\ 2 & 1 & 0 & 0 \\ 3 & 2 & 1 & 0 \\ 4 & 5 & 2 & 1 \end{vmatrix},
\quad
\begin{vmatrix} 1 & 1 & 1 & 1 \\ 2 & 2 & 2 & 2 \\ 1 & 0 & 1 & 0 \\ 5 & -1 & 6 & -1 \end{vmatrix},
$$

$$
\begin{vmatrix} 1 & -1 & 0 & 0 & 1 \\ 0 & 1 & 2 & 3 & 0 \\ 1 & 1 & 1 & 1 & 1 \\ 2 & 0 & -2 & 0 & 1 \\ 0 & 3 & 1 & 2 & 3 \end{vmatrix},
\quad
\begin{vmatrix} 0 & 0 & 1 & 1 & 1 \\ 0 & 0 & 2 & -2 & 3 \\ 0 & 0 & 4 & 4 & 9 \\ 0 & 5 & 1 & 0 & 1 \\ 3 & 2 & 0 & 1 & 0 \end{vmatrix}.
$$

3. The matrices

$$
L = \begin{bmatrix} a & 0 & 0 & 0 \\ u & b & 0 & 0 \\ x & v & c & 0 \\ z & y & w & d \end{bmatrix},
\quad
U = \begin{bmatrix} a & u & x & z \\ 0 & b & v & y \\ 0 & 0 & c & w \\ 0 & 0 & 0 & d \end{bmatrix}
$$

are, respectively, ***lower and upper triangular matrices***. Verify that $\det L = \det U = abcd$ irrespective of the values of u, v, w, x, y, and z. Extend this result to higher dimensions.

4. Show that if P is the product of m elementary matrices of the first kind, then $\det P = -1$ if m is odd and $\det P = 1$ if m is even.

5. Show that if A is the product of elementary matrices of the third kind, $\det A = 1$.

6. Show that if B is nonsingular, $\det BAB^{-1} = \det A$. *Hint:* Use the fact that $\det B \det B^{-1} = 1$.

7. Let A be a $n \times n$-matrix. Show that $\det sA = s^n \det A$ first for the case $n = 3$ and then for a general integer n. In particular, $\det (-A) = (-1)^n \det A$.

8. Find all values of x for which

(a) $\begin{vmatrix} x+1 & 3 & 4 \\ 0 & x-1 & 2 \\ 0 & 0 & x+3 \end{vmatrix} = 0$ (b) $\begin{vmatrix} x & 1 & 1 \\ 1 & x & 1 \\ 1 & 1 & x \end{vmatrix} = 0$

(c) $\begin{vmatrix} 5-x & 0 & 8 \\ -1 & 2-x & 4 \\ 2 & 0 & 5-x \end{vmatrix} = 0$ (d) $\begin{vmatrix} 5-x & 8 & 1 \\ 2 & 5-x & 3 \\ 0 & 0 & 3-x \end{vmatrix} = 0$

9. (a) Determinants of the form

$$
\begin{vmatrix} 1 & 1 & 1 \\ a & b & c \\ a^2 & b^2 & c^2 \end{vmatrix},
\quad
\begin{vmatrix} 1 & 1 & 1 & 1 \\ a & b & c & d \\ a^2 & b^2 & c^2 & d^2 \\ a^3 & b^3 & c^3 & d^3 \end{vmatrix}
$$

are called **Vandermonde determinants**. Find their values. For example,

$$\begin{vmatrix} 1 & 1 & 1 \\ a & b & c \\ a^2 & b^2 & c^2 \end{vmatrix} = \begin{vmatrix} 1 & 0 & 0 \\ a & b-a & c-a \\ a^2 & b^2-a^2 & c^2-a^2 \end{vmatrix} = \begin{vmatrix} b-a & c-a \\ b^2-a^2 & c^2-a^2 \end{vmatrix}$$

$$= (b-a)(c-a)\begin{vmatrix} 1 & 1 \\ b+a & c+a \end{vmatrix} = (b-a)(c-a)(c-b).$$

(b) Evaluate the Vandermonde determinants

$$\begin{vmatrix} 1 & 1 & 1 \\ 0 & 2 & 4 \\ 0 & 4 & 16 \end{vmatrix}, \qquad \begin{vmatrix} 1 & 1 & 1 & 1 \\ 1 & 3 & -1 & -3 \\ 1 & 9 & 1 & 9 \\ 1 & 27 & -1 & -27 \end{vmatrix}.$$

10. Let (x_1, y_1) and (x_2, y_2) be distinct points in the familiar xy-plane. Show that the equation

$$\begin{vmatrix} x & y & 1 \\ x_1 & y_1 & 1 \\ x_2 & y_2 & 1 \end{vmatrix} = 0$$

is an equation of the line throught these two points. Use this determinant to find an equation of the line passing through the points $(2, 3)$ and $(-1, 0)$. Also find the equation of the line passing through the points $(1, 1)$ and $(2, 9)$.

11. Let (x, y), (x_1, y_1), and (x_2, y_2) be three distinct points in the xy-plane not on a line. They are the vertices of a triangle T. Show that

$$d = \begin{vmatrix} x & y & 1 \\ x_1 & y_2 & 1 \\ x_2 & y_2 & 1 \end{vmatrix} = \pm 2 \text{ (area of } T).$$

Hint: $d = ax + by + c$ with $a = y_1 - y_2$, $b = x_2 - x_1$, $c = x_1 y_2 - x_2 y_1$.

Then

$$\sqrt{a^2 + b^2} = \text{distance from } (x_1, y_1) \text{ to } (x_2, y_2).$$

Moreover,

$$h = \frac{ax + by + c}{\sqrt{a^2 + b^2}} = \pm \text{ distance of } (x, y)$$

from the line passing through (x_1, y_1). So $\frac{1}{2} |h| \sqrt{a+b^2} = $ area of T. Hence

$$d = h \sqrt{a^2 + b^2} = \pm 2 \text{ (area of } T).$$

12. Find the area of the triangle whose vertices are

(a) $(1, 0)$, $(0, 1)$, $(5, 5)$ (b) $(1, 1)$, $(3, -6)$, $(-3, -2)$
(c) $(0, 0)$, $(5, 2)$, $(2, 6)$

13. Suppose that (x_1, y_1) and (x_2, y_2) are distinct points not on a line through the origin. show that the area of the triangle T having $(0, 0)$, (x_1, y_1), and (x_2, y_2) as vertices is given by the formula

$$\pm \frac{1}{2} \begin{vmatrix} x_1 & y_1 \\ x_2 & y_2 \end{vmatrix}$$

Hint: Set $(x, y) = (0, 0)$ in Exercise 11.

14. Show that if (x_1, y_1), (x_2, y_2), and (x_3, y_3) are points not on a line, then

$$\begin{vmatrix} x^2 + y^2 & x & y & 1 \\ x_1^2 + y_1^2 & x_1 & y_1 & 1 \\ x_2^2 + y_2^2 & x_2 & y_2 & 1 \\ x_3^2 + y_3^2 & x_3 & y_3 & 1 \end{vmatrix} = 0$$

is an equation of the circle passing through these three points. *Hint:* Expand by minors of the first row without evaluating these minors.

15. Find the equation of the circle passing through the points

   (a) $(0, 0)$, $(5, 0)$, $(0, 5)$ (b) $(1, 1)$ $(1, 7)$, $(7, 1)$ (c) $(0, 6)$, $(8, 0)$, $(2, 6)$

16. Let M be a matrix of the form

$$M = \begin{bmatrix} A & C \\ 0 & B \end{bmatrix},$$

where A and B are square matrices. Show that

$$\det M = \det A \det B.$$

Hint: Show first that

$$\begin{vmatrix} A & C \\ 0 & I \end{vmatrix} = \begin{vmatrix} A & 0 \\ 0 & I \end{vmatrix} = |A|$$

and then use the relation

$$\begin{vmatrix} A & C \\ 0 & B \end{vmatrix} = \begin{vmatrix} I & 0 \\ 0 & B \end{vmatrix} \begin{vmatrix} A & C \\ 0 & I \end{vmatrix}.$$

17. By the use of transposition, show by Exercise 16 that

$$\begin{vmatrix} A & 0 \\ D & B \end{vmatrix} = \det A \det B$$

when A and B are square.

18. Use the results described in Exercise 16 and 17 to evaluate the following determinants:

$$\begin{vmatrix} 3 & 4 & 0 & 0 \\ 2 & 3 & 0 & 0 \\ 5 & 6 & 7 & 1 \\ 0 & 1 & 1 & 2 \end{vmatrix}, \quad \begin{vmatrix} 5 & 2 & 6 & 7 \\ -2 & 1 & 2 & -1 \\ 0 & 0 & 1 & 3 \\ 0 & 0 & 1 & 1 \end{vmatrix}, \quad \begin{vmatrix} 8 & -3 & 0 & 0 \\ -5 & 2 & 0 & 0 \\ 0 & 0 & 3 & 2 \\ 0 & 0 & 4 & 3 \end{vmatrix},$$

$$\begin{vmatrix} 3 & 4 & 5 & 6 & 7 \\ 2 & 3 & 4 & 5 & 6 \\ 0 & 0 & 1 & 1 & 1 \\ 0 & 0 & 1 & 0 & 1 \\ 0 & 0 & 0 & 1 & 1 \end{vmatrix}, \qquad \begin{vmatrix} 3 & 4 & 0 & 0 & 0 & 0 \\ 2 & 3 & 0 & 0 & 0 & 0 \\ 1 & 1 & 1 & 1 & 1 & 1 \\ 2 & 2 & 1 & 0 & 2 & 2 \\ 3 & 3 & 0 & 0 & 2 & 3 \\ 4 & 4 & 0 & 0 & 3 & 4 \end{vmatrix}.$$

19. Let A, B, and C be square matrices. Use the results described in Exercises 16 and 17 to show that

$$\begin{vmatrix} A & 0 & 0 \\ D & B & 0 \\ F & E & C \end{vmatrix} = \begin{vmatrix} A & 0 & 0 \\ D & B & E \\ F & 0 & C \end{vmatrix} = \det A \det B \det C.$$

Supplementary Proofs

Fill in the details in the following proofs. Recall that we defined det A by its first column expansion. Consequently, we are restricted to a first column expansion for det A in our proofs of Properties 1, 2, and 5.

P1. Let B be the matrix obtained by A by interchanging rows j and k. Prove that det $B = -\det A$.

The proof will be made by induction. We have seen already that this result is true in the two-dimensional case. Assume that $n > 2$ and that it holds in the $(n-1)$-dimensional case. We shall show that it holds in the n-dimensional case, thereby completing our induction.

Consider first the case in which $k = j + 1$ so that we are interchanging adjacent rows of $A = [a_{ih}]$ to obtain $B = [b_{ih}]$. If i is neither j nor k, we have $b_{i1} = a_{i1}$. Moreover, $B(i|1)$ is obtained from $A(i|1)$ by interchanging two rows so that det $B(i|1) = -\det A(i|1)$ by our induction hypotheses. Hence we have

$$(-1)^{i+1}b_{i1} \det B(i|1) = -(-1)^{i+1}a_{i1} \det A(i|1)$$

when i is neither j nor $k = j + 1$. On the other hand,

$$b_{j1} = a_{k1}, \qquad B(j|1) = A(k|1)$$
$$b_{k1} = a_{j1}, \qquad B(k|1) = A(j|1),$$

so that, with $k = j + 1$,

$$(-1)^{j+1}b_{j1} \det B(j|1) = -(-1)_k{}^{+1}a_{k1} \det A(k|1)$$
$$(-1)^{k+1}b_{k1} \det B(k|1) = -(-1)^{j+1}a_{j1} \det A(j|1).$$

From these results we see that

$$\sum_{i=1}^{n} (-1)^{i+1} b_{i1} \det B\,(i|1) = -\sum_{i=1}^{n} (-1)^{i+1} a_{i1} \det A\,(i|1).$$

Hence det $B = -\det A$ when $k = j+1$, that is, when adjacent rows are interchanged. In case $k > j + 1$ we can transform A into B by $m = 2(j - i) - 1$ adjacent row interchanges. In this event det $B = (-1)^m \det A = -\det A$ since m is odd.

We will see presently that det $A^T = \det A$ for every matrix A. If B is obtained from A by interchanging two of its columns, then B^T is obtained from A^T by interchanging two rows. Hence det $B = \det B^T = -\det A^T = -\det A$, as stated in Property 1.　　■

■ **P2.** Let B be the matrix obtained from A by multiplying its kth row by s. Show that det $B = s \det A$.

By Property 1 we can suppose that $k = 1$. Why? Again we use induction. As was seen earlier, this result is true when $n = 2$. Suppose that $n > 2$ and that the result is true in the $(n - 1)$-dimensional case. Let $B = [b_{ij}]$ be obtained from $A = [a_{ij}]$ by multiplying its first row by s. Then

$$b_{11} = sa_{11}, \qquad B\,(1|1) = A(1|1).$$

For $i > 1$ the matrix $B\,(i|1)$ is obtained from $A(i|1)$ by multiplying its first row by s. Consequently, we have

$$b_{i1} = a_{i1}, \qquad \det B\,(i|1) = s \det A\,(i|1) \qquad (i > 1).$$

It follows that for all i

$$(-1)^{i+1} b_{i1} \det B\,(i|1) = s\,(-1)^{i+1} a_{i1} \det A\,(i|1).$$

Summing i from 1 to n we find that det $B = s \det A$, as was to be proved. Using this result together with det $A^T = \det A$, it follows that det $B = s \det A$ when B is obtained from A by multiplying a column of A by s.　　■

■ **P3.** Prove the row additivity property of matrices. That is, let M be the matrix obtained from a matrix A by replacing its kth row R_k by a row vector $R = [r_1 \quad r_2 \quad \cdots \quad r_n]$. Let B be the matrix obtained from A by adding R to R_k. Show that

$$\det B = \det A + \det M.$$

Again, by Property 1, it is sufficient to consider the case $k = 1$. Why? We have

$$b_{11} = a_{11} + r_1, \qquad m_{11} = r_1, \qquad B(1|1) = A(1|1) = M(1|1).$$

It follows that

$$b_{11} \det B(1|1) = a_{11} \det A(1|1) + m_{11} \det M(1|1).$$

On the other hand, for $i > 1$, we have $b_{i1} = a_{i1} = m_{i1}$. Moreover, $M(i|1)$ is obtained from $A(i|1)$ by replacing its first row by $S = [r_1 \quad r_2 \quad \cdots \quad r_n]$. In addition, $B(i|1)$ is obtained by adding S to the first row of $A(i|1)$. By our induction hypotheses we have

$$\det B(i|1) = \det A(i|1) + \det M(i|1) \qquad (i > 1).$$

Because $b_{i1} = a_{i1} = m_{i1}$ it follows that the equation

$$b_{i1} \det B(i|1) = a_{i1} \det A(i|1) + m_{i1} \det M(i|1)$$

holds for $i > 1$ as well as for $i = 1$. Multiplying by $(-1)^{i+1}$ and summing from 1 to n, we find that

$$\det B = \det A + \det M,$$

as was to be proved.   ▬

▬ **P4.** Proof of Property 8.

As a result of the proof given in Property 7 we see that the determinant of a nonsingular matrix A is the product of determinants of elementary matrices. Since determinants of elementary matrices do not vanish, it follows that det A does not vanish. On the other hand, det $A = 0$ when A is singular. Conclude therefore that det $A = 0$ if and only if A is singular and that A is nonsingular if and only if det A does not vanish. Since a matrix is invertible if and only if it is nonsingular, it follows that a matrix is invertible if and only if its determinant is not zero. This proves Property 8.   ▬

▬ **P5.** Proof of Property 9: det $A^T = $ det A.

This result holds when A is an elementarty matrix of the first and second kinds, because in these cases A is symmetric, that is, $A^T = A$. It also holds when A is an elementary matrix of the third kind because then det $A = 1$ and det $A^T = 1$, in view of the fact that A^T is also an elementary matrix of the third kind. If A is the product $A = FGH$ of three elementary matrices, then $A^T = H^T G^T F^T$, so that

$$\det A^T = \det H^T \det G^T \det F^T = \det H \det G \det F = \det A.$$

Similarly, det $A^T = $ det A when A is the product of any number of elementary matrices. Because a nonsingular matrix A is expressible as a product of elementary matrices, we have det A^T = det A when A is nonsingular. When A is singular, so also is A^T and their determinants are zero and hence equal. It follows that det A^T = det A in all cases, as was to be shown.

▬ P6. Show that det$(A + B)$ = det A + the sum of the determinants obtained from det A by replacing one or more rows of A by the corresponding rows of det B.

In the case $n = 3$ with R, S, and T as the rows of A and U, V, and W as the rows of B, the determinant of $A + B$ is

$$\begin{vmatrix} R + U \\ S + V \\ T + W \end{vmatrix} = \begin{vmatrix} R \\ S \\ T \end{vmatrix} + \begin{vmatrix} U \\ S \\ T \end{vmatrix} + \begin{vmatrix} R \\ V \\ T \end{vmatrix} + \begin{vmatrix} U \\ V \\ T \end{vmatrix} + \begin{vmatrix} R \\ S \\ W \end{vmatrix} + \begin{vmatrix} U \\ S \\ W \end{vmatrix} + \begin{vmatrix} R \\ V \\ W \end{vmatrix} + \begin{vmatrix} U \\ V \\ W \end{vmatrix}.$$

This result can be obtained by a repeated application of Property 5. For example, we have

$$\begin{vmatrix} R + U \\ S + V \\ T + W \end{vmatrix} = \begin{vmatrix} R + U \\ S + V \\ T \end{vmatrix} + \begin{vmatrix} R + U \\ S + V \\ W \end{vmatrix} + \begin{vmatrix} R + U \\ S \\ T \end{vmatrix}$$

$$+ \begin{vmatrix} R + U \\ V \\ T \end{vmatrix} + \begin{vmatrix} R + U \\ S \\ W \end{vmatrix} + \begin{vmatrix} R + U \\ V \\ W \end{vmatrix}.$$

Using Property 5 again we get the expansion given above. The proof in the general case can be carried out in this manner.

In the case $n = 2$ with

$$A = \begin{bmatrix} a & b \\ c & d \end{bmatrix}, \qquad B = \begin{bmatrix} -\lambda & 0 \\ 0 & -\lambda \end{bmatrix} = \lambda \begin{bmatrix} 1 & 0 \\ 0 & 1 \end{bmatrix} = -\lambda I$$

we have

$$\det (A + B) = \begin{vmatrix} a - \lambda & b \\ c & d - \lambda \end{vmatrix}$$

$$= \begin{vmatrix} a & b \\ c & d \end{vmatrix} + \begin{vmatrix} -\lambda & 0 \\ c & d \end{vmatrix} + \begin{vmatrix} a & b \\ 0 & -\lambda \end{vmatrix} + \begin{vmatrix} -\lambda & 0 \\ 0 & -\lambda \end{vmatrix}.$$

Hence det $(A - \lambda I)$ = det $A - (a + d)\lambda + \lambda^2$. This procedure is particularly useful in the computation of det $(A - \lambda I)$ in higher-dimensional cases. ▬

3.3 CLASSICAL ADJOINT AND CRAMER'S RULE

We continue to denote by $A(i|j)$, the submatrix of an $n \times n$-matrix

$$A = \begin{bmatrix} a_{11} & a_{12} & \cdots & a_{1n} \\ a_{21} & a_{22} & \cdots & a_{2n} \\ \vdots & \vdots & & \vdots \\ a_{n1} & a_{n2} & \cdots & a_{nn} \end{bmatrix}$$

obtained by deleting the ith row and jth column of A. Recall that the quantity

$$c_{ij} = (-1)^{i+j} \det A(i|j)$$

is called the **cofactor** of a_{ij} in A. The matrix defined by

$$\text{adj } A = [c_{ij}]^T = \begin{bmatrix} c_{11} & c_{21} & \cdots & c_{n1} \\ c_{12} & c_{22} & \cdots & c_{n2} \\ \vdots & \vdots & & \vdots \\ c_{1n} & c_{2n} & \cdots & c_{nn} \end{bmatrix}$$

is called the **classical adjoint**. It is also called the **adjugate** of A. Note the reversal of indices on c_{ij} in the definition of adj A.

■ **EXAMPLE 1** _____

Let A be the matrix

$$A = \begin{bmatrix} 3 & -4 & 2 \\ 2 & -3 & -1 \\ 1 & 2 & -4 \end{bmatrix}.$$

The cofactors of its entries are

$$c_{11} = \begin{vmatrix} -3 & -1 \\ 2 & -4 \end{vmatrix} = 14, \quad c_{21} = -\begin{vmatrix} -4 & 2 \\ 2 & -4 \end{vmatrix} = -12, \quad c_{31} = \begin{vmatrix} -4 & 2 \\ -3 & -1 \end{vmatrix} = 10,$$

$$c_{12} = -\begin{vmatrix} 2 & -1 \\ 1 & -4 \end{vmatrix} = 7, \quad c_{22} = \begin{vmatrix} 3 & 2 \\ 1 & -4 \end{vmatrix} = -14, \quad c_{32} = -\begin{vmatrix} 3 & 2 \\ 2 & -1 \end{vmatrix} = 7,$$

$$c_{13} = \begin{vmatrix} 2 & -3 \\ 1 & 2 \end{vmatrix} = 7, \quad c_{23} = -\begin{vmatrix} 3 & -4 \\ 1 & 2 \end{vmatrix} = -10, \quad c_{33} = \begin{vmatrix} 3 & -4 \\ 2 & -3 \end{vmatrix} = -1.$$

It follows that

$$\text{adj } A = \begin{bmatrix} 14 & -12 & 10 \\ 7 & -14 & 7 \\ 7 & -10 & -1 \end{bmatrix}.$$

Observe that

$$(\text{adj } A)A = \begin{bmatrix} 14 & -12 & 10 \\ 7 & -14 & 7 \\ 7 & -10 & -1 \end{bmatrix} \begin{bmatrix} 3 & -4 & 2 \\ 2 & -3 & -1 \\ 1 & 2 & -4 \end{bmatrix} = \begin{bmatrix} 28 & 0 & 0 \\ 0 & 28 & 0 \\ 0 & 0 & 28 \end{bmatrix}.$$

The reader should verify that $\det A = 28$. It follows that $(\text{adj } A)A = 28I = (\det A)I$. Similarly, $A(\text{adj } A) = 28I$. This illustrates the result given in the following:

PROPOSITION 1. An $n \times n$-matrix A and its classical adjoint adj A satisfy the relations

$$(\text{adj } A)A = A(\text{adj } A) = (\det A)I.$$

COROLLARY. If $\det A \neq 0$, then

$$A^{-1} = \left(\frac{1}{\det A}\right) \text{adj } A.$$

This follows from Proposition 1 because

$$\left[\left(\frac{1}{\det A}\right) \text{adj } A\right]A = \left(\frac{1}{\det A}\right)(\text{adj } A)A = \left(\frac{1}{\det A}\right)(\det A)I = I,$$

so that $\left(\dfrac{1}{\det A}\right) \text{adj } A$ is the inverse of A.

■ **EXAMPLE 1 (continued)** _____

In the case where the matrix A is given by

$$A = \begin{bmatrix} 3 & -4 & 2 \\ 2 & -3 & -1 \\ 1 & 2 & -4 \end{bmatrix},$$

whose determinant is $\det A = 28$, we have as its inverse

$$A^{-1} = \left(\frac{1}{\det A}\right) \text{adj } A = \frac{1}{28} \begin{bmatrix} 14 & -12 & 10 \\ 7 & -14 & 7 \\ 7 & -10 & -1 \end{bmatrix} = \begin{bmatrix} 1/2 & -3/7 & 5/14 \\ 1/4 & -1/2 & 1/4 \\ 1/4 & -5/14 & -1/28 \end{bmatrix}.$$

This formula for finding A^{-1}, while useful in the case when A is a two- or three-dimensional matrix, is not the method used for higher-dimensional matrices. In these cases, the Gauss elimination method is more appropriate.

There are many theoretical as well as practical results which are obtained using the adjoint of a matrix. An example is *Cramer's rule* for finding the solution of a system of linear equations. We motivate this idea by looking at the following system of equations:

$$a_{11}x + a_{12}y = h_1$$

$$a_{21}x + a_{22}y = h_2,$$

which is equivalent to $AX = H$:

$$\begin{bmatrix} a_{11} & a_{12} \\ a_{21} & a_{22} \end{bmatrix} \begin{bmatrix} x \\ y \end{bmatrix} = \begin{bmatrix} h_1 \\ h_2 \end{bmatrix}.$$

If A is nonsingular (det $A \neq 0$), then from the previous Corollary,

$$A^{-1} = \frac{\text{adj } A}{\text{det } A},$$

so that the unique solution

$$X = A^{-1}H = \frac{\text{adj } A}{\text{det } A} H = \frac{1}{\text{det } A} \begin{bmatrix} c_{11} & c_{21} \\ c_{12} & c_{22} \end{bmatrix} \begin{bmatrix} h_1 \\ h_2 \end{bmatrix},$$

where the c_{ij} are cofactors of the a_{ij} in the matrix A. Multiplying out the right side gives

$$X = A^{-1}H = \frac{1}{\text{det } A} \begin{bmatrix} c_{11}h_1 + c_{21}h_2 \\ c_{12}h_1 + c_{22}h_2 \end{bmatrix}.$$

We see that

$$x = \frac{c_{11}h_1 + c_{21}h_2}{\text{det } A}, \qquad y = \frac{c_{12}h_1 + c_{22}h_2}{\text{det } A}.$$

Computing the cofactors c_{ij} of the matrix A, we have $c_{11} = a_{22}$, $c_{21} = -a_{12}$, $c_{12} = -a_{21}$, $c_{22} = a_{11}$. It follows that

$$x = \frac{a_{22}h_1 - a_{12}h_2}{\text{det } A} = \frac{\begin{vmatrix} h_1 & a_{12} \\ h_2 & a_{22} \end{vmatrix}}{\begin{vmatrix} a_{11} & a_{12} \\ a_{21} & a_{22} \end{vmatrix}}, \qquad y = \frac{-a_{21}h_1 + a_{11}h_2}{\text{det } A} = \frac{\begin{vmatrix} a_{11} & h_1 \\ a_{21} & h_2 \end{vmatrix}}{\begin{vmatrix} a_{11} & a_{12} \\ a_{21} & a_{22} \end{vmatrix}}.$$

We observe that the numerator of the *first* unknown x is the determinant of the matrix obtained by *replacing the first* column of A by the column vector H. The denominator is just det A. Similarly, the numerator of the *second* unknown y is the determinant of the matrix obtained by *replacing the second* column of A by the column vector H. Again, the denominator is just det A.

This generalizes to systems of n linear equations in n unknowns and is known as **Cramer's rule**.

CRAMER'S RULE. To solve the equation $AX = H$, let A_j be the matrix obtained from A by replacing its jth column by H. Then if det A is not zero, the solution X of our equation $AX = H$ is given by the formula

$$X = \begin{bmatrix} x_1 \\ x_2 \\ \vdots \\ \vdots \\ x_n \end{bmatrix}, \quad \text{where } x_1 = \frac{\det A_1}{\det A}, \quad x_2 = \frac{\det A_2}{\det A}, \quad \cdots, \quad x_n = \frac{\det A_n}{\det A}.$$

The proof of this rule is given in the Supplementary Proofs. In the general three-dimensional case we have the equations

$$a_{11}x_1 + a_{12}x_2 + a_{13}x_3 = h_1$$

$$a_{21}x_1 + a_{22}x_2 + a_{23}x_2 = h_2$$

$$a_{31}x_1 + a_{32}x_2 + a_{33}x_3 = h_3.$$

These equations are of the form $AX = H$ with

$$A = \begin{bmatrix} a_{11} & a_{12} & a_{13} \\ a_{21} & a_{22} & a_{23} \\ a_{31} & a_{32} & a_{33} \end{bmatrix}, \quad X = \begin{bmatrix} x_1 \\ x_2 \\ x_3 \end{bmatrix}, \quad H = \begin{bmatrix} h_1 \\ h_2 \\ h_3 \end{bmatrix}.$$

We form the matrices

$$A_1 = \begin{bmatrix} h_1 & a_{12} & a_{13} \\ h_2 & a_{22} & a_{23} \\ h_3 & a_{32} & a_{33} \end{bmatrix}, \quad A_2 = \begin{bmatrix} a_{11} & h_1 & a_{13} \\ a_{21} & h_2 & a_{23} \\ a_{31} & h_3 & a_{33} \end{bmatrix}, \quad A_3 = \begin{bmatrix} a_{11} & a_{12} & h_1 \\ a_{21} & a_{22} & h_2 \\ a_{31} & a_{32} & h_3 \end{bmatrix}.$$

If det A is not zero, the solution to our equation is

$$x_1 = \frac{\det A_1}{\det A}, \quad x_2 = \frac{\det A_2}{\det A}, \quad x_3 = \frac{\det A_3}{\det A}.$$

■ **EXAMPLE 2** _____

A television manufacturer needs 72 resistors, 200 capacitors, and 80 diodes in order to build a particular model. The manufacturer can buy the components from three different electronic distributors. The first distributor packages 4 resistors, 10 capacitors, and 10 diodes in one grab bag. The second distributor packages 8 resistors, 20 capacitors, and 0 diodes in his grab bag. The third distributor's grab bag consists of 12 resistors, 40 capacitors, and 20 diodes. We can find all possible combinations of grab bags that will provide exactly the required amounts of resistors, capacitors, and diodes.

We begin by considering the following table:

	Resistors	Capacitors	Diodes
Distributor 1	4	10	10
Distributor 2	8	20	0
Distributor 3	12	40	20

If we let x, y, and z represent the amount of grab bags purchased from distributors 1, 2, and 3, respectively, we arrive at the following system of linear equations:

$$4x + 8y + 12z = 72$$

$$10x + 20y + 40z = 200$$

$$10x + 0y + 20z = 80.$$

We have the matrix equation

$$\begin{bmatrix} 4 & 8 & 12 \\ 10 & 20 & 40 \\ 10 & 0 & 20 \end{bmatrix} \begin{bmatrix} x \\ y \\ z \end{bmatrix} = \begin{bmatrix} 72 \\ 200 \\ 80 \end{bmatrix}, \quad \det A = \begin{vmatrix} 4 & 8 & 12 \\ 10 & 20 & 40 \\ 10 & 0 & 20 \end{vmatrix} = 800,$$

$$x = \frac{1}{800} \begin{vmatrix} 72 & 8 & 12 \\ 200 & 20 & 40 \\ 80 & 0 & 20 \end{vmatrix}, \quad y = \frac{1}{800} \begin{vmatrix} 4 & 72 & 12 \\ 10 & 200 & 40 \\ 10 & 80 & 20 \end{vmatrix}, \quad z = \frac{1}{800} \begin{vmatrix} 4 & 8 & 72 \\ 10 & 20 & 200 \\ 10 & 0 & 80 \end{vmatrix},$$

so $x = 4$, $y = 4$, $z = 2$.

■ **EXAMPLE 3** _____

As a final example, consider the system

$$3x_1 - 4x_2 + 2x_3 = 1$$

$$2x_1 - 3x_2 - x_3 = 1$$

$$x_1 + 2x_2 - 4x_3 = 1.$$

The matrix representation $AX = H$ is given by

$$\begin{bmatrix} 3 & -4 & 2 \\ 2 & -3 & -1 \\ 1 & 2 & -4 \end{bmatrix} \begin{bmatrix} x_1 \\ x_2 \\ x_3 \end{bmatrix} = \begin{bmatrix} 1 \\ 1 \\ 1 \end{bmatrix}.$$

The determinant of the coefficient matrix A is

$$\det A = \begin{vmatrix} 3 & -4 & 2 \\ 2 & -3 & -1 \\ 1 & 2 & -4 \end{vmatrix} = 28.$$

By Cramer's rule, the solution is

$$x_1 = \frac{1}{28}\begin{vmatrix} 1 & -4 & 2 \\ 1 & -3 & -1 \\ 1 & 2 & -4 \end{vmatrix}, \qquad x_2 = \frac{1}{28}\begin{vmatrix} 3 & 1 & 2 \\ 2 & 1 & -1 \\ 1 & 1 & -4 \end{vmatrix}, \qquad x_3 = \frac{1}{28}\begin{vmatrix} 3 & -4 & 1 \\ 2 & -3 & 1 \\ 1 & 2 & 1 \end{vmatrix}.$$

Notice the position in these formulas of the boldfaced vector $H = \begin{bmatrix} 1 \\ 1 \\ 1 \end{bmatrix}$.

Evaluating the determinants we find that

$$x_1 = 3/7, \qquad x_2 = 0, \qquad x_3 = -1/7.$$

Exercises

1. (a) Verify the relations

    $$A = \begin{bmatrix} a & b \\ c & d \end{bmatrix}, \qquad \operatorname{adj} A = \begin{bmatrix} d & -b \\ -c & a \end{bmatrix}.$$

    (b) Verify Proposition 1 when A and adj A are as above.

    (c) Show that if $\det A = 1$, then $A^{-1} = \operatorname{adj} A$. This result always holds.

2. Find the classical adjoints of the matrices

    $$A = \begin{bmatrix} a & 0 & 0 \\ 0 & b & 0 \\ 0 & 0 & c \end{bmatrix}, \qquad B = \begin{bmatrix} 1 & 0 & 0 \\ 0 & 0 & 1 \\ 0 & 1 & 0 \end{bmatrix}, \qquad C = \begin{bmatrix} 1 & 0 & 0 \\ a & 1 & 0 \\ b & 0 & 1 \end{bmatrix},$$

    $$D = \begin{bmatrix} 1 & 1 & 1 & 1 \\ -1 & 1 & -1 & 1 \\ -1 & -1 & 1 & 1 \\ 1 & -1 & -1 & 1 \end{bmatrix}, \qquad E = \begin{bmatrix} 1 & 2 & -1 & 1 \\ 2 & 5 & 0 & 2 \\ -1 & 0 & 6 & 0 \\ 1 & 2 & 0 & 3 \end{bmatrix},$$

    $$F = \begin{bmatrix} 3 & 4 & 0 & 0 \\ 2 & 3 & 0 & 0 \\ 5 & 6 & 7 & 1 \\ 0 & 1 & 1 & 2 \end{bmatrix}, \qquad G = \begin{bmatrix} 5 & 4 & 6 & 7 \\ -2 & 1 & 2 & -1 \\ 0 & 0 & 1 & 3 \\ 0 & 0 & 1 & 1 \end{bmatrix}, \qquad H = \begin{bmatrix} 8 & -3 & 0 & 0 \\ -5 & 2 & 0 & 0 \\ 0 & 0 & 3 & 2 \\ 0 & 0 & 4 & 3 \end{bmatrix}.$$

3. Show that if A is symmetric, so is adj A.

4. Show that $\det(\text{adj } A) = (\det A)^{n-1}$, where A is $n \times n$. *Hint:* Use Proposition 1. Conclude that adj A is nonsingular if and only if A is.

5. Show that if A is $n \times n$ and $\det A$ is not zero,
$$\text{adj (adj } A) = (\det A)^{n-2}A.$$

6. Given that
$$A = \begin{bmatrix} 1 & 0 & 1 \\ 2 & 1 & 1 \\ 3 & 1 & 2 \end{bmatrix}, \quad \text{adj } A = \begin{bmatrix} 1 & 1 & -1 \\ -1 & -1 & 1 \\ -1 & -1 & 1 \end{bmatrix},$$
   show that $(\text{adj } A)A = 0$ even though adj A is not zero. Conclude that $\det A = 0$.

7. Use Cramer's rule to solve the following systems of equations:

   (a) $3x - 4y = 7$ (b) $x - 7y = 18$
   $\ 2x - 6y = 8$ $\ 2x + 3y = 10$

   (c) $x_1 - 2x_2 + x_3 = 8$ (d) $x + y + z = 1$
   $\ 2x_1 + \ x_2 + x_3 = 9$ $\ x - y \ \ \ = 0$
   $\ x_1 - \ x_2 + x_3 = 1$ $\ 2x + y - z = 9$

8. Consider the system of two linear equations in the three unknowns x, y, and z:
$$a_{11}x + a_{12}y + a_{13}z = h_1$$
$$a_{21}x + a_{22}y + a_{23}z = h_2.$$

   Setting $z = k$, show that under the proper assumptions, one can determine x and y using Cramer's rule.

9. Let B be the bordered square matrix
$$B = \begin{bmatrix} A & U \\ V & c \end{bmatrix},$$

   where U is a column vector, V is a row vector, and c is a number. Show that
$$\det B = c\det A - V(\text{adj } A)U.$$

   Hint: Suppose first that A is nonsingular. Then
$$B = \begin{bmatrix} I & 0 \\ VA^{-1} & 1 \end{bmatrix} \begin{bmatrix} A & U \\ 0 & c - VA^{-1}U \end{bmatrix}.$$

   Conclude that $\det B = (\det A)(c - VA^{-1}U)$. Then use the relation $(\det A)A^{-1} = \text{adj } A$. When A is singular, the result holds when A is replaced by $A + bI$ for small nonzero values of b and hence for $b = 0$.

Supplementary Proofs

■ **P1.** Show that $(\text{adj } A)A = (\det A)I$.

To prove this result, fix j. Let A_{jk} be the matrix obtained from A by replacing its jth column by its kth column, keeping the remaining columns including the kth intact.

Then $A_{jk} = A$ when $j = k$ and A_{jk} has two identical columns otherwise. Hence
det A_{jj} = det A and det $A_{jk} = 0$ when k is different from j. Moreover, $A_{jk}(i|j)$. Using
the jth column expansion for det A_{jk}, we find that

$$\det A_{jk} = \sum_{i=1}^{n} (-1)^{i+j} a_{ik} \det A_{jk}(i|j) = \sum_{i=1}^{n} (-1)^{i+j} a_{ik} \det A(i|j)$$

$$\det A_{jk} = \sum_{i=1}^{n} c_{ji} a_{ik} = \det A \qquad \text{when } k = j \text{ and } = 0 \text{ otherwise.}$$

This quantity is the (j, k)-entry in the product (adj A)A so that (adj A)A = (det A)I.
In the same manner using rows in place of columns we find that A (adj A) =
(det A)I. This proves Proposition 1.   ▬

▬ **P2.** Establish Cramer's rule given the equation $AX = H$ with det A different from
zero.

Let A_j be the matrix obtained from A by replacing the jth column of A by $[h_i]$.
Then $A_j(i|j) = A\ (i|j)$ so that, by the jth column expansion,

$$\det A_j = \sum_{i=1}^{n} (-1)^{i+j} h_i \det A_j(i|j) = \sum_{i=1}^{n} (-1)^{i+j} h_i \det A\ (i|j).$$

Recalling the definition of the cofactor c_{ji}, we have

$$\det A_j = \sum_{i=1}^{n} c_{ji} h_i.$$

This quantity is the j-entry of the vector $Y = $ (adj A)H. It follows that the vector

$$X = \frac{1}{\det A} Y = \frac{1}{\det A}(\text{adj } A)H = A^{-1}H$$

is the solution of $AX = H$. Since the j-th component of Y is det A_j, the j-th component
of X is det A_j /det A, as was to be proved.   ▬

3.4   RANK OF A MATRIX

Most of our concern over the last few sections has been with the development
of the theory of determinants for square $n \times n$-matrices. Since many
applications involve $m \times n$-matrices, for which the concept of determinants
is not defined, we seek a generalization of sorts in order to gain insight into
the properties of nonsquare matrices. This will involve the study of square
submatrices of a given non-square matrix A. Much information about the
matrix A can be obtained by looking at the determinants of these square
submatrices.

We have met the concept of a submatrix many times. For example, the
matrix $A\ (i|\ j)$ used in the definition of determinant of A is a submatrix of
A. In the study of the linear equation $AX = H$, the matrices A and H are

submatrices of the augmented matrix [A | H]. A row of A is a submatrix of A. So also is a column of A. A submatrix is a matrix within a matrix. Recall that a submatrix can be obtained by deleting rows and columns. It also can be viewed as the intersection of rows and columns.

■ **EXAMPLE 1** _____

Consider the 4 × 6-matrix

$$A = \begin{bmatrix} 0 & 0 & -1 & 5 & 1 & -2 \\ 1 & 4 & 0 & 1 & -1 & 2 \\ 0 & -7 & 6 & 5 & -4 & 3 \\ 1 & 9 & 3 & -2 & 1 & -5 \end{bmatrix}.$$

The matrices

$$B = \begin{bmatrix} 4 & 1 \\ 9 & -2 \end{bmatrix}, \qquad C = \begin{bmatrix} -1 & 2 \\ 1 & -5 \end{bmatrix}, \qquad D = \begin{bmatrix} 0 & 5 & 1 \\ 9 & -2 & 1 \end{bmatrix},$$

$$E = \begin{bmatrix} 0 & 5 & 1 \\ 4 & 1 & -1 \\ 9 & -2 & 1 \end{bmatrix}, \qquad F = \begin{bmatrix} 0 & -1 & 5 & 1 \\ 4 & 0 & 1 & -1 \\ -7 & 6 & 5 & -4 \\ 9 & 3 & -2 & 1 \end{bmatrix}$$

are submatrices of A. There are many other submatrices of A. The matrix B is the intersection of rows 2 and 4 with columns 2 and 4. It can also be obtained from A by deleting rows 1, 3 and columns 1, 3, 5, and 6. The submatrix C is the intersection with rows 2 and 4 with columns 5 and 6. How can the submatrix D be obtained? The matrix E is the intersection of rows 1, 2, and 4 with columns 2, 4, and 5. The matrix F is obtained by deleting columns 1 and 6 of A. The matrix B is a submatrix of E, which in turn is a submatrix of F.

_____

When we are dealing with square matrices, it is often useful, especially in certain proofs, to have the concept of a principal submatrix. A *principal submatrix* of a square matrix S is a submatrix of S, which is the intersection of rows and columns of S having the same indices.

■ **EXAMPLE 2** _____

Consider the 4 × 4 matrix

$$S = \begin{bmatrix} 1 & 3 & -6 & 0 \\ 2 & 5 & 1 & 3 \\ 7 & -2 & 8 & 9 \\ 2 & 8 & 0 & 6 \end{bmatrix}.$$

The matrices

$$P = \begin{bmatrix} 1 & 3 \\ 2 & 5 \end{bmatrix}, \qquad Q = \begin{bmatrix} 5 & 1 \\ -2 & 8 \end{bmatrix}, \qquad R = \begin{bmatrix} 5 & 1 & 3 \\ -2 & 8 & 9 \\ 8 & 0 & 6 \end{bmatrix}$$

are principal submatrices of S. The matrix P is a submatrix of S, which is the intersection of rows 1 and 2 and columns 1 and 2 of S. It can also be obtained by deleting rows 3 and 4 and columns 3 and 4 of S. The matrix Q is a submatrix of S, which is the intersection of rows 2 and 3 and columns 2 and 3 of S. It can also be obtained by deleting rows 1 and 4 and columns 1 and 4 of S. Similarly, the matrix R is the intersection of rows 2, 3, and 4 and columns 2, 3, and 4 of S. It, too, can also be obtained by deleting row 1 and column 1 of S.

We see that a principal submatrix can therefore also be obtained from a square matrix by deleting rows and columns having the same indices. A principal submatrix of a square matrix S is square and so has a determinant. This determinant is called a ***principal minor*** of S. For the principal submatrices P, Q, and R of the matrix S in Example 2, det $P = -1$, det $Q = 42$, and det $R = 132$ are principal minors of S.

Some of the submatrices of the 4×6-matrix A in Example 1 are square matrices. In particular, the submatrices B, C, E, and F are square. Because they are square, their determinants are defined. We have

$$\det B = \begin{bmatrix} 4 & 1 \\ 9 & -2 \end{bmatrix} = -17, \qquad \det C = \begin{bmatrix} -1 & 2 \\ 1 & -5 \end{bmatrix} = 3,$$

$$\det E = \begin{bmatrix} 0 & 5 & 1 \\ 4 & 1 & -1 \\ 9 & -2 & 1 \end{bmatrix} = -82, \qquad \det F = \begin{bmatrix} 0 & -1 & 5 & 1 \\ 4 & 0 & 1 & -1 \\ -7 & 6 & 5 & -4 \\ 9 & 3 & -2 & 1 \end{bmatrix} = 908.$$

These determinants are called ***minors*** of the original matrix A. Summarizing, we have

> A matrix B obtained from a matrix A by deleting rows and columns is called a ***submatrix*** of A. If B is square, det B is called a ***minor*** of A. If B is $k \times k$, det B is a k-rowed minor of A.

> A submatrix B of a square matrix A is called a ***principal submatrix*** of A if B is obtained from A by deleting rows and columns having the same indices. Such a matrix B is square. Its determinant, det B, is called a ***principal minor*** of A.

The concept of minors of a matrix A enables us to introduce the concept of the ***determinantal rank*** r of A. It is the largest integer r for which there is a nonzero r-rowed minor of A. When $A = 0$, all of its minors are zero, so we call $r = 0$ its determinantal rank.

■ **EXAMPLE 3**

Consider the 2×3-matrix

$$A = \begin{bmatrix} 1 & 0 & 7 \\ 1 & 0 & 12 \end{bmatrix}.$$

Because A is a 2×3-matrix, the largest square submatrices are 2×2. The determinants of the 2×2-submatrices are the two-rowed minors of A. They are

$$\begin{vmatrix} 1 & 0 \\ 1 & 0 \end{vmatrix} = 0, \qquad \begin{vmatrix} 0 & 7 \\ 0 & 12 \end{vmatrix} = 0, \qquad \begin{vmatrix} 1 & 7 \\ 1 & 12 \end{vmatrix} = 5.$$

We notice that two of the two-rowed minors are zero, but one of them is nonzero. We say that the matrix A has *determinantal rank* 2 since we have found at least one two-rowed minor that was nonzero. We observe that since A is 2×3, there could be no larger square submatrix with a nonzero determinant.

■ **EXAMPLE 4**

Consider the 3×4-matrix

$$A = \begin{bmatrix} 1 & -1 & 2 & -2 \\ 0 & 1 & 2 & 3 \\ 1 & 0 & 4 & 1 \end{bmatrix}.$$

The 3×3-submatrices are

$$B = \begin{bmatrix} 1 & -1 & 2 \\ 0 & 1 & 2 \\ 1 & 0 & 4 \end{bmatrix}, \quad C = \begin{bmatrix} 1 & -1 & -2 \\ 0 & 1 & 3 \\ 1 & 0 & 1 \end{bmatrix},$$

$$D = \begin{bmatrix} 1 & 2 & -2 \\ 0 & 2 & 3 \\ 1 & 4 & 1 \end{bmatrix}, \quad E = \begin{bmatrix} -1 & 2 & -2 \\ 1 & 2 & 3 \\ 0 & 4 & 1 \end{bmatrix}.$$

Their determinants are all 0. We see that all three-rowed minors of A are 0. The ***determinantal rank*** of A *cannot* be 3. We therefore look for at least one two-rowed minor of A. There are many. For example, the 2×2-submatrix

$$F = \begin{bmatrix} 1 & 2 \\ 0 & 4 \end{bmatrix}$$

has det $F = 4$. It follows that the 3×4-matrix A has determinantal rank equal to 2.

The definition of determinantal rank given above can be restated as follows.

A $m \times n$-matrix A has **determinantal rank r** if both of the following conditions hold:

   1. There is at least one $r \times r$-submatrix of A whose determinant is nonzero.
   2. There is no square submatrix of A whose size is larger than $r \times r$ and whose determinant is nonzero.

If $A = 0$, its determinantal rank is $r = 0$.

Proceeding, let us suppose that we are given a 6×8-matrix A. Suppose further that all of the three-rowed minors happen to be 0. Since a four-rowed minor of A is a determinant of a 4×4-submatrix of A, its expansion about a row or column will be in terms of determinants of 3×3-submatrices of A. But we have assumed that all 3×3-submatrices have determinant 0. Therefore all four-rowed minors of A are 0. By the same argument, all five-rowed and six-rowed minors of A are 0. This illustrates the following general property:

---

   1. Given an $m \times n$-matrix A, such that all k-rowed minors of A are zero where $k < m$ and $k < n$, then all $(k + 1)$-rowed minors of A are also zero. Hence every q-rowed minor of A with $q > k$ is zero.

By similar reasoning we have

---

   2. If there is a nonzero k-rowed minor of A, then for each integer $j < k$, there are one or more nonzero j-rowed minors of A.

We shall establish three basic properties of the determinantal rank of a matrix, the first of which is the following.

PROPERTY 1. The determinantal ranks of A and A^T are the same.

This follows because if B is a $q \times q$-submatrix of A, then B^T is a $q \times q$-submatrix of A^T. Moreover, det B = det B^T. It follows that if det B is not zero, then neither is det B^T. Hence B is the largest (in size) nonsingular submatrix of A if and only if B^T is the largest nonsingular submatrix of A^T. The matrices A and A^T therefore have the same determinantal rank.

PROPERTY 2. The determinantal rank of a submatrix B of a matrix A cannot exceed the determinantal rank of A.

The third property of determinantal rank in a sense generalizes the idea of "nonsingularity" of $n \times n$-square matrices to $m \times m$-nonsquare matrices.

Recall that a square matrix A was nonsingular if and only if $AX = 0$ implies that $X = 0$. This was also equivalent to the det $A \neq 0$ and to A being invertible.

While a nonsquare matrix A does not have an inverse or a determinant, a generalization exists in the following.

PROPERTY 3. Let A be a $m \times n$-matrix. The following statements are equivalent.

   (a) The columns of A are linearly independent.
   (b) $AX = 0$ implies that $X = 0$.
   (c) The determinantal rank of A is n.

This property is established in the Supplementary Proofs at the end of the section. The equivalence of statements (a) and (b) was established in Section 1.5.

The reader should be careful not to confuse the criterion above with the assumption that A^{-1} exists, an assumption that is impossible when A is not square.

■ **EXAMPLE 5** _____

Consider the matrix

$$A = \begin{bmatrix} 1 & 0 & -2 & 1 & 3 \\ 0 & 1 & 0 & 1 & -4 \\ 1 & 0 & -2 & 1 & 3 \end{bmatrix} = [C_1 \quad C_2 \quad C_3 \quad C_4 \quad C_5].$$

It is easily seen that A has determinantal rank 2. Observe that the first two columns C_1 and C_2 are linearly independent. Moreover,

$$C_3 = -2C_1, \qquad C_4 = C_1 + C_2, \qquad C_5 = 3C_1 - 4C_2,$$

so that the columns of A are linear combinations of C_1 and C_2. Similarly, C_2 and C_4 are linearly independent and again the remaining columns of A are linear combinations

$$C_1 = -C_2 + C_4, \qquad C_3 = 2C_2 - 2C_4, \qquad C_5 = -7C_2 + 3C_4$$

of C_2 and C_4. Note that in each case the number of linearly independent columns is equal to the determinantal rank of A.

This suggests the following property of determinantal rank.

> **PROPERTY 4.** Let A be a $m \times n$-matrix of determinantal rank r. Let $C_1, C_2, \ldots, C_s$ be s columns of A such that:
>
> **(a)** They are linearly independent.
> **(b)** Each column of A is a linear combination of these s columns of A.
>
> Then $s = r$, the determinantal rank of A.

In view of this result, the number s of columns of A having properties (a) and (b), being equal to the determinantal rank of A, is independent of the choice of these vectors. The number s of columns of A having properties (a) and (b) is called the **column rank** of A. Property 4 therefore states that the column rank of A is equal to the determinantal rank of A. The proof of Property 4 is given in the Supplementary Proofs found at the end of the section.

Recall that the rows of A are the columns of its transpose A^T. Applying the results described in Properties 1 and 4 to A^T, we obtain the following additional property of determinantal rank of A.

PROPERTY 5. Let A be a $m \times n$-matrix of determinantal rank r. Let $R_1, R_2, \ldots, R_t$ be t rows of A such that:

(a) they are linearly independent

(b) Each row of A is a linear combination of these t rows.

Then $t = r$, the determinantal rank of A.

It follows that the number t of rows having properties (a) and (b) is independent of our choice of these rows. The number t of rows of A having properties (a) and (b) is called the **row rank** of A. Accordingly, we have the following result:

column rank of A = row rank of A = determinantal rank of A.

The common value of these three ranks of A will be called the **rank** of A. Thus when we speak of the rank of A, we interpret it to be the determinantal rank, the column rank, or the row rank, whichever is appropriate in context. An alternative definition of rank is given in Section 6.2. By Property 1, the rank of A is the same as the rank of A^T for each of these three concepts of rank.

Summarizing, we associate with the term **rank** the adjective **largest** as follows:

The rank of A is the largest integer r such that:

1. There are r linearly independent columns of A.
2. There are r linearly independent rows of A.
3. There is an r-rowed nonzero minor of A.

If $A = 0$, the rank r of A is 0.

The following further property of rank is useful:

PROPERTY 6. Let A be an $m \times n$-matrix and let P be a nonsingular $m \times m$-matrix. Then the matrices A and PA have the same rank and so have the same determinantal rank. In addition, if Q is a nonsingular $n \times n$-matrix, the matrices AQ and PAQ have the same rank as A.

The proof of this property is given in the Supplementary Proofs at the end of the section. To illustrate this result, consider the following.

■ **EXAMPLE 6** _____

Let A be the 3×5-matrix

$$A = \begin{bmatrix} 1 & 0 & -2 & 1 & 3 \\ 0 & 1 & 0 & 1 & -4 \\ 1 & 0 & -2 & 1 & 3 \end{bmatrix}$$

considered earlier. Let P be the elementary 3×3-matrix that subtracts row 1 from row 3.

Then $F = PA$ is the matrix

$$\begin{bmatrix} 1 & 0 & 0 \\ 0 & 1 & 0 \\ -1 & 0 & 1 \end{bmatrix} \begin{bmatrix} 1 & 0 & -2 & 1 & 3 \\ 0 & 1 & 0 & 1 & -4 \\ 1 & 0 & -2 & 1 & 3 \end{bmatrix} = \begin{bmatrix} 1 & 0 & -2 & 1 & 3 \\ 0 & 1 & 0 & 1 & -4 \\ 0 & 0 & 0 & 0 & 0 \end{bmatrix} = F.$$

Clearly, the determinantal rank of F is two, as is the determinantal rank of A. Similarly, let Q be the nonsingular 5×5-matrix

$$Q = \begin{bmatrix} 1 & 0 & 2 & -1 & -3 \\ 0 & 1 & 0 & -1 & 4 \\ 0 & 0 & 1 & 0 & 0 \\ 0 & 0 & 0 & 1 & 0 \\ 0 & 0 & 0 & 0 & 1 \end{bmatrix}.$$

It is easily seen that the matrices AQ and $FQ = PAQ$ are the matrices

$$AQ = \begin{bmatrix} 1 & 0 & 0 & 0 & 0 \\ 0 & 1 & 0 & 0 & 0 \\ 1 & 0 & 0 & 0 & 0 \end{bmatrix}, \quad PAQ = \begin{bmatrix} 1 & 0 & 0 & 0 & 0 \\ 0 & 1 & 0 & 0 & 0 \\ 0 & 0 & 0 & 0 & 0 \end{bmatrix}.$$

Again these matrices have rank 2. The matrices A, PA, AQ, and PAQ therefore have the same rank, as stated in Property 6.

_____

In our example the matrix $F = PA$ happens to be the reduced row echelon matrix for the matrix A. The first two rows of F are the nonzero rows of F. They are linearly independent. It follows that the number of nonzero rows of F is the rank of A. This result illustrates the following result, which is often quite useful in calculating the rank of a matrix A.

> **COROLLARY 1.** The rank of a matrix A is equal to the number of nonzero rows in the reduced row echelon matrix F associated with A.

In Proposition 3 of Section 2.4, it was shown that a matrix linear equation $AX = H$ is solvable if and only if the reduced row echelon matrices F and $[F \mid K]$ of A and the augmented matrix $[A \mid H]$ had the same number r of nonzero rows. By Corollary 1, this number r is the rank of A and of $[A \mid H]$. This corollary tells us that we can find the rank of A by finding the reduced row echelon form of A and counting its nonzero rows.

COROLLARY 2. A linear equation $AX = H$ is solvable if and only if the matrix A and the augmented matrix $[A \mid H]$ have the same rank.

Referring to Proposition 1 in Section 2.4 and the remarks following it, we see that the nullity of A is the largest number of linearly independent solutions of $AX = 0$. It is given by the formula $q = n - r$ where r is the number of nonzero rows of the reduced row echelon matrix F for A. By Corollary 1 this number r is the rank of A. This gives the following:

PROPERTY 7. The rank r of a $m \times n$-matrix A and the nullity q of A are connected by the relation
$$q + r = n.$$

Accordingly, $r = n - q$ and $q = n - r$.

The final property of rank, we consider, is the following:

PROPERTY 8. Let A be the product $A = BC$ of a $m \times k$-matrix B and a $k \times n$-matrix C of ranks s and t, respectively. The rank r of A satisfies the inequalities

    (a) $r < s$ with $r = s$ when $t = k$,

    (b) $r < t$ with $r = t$ when $s = k$.

There are situations in which $r < s$ and $r < t$. For example, this is true when

$$A = \begin{bmatrix} 0 & 0 \\ 0 & 0 \end{bmatrix} = \begin{bmatrix} 1 & 1 & 1 & 1 \\ 1 & -1 & 1 & -1 \end{bmatrix} \begin{bmatrix} 1 & 1 \\ 1 & 1 \\ -1 & -1 \\ -1 & -1 \end{bmatrix} = BC.$$

The rank of A is $r = 0$, the rank of B is $s = 2$, and the rank of C is $t = 1$.

The proof of Property 8 is given in the Supplementary Proofs found at the end of the section.

Exercises

1. Find the determinantal ranks r of the following matrices.

$$A = \begin{bmatrix} 2 & 1 \\ 1 & 0 \end{bmatrix}, \qquad B = \begin{bmatrix} 2 & 1 \\ 4 & 2 \end{bmatrix}, \qquad C = \begin{bmatrix} 1 & 0 & 3 \\ 0 & 1 & 1 \end{bmatrix},$$

$$D = \begin{bmatrix} 1 & 0 & 3 \\ 0 & 1 & 1 \\ 1 & 1 & 4 \end{bmatrix}, \qquad E = \begin{bmatrix} 1 & 1 & -2 \\ -2 & -2 & 4 \\ 3 & 3 & -6 \end{bmatrix}, \qquad F = \begin{bmatrix} 1 & 1 & 0 \\ 0 & 1 & 2 \\ 0 & 0 & 1 \end{bmatrix},$$

$$G = \begin{bmatrix} 3 & 4 & 1 & 1 \\ 2 & 3 & 0 & 2 \\ 5 & 6 & 7 & 1 \\ 0 & 1 & -6 & 2 \end{bmatrix}, \qquad H = \begin{bmatrix} 5 & 4 & 6 & 7 \\ -2 & 1 & 2 & -1 \\ 0 & 0 & 1 & 3 \\ 3 & 5 & 9 & 9 \end{bmatrix}, \qquad K = \begin{bmatrix} 8 & -3 & 0 & 0 \\ -5 & 2 & 0 & 0 \\ 0 & 0 & 3 & 2 \\ 0 & 0 & 4 & 3 \end{bmatrix}.$$

2. Find the rank of each of the matrices in Exercise 1 by first computing their reduced row echelon matrices. In each case exhibit:

   (a) A nonsingular $r \times r$-submatrix L.

   (b) A set of r linearly independent columns of the matrix.

   (c) A set of r linearly independent rows of the matrix.

3. Use the term *column-submatrix* to designate a submatrix obtained by deleting columns only. Let U be a column submatrix of an $m \times n$-matrix A of maximum size whose columns are linearly independent. Show that:

   (a) Every column of A is a linear combination of the columns of U.

   (b) The number r of columns of U is the rank of A.

   (c) There is an $r \times n$-matrix W such that $A = UW$.

   What is the rank of W? Exhibit such a column submatrix U for each of the matrices given in Exercise 1.

4. Show that the matrices

$$A = \begin{bmatrix} 0 & 1 & -2 & 1 & 0 & 4 & 1 & 5 \\ 0 & 1 & -2 & -1 & -2 & 2 & -1 & -1 \\ 0 & 1 & -2 & 1 & 0 & 4 & -1 & 3 \\ 0 & 1 & -2 & -1 & -2 & 2 & 1 & 1 \end{bmatrix}, \qquad U = \begin{bmatrix} 1 & 1 & 1 \\ 1 & -1 & -1 \\ 1 & 1 & -1 \\ 1 & -1 & 1 \end{bmatrix}$$

   are related as described in Exercise 3. Show that $A = UW$, where

$$W = \begin{bmatrix} 0 & 1 & -2 & 0 & -1 & 3 & 0 & 2 \\ 0 & 0 & 0 & 1 & 1 & 1 & 0 & 2 \\ 0 & 0 & 0 & 0 & 0 & 0 & 1 & 1 \end{bmatrix}.$$

   Why do A, U, and W have the same rank?

5. Let A and U be the matrices described in Exercise 4. Show that the column submatrix U of A can be obtained by the following deletions:

(a) Delete all zero columns.

(b) Delete all columns which are linear combinations of the columns that precede it.

Can these rules for deletions be applied to any matrix A of rank r to produce a column submatrix U of maximum size whose columns are linearly independent?

6. Use the term *row submatrix* to designate a submatrix obtained by deleting rows only. Let V be a row submatrix of an $m \times n$-matrix A of maximum size whose rows are linearly independent. Show that:

   (a) Every row of A is a linear combination of the rows of V.

   (b) The number r of rows of V is the rank of A.

   (c) There is an $m \times r$-matrix Z such that $A = ZV$.

   Why is the rank of Z the same as the rank of A and hence also of V? Exhibit such a row submatrix V for each of the matrices given in Exercise 1.

7. Let V be the row submatrix of the matrix A exhibited in Exercise 4 obtained by deleting the last row. Show that V is related to A as described in Exercise 6. What is the rank of V?

8. Let U be a column submatrix related to A as described in Exercise 3 and let V be a row submatrix related to A as described in Exercise 6. The matrices A, U, and V have the same rank r. Show that the intersection G of U and V is a nonsingular $r \times r$-submatrix of A. *Hint:* Use the fact that because the columns of A are linear combinations of the columns of U, the columns of V are linear combinations of the columns of G. Since V and hence G has rank r, G is nonsingular.

9. Let A be a matrix of rank r. Show that

   (a) $A^T A X = 0$ if and only if $A X = 0$ (see Exercise 18 of Section 1.4).

   (b) $A A^T A X = 0$ if and only if $A^T A X = 0$.

   (c) A, $A^T A$, $A A^T$,   . . . have the same nullity.

   (d) A, $A^T A$, $A A^T$,   . . . have the same rank r.

   (e) A^T, $A A^T$, $A^T A A^T$,   . . . also have rank r.

   (f) If A is a square *symmetric* matrix, the powers of A have the same rank and nullity as A.

10. Let A be an $n \times n$-matrix and adj A be its classical adjoint.

    (a) adj $A = 0$ if and only if the rank of A is less than $n - 1$.

    (b) If A has rank n, then adj A has rank n.

    (c) If A has rank $n - 1$, then adj A has rank 1.

    Conclude that the rank of adj A is either 0, 1, or n.

11. Let B be a matrix obtained from a matrix A by permuting (rearranging) its columns. Let C be a matrix obtained from A by permuting its rows. Show that

A, *B*, and *C* have the same rank and the same nullity. *Hint:* This follows from Property 6 because there are permutation matrices *P* and *Q* such that $C = PA$ and $B = AQ$.

12. (a) Suppose that the nonzero columns of a matrix *A* are linearly independent. Show that the number of zero columns of *A* is the nullity of *A*.

    (b) Suppose that the nonzero rows of *A* are linearly independent. Show that the number of zero rows of *A* is equal to the nullity of A^T. *Hint:* In part (a) use the fact that in this case the number of nonzero columns of *A* is the rank of *A*. Part (b) is obtained from (a) by applying (a) to A^T.

13. Let *q* be the nullity of an $m \times n$-matrix *A*. Show that there is an $n \times q$-matrix *Z* of rank *q* such that

    (a) $AZ = 0$.

    (b) If *X* solves $AX = 0$, it is expressible in the form $X = ZC$.

    This is a restatement in matrix form of Proposition 1 in Section 2.4. As an example, we have

    $$A = \begin{bmatrix} 3 & 0 & 3 \\ 6 & 0 & 6 \\ 3 & 0 & 3 \end{bmatrix}, \quad Z = \begin{bmatrix} 0 & 1 \\ 1 & 0 \\ 0 & -1 \end{bmatrix}, \quad X = \begin{bmatrix} 1 \\ 2 \\ -1 \end{bmatrix} = \begin{bmatrix} 0 & 1 \\ 1 & 0 \\ 0 & -1 \end{bmatrix} \begin{bmatrix} 2 \\ 1 \end{bmatrix}.$$

14. Let *A* be an $m \times n$-matrix of nullity *q*. As in Problem 13, select an $n \times q$-matrix *Z* of rank *q* such that $AZ = 0$. Let *C* be a $k \times n$-matrix of nullity *p* such that $AX = 0$ whenever $CX = 0$. Conclude that the rank *d* of $D = CZ$ is given by the formula

    $$d = q - p = t - r,$$

    where $r = n - q$ is the rank of *A* and $t = n - p$ is the rank of *C*. As an example, select *A* and *Z* as in Exercise 13. We have $r = 1$ and $q = 2$. Select *C* and *D* as follows.

    $$C = \begin{bmatrix} 2 & 1 & 3 \\ 1 & 1 & 2 \\ 0 & 1 & 1 \end{bmatrix}, \quad Z = \begin{bmatrix} 0 & 1 \\ 1 & 0 \\ 0 & -1 \end{bmatrix}, \quad D = CZ = \begin{bmatrix} 1 & -1 \\ 1 & -1 \\ 1 & -1 \end{bmatrix}.$$

    We have $t = 2$, $p = 1$, $d = 2 - 1 = 1$.

15. Let *B* be an $m \times k$-matrix of rank *s*. Let *D* be a $k \times q$-matrix of rank *d* such that $BD = 0$. Show that $d \leq k - s$, the nullity of *B*. Conclude that $s + d \leq k$. As an example we have

    $$B = \begin{bmatrix} 1 & 1 & -2 \\ 2 & 2 & -4 \\ 1 & 1 & -2 \end{bmatrix}, \quad D = \begin{bmatrix} 1 & -1 \\ 1 & -1 \\ 1 & -1 \end{bmatrix}, \quad BD = 0.$$

    We have $m = k = 3$, $s = 1$, $d = 1$, $d + s = 2 \leq 3$.

16. Referring to Proposition 8, in which $A = BC$, where *A* is $m \times n$, *B* is $m \times k$, and *C* is $k \times n$. Show that the ranks *r*, *s*, *t* of *A*, *B*, and *C* satisfy the inequalities

$$r \le s, \qquad r \le t, \qquad 2r \le s + t \le k + r.$$

Hint: Choose Z related to A as described in Exercise 14. Then the rank d of $D = CZ$ is given by the formula $d = t - r$. We have

$$AZ = BCZ = BD = 0.$$

By Exercise 15, $s + d \le k$ so that $s + t - r \le k$ and $s + k + r$. As an example choose B and C as exhibited in the preceding two exercises. We have $m = k = n = 3$, $s = 1$, $t = 2$. Forming the product $A = BC$, we obtain the matrix A exhibited in Exercise 13. It has rank $r = 1$. This gives

$$2r = 2 \le s + t = 3 \le k + r = 4.$$

17. Verify that the matrices

$$A = \begin{bmatrix} 2 & 1 & 3 & 4 \\ 4 & 2 & 6 & 8 \end{bmatrix}, \qquad U = \begin{bmatrix} 1 \\ 2 \end{bmatrix}$$

have rank 1. Note that U is column 2 of A. Verify that A is the column–row product

$$A = \begin{bmatrix} 1 \\ 2 \end{bmatrix} [2 \quad 1 \quad 3 \quad 4] = UV, \qquad V = [2 \quad 1 \quad 3 \quad 4].$$

18. Let A be a nonzero $m \times n$-matrix. Show that A has rank 1 if and only if there is an m-dimensional column vector U such that the columns $C_1, C_2, \ldots, C_n$ of A are multiples

$$C_1 = Uk_1, \quad C_2 = Uk_2, \ldots, C_n = Uk_n$$

of U. Verify that, in this case,

$$A = [Uk_1 \quad Uk_2 \quad \cdots \quad Uk_n] = U[k_1 \quad k_2 \quad \cdots \quad k_n] = UV.$$

19. Let A be the first of the matrices

$$A = \begin{bmatrix} 1 & 0 & -2 & 1 & 3 \\ 0 & 1 & 0 & 1 & -4 \\ 1 & 0 & -2 & 1 & 3 \end{bmatrix}, \qquad U = \begin{bmatrix} 1 & 0 \\ 0 & 1 \\ 1 & 0 \end{bmatrix}.$$

It has rank 2. The matrix U is a submatrix of A of rank 2. Verify that

$$A = \begin{bmatrix} 1 & 0 & -2 & 1 & 3 \\ 0 & 1 & 0 & 1 & -4 \\ 1 & 0 & -2 & 1 & 3 \end{bmatrix} = \begin{bmatrix} 1 & 0 \\ 0 & 1 \\ 1 & 0 \end{bmatrix} \begin{bmatrix} 1 & 0 & -2 & 1 & 3 \\ 0 & 1 & 0 & 1 & -4 \end{bmatrix}.$$

Thus A is the product of U and a matrix V. Show that V has rank 2. Show that in this case, V is comprised of the nonzero rows of the reduced row echelon form for A.

20. Let r be the rank of an $m \times n$-matrix A. Select an $m \times r$-submatrix U of A whose columns are linearly independent. It too has rank r. Why? Next select an $r \times n$-submatrix V of A whose rows are linearly independent. Its rank is also r. The intersection G of U and V is an $r \times r$-submatrix of A. Show that $\det G \ne 0$. *Hint:* Add linear combinations of the column U to each of the remaining columns of U so that they will be zero. This does not alter the ranks of A or of V. Moreover, G is unaltered.

21. Find several principal two-rowed minors of the tridiagonal matrix

$$G = \begin{bmatrix} 2 & 1 & 0 & 0 & 0 \\ 1 & 2 & 1 & 0 & 0 \\ 0 & 1 & 2 & 1 & 0 \\ 0 & 0 & 1 & 2 & 1 \\ 0 & 0 & 0 & 1 & 2 \end{bmatrix}.$$

What values do they have? What is the value of the two-rowed minor d_2 determined by rows 1 and 2 of G? The value of the one-rowed principal minors are $d_2 = 2$. Show by the last column determinantal expansion that the three-rowed minor d_3 determined by rows 1, 2, and 3 can be computed by the formula $d_3 = 2d_2 - d_1$. Show, by the last column determinantal expansion, that the four-rowed minor d_4 determined by rows 1, 2, 3, and 4 has the value $d_4 = 2d_3 - d_2$. Show that $d_5 = \det G = 2d_4 - d_3$. Let H be the matrix obtained from G by replacing 1's by (-1)'s. Show that corresponding principal minors of G and H have the same value.

22. Let T_5 be the tridiagonal matrix

$$T_5 = \begin{bmatrix} a_1 & b_1 & 0 & 0 & 0 \\ c_1 & a_2 & b_2 & 0 & 0 \\ 0 & c_2 & a_3 & b_3 & 0 \\ 0 & 0 & c_3 & a_4 & b_4 \\ 0 & 0 & 0 & c_4 & a_5 \end{bmatrix}.$$

Let d_i be the principal minor of T_5 determined by rows 1, . . . , i of T_5. Then $d_1 = a_1$ and $d_2 = a_2 d_1 - b_1 c_1$. Show, by the last column determinantal expansion that for $i > 1$, we have $d_{i+1} = a_{i+1} d_i - b_i c_i d_{i-1}$. This algorithm can be used to compute the determinant of a tridiagonal matrix, such as T_5. Verify that this formula was used in Exercise 21. Generalize to higher-dimensional tridiagonal matrices T_6, T_7, ...

23. Suppose that $a_i = 4$, $b_i = c_i = -1$ for $i = 1, \ldots$ in the tridiagonal matrix T_n of the type described in Exercise 22. By the use of the formula given in Exercise 22, compute the determinants $d_n = \det T_n$ for $n = 3, 4, \ldots$, as far as you like. Of course, in this case, $d_1 = 4$ and $d_2 = 15$. Tridiagonal matrices arise in the study of numerical methods for solving ordinary and partial differential equations. They also arise in the study of numerical methods for solving linear equations and for finding eigenvalues of symmetric matrices.

Supplementary Proofs

➤ **P1.** Property 3 of determinantal rank of a matrix A follows from the following result.

Let A be a m $\times$ n-matrix of determinantal rank r. Then

1. *If r = n, the columns of A are linearly independent; that is, AX = 0 holds only if X = 0.*

2. *If r < n, the columns of A are linearly dependent; that is, AX = 0 has a nonzero solution X.*

Because r is the determinantal rank of A, there is a nonsingular $r \times r$-submatrix G of A. Without loss of generality we can suppose that G lies in the first r rows of A and in the first r columns of A, since this result can be obtained by interchanges of rows and by interchanges of columns of A.

Consider first the case in which $r = n$, the number of columns of A. Then $AX = 0$ implies that $GX = 0$ and hence that $X = 0$ because G is nonsingular. Hence conclusion 1 holds.

Consider next the case in which $r < n$. Let B be the $m \times (r + 1)$-submatrix of A comprised of the first $r + 1$ columns of A. If $r < m$, B is of the form

$$B = \begin{bmatrix} G & H \\ M & N \end{bmatrix}.$$

The matrix B has determinantal rank r. Its $(r + 1)$-minors are therefore all zero. Set $Y = G^{-1}H$ so that $GY = H$. The matrix

$$C = \begin{bmatrix} G & H - GY \\ M & N - MY \end{bmatrix} = \begin{bmatrix} G & 0 \\ M & N - MY \end{bmatrix}$$

is obtained from B by subtracting a linear combination of the first r columns of B to the last column of B. This does not alter the $(r + 1)$-rowed minors of B so that the $(r + 1)$-rowed minors of B and C are the same. These minors are all zero. Because G is nonsingular, this is possible only if we have $N - MY = 0$ as well as $H - GY = 0$. The $(r + 1)$st column.

$$\begin{bmatrix} H \\ N \end{bmatrix}$$

of B and hence of A is therefore a linear combination of the first r columns of A. The columns of A are therefore linearly dependent, so that $AX = 0$ has a nonzero solution X. The same result holds when $r = m$. Hence conclusion 2 holds. This proves Property 3 of determinantal ranks.   ▬

▬ **P2.** Establish Property 4.

It will be convenient to rewrite Property 4 as follows.

PROPERTY 4. *Let U be a m × n-submatrix of an m × n-matrix A having the following two properties:*

(a) *The columns of U are linearly independent.*

(b) *Each column of A is a linear combination of the columns of U. Then the number s of columns of U is equal to the determinantal rank r of A.*

Let V be a second $m \times t$-submatrix of A whose columns are linearly independent and which has the property that the columns of A are linear combinations of the columns of V. Since the columns of V are linear combinations of the columns of U, it follows that there is a $s \times t$-matrix B such that $V = UB$. If $t > s$ there is a nonzero vector X such that $BX = 0$ and hence such that $VX = UBC = 0$. But this would imply that the columns of V are linearly dependent, which is not the case. Hence $t \leq s$. Interchanging the roles of U and V, we see that $s \leq t$ so that $s = t$, as was to be proved.

It remains to show that our second matrix V can be chosen so that $t = r$, the determinantal rank of A. To this end let G be a nonsingular $r \times r$-submatrix. The $m \times r$-matrix V, whose columns contain the columns of G, has the desired property. Since G is a nonsingular submatrix of V, it follows from Property 3 that the columns of V are linearly independent. Moreover, every column of A is a linear combination of the columns of V. Otherwise, there would exist a column C of A that is not a linear combination of the columns of V. The column C and the columns of V would form a $m \times (r + 1)$-submatrix W of A whose columns are linearly independent. By Property 3, W would have a nonsingular $(r + 1) \times (r + 1)$-submatrix and r could not be the rank of A. The matrix V therefore has the desired property and we have the relations $s = r$, by the result given in the preceding paragraph.   ▬

▬ **P3.** Proof of Property 6.

Let r be the rank of A. By Property 4 there is a $m \times r$-submatrix V of A whose columns are linearly independent. The columns of the product $U = PV$ are columns of $F = PA$. We shall show that they are linearly independent. Since the columns of P and V are linearly independent, we see that

$$UX = PVX = 0 \text{ implies that } VX = 0 \text{ implies that } X = 0.$$

The columns of U are therefore linearly independent. Hence

$$\text{rank of } A = P^{-1}F \geq \text{rank of } F.$$

It follows that A and $F = PA$ have the same rank, as was to be shown.

Let Q be a nonsingular $n \times n$-matrix and set $C = AQ$. Applying the result just obtained to the product $C^T = Q^T A^T$, we see that C^T and A^T have the same rank. It follows that A and $C = AQ$ have the same rank. Similarly, $PC = PAQ$ has the same rank as C and hence has the same rank as A. This completes the proof of Property 6.   ▬

▬ **P4.** Proof of Property 8.

PROPERTY 8. *Let A be the product $A = BC$ of a $m = k$-matrix B and $k \times n$-matrix C of ranks s and t, respectively. The rank r of A satisfies the inequalities*

(a) $r \le s$ *with* $r = s$ *when* $t = k$

(b) $r \le t$ *with* $r = t$ *when* $s = k$.

To establish Property 8, observe that the equation $AX = BCX = 0$ holds whenever $CX = 0$. It follows that the nullity $n - t$ of C cannot exceed the nullity $n - r$ of A. Hence $n - t \le n - r$, so that $r \le t$. If $s = k$, the columns of B are linearly independent. In this case $AX = BCX = 0$ if and only if $CX = 0$. This implies that $n - r = n - t$ and hence that $r = t$. Applying the result just obtained to the transpose $A^T = C^T B^T$ of $A = BC$, we see that $r \le s$, the equality holding when $t = k$. ▬

▬ **P5.** Establish the following additional property of rank.

PROPERTY 9. *Suppose that the columns of a $m \times n$-matrix A of rank r are linear combinations of the columns of a $m \times r$-matrix U of rank r. Then there exists a $r \times n$-matrix V of rank r such that $A = UV$. A matrix A has rank r if and only if it is the product $A = UV$ of a $m \times r$-matrix U of rank r and a $r \times n$-matrix V of rank r.*

Suppose that A has rank r and that its columns $C_1, C_2, \ldots, C_n$ are linear combinations

$$C_1 = UK_1, \ C_2 = UK_2, \ldots, \ C_n = UK_n$$

of an $m \times r$-matrix of rank r. Here the K's are r-dimensional column vectors. We have

$$A = [UK_1 \quad UK_2 \quad \cdots \quad UK_n] = U[K_1 \quad K_2 \quad \cdots \quad K_n] = UV,$$

where $V = [K_1 \quad K_2 \quad \cdots \quad K_n]$ is an r-rowed matrix. By Property 8 the matrix V has the same rank as A and so is of rank r. By Property 4 the matrix U can be chosen to be a submatrix of A.

Conversely, if A is the product $A = UV$ of an $m \times r$-matrix U of rank r and an $r \times n$-matrix V of rank r, then, by Property 8, A has rank r. ▬

4

Eigenvalues and Eigenvectors of a Square Matrix

4.1 CHARACTERISTIC POLYNOMIAL AND EIGENVALUES OF A SQUARE MATRIX

In the study of the characteristic polynomial and eigenvalues of a square matrix A, it is customary to use the Greek lowercase letter λ, lambda, to denote a scalar playing a particular role. As we proceed, the Greek lowercase letter μ, mu, is also used. To this end, let A be an $n \times n$-matrix and let I be the corresponding n-dimensional identity matrix. It turns out that much information about A can be obtained from the properties of the matrix $A - \lambda I$, where λ is an arbitrary scalar. In particular, its determinant,

$$|A - \lambda I|,$$

plays a significant role in the study of the properties of the matrix A. The matrix $A - \lambda I$ is obtained from A by subtracting λ from each of its main diagonal entries. For example, in a numerical two-dimensional case, we have

$$A = \begin{bmatrix} 5 & 8 \\ 2 & 5 \end{bmatrix}, \quad I = \begin{bmatrix} 1 & 0 \\ 0 & 1 \end{bmatrix}, \quad A - \lambda I = \begin{bmatrix} 5 - \lambda & 8 \\ 2 & 5 - \lambda \end{bmatrix}.$$

The determinant of $A - \lambda I$ is

$$|A - \lambda I| = \begin{vmatrix} 5 - \lambda & 8 \\ 2 & 5 - \lambda \end{vmatrix} = (5 - \lambda)^2 - 16 = 9 - 10\lambda + \lambda^2.$$

In a 3×3 case, the matrix

$$A = \begin{bmatrix} 2 & 3 & -7 \\ 0 & 5 & 8 \\ 0 & 2 & 5 \end{bmatrix}$$

has determinant of $A - \lambda I$ given by

$$|A - \lambda I| = \begin{vmatrix} 2 - \lambda & 3 & -7 \\ 0 & 5 - \lambda & 8 \\ 0 & 2 & 5 - \lambda \end{vmatrix} = 18 - 29\lambda + 12\lambda^2 - \lambda^3.$$

In each of the examples above, observe that $|A - \lambda I|$ is a polynomial in λ. When A is a 2×2-matrix, the polynomial $|A - \lambda I|$ is of degree 2, and when A is a 3×3-matrix, the polynomial $|A - \lambda I|$ is of degree 3.

In the $n \times n$ case, the determinant $|A - \lambda I|$ is a polynomial

$$|A - \lambda I| = a_0 + a_1(-\lambda) + \cdots + a_{n-1}(-\lambda)^{n-1} + (-\lambda)^n$$

of degree n in the variable $-\lambda$ and hence in λ. This polynomial is called the *characteristic polynomial* of A, which we shall denote by $p(\lambda)$.

Setting the characteristic polynomial $p(\lambda)$ equal to 0, we have

$$p(\lambda) = |A - \lambda I| = a_0 + a_1(-\lambda) + \cdots + a_{n-1}(-\lambda)^{n-1} + (-\lambda)^n = 0.$$

This equation is called the ***characteristic equation*** of A. The roots of the polynomial $p(\lambda)$ [i.e., the values of λ for which $p(\lambda) = 0$] are called by various names, such as *eigenvalues, proper values, characteristic values, characteristic roots*, and *latent roots*. We shall use the term ***eigenvalue***. An eigenvalue can be real or complex.

We begin the study of eigenvalues by examining the case when A is a 2×2-matrix.

■ **EXAMPLE 1** —————————————————————————————————

Suppose that

$$A = \begin{bmatrix} 0 & 1 \\ 1 & 0 \end{bmatrix}.$$

The characteristic polynomial

$$p(\lambda) = |A - \lambda I| = \begin{vmatrix} -\lambda & 1 \\ 1 & -\lambda \end{vmatrix} = \lambda^2 - 1,$$

and therefore the characteristic equation $p(\lambda) = \lambda^2 - 1 = 0$ has as its solutions $\lambda = \pm 1$. The eigenvalues of A are therefore 1 and -1. In this example, the eigenvalues of A are *distinct* and *real*.

■ **EXAMPLE 2** —————————————————————————————————

Suppose that

$$A = \begin{bmatrix} 0 & 1 \\ -1 & 0 \end{bmatrix}.$$

The characteristic polynomial $p(\lambda)$ is

$$p(\lambda) = |A - \lambda I| = \begin{vmatrix} -\lambda & 1 \\ -1 & -\lambda \end{vmatrix} = \lambda^2 + 1.$$

The characteristic equation $p(\lambda) = \lambda^2 + 1 = 0$ has as its solutions $\lambda = \pm\sqrt{-1} = \pm i$. The eigenvalues of A are in this case the ***complex numbers*** i and $-i$.

■ **EXAMPLE 3** —————————————————————————————————

Let

$$A = \begin{bmatrix} 1 & 0 \\ 0 & 1 \end{bmatrix}.$$

The characteristic polynomial $p(\lambda)$ is

$$p(\lambda) = |A - \lambda I| = \begin{vmatrix} 1 - \lambda & 0 \\ 0 & 1 - \lambda \end{vmatrix} = (1 - \lambda)^2.$$

The eigenvalues of A are therefore 1, 1. In this case, the eigenvalues of A are **real** and **repeated**. The eigenvalue $\lambda = 1$ is said to have **multiplicity** 2.

The three examples above illustrate the general situation for a real 2 × 2-matrix A. The eigenvalues of A will fall into one of the following three cases:

1. λ_1, λ_2 real; $\lambda_1 \neq \lambda_2$ (distinct)
2. λ_1, λ_2 complex; $\overline{\lambda_1} = \lambda_2$ (complex conjugates)
3. λ_1, λ_2 real; $\lambda_1 = \lambda_2$ (multiple root)

We can justify this by considering the general two-dimensional case, in which

$$A = \begin{bmatrix} a & b \\ c & d \end{bmatrix}, \qquad A - \lambda I = \begin{bmatrix} a - \lambda & b \\ c & d - \lambda \end{bmatrix}.$$

The characteristic polynomial is

$$\begin{aligned} p(\lambda) &= \begin{vmatrix} a - \lambda & b \\ c & d - \lambda \end{vmatrix} \\ &= (a - \lambda)(d - \lambda) - bc \\ &= (ad - bc) - (a + d)\lambda + \lambda^2. \end{aligned}$$

This polynomial is quadratic.

Applying the quadratic formula to the characteristic equation

$$p(\lambda) = \lambda^2 - (a + d)\lambda + (ad - bc) = 0,$$

we find that the eigenvalues of A are

$$\lambda = \frac{(a + d) \pm \sqrt{(a + d)^2 - 4(ad - bc)}}{2}.$$

Recall that the value of the discriminant

$$(a + d)^2 - 4(ad - bc)$$

of $p(\lambda)$ determines whether the eigenvalues λ_1 and λ_2 of A are real, complex, or repeated.

The following list illustrates the three possibilities:

1. If $(a + d)^2 - 4(ad - bc) > 0$,   then λ_1, λ_2 are real and distinct.
2. If $(a + d)^2 - 4(ad - bc) < 0$,   then λ_1, λ_2 are complex conjugates.
3. If $(a + d)^2 - 4(ad - bc) = 0$,   then λ_1, λ_2 are real and equal (repeated).

Observe that in the characteristic polynomial $p(\lambda) = |A - \lambda I| = \lambda^2 - (a + d)\lambda + (ad - bc)$ of A, the constant term $(ad - bc)$ is the **determinant** det A of A. The quantity $(a + d)$ which is the sum of the main diagonal entries of A is called the **trace** of A. Accordingly, $p(\lambda)$ is of the form

$$p(\lambda) = |A - \lambda I| = \lambda^2 - (\text{trace } A)\lambda + (\det A).$$

The discriminant of $p(\lambda)$ can be written as

$$(\text{trace } A)^2 - 4(\det A).$$

We see that the nature of the eigenvalues of A can be determined by calculating the trace of A and the determinant of A. Using the formulas for λ_1 and λ_2 obtained from the quadratic formula above, a simple calculation yields

$$\lambda_1\lambda_2 = ad - bc = \det A, \qquad \lambda_1 + \lambda_2 = a + d = \text{trace of } A.$$

We turn now to the three-dimensional case in which

$$A = \begin{bmatrix} a_{11} & a_{12} & a_{13} \\ a_{21} & a_{22} & a_{23} \\ a_{31} & a_{32} & a_{33} \end{bmatrix}, \qquad A - \lambda I = \begin{bmatrix} a_{11} - \lambda & a_{12} & a_{13} \\ a_{21} & a_{22} - \lambda & a_{23} \\ a_{31} & a_{32} & a_{33} - \lambda \end{bmatrix}.$$

The characteristic polynomial is a cubic of the form

$$p(\lambda) = \det (A - \lambda I) = a_0 - a_1\lambda + a_2\lambda^2 - \lambda^3.$$

Clearly, $p(0) = \det A = a_0$. In addition, as will be seen in Exercise 7, we have

$$a_1 = \begin{vmatrix} a_{22} & a_{23} \\ a_{32} & a_{33} \end{vmatrix} + \begin{vmatrix} a_{11} & a_{13} \\ a_{31} & a_{33} \end{vmatrix} + \begin{vmatrix} a_{11} & a_{12} \\ a_{21} & a_{22} \end{vmatrix},$$

$$a_2 = a_{11} + a_{22} + a_{33} = \text{trace of } A.$$

The cubic $p(\lambda)$ has three zeros $\lambda_1, \lambda_2,$ and λ_3. From the factoring property of polynomials,

$$p(\lambda) = (\lambda_1 - \lambda)(\lambda_2 - \lambda)(\lambda_3 - \lambda).$$

Multiplying out the expression on the right, it follows that

$$p(\lambda) = \lambda_1\lambda_2\lambda_3 - (\lambda_1\lambda_2 + \lambda_2\lambda_3 + \lambda_3\lambda_1)\lambda + (\lambda_1 + \lambda_2 + \lambda_3)\lambda^2 - \lambda^3.$$

We see that the coefficients a_0, a_1, a_2 in $p(\lambda)$ are given by the formulas

$$a_0 = \det A = \lambda_1\lambda_2\lambda_3, \qquad a_1 = \lambda_1\lambda_2 + \lambda_2\lambda_3 + \lambda_3\lambda_1,$$

$$a_2 = \text{trace } A = \lambda_1 + \lambda_2 + \lambda_3.$$

■ **EXAMPLE 4** _____
Consider

$$A = \begin{bmatrix} 5 & 0 & 8 \\ -1 & 2 & 4 \\ 2 & 0 & 5 \end{bmatrix}, \qquad p(\lambda) = |A - \lambda I| = \begin{vmatrix} 5 - \lambda & 0 & 8 \\ -1 & 2 - \lambda & 4 \\ 2 & 0 & 5 - \lambda \end{vmatrix}.$$

Expanding $p(\lambda)$ by its second column, we find that

$$p(\lambda) = (2 - \lambda) \begin{vmatrix} 5 - \lambda & 8 \\ 2 & 5 - \lambda \end{vmatrix} = (2 - \lambda)[(5 - \lambda)^2 - 16]$$

and hence that

$$p(\lambda) = (2 - \lambda)(9 - 10\lambda + \lambda^2) = 18 - 29\lambda + 12\lambda^2 - \lambda^3.$$

Factoring, we have

$$p(\lambda) = (2 - \lambda)(9 - \lambda)(1 - \lambda) = 0$$

as the characteristic equation of A. It follows that $\lambda_1 = 2$, $\lambda_2 = 9$, and $\lambda_3 = 1$ are the eigenvalues of A.

When A is a diagonal matrix, such as

$$A = \begin{bmatrix} 2 & 0 & 0 \\ 0 & 3 & 0 \\ 0 & 0 & 4 \end{bmatrix},$$

its main diagonal entries are its eigenvalues. This follows because its characteristic polynomial is

$$p(\lambda) = \begin{vmatrix} 2 - \lambda & 0 & 0 \\ 0 & 3 - \lambda & 0 \\ 0 & 0 & 4 - \lambda \end{vmatrix} = (2 - \lambda)(3 - \lambda)(4 - \lambda).$$

The roots or zeros of $p(\lambda)$ are obviously 2, 3, and 4, the main diagonal entries of A.

In another case,

$$A = \begin{bmatrix} 1 & 0 & 0 \\ 0 & 0 & 1 \\ 0 & 1 & 0 \end{bmatrix}, \qquad p(\lambda) = \begin{vmatrix} 1 - \lambda & 0 & 0 \\ 0 & -\lambda & 1 \\ 0 & 1 & -\lambda \end{vmatrix} = (1 - \lambda)(\lambda^2 - 1).$$

The characteristic polynomial of A is therefore

$$p(\lambda) = -1 + \lambda + \lambda^2 - \lambda^3 = (1 - \lambda)(\lambda - 1)(\lambda + 1)$$

and its eigenvalues are 1, 1, and -1. The eigenvalue $\lambda = 1$ has multiplicity 2.

The concept of **multiplicity** of an eigenvalue λ of A is illustrated further in the following example. Let A be a 6×6 matrix whose eigenvalues are 1, 2, 3, 1, 2, and 1. Its characteristic polynomial is $p(\lambda) = (1 - \lambda)^3 \times (2 - \lambda)^2(3 - \lambda)$. The eigenvalue 1 is repeated three times and we say that its multiplicity is 3. The eigenvalue 2 is repeated twice, so its multiplicity is 2. The eigenvalue 3 is of multiplicity 1. In general, if $\lambda_1, \lambda_2, \ldots, \lambda_n$ are the n eigenvalues of an $n \times n$-matrix A, the number of times a particular eigenvalue is repeated is called its **multiplicity**.

Motivated by the 2×2 and 3×3 cases above, and from our knowledge of polynomials, we list the following properties of the characteristic polynomial

$$p(\lambda) = |A - \lambda I| = a_0 + a_1(-\lambda) + \cdots + a_{n-1}(-\lambda)^{n-1} + (-\lambda)^n$$

of a general $n \times n$-matrix A:

1. It has n zeros $\lambda_1, \lambda_2, \ldots, \lambda_n$, not necessarily distinct. These are the eigenvalues of A, real or complex.
2. $p(\lambda) = (\lambda_1 - \lambda)(\lambda_2 - \lambda) \cdots (\lambda_n - \lambda)$
3. When the coefficients $a_0, a_1, \ldots, a_{n-1}$ are real, as is the case when A is real, any of the complex roots of $p(\lambda)$ come in pairs; that is, if λ is a complex zero of $p(\lambda)$, so is its conjugate $\bar{\lambda}$.
4. $p(0) = a_0 = \det A = \lambda_1\lambda_2 \cdots \lambda_n$.
5. $a_{n-1} = \lambda_1 + \lambda_2 + \cdots + \lambda_n = \text{trace of } A = \text{sum of main diagonal}$ entries of A.

There are, of course, formulas for the remaining coefficients $a_1, a_2, \ldots, a_{n-2}$ in $p(\lambda)$, but we shall not be concerned with them at this time.

It should be noted that some authors define the determinant $|\lambda I - A|$ to be the characteristic polynomial of A. Because $|\lambda I - A| = (-1)^n |A - \lambda I|$, the two polynomials $|\lambda I - A|$ and $|A - \lambda I|$ are the same when n is even and differ only in sign when n is odd. It is immaterial which one is used. We prefer to use $p(\lambda) = |A - \lambda I|$ because it involves fewer minus signs in the applications that we discuss.

At this stage we shall be concerned mainly with real eigenvalues. However, complex eigenvalues play an important role in the theory of matrices and their applications and so should not be disregarded.

We now present a list of properties of the characteristic polynomials and eigenvalues of general $n \times n$-matrices $A, B, C, \ldots$.

PROPERTY 1. λ is an eigenvalue of A if and only if the matrix $A - \lambda I$ is singular.

This follows because $A - \lambda I$ is singular if and only if $p(\lambda) = \det(A - \lambda I) = 0$, that is, if and only if λ is an eigenvalue of A.

PROPERTY 2. $\det A = 0$ if and only if $\lambda = 0$ is an eigenvalue of A.

This follows from Property 1 with $\lambda = 0$.

PROPERTY 3. A and A^T have the same characteristic polynomial and hence the same eigenvalues.

The transpose of $B = A - \lambda I$ is $B^t = (A - \lambda I)^t = A^t - \lambda I$. Since $\det B = \det B^T$, we have $\det(A - \lambda I) = \det(A^T - \lambda I)$; that is, the characteristic polynomials of A and A^T are equal.

PROPERTY 4. If B is a nonsingular matrix and $C = B^{-1}AB$, then A and C have the same characteristic polynomial and hence the same eigenvalues.

Observe that

$$C - \lambda I = B^{-1}AB - \lambda I = B^{-1}AB - \lambda B^{-1}IB = B^{-1}(A - \lambda I)B.$$

Since the determinant of products is a product of determinants we have

$$\det (C - \lambda I) = (\det B^{-1})(\det (A - \lambda I))(\det B) = \det (A - \lambda I)$$

in view of the relation $(\det B^{-1})(\det B) = 1$. Hence C and A have the same characteristic polynomial and the same eigenvalues.

PROPERTY 5. AB and BA have the same characteristic polynomial and the same eigenvalues.

When B is nonsingular we have $AB = B^{-1}(BA)B$. By Property 4, the matrices BA and AB have the same characteristic polynomial and eigenvalues. When B is singular, $B - cI$ is nonsingular for small nonzero values of c. Consequently,

$$A(B - cI) = AB - cA \quad \text{and} \quad (B - cI)A = BA - cA$$

have the same characteristic polynomial for small nonzero values of c and hence also for $c = 0$, as was to be proved.

Choosing $B = A^T$, we obtain from Property 5:

PROPERTY 6. $A^T A$ and AA^T have the same characteristic polynomial. Hence they have the same eigenvalues.

In order to motivate a final result for general $n \times n$-matrices, we once again consider the general 2×2-matrix

$$A = \begin{bmatrix} a & b \\ c & d \end{bmatrix}$$

and its characteristic polynomial

$$p(\lambda) = |A - \lambda I| = (ad - bc) - (a + d)\lambda + \lambda^2.$$

If in the polynomial $p(\lambda)$, we replace 1 by I and λ by A, the result is a polynomial

$$p(A) = (ad - bc)I - (a + d)A + A^2$$

in A. Using the fact that

$$I = \begin{bmatrix} 1 & 0 \\ 0 & 1 \end{bmatrix}, \quad A = \begin{bmatrix} a & b \\ c & d \end{bmatrix}, \quad A^2 = \begin{bmatrix} a^2 + bc & ab + bd \\ ca + cd & cb + d^2 \end{bmatrix},$$

we see that $p(A)$ is given by the sum

$$p(A) = \begin{bmatrix} ad - bc & 0 \\ 0 & ad - bc \end{bmatrix} - \begin{bmatrix} a(a + d) & b(a + d) \\ c(a + d) & d(a + d) \end{bmatrix}$$

$$+ \begin{bmatrix} a^2 + bc & ab + bd \\ ca + cd & cb + d^2 \end{bmatrix} = \begin{bmatrix} 0 & 0 \\ 0 & 0 \end{bmatrix}.$$

Consequently, $p(A) = 0$. That is, the matrix A satisfies its own characteristic equation. This important result, which holds for any $n \times n$-matrix A, is known as the *Cayley–Hamilton theorem*. We have:

PROPERTY 7. CAYLEY–HAMILTON THEOREM. Let

$$p(\lambda) = a_0 + a_1(-\lambda) + \cdots + a_{n-1}(-\lambda)^{n-1} + (-\lambda)^n$$

be the characteristic polynomial of A. Then

$$p(A) = a_0 I + a_1(-A) + \cdots + a_{n-1}(-A)^{n-1} + (-A)^n = 0;$$

that is, a matrix A satisfies its characteristic equation.

■ **EXAMPLE 5**

Consider the 3×3 case in which

$$A = \begin{bmatrix} 1 & 0 & 0 \\ 0 & 0 & 1 \\ 0 & 1 & 0 \end{bmatrix}.$$

As was seen earlier its characteristic polynomial is

$$p(\lambda) = -1 + \lambda + \lambda^2 - \lambda^3.$$

Because $A^2 = I$ and $A = A^3$ we have

$$p(A) = -I + A + A^2 - A^3 = 0,$$

as stated in Property 7.

The proof of Property 7 in the general case is like that of the case $n = 3$. Observe that the cofactors of the entries of $A - \lambda I$ are polynomials in λ of degree $\leq n - 1 = 3 - 1 = 2$. Hence the classical adjoint, adj $(A - \lambda I)$ of $A - \lambda I$ is expressible in the form

$$\text{adj } (A - \lambda I) = C_0 - C_1 \lambda + C_2 \lambda^2,$$

where C_0, C_1, and C_2 are constant matrices whose values need not be determined. Recall that

$$(A - \lambda I) \text{ adj } (A - \lambda I) = p(\lambda)I,$$

where $p(\lambda) = a_0 - a_1 \lambda + a_2 \lambda^2 - \lambda^3$ is the characteristic polynomial of A. This relation can be put in the form

$$(A - \lambda I)(C_0 - C_1 \lambda + C_2 \lambda^2) = (a_0 - a_1 \lambda + a_2 \lambda^2 - \lambda^3)I.$$

Equating the coefficients of like powers of λ, we find that

$$AC_0 = a_0 I, \quad AC_1 + C_0 = a_1 I, \quad AC_2 + C_1 = a_2 I, \quad C_2 = I.$$

Multiplying the first equation by I, the second by $-A$, the third by A^2, and the last by $-A^3$, we get

$$AC_0 = a_0 I, \quad -A^2 C_1 - AC_0 = -a_1 A, \quad A^3 C_2 + A^2 C_1 = a_2 A^2,$$
$$-A^3 C_2 = -A^3.$$

Adding these equations, we have

$$0 = a_0 I - a_1 A + a_2 A^2 - A^3 = p(A),$$

as was to be proved.

Exercises

1. Find the characteristic polynomial and the eigenvalues of each of the following matrices.

$$H = \begin{bmatrix} 5 & 4 \\ 4 & 5 \end{bmatrix}, \qquad K = \begin{bmatrix} 0 & 2 \\ -3 & 5 \end{bmatrix}, \qquad L = \begin{bmatrix} 4 & 3 \\ -3 & 4 \end{bmatrix},$$

$$M = \begin{bmatrix} 12 & -5 \\ 5 & 12 \end{bmatrix}, \qquad N = \begin{bmatrix} 5 & 4 & 0 \\ 4 & 5 & 0 \\ 2 & -3 & 0 \end{bmatrix}, \qquad P = \begin{bmatrix} 0 & -1 & 0 \\ 0 & 0 & -1 \\ 1 & 0 & 0 \end{bmatrix},$$

$$Q = \begin{bmatrix} 0 & -1 & 0 \\ 0 & 0 & -1 \\ -9 & -9 & 1 \end{bmatrix}, \qquad R = \begin{bmatrix} 0 & 1 & 0 & 0 \\ -3 & 4 & 0 & 0 \\ 5 & 6 & 5 & 1 \\ -1 & 1 & -9 & -1 \end{bmatrix},$$

$$S = \begin{bmatrix} 0 & 1 & 0 & 0 \\ 0 & 0 & 1 & 0 \\ 1 & 0 & 0 & 0 \\ 3 & 2 & 4 & 5 \end{bmatrix}, \qquad T = \begin{bmatrix} 0 & 0 & 0 & 2 \\ 0 & 0 & 3 & 0 \\ 0 & 3 & 0 & 0 \\ 2 & 0 & 0 & 0 \end{bmatrix}.$$

2. If A is the 2×2-matrix

$$A = \begin{bmatrix} a & b \\ c & d \end{bmatrix},$$

whose characteristic equation is

$$P(\lambda) = \lambda^2 - (a + d)\lambda + (ad - bc) = 0,$$

show that the value of the discriminant

$$(a + d)^2 - 4(ad - bc)$$

is also given by the expression

$$(a - d)^2 + 4bc.$$

Conclude that, if a, b, c, and d are real, the eigenvalues of A are real whenever b and c have the same sign.

3. Show that $p(\lambda) = a - b\lambda + c\lambda^2 - d\lambda^3 + \lambda^4$ is the characteristic polynomial of the 4×4-matrix

$$A = \begin{bmatrix} 0 & -1 & 0 & 0 \\ 0 & 0 & -1 & 0 \\ 0 & 0 & 0 & -1 \\ a & b & c & d \end{bmatrix}.$$

4. Use the result described in Exercise 3 to construct a 4×4-matrix A whose eigenvalues are 1, -1, 2, and -2.

5. Find the characteristic polynomials of the matrices

$$U = \begin{bmatrix} 0 & 1 & 0 \\ 0 & 0 & 1 \\ 1 & 0 & 0 \end{bmatrix}, \qquad V = \begin{bmatrix} 0 & 1 & 0 \\ 0 & 0 & 1 \\ -1 & 0 & 0 \end{bmatrix},$$

$$W = \begin{bmatrix} 0 & -1 & 0 \\ 0 & 0 & -1 \\ -1 & 0 & 0 \end{bmatrix}, \qquad X = \begin{bmatrix} 0 & 1 & 0 & 0 \\ 0 & 0 & 1 & 0 \\ 0 & 0 & 0 & 1 \\ 1 & 0 & 0 & 0 \end{bmatrix},$$

$$Y = \begin{bmatrix} 0 & 1 & 0 & 0 \\ 0 & 0 & 1 & 0 \\ 0 & 0 & 0 & 1 \\ -1 & 0 & 0 & 0 \end{bmatrix}, \qquad Z = \begin{bmatrix} 0 & -1 & 0 & 0 \\ 0 & 0 & -1 & 0 \\ 0 & 0 & 0 & -1 \\ -1 & 0 & 0 & 0 \end{bmatrix}.$$

6. There are many $n \times n$-matrices which have $(-1)^n(\lambda^n - 1)$ as their characteristic polynomial. Verify that the characteristic polynomials of the matrices

$$\begin{bmatrix} 0 & 1 \\ 1 & 0 \end{bmatrix}, \quad \begin{bmatrix} 0 & 1 & 0 \\ 0 & 0 & 1 \\ 1 & 0 & 0 \end{bmatrix}, \quad \begin{bmatrix} 0 & 1 & 0 & 0 \\ 0 & 0 & 1 & 0 \\ 0 & 0 & 0 & 1 \\ 1 & 0 & 0 & 0 \end{bmatrix}, \quad \begin{bmatrix} 0 & 1 & 0 & 0 & 0 \\ 0 & 0 & 1 & 0 & 0 \\ 0 & 0 & 0 & 1 & 0 \\ 0 & 0 & 0 & 0 & 1 \\ 1 & 0 & 0 & 0 & 0 \end{bmatrix}$$

are, respectively, $\lambda^2 - 1$, $1 - \lambda^3$, $\lambda^4 - 1$, and $1 - \lambda^5$. To find the eigenvalues for matrices of this type, we use *DeMoivre's theorem.* It states that with $i^2 = -1$,

$$(\cos \theta + i \sin \theta)^n = \cos n\theta + i \sin n\theta$$

for every integer n. Recall that $\cos 2k\pi = 1$, $\sin 2k\pi = 0$ for every integer k. Use this property to show that

$$w_k = \cos \frac{2k\pi}{n} + i \sin \frac{2k\pi}{n} \qquad (k = 1, 2, \ldots, n)$$

is an nth root of unity; that is, $w_k^n = 1$. Verify further that

$$w_k = w_1^k.$$

Conclude that w_1, $w_2 = w_1^3$, . . . , $w_n = 1$ are the eigenvalues of an $n \times n$-matrix A having $(-1)^n(\lambda^n - 1)$ as its characteristic polynomial. Find the eigenvalues of A when the characteristic polynomial of A is

(a) $\lambda^2 - 1$ **(b)** $1 - \lambda^3$ **(c)** $\lambda^4 - 1$ **(d)** $1 - \lambda^5$

7. Given that

$$A = \begin{bmatrix} 2 & 0 & 0 & 0 \\ 0 & -2 & 0 & 0 \\ 0 & 0 & 3 & 0 \\ 0 & 0 & 0 & -3 \end{bmatrix}, \qquad B = \begin{bmatrix} 1 & 1 & 1 & 1 \\ -1 & 1 & -1 & 1 \\ -1 & -1 & 1 & 1 \\ 1 & -1 & -1 & 1 \end{bmatrix},$$

show that $B^{-1} = (1/4)B^T$. Compute $C = B^{-1}AB$. What are its eigenvalues? This technique is a standard technique for constructing nondiagonal matrices with prescribed eigenvalues.

8. Let B be the matrix displayed in Exercise 7. Its characteristic equation is of the form

$$p(\lambda) = a - b\lambda + c\lambda^2 - d\lambda^3 + \lambda^4.$$

Compute $a = \det B$. Compute b as the sum of the three-rowed principal minors of B. Compute c as the sum of the two-rowed principal minors of B. Compute d as the trace of B, the sum of the one-rowed principal minors of B.

9. Let A be a 3×3-matrix having 2, 4, and 6 as its eigenvalues. In each of the following cases, find the eigenvalues of the matrix B related to A as follows:

(a) $B = A - 5I$ **(b)** $B = A + 2I$ **(c)** $B = A + cI$ **(d)** $B = -A$

(e) $B = 2A$ **(f)** $B = cA$ **(g)** $B = 2A + 3I$

10. In our examples and exercises we chose matrices A whose entries are positive or negative integers or zero. As a consequence, the coefficients in the characteristic polynomial of our matrix A are also positive or negative integers or zero. In particular, $\det A$ is an integer. In many cases at least one of the eigenvalues is an integer. Such an eigenvalue r must divide $\det A$. For example, when $\det A = 6$, the possible integer eigenvalues are ± 1, ± 2, ± 3, and ± 6. In case the characteristic polynomial is $6 - 5\lambda + \lambda^2$, 2 and 3 are eigenvalues. However, when $6 + \lambda^2$ is the characteristic polynomial, there are no integer eigenvalues. List the possible integer eigenvalues in the following cases:

(a) $\det A = 1$ **(b)** $\det A = 3$ **(c)** $\det A = -8$ **(d)** $\det A = 11$

11. Verify the result described in Exercise 10 for the matrices given in Exercise 1.

12. Let A and $A - \lambda I$ be the matrices

$$A = \begin{bmatrix} a_{11} & a_{12} & a_{13} \\ a_{21} & a_{22} & a_{23} \\ a_{31} & a_{32} & a_{33} \end{bmatrix}, \qquad A - \lambda I = \begin{bmatrix} a_{11} - \lambda & a_{12} & a_{13} \\ a_{21} & a_{22} - \lambda & a_{23} \\ a_{31} & a_{32} & a_{33} - \lambda \end{bmatrix}.$$

Set $B = -\lambda I$ so that $A + B = A - \lambda I$. Let

$$A_1 = \begin{bmatrix} -\lambda & 0 & 0 \\ a_{21} & a_{22} & a_{23} \\ a_{31} & a_{32} & a_{33} \end{bmatrix}, \qquad A_2 = \begin{bmatrix} a_{11} & a_{12} & a_{13} \\ 0 & -\lambda & 0 \\ a_{31} & a_{32} & a_{33} \end{bmatrix},$$

$$A_3 = \begin{bmatrix} a_{11} & a_{12} & a_{13} \\ a_{21} & a_{22} & a_{23} \\ 0 & 0 & -\lambda \end{bmatrix}$$

be the matrices obtained from A by replacing a row of A by the corresponding row of $B = -\lambda I$. Let

$$A_4 = \begin{bmatrix} -\lambda & 0 & 0 \\ 0 & -\lambda & 0 \\ a_{31} & a_{32} & a_{33} \end{bmatrix}, \qquad A_5 = \begin{bmatrix} a_{11} & a_{12} & a_{13} \\ 0 & -\lambda & 0 \\ 0 & 0 & -\lambda \end{bmatrix},$$

$$A_6 = \begin{bmatrix} -\lambda & 0 & 0 \\ a_{21} & a_{22} & a_{23} \\ 0 & 0 & -\lambda \end{bmatrix}$$

be the matrices obtained from A by replacing two rows of A by the corresponding two rows of $B = -\lambda I$. According to Exercise P6 of Section 3.2,

$$\det (A - \lambda I) = \det A + [\det A_1 + \det A_2 + \det A_3]$$
$$+ [\det A_4 + \det A_5 + \det A_6] + \det (-\lambda I).$$

The quantity in the first set of brackets is equal to $-b\lambda$, where

$$b = \begin{vmatrix} a_{22} & a_{23} \\ a_{32} & a_{33} \end{vmatrix} + \begin{vmatrix} a_{11} & a_{13} \\ a_{31} & a_{33} \end{vmatrix} + \begin{vmatrix} a_{11} & a_{12} \\ a_{21} & a_{22} \end{vmatrix}.$$

The quantity in the second set of brackets equals $c\lambda^2$, where

$$c = a_{11} + a_{22} + a_{33} = \text{trace of } A.$$

Consequently, the characteristic polynomial of A takes the form

$$\det (A - \lambda I) = \det A - b\lambda + c\lambda^2 - \lambda^3,$$

where b is the sum of the two-rowed principal minors of A and c is the trace of A.

13. Let A be an $n \times n$-matrix and let

$$p(\lambda) = a_0 + a_1(-\lambda) + a_2(-\lambda)^2 + \cdots + a_{n-1}(-\lambda)^{n-1} + (-\lambda)^n$$

be its characteristic polynomial. By the technique described in Exercise 12 show that the coefficient a_k is given by the formula

$$a_k = \text{sum of the } (n - k)\text{-rowed principal minors of } A.$$

14. Let A be a 4×4-matrix and let

$$p(\lambda) = a - b\lambda + c\lambda^2 - d\lambda^3 + \lambda^4$$

be its characteristic polynomial. Suppose that $a = \det A$ is not zero. By the use of the Cayley–Hamilton theorem show that the matrix

$$B = \frac{1}{a}[bI - cA + dA^2 - A^3]$$

is the inverse of A. Generalize to the n-dimensional case.

15. Consider the case in which

$$A = \begin{bmatrix} 1 & 0 & 0 & 1 \\ 0 & 1 & 1 & 0 \\ 0 & 1 & 1 & 0 \\ 1 & 0 & 0 & 1 \end{bmatrix}, \qquad A - \lambda I = \begin{bmatrix} 1-\lambda & 0 & 0 & 1 \\ 0 & 1-\lambda & 1 & 0 \\ 0 & 1 & 1-\lambda & 0 \\ 1 & 0 & 0 & 1-\lambda \end{bmatrix}.$$

Show that $p(\lambda) = \det(A - \lambda I) = \lambda^2 (2 - \lambda)^2 = q(\lambda)^2$, where $q(\lambda) = 2\lambda - \lambda^2$. Show that $A^2 = 2A$. Conclude that we have $q(A) = 0$, as well as the relation $p(A) = 0$. It follows that a special matrix A can satisfy a polynomial equation $q(\lambda) = 0$ of degree less than that of its characteristic equation $p(\lambda) = 0$. However, do not conclude that if the characteristic polynomial of a matrix A is of the form $p(\lambda) = q(\lambda)^2$, then the relation $q(A) = 0$ necessarily holds. Show that this is not the case when

$$A = \begin{bmatrix} 0 & 1 \\ 0 & 0 \end{bmatrix}.$$

16. Let A be an $m \times n$-matrix with $m \geq n$. Let $p(\lambda)$ be the characteristic polynomial of the $n \times n$-matrix $A^T A$. Let $q(\lambda)$ be the characteristic polynomial of the $m \times m$-matrix AA^T. Show that $q(\lambda) = (-\lambda)^{m-n} p(\lambda)$. Conclude that the matrices $A^T A$ and AA^T have the same **nonzero** eigenvalues with the same multiplicities. *Hint:* Augment the matrix A by adding $m - n$ zero columns to form a $m \times m$-matrix $B = [A \quad 0]$. We then have $BB^T = AA^T$ and

$$B^T B = \begin{bmatrix} A^T A & 0 \\ 0 & 0 \end{bmatrix}.$$

Denote the $r \times r$-identity matrix by I_r. Since B is a square matrix, the matrices BB^T and $B^T B$ have the same characteristic polynomial. Hence

$$q(\lambda) = \det(AA^T - \lambda I_m) = \det(BB^T - \lambda I_m) = \det(B^T B - \lambda I_m)$$

and since

$$\det(B^T B - \lambda I_m) = \begin{vmatrix} A^T A - \lambda I_n & 0 \\ 0 & -\lambda I_{m-n} \end{vmatrix} = (-\lambda)^{m-n} p(\lambda),$$

we have $q(\lambda) = (-\lambda)^{m-n} p(\lambda)$.

17. Continuing with Exercise 16, show that $\det A^T A$ is the sum of the n-rowed principal minors of AA^T. *Hint:* Observe that $B^T B$ has but one nonzero n-rowed principal minor, $\det A^T A$. Because the coefficient a_n of $(-\lambda)^{m-n}$ in the polynomial

$$q(\lambda) = \det(AA^T - \lambda I) = \det(B^T B - \lambda I)$$

is equal to the sum of the n-rowed principal minors of AA^T on the one hand, and is equal to the sum $\det A^T A$ of the n-rowed principal minors of $B^T B$ on the other hand, we have $\det A^T A$ equal to the sum of the principal minors of AA^T.

18. Continuing with Exercises 16 and 17, establish the following generalized Lagrangian identity:

    $\det A^T A =$ the sum of the squares of the n-rowed principal minors of A.

    Hint: Show that an n-rowed principal submatrix P of AA^T is the product $P = NN^T$ of the $n \times n$-submatrix N of A having the same row indices as P. Conclude that

    $$\det P = \det N \det N^T = (\det N)^2.$$

19. Use the result given in Exercise 18 to establish the following Lagrangian identity. For j and k summed on the range $1, 2, \ldots, m$, the equation

$$(\Sigma a_j^2)(\Sigma b_k^2) - (\Sigma a_k b_k)^2 = \Sigma_{j<k}(a_j b_k - b_j a_k)^2$$

holds for all choices of the numbers $a_1, \ldots, a_m, b_1, \ldots, b_m$. *Hint:* Let A be the $m \times 2$-matrix whose first column has $a_1, \ldots, a_m$ as its entries and whose second column has $b_1, \ldots, b_m$ as its entries. Then the left member of the identity is the determinant det $A^T A$ and the right member is the sum of the squares of the two-rowed minors of A.

4.2   EIGENVECTORS OF A SQUARE MATRIX

In the preceding section we saw that much information can be obtained about a square matrix A by the study of its eigenvalues. In this section we show that further information about A can be obtained by the study of its special vectors, called *eigenvectors*.

Recall that λ is an eigenvalue of A if and only if $|A - \lambda I| = 0$. This is true if and only if the matrix $A - \lambda I$ is singular, that is, if and only if there is a *nonzero* vector $\mathbf{x}$ such that

$$(A - \lambda I)\mathbf{x} = \mathbf{0}.$$

Multiplying out, we see that this nonzero vector $\mathbf{x}$ also satisfies

$$A\mathbf{x} = \lambda\mathbf{x}.$$

We see that an equivalent definition of an eigenvalue of A is a number λ for which there exists a nonzero vector $\mathbf{x}$ such that $A\mathbf{x} = \lambda\mathbf{x}$. This is the definition usually used for an eigenvalue λ of A. The corresponding nonzero vector $\mathbf{x}$ is called an *eigenvector* of A.

So an *eigenvector* of A is a *nonzero* vector $\mathbf{x}$ for which there is a number λ called an eigenvalue of A such that $\mathbf{x}$ and λ both satisfy the equation

$$A\mathbf{x} = \lambda\mathbf{x}.$$

One can think of the left side of this equation as expressing what happens to $\mathbf{x}$ when it is multiplied on the left side by A (i.e., "transformed by A"). The right side of the equation is simply a multiple of $\mathbf{x}$.

We see that an eigenvector $\mathbf{x}$ when multiplied on the left by its matrix A yields a multiple of $\mathbf{x}$, the multiple being the eigenvalue λ of A.

■ **EXAMPLE 1**

Let

$$A = \begin{bmatrix} 5 & 0 & 8 \\ -1 & 2 & 4 \\ 2 & 0 & 5 \end{bmatrix}.$$

In order to calculate the eigenvalues and eigenvectors of A, we begin by finding the eigenvalues. Expanding the determinant of $A - \lambda I$ by minors of its second column, we see that the characteristic polynomial $p(\lambda)$ of A is

$$p(\lambda) = |A - \lambda I| = (2 - \lambda)(\lambda - 9)(\lambda - 1).$$

It follows that the eigenvalues of A are

$$\lambda_1 = 2, \qquad \lambda_2 = 9, \qquad \lambda_3 = 1.$$

We will see that the vectors

$$\mathbf{x}_1 = \begin{bmatrix} 0 \\ 1 \\ 0 \end{bmatrix}, \qquad \mathbf{x}_2 = \begin{bmatrix} 14 \\ 2 \\ 7 \end{bmatrix}, \qquad \mathbf{x}_3 = \begin{bmatrix} -2 \\ -6 \\ 1 \end{bmatrix}$$

are eigenvectors of A corresponding to the eigenvalues $\lambda_1 = 2$, $\lambda_2 = 9$, $\lambda_3 = 1$, respectively. We proceed to calculate the eigenvectors of A. To do so, we return to the defining equation

$$(A - \lambda I)\mathbf{x} = \mathbf{0}.$$

We have

$$(A - \lambda I)\mathbf{x} = \begin{bmatrix} 5 - \lambda & 0 & 8 \\ -1 & 2 - \lambda & 4 \\ 2 & 0 & 5 - \lambda \end{bmatrix} \begin{bmatrix} x_1 \\ x_2 \\ x_3 \end{bmatrix} = \begin{bmatrix} 0 \\ 0 \\ 0 \end{bmatrix}.$$

To find a nonzero eigenvector $\mathbf{x}$ corresponding to each eigenvalue λ, we substitute the value of λ in the equation above and calculate the components of $\mathbf{x}$. In this example we must do this three times since there are three eigenvalues.

For $\lambda_1 = 2$, our equation becomes

$$(A - 2I)\mathbf{x} = \begin{bmatrix} 3 & 0 & 8 \\ -1 & 0 & 4 \\ 2 & 0 & 3 \end{bmatrix} \begin{bmatrix} x_1 \\ x_2 \\ x_3 \end{bmatrix} = \begin{bmatrix} 0 \\ 0 \\ 0 \end{bmatrix}.$$

It follows that

$$\begin{aligned} 3x_1 + 0x_2 + 8x_3 &= 0 \\ -1x_1 + 0x_2 + 4x_3 &= 0 \\ 2x_1 + 0x_2 + 3x_3 &= 0. \end{aligned}$$

The solution is $x_1 = x_3 = 0$, x_2 arbitrary. Choosing $x_2 = 1$, we obtain the eigenvector

$$\mathbf{x}_1 = \begin{bmatrix} 0 \\ 1 \\ 0 \end{bmatrix}.$$

If we let $x_2 = a$, where a is any real number, all eigenvectors corresponding to the eigenvalue $\lambda = 2$ are of the form

$$\mathbf{x} = \begin{bmatrix} 0 \\ a \\ 0 \end{bmatrix} = a \begin{bmatrix} 0 \\ 1 \\ 0 \end{bmatrix}.$$

For $\lambda_2 = 9$, our equation becomes

$$\begin{bmatrix} -4 & 0 & 8 \\ -1 & -7 & 4 \\ 2 & 0 & -4 \end{bmatrix} \begin{bmatrix} x_1 \\ x_2 \\ x_3 \end{bmatrix} = \begin{bmatrix} 0 \\ 0 \\ 0 \end{bmatrix}.$$

We have

$$\begin{aligned} -4x_1 + 0x_2 + 8x_3 &= 0 \\ -1x_1 - 7x_2 + 4x_3 &= 0 \\ 2x_1 + 0x_2 - 4x_3 &= 0. \end{aligned}$$

Solving, we get $x_1 = 2a$, $x_2 = \frac{2}{7}a$, $x_3 = a$, where a is any real number. Choosing $a = 7$ yields the eigenvector

$$\mathbf{x}_2 = \begin{bmatrix} 14 \\ 2 \\ 7 \end{bmatrix}.$$

All eigenvectors corresponding to $\lambda = 9$ are of the form

$$\mathbf{x} = \begin{bmatrix} 2a \\ \frac{2}{7}a \\ a \end{bmatrix} = a \begin{bmatrix} 2 \\ \frac{2}{7} \\ 1 \end{bmatrix}.$$

Finally, for $\lambda_3 = 1$, we have

$$\begin{bmatrix} 4 & 0 & 8 \\ -1 & 1 & 4 \\ 2 & 0 & 4 \end{bmatrix} \begin{bmatrix} x_1 \\ x_2 \\ x_3 \end{bmatrix} = \begin{bmatrix} 0 \\ 0 \\ 0 \end{bmatrix},$$

which yields the solution $x_1 = -2a$, $x_2 = -6a$, $x_3 = a$ for any real number a. Choosing $a = 1$, we get as a specific eigenvector

$$\mathbf{x} = \begin{bmatrix} -2 \\ -6 \\ 1 \end{bmatrix}.$$

All of the eigenvectors corresponding to $\lambda_3 = 1$ are of the form

$$\mathbf{x} = \begin{bmatrix} -2a \\ -6a \\ a \end{bmatrix} = a \begin{bmatrix} -2 \\ -6 \\ 1 \end{bmatrix}.$$

In applications, we will usually only need one eigenvector corresponding to each eigenvalue. However, as we have seen above, once we have obtained one eigenvector corresponding to a particular eigenvalue, any nonzero multiple of this eigenvector is again an eigenvector corresponding to the same eigenvalue. In general,

---

If $\mathbf{x}$ is an eigenvector of A corresponding to an eigenvalue λ, then $c\mathbf{x}$ is an eigenvector of A corresponding to the same eigenvalue λ, for all $c \neq 0$.

This follows from the defining equation since

$$A(c\mathbf{x}) = c(A\mathbf{x}) = c(\lambda\mathbf{x}) = \lambda(c\mathbf{x}).$$

It turns out that when λ is a nonrepeating eigenvalue, then all eigenvectors of A corresponding to λ will always be of the form $c\mathbf{x}$, where $\mathbf{x}$ is an eigenvector of A and $c \neq 0$.

In finding the eigenvectors of a matrix A, we have seen that we must substitute each eigenvalue of A into the matrix equation $(A - \lambda I)\mathbf{x} = \mathbf{0}$, which involves solving a system of linear equations in each case.

In Section 3.4 we learned a method for solving the homogeneous equation $BX = 0$. This method consists of reducing B to its row echelon form. This method always works. It is perhaps the most effective way for solving $BX = 0$. However, there are other methods that frequently require less computation. For example, one might guess at a solution and then verify it. In the case A is $n \times n$, and $B = A - \lambda I$, the rank of B is often $n - 1$. In this event we can use the properties of the classical adjoint of $A - \lambda I$ to assist us in finding the solution of $(A - \lambda I)X = 0$. This technique for computing the eigenvectors of a matrix A often requires less computation for small matrices. We proceed as follows: Recall the relationship between a matrix A and its adjoint, adj A:

$$A(\text{adj } A) = (\det A)I.$$

Since this relationship holds for all $n \times n$-matrices A, we may replace A by the matrix $A - \lambda I$ above to get

$$(A - \lambda I) \text{ adj } (A - \lambda I) = \det (A - \lambda I)I.$$

If λ is an eigenvalue of A, then det $(A - \lambda I) = 0$. It follows that when λ is an eigenvalue of A,

$$(A - \lambda I) \text{ adj } (A - \lambda I) = 0.$$

The equation above is of the form

$$(A - \lambda I)X = 0.$$

Here $X = $ adj $(A - \lambda I)$ is an $n \times n$-*matrix*. For the case $n = 3$, with C_1, C_2, and C_3 as the columns of adj $(A - \lambda I)$, the equation

$$(A - \lambda I) \text{ adj } (A - \lambda I) = 0$$

becomes

$$(A - \lambda I)[C_1 \quad C_2 \quad C_3] = [(A - \lambda I)C_1 \quad (A - \lambda I)C_2 \quad (A - \lambda I)C_3]$$
$$= [0 \quad 0 \quad 0],$$

so that

$$(A - \lambda I)C_1 = 0, \quad (A - \lambda I)C_2 = 0, \quad (A - \lambda I)C_3 = 0.$$

The column C_1, if nonzero, is therefore an eigenvector of A corresponding to the eigenvalue λ of A. Similarly, C_2 and C_3, if nonzero, are eigenvectors of A corresponding to λ. It can be shown that if one of these columns is nonzero, the other columns are multiples of it. To find an eigenvector of A corresponding to an eigenvalue λ, look for a nonzero column of adj $(A - \lambda I)$. This result holds for any $n \times n$-matrix.

■ **EXAMPLE 2**

Consider the example already encountered,

$$A = \begin{bmatrix} 5 & 0 & 8 \\ -1 & 2 & 4 \\ 2 & 0 & 5 \end{bmatrix}, \quad A - \lambda I = \begin{bmatrix} 5 - \lambda & 0 & 8 \\ -1 & 2 - \lambda & 4 \\ 2 & 0 & 5 - \lambda \end{bmatrix}.$$

As was seen earlier, the eigenvalues of A are $\lambda_1 = 2$, $\lambda_2 = 9$, and $\lambda_3 = 1$. The matrix adj $(A - \lambda I)$ is seen to be

$$\begin{bmatrix} \begin{vmatrix} 2 - \lambda & 4 \\ 0 & 5 - \lambda \end{vmatrix} & -\begin{vmatrix} 0 & 8 \\ 0 & 5 - \lambda \end{vmatrix} & \begin{vmatrix} 0 & 8 \\ 2 - \lambda & 4 \end{vmatrix} \\ -\begin{vmatrix} -1 & 4 \\ 2 & 5 - \lambda \end{vmatrix} & \begin{vmatrix} 5 - \lambda & 8 \\ 2 & 5 - \lambda \end{vmatrix} & -\begin{vmatrix} 5 - \lambda & 8 \\ -1 & 4 \end{vmatrix} \\ \begin{vmatrix} -1 & 2 - \lambda \\ 2 & 0 \end{vmatrix} & -\begin{vmatrix} 5 - \lambda & 0 \\ 2 & 0 \end{vmatrix} & \begin{vmatrix} 5 - \lambda & 0 \\ -1 & 2 - \lambda \end{vmatrix} \end{bmatrix}.$$

Since each nonzero column of adj $(A - \lambda I)$ is an eigenvector of A when λ is an eigenvalue, we consider the first column,

$$C_1 = \begin{bmatrix} (2 - \lambda)(5 - \lambda) \\ 13 - \lambda \\ -4 + 2\lambda \end{bmatrix}.$$

To obtain an eigenvector corresponding to a particular eigenvector of A, all that is needed is to substitute the value of λ into the expression for C_1 above.

When $\lambda_1 = 2$, a corresponding eigenvector is

$$\mathbf{x}_1 = \begin{bmatrix} (2 - 2)(5 - 2) \\ 13 - 2 \\ -4 + 2(2) \end{bmatrix} = \begin{bmatrix} 0 \\ 11 \\ 0 \end{bmatrix} = 11 \begin{bmatrix} 0 \\ 1 \\ 0 \end{bmatrix}.$$

When $\lambda_2 = 9$, a corresponding eigenvector is

$$\mathbf{x}_2 = \begin{bmatrix} (2 - 9)(5 - 9) \\ 13 - 9 \\ -4 + 2(9) \end{bmatrix} = \begin{bmatrix} 28 \\ 4 \\ 14 \end{bmatrix} = 2 \begin{bmatrix} 14 \\ 2 \\ 7 \end{bmatrix}.$$

When $\lambda_3 = 1$, a corresponding eigenvector is

$$\mathbf{x}_3 = \begin{bmatrix} (2 - 1)(5 - 1) \\ 13 - 1 \\ -4 + 2(1) \end{bmatrix} = \begin{bmatrix} 4 \\ 12 \\ -2 \end{bmatrix} = -2 \begin{bmatrix} -2 \\ -6 \\ 1 \end{bmatrix}.$$

The reader may wonder what would have occurred had we chosen the second column, C_2, of adj $(A - \lambda I)$ rather than the first column, C_1. Since

$$C_2 = \begin{bmatrix} 0 \\ (5 - \lambda)^2 - 16 \\ 0 \end{bmatrix},$$

we immediately observe that for $\lambda_1 = 2$, a corresponding eigenvector is

$$\mathbf{x}_1 = \begin{bmatrix} 0 \\ -7 \\ 0 \end{bmatrix},$$

a multiple of $\begin{bmatrix} 0 \\ 1 \\ 0 \end{bmatrix}$.

However, when $\lambda_2 = 9, C_2 = \begin{bmatrix} 0 \\ 0 \\ 0 \end{bmatrix}$, and when $\lambda_3 = 1$,

$$C_2 = \begin{bmatrix} 0 \\ 0 \\ 0 \end{bmatrix}.$$

In these last two cases we do not get eigenvectors since eigenvectors must be nonzero.

As we have seen, all the columns of adj $(A - \lambda I)$ are not needed in order to find an eigenvector of A corresponding to an eigenvalue λ. In actual computation it often saves time to find just one of the columns of adj $(A - \lambda I)$.

We can now state the general result.

> Let λ be an eigenvalue of A. If adj $(A - \lambda I)$ is not zero, a nonzero column **x** of adj $(A - \lambda I)$ is a corresponding eigenvector of A.

This result is useful computationally only for "small" matrices, because of the excessive computations involved in finding a column of adj $(A - \lambda I)$ in higher-dimensional cases.

■ **EXAMPLE 3** _____

We now consider a case in which the technique used above for finding the eigenvectors of A is not effective because adj $(A - \lambda I) = 0$ for all of the eigenvalues λ of A. This is the situation when

$$
A = \begin{bmatrix} 1 & 0 & 0 & 1 \\ 0 & 1 & 1 & 0 \\ 0 & 1 & 1 & 0 \\ 1 & 0 & 0 & 1 \end{bmatrix}, \quad A - \lambda I = \begin{bmatrix} 1 - \lambda & 0 & 0 & 1 \\ 0 & 1 - \lambda & 1 & 0 \\ 0 & 1 & 1 - \lambda & 0 \\ 1 & 0 & 0 & 1 - \lambda \end{bmatrix}.
$$

The four eigenvalues of A are 0, 0, 2, and 2. When $\lambda = 0$, we have $A - \lambda I = A$. It is easily verified that adj $A = 0$. Similarly, adj $(A - 2I) = 0$. Consequently, the classical adjoint of A cannot be used for finding *any* of the eigenvectors of A.

_____

In the example above, the matrix A had four eigenvalues: 0, 0, 2, and 2. We see that 0 and 2 are each eigenvalues of multiplicity two. In general, if λ is a multiple root of $|A - \lambda I| = 0$, then normally adj $(A - \lambda I)$ is identically zero, so that this technique for finding eigenvectors is useful mainly when λ is a simple (nonrepeating) eigenvalue of A.

■ **EXAMPLE 4** _____

We consider another example of a matrix A with multiple eigenvalues.

$$
A = \begin{bmatrix} 1 & 1 \\ 0 & 1 \end{bmatrix}, \quad A - \lambda I = \begin{bmatrix} 1 - \lambda & 1 \\ 0 & 1 - \lambda \end{bmatrix}.
$$

The eigenvalues are $\lambda = 1$ and 1. We see that 1 is a multiple eigenvalue. To find eigenvectors of A corresponding to $\lambda = 1$, we look at

$$(A - \lambda I)\mathbf{x} = \begin{bmatrix} 0 & 1 \\ 0 & 0 \end{bmatrix} \begin{bmatrix} x_1 \\ x_2 \end{bmatrix} = \begin{bmatrix} 0 \\ 0 \end{bmatrix}.$$

It follows that $x_2 = 0$, x_1 arbitrary. We see that if $\mathbf{x}$ is an eigenvector of A corresponding to $\lambda = 1$, then $\mathbf{x}$ is of the form $a\begin{bmatrix} 1 \\ 0 \end{bmatrix}$ where a is any nonzero real number. In this example, all eigenvectors of A are multiples of the eigenvector $\begin{bmatrix} 1 \\ 0 \end{bmatrix}$. There is just one linearly independent eigenvector corresponding to the eigenvalue $\lambda = 1$.

■ **EXAMPLE 5**

Let

$$A = \begin{bmatrix} 1 & 0 & 0 \\ 0 & 1 & 0 \\ 0 & 0 & 2 \end{bmatrix}, \qquad A - \lambda I = \begin{bmatrix} 1 - \lambda & 0 & 0 \\ 0 & 1 - \lambda & 0 \\ 0 & 0 & 2 - \lambda \end{bmatrix}.$$

Because $\det(A - \lambda I) = (1 - \lambda)^2(2 - \lambda)$, it follows that the eigenvalue $\lambda = 1$ is a double eigenvalue of A. To find eigenvectors of A corresponding to $\lambda = 1$, we compute $(A - 1I)\mathbf{x} = \mathbf{0}$.

$$(A - 1I)\mathbf{x} = \begin{bmatrix} 0 & 0 & 0 \\ 0 & 0 & 0 \\ 0 & 0 & 1 \end{bmatrix} \begin{bmatrix} x_1 \\ x_2 \\ x_3 \end{bmatrix} = \begin{bmatrix} 0 \\ 0 \\ 0 \end{bmatrix}$$

yields $x_3 = 0$. There is no restriction on x_1 and x_2.

The eigenvectors of A corresponding to $\lambda = 1$ are of the form

$$\mathbf{x} = \begin{bmatrix} a \\ b \\ 0 \end{bmatrix},$$

where a and b are any real numbers. We write this as

$$\mathbf{x} = \begin{bmatrix} a \\ b \\ 0 \end{bmatrix} = a \begin{bmatrix} 1 \\ 0 \\ 0 \end{bmatrix} + b \begin{bmatrix} 0 \\ 1 \\ 0 \end{bmatrix}.$$

Since $\begin{bmatrix} 1 \\ 0 \\ 0 \end{bmatrix}$ and $\begin{bmatrix} 0 \\ 1 \\ 0 \end{bmatrix}$ are each eigenvectors of A corresponding to $\lambda = 1$, and since they are not multiples of each other, we have found two linearly independent eigenvectors corresponding to a repeated eigenvalue.

The reader should refer back to the preceding example, where this was not possible.

In the situation when the eigenvectors correspond to the same eigenvalue λ, we have the following result:

> Let $\mathbf{y}$ and $\mathbf{z}$ be eigenvectors of A corresponding to an eigenvalue λ. Then every nonzero linear combination $\mathbf{x} = a\mathbf{y} + b\mathbf{z}$ of $\mathbf{y}$ and $\mathbf{z}$ is an eigenvector of A corresponding to λ.

In the general case this follows because the relations $A\mathbf{y} = \lambda\mathbf{y}$ and $A\mathbf{z} = \lambda\mathbf{z}$ imply the relations

$$aA\mathbf{y} + bA\mathbf{z} = a\lambda\mathbf{y} + b\lambda\mathbf{z} = \lambda(a\mathbf{y} + b\mathbf{z})$$

so that

$$A(a\mathbf{y} + b\mathbf{z}) = \lambda(a\mathbf{y} + b\mathbf{z}).$$

In the preceding examples we have seen that if an $n \times n$-matrix A has repeating eigenvalues, we may or may not get n linearly independent eigenvectors. This topic is a theme of the next section.

As a final result, we have the following:

> If A is nonsingular (invertible), then A and A^{-1} have the same eigenvectors. If λ is an eigenvalue of A, then $1/\lambda$ is an eigenvalue of A^{-1}.

For if $A\mathbf{x} = \lambda\mathbf{x}$, multiplying both sides by A^{-1} yields $A^{-1}A\mathbf{x} = \lambda A^{-1}\mathbf{x}$, so that

$$\mathbf{x} = \lambda A^{-1}\mathbf{x}.$$

Dividing by λ, we have

$$A^{-1}\mathbf{x} = \frac{1}{\lambda}\mathbf{x}.$$

The preceding result can also be stated as follows:

> If $\mathbf{x}$ is an eigenvector of a nonsingular (invertible) matrix A and λ is its corresponding eigenvalue, then:
>
> 1. $\lambda \neq 0$.
> 2. $\mathbf{x}$ is an eigenvector of A^{-1} with $1/\lambda$ as its corresponding eigenvalue.

Exercises

1. Find real eigenvectors, if any, of the following matrices. Give the corresponding eigenvalue.

$$H = \begin{bmatrix} 5 & 4 \\ 4 & 5 \end{bmatrix}, \qquad K = \begin{bmatrix} 0 & 2 \\ -3 & 5 \end{bmatrix},$$

$$L = \begin{bmatrix} 4 & 3 \\ -3 & 4 \end{bmatrix}, \qquad M = \begin{bmatrix} 12 & -5 \\ 5 & 12 \end{bmatrix},$$

$$N = \begin{bmatrix} 5 & 4 & 0 \\ 4 & 5 & 0 \\ 2 & -3 & 9 \end{bmatrix}, \qquad P = \begin{bmatrix} 0 & -1 & 0 \\ 0 & 0 & -1 \\ 1 & 0 & 0 \end{bmatrix},$$

$$Q = \begin{bmatrix} 0 & -1 & 0 \\ 0 & 0 & -1 \\ -9 & -9 & 1 \end{bmatrix}, \qquad R = \begin{bmatrix} 0 & 1 & 0 & 0 \\ -3 & 4 & 0 & 0 \\ 5 & 6 & 5 & 1 \\ -1 & 1 & -9 & -1 \end{bmatrix},$$

$$S = \begin{bmatrix} 0 & 1 & 0 & 0 \\ 0 & 0 & 1 & 0 \\ 1 & 0 & 0 & 0 \\ 3 & 2 & 4 & 5 \end{bmatrix}, \qquad T = \begin{bmatrix} 0 & 0 & 0 & 2 \\ 0 & 0 & 3 & 0 \\ 0 & 3 & 0 & 0 \\ 2 & 0 & 0 & 0 \end{bmatrix}.$$

2. Show that 2, -2, 4, and -4 are the eigenvalues of the symmetric matrix

$$A = \begin{bmatrix} 0 & 3 & 0 & -1 \\ 3 & 0 & -1 & 0 \\ 0 & -1 & 0 & 3 \\ -1 & 0 & 3 & 0 \end{bmatrix}.$$

   Find the corresponding eigenvectors of A.

3. Let B be a nonsingular matrix. Let $\mathbf{x}$ be an eigenvector of A corresponding to an eigenvalue λ. Show that λ is an eigenvalue of $C = BAB^{-1}$ and that $\mathbf{y} = B\mathbf{x}$ is a corresponding eigenvector of C.

4. If $\mathbf{x}$ is an eigenvector of A, show that its corresponding eigenvalue λ is given by the formula

$$\lambda = \frac{\mathbf{x}^T A \mathbf{x}}{\mathbf{x}^T \mathbf{x}}.$$

5. Let λ and μ be distinct eigenvalues of A. Let $\mathbf{x}$ be an eigenvector of A corresponding to λ, and let $\mathbf{u}$ be an eigenvector of A^T corresponding to μ. Show that $\mathbf{x}$ and $\mathbf{u}$ are orthogonal. *Hint:* Use the relation

$$\mathbf{u}^T A \mathbf{x} = \lambda \mathbf{u}^T \mathbf{x} = \mu \mathbf{u}^T \mathbf{x}.$$

   Since $\lambda \neq \mu$, we have $\mathbf{u}^T \mathbf{x} = 0$.

6. Let A be a symmetric matrix. Let $\mathbf{x}$ and $\mathbf{y}$ be eigenvectors of A corresponding to distinct eigenvalues λ and μ. Show that $\mathbf{x}$ and $\mathbf{y}$ are orthogonal. *Hint:* Here $A^T = A$. Use the result given in Exercise 5.

7. Consider the vectors

$$\mathbf{u}_1 = \begin{bmatrix} 1 \\ 0 \\ 1 \end{bmatrix}, \qquad \mathbf{u}_2 = \begin{bmatrix} 0 \\ 1 \\ 1 \end{bmatrix}, \qquad \mathbf{v}_1 = \begin{bmatrix} 1 \\ -1 \\ 1 \end{bmatrix}, \qquad \mathbf{v}_2 = \begin{bmatrix} 0 \\ 2 \\ 0 \end{bmatrix}.$$

(a) Show that $\mathbf{u}_1$ is orthogonal to $\mathbf{v}_2$ and $\mathbf{u}_2$ is orthogonal to $\mathbf{v}_1$.

(b) Show that $\mathbf{u}_1$ and $\mathbf{u}_2$ are eigenvectors of the matrix

$$A = \mathbf{u}_1 \mathbf{v}_1^T + \mathbf{u}_2 \mathbf{v}_2^T$$

(c) What are the corresponding eigenvalues of A?

(d) Show that $\mathbf{v}_1$ and $\mathbf{v}_2$ are eigenvectors of A^T.

(e) What are the corresponding eigenvalues of A^T?

(f) Show that

$$\mathbf{u}_3 = \begin{bmatrix} 1 \\ 0 \\ -1 \end{bmatrix}, \qquad \mathbf{v}_3 = \begin{bmatrix} 1 \\ 1 \\ -1 \end{bmatrix}$$

are eigenvectors of A and A^T, respectively. Find the corresponding eigenvalue.

8. Consider two pairs $\mathbf{u}_1$, $\mathbf{u}_2$, and $\mathbf{v}_1$, $\mathbf{v}_2$ of linearly independent vectors having the property that

$$\mathbf{v}_1^T \mathbf{u}_2 = 0, \qquad \mathbf{v}_2^T \mathbf{u}_1 = 0.$$

(a) Show that $\mathbf{u}_1$ and $\mathbf{u}_2$ are eigenvectors of

$$A = \mathbf{u}_1 \mathbf{v}_1^T + \mathbf{u}_2 \mathbf{v}_2^T$$

with $\lambda_1 = \mathbf{v}_1^T \mathbf{u}_1$ and $\lambda_2 = \mathbf{v}_2^T \mathbf{u}_2$ as corresponding eigenvalues.

(b) Show that $\mathbf{v}_1$ and $\mathbf{v}_2$ are eigenvectors of A^T. What are the corresponding eigenvalues?

(c) Show that a nonzero vector $\mathbf{x}$ orthogonal to $\mathbf{v}_1$ and $\mathbf{v}_2$ is an eigenvector of A with $\lambda = 0$ as the corresponding eigenvalue.

(d) Show that a nonzero vector $\mathbf{y}$ orthogonal to $\mathbf{u}_1$ and $\mathbf{u}_2$ is an eigenvector of A^T with $\lambda = 0$ as the corresponding eigenvalue.

9. Verify the following statements for an $n \times n$-matrix A.

(a) Suppose that $\mathbf{x}$ is a nonzero solution of $A\mathbf{x} = 0$. Show that $\mathbf{x}$ is an eigenvector of A with $\lambda = 0$ as the corresponding eigenvalue.

(b) Suppose that A is of nullity $q > 0$. Recall that there is an $n \times q$-matrix X of rank q such that $AX = 0$. Show that the columns of X are eigenvectors of A. What are the corresponding eigenvalues of A?

(c) Continuing with statement (b), let C be a nonsingular $q \times q$-matrix. Show that the columns of $Y = XC$ are eigenvectors of A.

10. Let A be a nonsingular $n \times n$-matrix.

(a) Let $\mathbf{x}$ be an eigenvector of A. Why is $\mathbf{y} = A\mathbf{x}$ an eigenvector of A? Why is

$\mathbf{z} = A^2\mathbf{x}$ also an eigenvector of A? Is $A^k\mathbf{x}$ an eigenvector of A for every positive integer k?

(b) Let X be an $n \times r$-matrix whose columns are eigenvectors of A. Let D be a nonsingular $r \times r$ diagonal matrix. Show that the columns of $Y = XD$ are eigenvectors of A. Show also that the columns of $Z = AX$ are eigenvectors of A. If k is a positive integer, are the columns of A^kX eigenvectors of A?

11. Consider the 4×4-matrices

$$A = \begin{bmatrix} 0 & 1 & 0 & 0 \\ 0 & 0 & 1 & 0 \\ 0 & 0 & 0 & 1 \\ 0 & 0 & 0 & 0 \end{bmatrix}, \quad B = \begin{bmatrix} 0 & 0 & 1 & 0 \\ 0 & 0 & 0 & 1 \\ 0 & 0 & 0 & 0 \\ 0 & 0 & 0 & 0 \end{bmatrix},$$

$$C = \begin{bmatrix} 0 & 0 & 0 & 1 \\ 0 & 0 & 0 & 0 \\ 0 & 0 & 0 & 0 \\ 0 & 0 & 0 & 0 \end{bmatrix}.$$

(a) Show that $B = A^2$ and that $C = A^3$.

(b) Show that $\lambda = 0$ is the only eigenvalue of these matrices.

(c) For each matrix find a maximal set of linearly independent eigenvectors of the matrix.

(d) Show that an eigenvector of A is also an eigenvector of B and of C.

(e) Show that B has an eigenvector which is not an eigenvector of A.

(f) Show that C has an eigenvector that is neither an eigenvector of B nor of A.

(g) What are the eigenvectors of A^4?

12. Consider the 3×3-matrices

$$A = \begin{bmatrix} 1 & 1 & 0 \\ 0 & 1 & 1 \\ 0 & 0 & 1 \end{bmatrix}, \quad B = \begin{bmatrix} 1 & 2 & 1 \\ 0 & 1 & 2 \\ 0 & 0 & 1 \end{bmatrix}, \quad C = \begin{bmatrix} 1 & 3 & 3 \\ 0 & 1 & 3 \\ 0 & 0 & 1 \end{bmatrix}.$$

(a) Show that $B = A^2$ and the $C = A^3$.

(b) Show that $\lambda = 1$ is the only eigenvalue of these matrices.

(c) Show that these matrices have the same eigenvectors. What are they?

13. Let $\mathbf{x}$ be an eigenvector of A corresponding to an eigenvalue λ. Show that λ^2 is an eigenvalue of A^2 with $\mathbf{x}$ as a corresponding eigenvector. Is λ^3 an eigenvalue of A^3 with $\mathbf{x}$ as a corresponding eigenvector? Generalize to A^k. *Hint:* Use the relation $A^2\mathbf{x} = \lambda A\mathbf{x} = \lambda^2\mathbf{x}$.

14. Let $p(\lambda)$ be the characteristic polynomial of an $n \times n$-matrix A. Let $q(\lambda)$ be a second polynomial. We have $p(A) = 0$.

(a) Show that if $A\mathbf{x}_1 = \lambda_1\mathbf{x}_1$, then $q(A)\mathbf{x}_1 = q(\lambda_1)\mathbf{x}_1$.

(b) Conclude that if λ_1 is an eigenvalue of A then $q(\lambda_1)$ is an eigenvalue of $q(A)$ with the same corresponding eigenvector $\mathbf{x}_1$.

(c) Show that if $p(\lambda)$ and $q(\lambda)$ are relatively prime (i.e., have no common nonconstant factor), then $q(A)$ is nonsingular.

(d) Show that if λ_1 is a common zero of $p(\lambda)$ and $q(\lambda)$, then $q(A)$ is singular.

15. Continue with Exercise 14.

(e) Let $r(\lambda)$ be a GCD (greatest common divisor) of $p(\lambda)$ and $q(\lambda)$. Show that if $q(A) = 0$, then $r(A) = 0$.

(f) Let $m(\lambda)$ be a polynomial of lowest degree such that $m(A) = 0$. Show that if $q(A) = 0$, then $m(\lambda)$ divides $q(\lambda)$.

16. Show that the eigenvalues of A^TA are nonnegative. *Hint:* Use the relation $x^TA^TAx = \lambda x^Tx \geq 0$, which holds for an eigenvalue λ of A^TA and its corresponding eigenvector **x**.

17. A symmetric matrix A is said to be positive definite if $x^TAx > 0$ except when $x = 0$. Use the result given in Exercise 4 to show that the eigenvalues of A are positive.

18. In the two-dimensional case interpret a vector

$$\mathbf{x} = \begin{bmatrix} x \\ y \end{bmatrix}$$

to be the point (x, y) in the xy-plane. Let A be the matrix

$$A = \begin{bmatrix} 4 & 0 \\ 0 & 9 \end{bmatrix}.$$

Show that the equation

$$\mathbf{x}^T A \mathbf{x} = 36$$

represents the ellipse

$$4x^2 + 9y^2 = 36.$$

How are the eigenvectors and eigenvalues of A related to the major and minor axes of the ellipse?

19. Continuing with the notations used in Exercise 10, let A be the matrix

$$A = \begin{bmatrix} 4 & 0 \\ 0 & -9 \end{bmatrix}.$$

Show that the equation

$$\mathbf{x}^T A \mathbf{x} = 36$$

represents the hyperbola $4x^2 - 9y^2 = 36$. How are the eigenvectors and eigenvalues of A related to the axes of the hyperbola?

20. In the three-dimensional case, identify a vector

$$\mathbf{x} = \begin{bmatrix} x \\ y \\ z \end{bmatrix}$$

with the point (x, y, z) in xyz-space. Let A be the matrix

$$A = \begin{bmatrix} 1 & 0 & 0 \\ 0 & 4 & 0 \\ 0 & 0 & 9 \end{bmatrix}.$$

Show that the equation $x^T A x = 36$ represents the ellipsoid $x^2 + 4y^2 + 9z^2 = 36$. How are the eigenvalues and eigenvectors of A related to the axes of the ellipsoid?

4.3   PROPERTIES OF EIGENVECTORS

In this section we present certain fundamental properties of the eigenvectors of an $n \times n$-matrix A. We continue to restrict ourselves to real matrices and vectors unless otherwise specified. Most of the results are valid in the complex case. The modifications that need to be made in the complex case are described at the end of the section. We shall see that in the theory that follows, the linear independence of eigenvectors will play a key role.

Consider the matrix

$$A = \begin{bmatrix} 2 & 2 \\ 0 & 3 \end{bmatrix}.$$

Its distinct eigenvalues are $\lambda = 2$ and $\mu = 3$. The column vectors

$$\mathbf{x} = \begin{bmatrix} 1 \\ 0 \end{bmatrix} \quad \text{and} \quad \mathbf{y} = \begin{bmatrix} 2 \\ 1 \end{bmatrix}$$

are eigenvectors of A corresponding to λ and μ, respectively, since

$$\begin{bmatrix} 2 & 2 \\ 0 & 3 \end{bmatrix}\begin{bmatrix} 1 \\ 0 \end{bmatrix} = 2\begin{bmatrix} 1 \\ 0 \end{bmatrix} \quad \text{and} \quad \begin{bmatrix} 2 & 2 \\ 0 & 3 \end{bmatrix}\begin{bmatrix} 2 \\ 1 \end{bmatrix} = 3\begin{bmatrix} 2 \\ 1 \end{bmatrix}.$$

These eigenvectors $\mathbf{x}$ and $\mathbf{y}$ are also linearly independent since

$$c_1\begin{bmatrix} 1 \\ 0 \end{bmatrix} + c_2\begin{bmatrix} 2 \\ 1 \end{bmatrix} = \begin{bmatrix} 0 \\ 0 \end{bmatrix}$$

yields $c_1 = c_2 = 0$.

This example illustrates the following property of eigenvectors when the corresponding eigenvalues are distinct.

PROPERTY 1. If $\mathbf{x}$ and $\mathbf{y}$ are eigenvectors of A corresponding to distinct eigenvalues λ and μ, then $\mathbf{x}$ and $\mathbf{y}$ are linearly independent.

For suppose that $\mathbf{x}$ and $\mathbf{y}$ are linearly dependent. Then $\mathbf{y}$ is a multiple $\mathbf{y} = m\mathbf{x}$ of $\mathbf{x}$. Using the two relations

$$A\mathbf{y} = \mu\mathbf{y} = \mu m\mathbf{x}, \qquad A\mathbf{y} = (Am\mathbf{x}) = mA\mathbf{x} = m\lambda\mathbf{x},$$

we find that $m\mu = m\lambda$. This is impossible because λ and μ are distinct and m is not zero. It follows that $\mathbf{x}$ and $\mathbf{y}$ are linearly independent, as was to be shown.

As a generalization of Property 1 we have:

PROPERTY 2. Let $\lambda_1, \lambda_2, \ldots, \lambda_q$ be distinct eigenvalues of a matrix A and let $\mathbf{x}_1, \mathbf{x}_2, \ldots, \mathbf{x}_q$ be corresponding eigenvectors of A. The eigenvectors $\mathbf{x}_1, \mathbf{x}_2, \ldots, \mathbf{x}_q$ are linearly independent.

We proceed to show it true for $q = 3$; that is, if λ_1, λ_2, and λ_3 are distinct eigenvalues of A and $\mathbf{x}_1, \mathbf{x}_2$, and $\mathbf{x}_3$ are their corresponding eigenvectors, then $\mathbf{x}_1, \mathbf{x}_2$, and $\mathbf{x}_3$ are linearly independent. Since the result holds for $q = 2$ by Property 1, we know from the start that $\mathbf{x}_1$ and $\mathbf{x}_2$ are linearly independent. Arguing by contradiction, suppose that $\mathbf{x}_3$ is linearly dependent on $\mathbf{x}_1$ and $\mathbf{x}_2$. Then there exist constants c_1 and c_2 such that

$$\mathbf{x}_3 = c_1\mathbf{x}_1 + c_2\mathbf{x}_2.$$

It follows that

$$A\mathbf{x}_3 = A(c_1\mathbf{x}_1 + c_2\mathbf{x}_2) = c_1(A\mathbf{x}_1) + c_2(A\mathbf{x}_2) = c_1\lambda_1\mathbf{x}_1 + c_2\lambda_2\mathbf{x}_2.$$

Furthermore, computing $A\mathbf{x}_3$ still another way, we have

$$A\mathbf{x}_3 = \lambda_3\mathbf{x}_3 = \lambda_3(c_1\mathbf{x}_1 + c_2\mathbf{x}_2) = c_1\lambda_3\mathbf{x}_1 + c_2\lambda_3\mathbf{x}_2.$$

Equating the two expressions of $A\mathbf{x}_3$ yields

$$c_1\lambda_1\mathbf{x}_1 + c_2\lambda_2\mathbf{x}_2 = c_1\lambda_3\mathbf{x}_1 + c_2\lambda_3\mathbf{x}_2.$$

Bringing everything to one side, we get

$$c_1(\lambda_1 - \lambda_3)\mathbf{x}_1 + c_2(\lambda_2 - \lambda_3)\mathbf{x}_2 = \mathbf{0}.$$

Since $\mathbf{x}_1$ and $\mathbf{x}_2$ were assumed to be linearly independent,

$$c_1(\lambda_1 - \lambda_3) = 0 \quad \text{and} \quad c_2(\lambda_2 - \lambda_3) = 0.$$

But λ_1, λ_2, and λ_3 are all distinct (i.e., $\lambda_1 \neq \lambda_3$ and $\lambda_2 \neq \lambda_3$). Therefore, $c_1 = c_2 = 0$. We see that $\mathbf{x}_3 = \mathbf{0}$, which cannot be the case since $\mathbf{x}_3$ was an eigenvector of A. Therefore, $\mathbf{x}_1, \mathbf{x}_2$, and $\mathbf{x}_3$ must be linearly independent. Assuming that Property 2 is true when $q = 2$, we have seen that Property 2 is true when $q = 3$. This same inductive reasoning can now be continued to show that the $q = 3$ case implies the $q = 4$ case, and so on.

■ **EXAMPLE 1**

The 3×3 matrix

$$A = \begin{bmatrix} 2 & 0 & 1 \\ -1 & 2 & 0 \\ 1 & 0 & 2 \end{bmatrix}$$

has distinct eigenvalues $\lambda_1 = 1$, $\lambda_2 = 2$, and $\lambda_3 = 3$. The corresponding eigenvectors are the column vectors $\mathbf{x}_1$, $\mathbf{x}_2$, and $\mathbf{x}_3$ of the matrix

$$X = [\mathbf{x}_1 \quad \mathbf{x}_2 \quad \mathbf{x}_3] = \begin{bmatrix} 1 & 0 & 1 \\ 1 & 1 & -1 \\ -1 & 0 & 1 \end{bmatrix}.$$

To verify this result, we compute the product AX, thereby obtaining the relation

$$AX = A[\mathbf{x}_1 \quad \mathbf{x}_2 \quad \mathbf{x}_3] = \begin{bmatrix} 1 & 0 & 3 \\ 1 & 2 & -3 \\ -1 & 0 & 3 \end{bmatrix} = [\mathbf{x}_1 \quad 2\mathbf{x}_2 \quad 3\mathbf{x}_3].$$

From this relation we conclude that

$$A\mathbf{x}_1 = \mathbf{x}_1, \qquad A\mathbf{x}_2 = 2\mathbf{x}_2, \qquad A\mathbf{x}_3 = 3\mathbf{x}_3$$

and hence that $\mathbf{x}_1$, $\mathbf{x}_2$, and $\mathbf{x}_3$ are eigenvectors of A with $\lambda_1 = 1$, $\lambda_2 = 2$, and $\lambda_3 = 3$ as corresponding eigenvalues. Because these eigenvalues are distinct, the eigenvectors $\mathbf{x}_1$, $\mathbf{x}_2$, and $\mathbf{x}_3$ are linearly independent, by Property 2. The matrix X whose columns are the eigenvectors of A therefore is nonsingular and hence invertible.

As we shall see in the next section and in Chapter 5, a given matrix A often is in such a form that it does not easily yield information about its properties. However, a "restructuring" of the matrix can sometimes simplify the form of the matrix without losing many of its desired properties. One such restructuring will now be discussed. It involves transforming a given matrix A, in a very special way, into a diagonal matrix D, which usually is much easier to work with but retains many of the desired properties of the original matrix A. We begin with the following definition.

A matrix A is said to be **_diagonalizable_** if there exists a nonsingular matrix X such that the matrix

$$D = X^{-1}AX$$

is a diagonal matrix.

■ **EXAMPLE 2**

To illustrate this idea, consider again the matrix

$$A = \begin{bmatrix} 2 & 0 & 1 \\ -1 & 2 & 0 \\ 1 & 0 & 2 \end{bmatrix}$$

encountered in Example 1. Associated with A was the matrix X, whose columns were three linearly independent eigenvectors of A:

$$X = \begin{bmatrix} 1 & 0 & 1 \\ 1 & 1 & -1 \\ -1 & 0 & 1 \end{bmatrix}.$$

This matrix X turns out to be the matrix X that will "restructure" A.

It is readily verified that the inverse of X is

$$X^{-1} = \begin{bmatrix} 1/2 & 0 & -1/2 \\ 0 & 1 & 1 \\ 1/2 & 0 & 1/2 \end{bmatrix}.$$

Carrying out the indicated multiplications, we see that

$$D = X^{-1}AX$$

$$= \begin{bmatrix} 1/2 & 0 & -1/2 \\ 0 & 1 & 1 \\ 1/2 & 0 & 1/2 \end{bmatrix} \begin{bmatrix} 2 & 0 & 1 \\ -1 & 2 & 0 \\ 1 & 0 & 2 \end{bmatrix} \begin{bmatrix} 1 & 0 & 1 \\ 1 & 1 & -1 \\ -1 & 0 & 1 \end{bmatrix}$$

$$= \begin{bmatrix} 1 & 0 & 0 \\ 0 & 2 & 0 \\ 0 & 0 & 3 \end{bmatrix}$$

is a diagonal matrix whose main diagonal entries are the eigenvalues of A.

It is a pleasant surprise to see that the main diagonal entries of D are in fact the eigenvalues of A. We have shown that the matrix A is diagonalizable. It turns out that A and D share many "similar" properties, as will be seen in the next section. For now, we are concerned mainly with the question: *When* can A be diagonalized?

We motivate the general result by returning to Example 2. We observe that in trying to find a diagonal matrix D such that

$$X^{-1}AX = D,$$

one must first find the appropriate nonsingular matrix X.

As we have seen, this nonsingular matrix X is constructed by making its columns the *eigenvectors* of A. To guarantee that this X is nonsingular (i.e.,

X^{-1} exists), these ***columns must be linearly independent.*** In this example we were able to find three linearly independent eigenvectors of A since A had three distinct eigenvalues (Property 2).

The general restructuring is expressed in the following.

PROPERTY 3. Let A be an $n \times n$-matrix possessing n linearly independent eigenvectors $\mathbf{x}_1 \ldots, \mathbf{x}_n$. Let X be the matrix

$$X = [\mathbf{x}_1 \quad \mathbf{x}_2 \cdots \mathbf{x}_n],$$

whose columns are these eigenvectors of A. Then the matrix

$$D = X^{-1}AX$$

is a diagonal matrix. Furthermore, the main diagonal entries λ_1, $\lambda_2, \ldots, \lambda_n$ of D are the eigenvalues of A corresponding, respectively, to the eigenvectors $\mathbf{x}_1, \mathbf{x}_2, \ldots, \mathbf{x}_n$.

Before proving this, we look at another example.

■ **EXAMPLE 3** _____

Suppose that

$$A = \begin{bmatrix} 5 & 0 & 8 \\ -1 & 2 & 4 \\ 2 & 0 & 5 \end{bmatrix}.$$

The eigenvalues are calculated to be $\lambda_1 = 2$, $\lambda_2 = 9$, and $\lambda_3 = 1$.

Three corresponding eigenvectors are, respectively,

$$\mathbf{x}_1 = \begin{bmatrix} 0 \\ 11 \\ 0 \end{bmatrix}, \qquad \mathbf{x}_2 = \begin{bmatrix} 28 \\ 4 \\ 14 \end{bmatrix}, \qquad \mathbf{x}_3 = \begin{bmatrix} 4 \\ 12 \\ -2 \end{bmatrix}.$$

Since the ***eigenvalues of A are distinct***, we are guaranteed by Property 2 that these eigenvectors are linearly independent. We now construct the matrix X by making its columns these eigenvectors. We have

$$X = \begin{bmatrix} 0 & 28 & 4 \\ 11 & 4 & 12 \\ 0 & 14 & -2 \end{bmatrix}.$$

A straightforward calculation shows that

$$X^{-1} = \begin{bmatrix} -1/7 & 1/11 & 20/77 \\ 1/56 & 0 & 1/28 \\ 1/8 & 0 & -1/4 \end{bmatrix}.$$

It follows that

$$X^{-1}AX = \begin{bmatrix} -1/7 & 1/11 & 20/77 \\ 1/56 & 0 & 1/28 \\ 1/8 & 0 & -1/4 \end{bmatrix} \begin{bmatrix} 5 & 0 & 8 \\ -1 & 2 & 4 \\ 2 & 0 & 5 \end{bmatrix} \begin{bmatrix} 0 & 28 & 4 \\ 11 & 4 & 12 \\ 0 & 14 & -2 \end{bmatrix}$$

$$= \begin{bmatrix} 2 & 0 & 0 \\ 0 & 9 & 0 \\ 0 & 0 & 1 \end{bmatrix} = D,$$

where once again we see that the main diagonal entries of D are the eigenvalues of A.

We prove Property 3 in the case $n = 3$. The general case is proven in exactly the same way. The matrix X is nonsingular because its columns are linearly independent. Consider

$$AX = A[\mathbf{x}_1 \quad \mathbf{x}_2 \quad \mathbf{x}_3] = [A\mathbf{x}_1 \quad A\mathbf{x}_2 \quad A\mathbf{x}_3] = [\lambda_1\mathbf{x}_1 \quad \lambda_2\mathbf{x}_2 \quad \lambda_3\mathbf{x}_3]$$

$$= [\mathbf{x}_1 \quad \mathbf{x}_2 \quad \mathbf{x}_3] \begin{bmatrix} \lambda_1 & 0 & 0 \\ 0 & \lambda_2 & 0 \\ 0 & 0 & \lambda_3 \end{bmatrix} = XD.$$

Hence

$$X^{-1}AX = X^{-1}XD = ID = D$$

is a diagonal matrix, as stated in Property 3.

There are, of course, matrices that are not diagonalizable. For example, the matrix

$$A = \begin{bmatrix} 0 & 1 \\ 0 & 0 \end{bmatrix}$$

is not diagonalizable because it *fails to have two linearly independent eigenvectors*. Its eigenvectors are all multiples of the single eigenvector

$$\mathbf{x} = \begin{bmatrix} 1 \\ 0 \end{bmatrix}.$$

We will see shortly that there is a certain class of matrices that are always diagonalizable. These are the symmetric matrices. Recall that a square matrix A is symmetric if and only if $A = A^T$.

■ **EXAMPLE 4** _____

Consider the real symmetric matrix:

$$A = \begin{bmatrix} 5 & 4 \\ 4 & 5 \end{bmatrix}.$$

The eigenvalues of A are the real numbers $\lambda_1 = 9$ and $\lambda_2 = 1$. Calculating, we find that the vectors

$$\mathbf{x}_1 = \begin{bmatrix} 1 \\ 1 \end{bmatrix} \quad \text{and} \quad \mathbf{x}_2 = \begin{bmatrix} 1 \\ -1 \end{bmatrix}$$

are linearly independent eigenvectors of A corresponding respectively to the eigenvalues $\lambda_1 = 9$ and $\lambda_2 = 1$.

Forming the matrices

$$X = \begin{bmatrix} 1 & 1 \\ 1 & -1 \end{bmatrix}, \quad X^{-1} = \begin{bmatrix} 1/2 & 1/2 \\ 1/2 & -1/2 \end{bmatrix}$$

we see that

$$X^{-1}AX = \begin{bmatrix} 1/2 & 1/2 \\ 1/2 & -1/2 \end{bmatrix} \begin{bmatrix} 5 & 4 \\ 4 & 5 \end{bmatrix} \begin{bmatrix} 1 & 1 \\ 1 & -1 \end{bmatrix} = \begin{bmatrix} 9 & 0 \\ 0 & 1 \end{bmatrix} = D.$$

It follows that A is diagonalizable.

The reader should observe that the eigenvectors,

$$\mathbf{x}_1 = \begin{bmatrix} 1 \\ 1 \end{bmatrix} \quad \text{and} \quad \mathbf{x}_2 = \begin{bmatrix} 1 \\ -1 \end{bmatrix}$$

of A, besides being linearly independent, are also *orthogonal*,

$$\mathbf{x}_1^T \mathbf{x}_2 = \begin{bmatrix} 1 & 1 \end{bmatrix} \begin{bmatrix} 1 \\ -1 \end{bmatrix} = (1)(1) + (1)(-1) = 0.$$

In general, as we have seen in Section 1.6,

> Two vectors $\mathbf{x}$ and $\mathbf{y}$ are *orthogonal* if they satisfy the relation $\mathbf{y}^T\mathbf{x} = 0$.

That the eigenvectors $\mathbf{x}_1$ and $\mathbf{x}_2$ of the real symmetric matrix A, in the example above, are orthogonal is no accident. One in fact has:

> **PROPERTY 4.** Let $\mathbf{x}$ and $\mathbf{y}$ be eigenvectors of a real symmetric matrix A corresponding to distinct eigenvalues λ and μ. Then $\mathbf{x}$ and $\mathbf{y}$ are orthogonal; that is, they satisfy the relation
> $$\mathbf{y}^T\mathbf{x} = 0$$

To prove this, we are given that

$$A\mathbf{x} = \lambda\mathbf{x} \quad \text{and} \quad A\mathbf{y} = \mu\mathbf{y},$$

where $\lambda \neq \mu$. Taking the transpose of the second equation above yields

$$\mathbf{y}^T A^T = \mu \mathbf{y}^T.$$

But A symmetric means that $A = A^T$, so that

$$\mathbf{y}^T A = \mu \mathbf{y}^T.$$

From this we have that

$$\mathbf{y}^T A \mathbf{x} = \mu \mathbf{y}^T \mathbf{x},$$

and since $A\mathbf{x} = \lambda \mathbf{x}$ it follows that

$$\lambda \mathbf{y}^T \mathbf{x} = \mu \mathbf{y}^T \mathbf{x}.$$

Therefore, $(\lambda - \mu)\mathbf{y}^T\mathbf{x} = 0$. Since $\lambda \neq \mu$, $\mathbf{y}^T\mathbf{x} = 0$, as was to be shown.

One can generalize the idea of orthogonality to more than two vectors as follows:

A set of nonzero vectors $\mathbf{x}_1, \mathbf{x}_2, \ldots, \mathbf{x}_q$ is said to be ***mutually orthogonal*** if the relation $\mathbf{x}_i^T\mathbf{x}_j = 0$ holds unless $i = j$. If, in addition, $\mathbf{x}_i^T\mathbf{x}_i = 1$, the vectors are said to be ***orthonormal***.

Since $\mathbf{x}_i \neq \mathbf{0}$, we have $\mathbf{x}_i^T\mathbf{x}_i > 0$.

■ **EXAMPLE 5** _____

Consider the three vectors

$$\mathbf{x}_1 = \begin{bmatrix} 1 \\ 0 \\ -1 \end{bmatrix}, \qquad \mathbf{x}_2 = \begin{bmatrix} 0 \\ 1 \\ 0 \end{bmatrix}, \qquad \mathbf{x}_3 = \begin{bmatrix} -2 \\ 0 \\ -2 \end{bmatrix}.$$

These vectors are mutually orthogonal since

$$\mathbf{x}_1^T\mathbf{x}_2 = 0, \qquad \mathbf{x}_2^T\mathbf{x}_3 = 0, \qquad \mathbf{x}_1^T\mathbf{x}_3 = 0.$$

If we divide each of the vectors $\mathbf{x}_1$, $\mathbf{x}_2$, and $\mathbf{x}_3$ by their respective lengths, we get

$$\mathbf{y}_1 = \begin{bmatrix} 1/\sqrt{2} \\ 0 \\ -1/\sqrt{2} \end{bmatrix}, \qquad \mathbf{y}_2 = \begin{bmatrix} 0 \\ 1 \\ 0 \end{bmatrix}, \qquad \mathbf{y}_3 = \begin{bmatrix} -1/\sqrt{2} \\ 0 \\ -1\sqrt{2} \end{bmatrix}.$$

These vectors are orthonormal since they are mutually orthogonal and each has length 1.

One of the really interesting properties of real $n \times n$ symmetric matrices is the fact that they always have n mutually orthogonal eigenvectors. For example, we have seen that the real symmetric matrix $A = \begin{bmatrix} 5 & 4 \\ 4 & 5 \end{bmatrix}$ had two eigenvectors $\mathbf{x}_1 = \begin{bmatrix} 1 \\ 1 \end{bmatrix}$

and $\mathbf{x}_2 = \begin{bmatrix} 1 \\ -1 \end{bmatrix}$,

which were orthogonal.

■ **EXAMPLE 6** _____

Consider the real 3×3 symmetric matrix

$$A = \begin{bmatrix} 0 & 0 & 0 \\ 0 & 0 & 1 \\ 0 & 1 & 0 \end{bmatrix},$$

whose eigenvalues are $\lambda_1 = 0$, $\lambda_2 = 1$, and $\lambda_3 = -1$. Calculating, we find three eigenvectors, one corresponding to $\lambda_1 = 0$, $\lambda_2 = 1$, and $\lambda_3 = -1$, respectively. They are

$$\mathbf{x}_1 = \begin{bmatrix} 1 \\ 0 \\ 0 \end{bmatrix}, \qquad \mathbf{x}_2 = \begin{bmatrix} 0 \\ 1 \\ 1 \end{bmatrix}, \qquad \mathbf{x}_3 = \begin{bmatrix} 0 \\ 1 \\ -1 \end{bmatrix}.$$

Clearly, these three eigenvectors are mutually orthogonal. They can be made orthonormal by dividing each by its length.

We now state the general property.

> **PROPERTY 5.** A real symmetric $n \times n$-matrix A possesses a set of n mutually orthogonal eigenvectors $\mathbf{x}_1, \mathbf{x}_2, \ldots, \mathbf{x}_n$. They can be chosen to be orthonormal.

The proof of this result is a simple consequence of Exercise P2 in the Supplementary Proofs at the end of the section.

Recall that an $n \times n$-matrix A was diagonalizable whenever A had n linearly independent eigenvectors. Since $n \times n$ real symmetric matrices always have n mutually orthogonal eigenvectors, we can deduce that they are diagonalizable provided that the n mutually orthogonal eigenvectors are *linearly independent*. This is, in fact, the case irrespective of whether the vectors are eigenvectors. We have:

> **PROPERTY 6.** Any set of n mutually orthogonal nonzero vectors are linearly independent.

We present the proof in the case $n = 3$. The general case follows in exactly the same way.

Suppose, then, that $\mathbf{x}_1$, $\mathbf{x}_2$ and $\mathbf{x}_3$ are mutually orthogonal. Consider

$$c_1\mathbf{x}_1 + c_2\mathbf{x}_2 + c_3\mathbf{x}_3 = \mathbf{0}.$$

Multiplying this last equation on the left by $\mathbf{x}_1^T$ yields

$$c_1\mathbf{x}_1^T\mathbf{x}_1 + c_2\mathbf{x}_1^T\mathbf{x}_2 + c_3\mathbf{x}_1^T\mathbf{x}_3 = \mathbf{x}_1^T\mathbf{0} = 0.$$

But $\mathbf{x}_1^T\mathbf{x}_2 = \mathbf{x}_1^T\mathbf{x}_3 = 0$. Therefore, $c_1\mathbf{x}_1^T\mathbf{x}_1 = 0$. But $\mathbf{x}_1^T\mathbf{x}_1 > 0$. Hence $c_1 = 0$. Multiplying the original equation

$$c_1\mathbf{x}_1 + c_2\mathbf{x}_2 + c_3\mathbf{x}_3 = \mathbf{0}$$

on the left by $\mathbf{x}_2^T$ yields, in a similar fashion, $c_2 = 0$. Finally, a multiplication on the left by $\mathbf{x}_3^T$ gives $c_3 = 0$. We have shown that the vectors $\mathbf{x}_1$, $\mathbf{x}_2$, and $\mathbf{x}_3$ are linearly independent.

Using Properties 3, 5, and 6 we now obtain:

PROPERTY 7. Every real $n \times n$-symmetric matrix A is diagonalizable.

■ **EXAMPLE 7**

Consider the real symmetric matrix

$$A = \begin{bmatrix} 0 & 1 & 1 \\ 1 & 0 & 1 \\ 1 & 1 & 0 \end{bmatrix}$$

whose eigenvalues are $\lambda_1 = 2$, $\lambda_2 = -1$, and $\lambda_3 = -1$.

We have seen that if A is an arbitrary matrix, we cannot always guarantee enough linearly independent eigenvectors unless the eigenvalues are distinct. Even though $\lambda_2 = \lambda_3 = -1$ is an eigenvalue of multiplicity 2, in this case, since A is real and symmetric, we can find two mutually orthogonal eigenvectors corresponding to the multiple eigenvalue -1. We have

$$(A - (-1)I)\mathbf{x} = \begin{bmatrix} 1 & 1 & 1 \\ 1 & 1 & 1 \\ 1 & 1 & 1 \end{bmatrix}\begin{bmatrix} x_1 \\ x_2 \\ x_3 \end{bmatrix} = \begin{bmatrix} 0 \\ 0 \\ 0 \end{bmatrix},$$

from which it follows that $x_1 + x_2 + x_3 = 0$. A simple calculation shows that

$$\mathbf{x}_2 = \begin{bmatrix} 1 \\ -1 \\ 0 \end{bmatrix}, \quad \mathbf{x}_3 = \begin{bmatrix} 1 \\ 1 \\ -2 \end{bmatrix}$$

are each eigenvectors corresponding to $\lambda = -1$. Furthermore, they are orthogonal. Another calculation yields

$$\mathbf{x}_1 = \begin{bmatrix} 1 \\ 1 \\ 1 \end{bmatrix}$$

as an eigenvector corresponding to the eigenvalue $\lambda_1 = 2$. The three eigenvectors

$$\mathbf{x}_1 = \begin{bmatrix} 1 \\ 1 \\ 1 \end{bmatrix}, \qquad \mathbf{x}_2 = \begin{bmatrix} 1 \\ -1 \\ 0 \end{bmatrix}, \qquad \text{and} \qquad \mathbf{x}_3 = \begin{bmatrix} 1 \\ 1 \\ -2 \end{bmatrix}$$

are mutually orthogonal and hence linearly independent. The matrix X whose columns are constructed from these eigenvectors is given by

$$X = \begin{bmatrix} 1 & 1 & 1 \\ 1 & -1 & 1 \\ 1 & 0 & -2 \end{bmatrix}.$$

If D is the diagonal matrix whose main diagonal entries are the eigenvalues $\lambda_1 = 2, \lambda_2 = -1, \lambda_3 = -1$ of A, it follows that

$$AX = \begin{bmatrix} 0 & 1 & 1 \\ 1 & 0 & 1 \\ 1 & 1 & 0 \end{bmatrix} \begin{bmatrix} 1 & 1 & 1 \\ 1 & -1 & 1 \\ 1 & 0 & -2 \end{bmatrix} = \begin{bmatrix} 2 & -1 & -1 \\ 2 & 1 & -1 \\ 2 & 0 & 2 \end{bmatrix},$$

$$XD = \begin{bmatrix} 1 & 1 & 1 \\ 1 & -1 & 1 \\ 1 & 0 & -2 \end{bmatrix} \begin{bmatrix} 2 & 0 & 0 \\ 0 & -1 & 0 \\ 0 & 0 & -1 \end{bmatrix} = \begin{bmatrix} 2 & -1 & -1 \\ 2 & 1 & -1 \\ 2 & 0 & 2 \end{bmatrix},$$

so that $AX = XD$. Consequently, $X^{-1}AX = D$. We see that A is diagonalizable.

We can gain even more information about A since it is symmetric. In the example above, we found three mutually orthogonal eigenvectors of A. Suppose that we now divide each eigenvector

$$\mathbf{x}_1 = \begin{bmatrix} 1 \\ 1 \\ 1 \end{bmatrix}, \qquad \mathbf{x}_2 = \begin{bmatrix} 1 \\ -1 \\ 0 \end{bmatrix}, \qquad \mathbf{x}_3 = \begin{bmatrix} 1 \\ 1 \\ -2 \end{bmatrix}$$

by its corresponding length. We then obtain three *orthonormal* eigenvectors

$$\mathbf{y}_1 = \begin{bmatrix} 1/\sqrt{3} \\ 1/\sqrt{3} \\ 1/\sqrt{3} \end{bmatrix}, \qquad \mathbf{y}_2 = \begin{bmatrix} 1/\sqrt{2} \\ -1/\sqrt{2} \\ 0 \end{bmatrix}, \qquad \mathbf{y}_3 = \begin{bmatrix} 1/\sqrt{6} \\ 1/\sqrt{6} \\ -2/\sqrt{6} \end{bmatrix}.$$

If we form the matrix Y whose columns are made up of these **orthonormal** eigenvectors, we get

$$Y = \begin{bmatrix} 1/\sqrt{3} & 1/\sqrt{2} & 1/\sqrt{6} \\ 1/\sqrt{3} & -1/\sqrt{2} & 1/\sqrt{6} \\ 1/\sqrt{3} & 0 & -2/\sqrt{6} \end{bmatrix}.$$

This matrix Y is an **orthogonal** matrix. In general,

An $n \times n$-matrix Y is said to be an **orthogonal** matrix if the columns of Y form an **orthonormal** set in R^n.

We observe that

$$YY^T = \begin{bmatrix} 1/\sqrt{3} & 1/\sqrt{2} & 1/\sqrt{6} \\ 1/\sqrt{3} & -1/\sqrt{2} & 1/\sqrt{6} \\ 1/\sqrt{3} & 0 & -2/\sqrt{6} \end{bmatrix} \begin{bmatrix} 1/\sqrt{3} & 1/\sqrt{3} & 1/\sqrt{3} \\ 1/\sqrt{2} & -1/\sqrt{2} & 0 \\ 1/\sqrt{6} & 1/\sqrt{6} & -2/\sqrt{6} \end{bmatrix}$$

$$= \begin{bmatrix} 1 & 0 & 0 \\ 0 & 1 & 0 \\ 0 & 0 & 1 \end{bmatrix} = I.$$

A similar calculation shows that $Y^T Y = I$. It follows that $Y^T = Y^{-1}$. In general, the following statements are all equivalent:

---

1. Y is an $n \times n$-orthogonal matrix.
2. The columns of Y form an orthonormal set in R^n.
3. The rows of Y form an orthonormal set in R^n.
4. $YY^T = Y^T Y = I$.
5. $Y^T = Y^{-1}$.

In the example above, we see that

$$AY = \begin{bmatrix} 0 & 1 & 1 \\ 1 & 0 & 1 \\ 1 & 1 & 0 \end{bmatrix} \begin{bmatrix} 1/\sqrt{3} & 1/\sqrt{2} & 1/\sqrt{6} \\ 1/\sqrt{3} & -1/\sqrt{2} & 1/\sqrt{6} \\ 1/\sqrt{3} & 0 & -2/\sqrt{6} \end{bmatrix}$$

$$= \begin{bmatrix} 2/\sqrt{3} & -1/\sqrt{2} & -1/\sqrt{6} \\ 2/\sqrt{3} & 1/\sqrt{2} & -1\sqrt{6} \\ 2/\sqrt{3} & 0 & 2/\sqrt{6} \end{bmatrix},$$

$$YD = \begin{bmatrix} 1/\sqrt{3} & 1/\sqrt{2} & 1/\sqrt{6} \\ 1/\sqrt{3} & -1/\sqrt{2} & 1/\sqrt{6} \\ 1/\sqrt{3} & 0 & -2\sqrt{6} \end{bmatrix} \begin{bmatrix} 2 & 0 & 0 \\ 0 & -1 & 0 \\ 0 & 0 & -1 \end{bmatrix}$$

$$= \begin{bmatrix} 2/\sqrt{3} & -1/\sqrt{2} & -1/\sqrt{6} \\ 2/\sqrt{3} & 1/\sqrt{2} & -1/\sqrt{6} \\ 2/\sqrt{3} & 0 & 2/\sqrt{6} \end{bmatrix},$$

so that $AY = YD$. It follows that

$$Y^{-1}AY = D.$$

We see that the symmetric matrix A is diagonalizable. Furthermore, the matrix Y that diagonalizes A is orthogonal. We have:

> **PROPERTY 8.** If A is a real symmetric matrix, there is an orthogonal matrix Y such that $Y^{-1}AY = Y^{T}AY = D$, a diagonal matrix.

Alternatively,

> Every real symmetric matrix A is ***orthogonally diagonalizable*** in the sense that there is an orthogonal matrix Y such that
>
> $$D = Y^{-1}AY = Y^{T}AY$$
>
> is a diagonal matrix.

All of the examples that we have looked at for *real symmetric matrices* have had *real eigenvalues*. This is, in fact, true for real symmetric matrices in general. We have:

> **PROPERTY 9.** The eigenvalues of a real symmetric matrix A are real and its eigenvectors can be chosen to be real.

A proof goes as follows:

Let λ be an eigenvalue of A. Let $\mathbf{z} = \mathbf{x} + i\mathbf{y}$ be a corresponding eigenvector of A. Here $\mathbf{x}$ and $\mathbf{y}$ are real vectors, not both $\mathbf{0}$. We have

$$A(\mathbf{x} + i\mathbf{y}) = \lambda(\mathbf{x} + i\mathbf{y}).$$

Also, since A is symmetric, $\mathbf{x}^{T}A\mathbf{y} = \mathbf{y}^{T}A\mathbf{x}$. Hence

$$(\mathbf{x}^{T} - i\mathbf{y}^{T})A(\mathbf{x} + i\mathbf{y}) = \mathbf{x}^{T}A\mathbf{x} + i\mathbf{x}^{T}A\mathbf{y} - i\mathbf{y}^{T}A\mathbf{x} - i^{2}\mathbf{y}^{T}A\mathbf{y}$$

$$= \mathbf{x}^{T}A\mathbf{x} + \mathbf{y}^{T}A\mathbf{y}$$

$$= \lambda(\mathbf{x}^{T} - i\mathbf{y}^{T})(\mathbf{x} + i\mathbf{y}) = \lambda(\mathbf{x}^{T}\mathbf{x} + \mathbf{y}^{T}\mathbf{y}).$$

The quantity in the last set of parentheses is a positive number. The quantities on the second line are real. We have, therefore,

$$(\text{real number}) = \lambda(\text{positive number}).$$

It follows that the eigenvalue λ of A is real. Because λ is real, the equation $A(\mathbf{x} + i\mathbf{y}) = \lambda(\mathbf{x} + i\mathbf{y})$ tells us that $A\mathbf{x} = \lambda\mathbf{x}$ and $A\mathbf{y} = \lambda\mathbf{y}$. It follows that if $\mathbf{x}$ is not zero, it is an eigenvector of A. Similarly, if $\mathbf{y}$ is not zero, it too is an eigenvector of A, perhaps a multiple of $\mathbf{x}$, but not necessarily so when λ is a repeated eigenvalue of A. We can therefore choose our eigenvectors of A to be real.

Exercises

1. Without computation, determine whether or not the matrix

$$A = \begin{bmatrix} 0 & 3 & -2 \\ 3 & 1 & 4 \\ -2 & 4 & 1 \end{bmatrix}$$

is diagonalizable.

2. Show that the following matrices are diagonalizable by showing that their eigenvalues are real and distinct.

$$A = \begin{bmatrix} 7 & 8 \\ 2 & 1 \end{bmatrix}, \quad B = \begin{bmatrix} 3 & 2 \\ 8 & -3 \end{bmatrix}, \quad C = \begin{bmatrix} 10 & 9 \\ -1 & 2 \end{bmatrix}, \quad D = \begin{bmatrix} 12 & -5 \\ -5 & -12 \end{bmatrix},$$

$$E = \begin{bmatrix} 5 & 0 & 8 \\ 0 & 3 & 0 \\ 2 & 0 & 5 \end{bmatrix}, \quad F = \begin{bmatrix} 5 & 0 & -8 \\ 6 & 3 & -6 \\ -2 & 0 & 5 \end{bmatrix}, \quad G = \begin{bmatrix} 0 & 0 & 4 \\ 0 & 7 & 0 \\ 1 & 0 & 0 \end{bmatrix},$$

$$H = \begin{bmatrix} A & B \\ 0 & C \end{bmatrix}, \quad K = \begin{bmatrix} B & 0 \\ A & D \end{bmatrix}, \quad L = \begin{bmatrix} E & 0 \\ F & G \end{bmatrix}.$$

3. Show that the following matrices are diagonalizable.

   (a) The matrices

$$A = \begin{bmatrix} 1 & 2 & 2 \\ 0 & 3 & 4 \\ 0 & 0 & 5 \end{bmatrix}, \quad B = \begin{bmatrix} 0 & 0 & 0 \\ 2 & 1 & 0 \\ -2 & 3 & -1 \end{bmatrix}.$$

   (b) An upper triangular matrix whose main diagonal entries are distinct.

   (c) A lower triangular matrix whose main diagonal entries are distinct.

4. If A is diagonalizable, show that:

   (a) A^2 is diagonalizable.

   (b) A^n is diagonalizable, n any positive integer.

5. Let

$$A = \begin{bmatrix} 2 & 0 & 0 \\ 1 & 0 & 0 \\ 3 & 2 & 1 \end{bmatrix}.$$

Find the eigenvalues of the matrices A, A^2, and A^n, where n is any positive integer. *Hint:* Use the fact that A is a lower triangular matrix.

6. Show directly that every real symmetric 2×2-matrix A has only real eigenvalues.

7. If $A = \begin{bmatrix} 1 & 1 \\ 1 & -1 \end{bmatrix}$:

    (a) Show that A *is symmetric.*

    (b) Find the eigenvalues of A.

    (c) Find two mutually orthogonal eigenvectors of A.

    (d) Find an orthogonal matrix X such that $X^{-1}AX$ is a diagonal matrix D.

8. Consider the matrix

$$A = \begin{bmatrix} 0 & 1 & -1 \\ 1 & 0 & 0 \\ -1 & 0 & 0 \end{bmatrix}.$$

    (a) Find the eigenvalues of A.

    (b) Find three mutually orthogonal eigenvectors of A.

    (c) Find a nonsingular matrix X such that $X^{-1}AX$ is a diagonal matrix D.

    (d) Find an orthogonal matrix Y such that $Y^{-1}AY$ is a diagonal matrix D.

9. If A is diagonalizable by X (i.e., $X^{-1}AX = D$), show that A^T is diagonalizable by $Y = (X^{-1})^T = (X^T)^{-1}$ with the same D. *Hint:* Take transposes.

10. Let A be a diagonalizable matrix. If X is a nonsingular matrix such that

$$D = X^{-1}AX$$

is a diagonal matrix, conclude that the columns of X are eigenvectors of A. *Hint:* Use the relation $AX = XD$.

11. In Exercise 10, use the relation $D = D^T$ to show that

$$D = X^T A^T (X^T)^{-1}.$$

Conclude that A^T is diagonalizable and that the columns of $(X^T)^{-1}$ are eigenvectors of A^T. Hence A^T is diagonalizable whenever A is.

12. Consider the matrices

$$A, \quad B = A + bI, \quad C = cA,$$

where b and c are arbitrary numbers with $c = 0$ excluded.

    (a) Show that A is diagonalizable by X if and only if B is diagonalizable by X.

    (b) Show that A is diagonalizable by X if and only if C is diagonalizable by X.

    (c) Conclude that if one of the matrices A, B, or C is diagonalizable by X, so are the other two.

Hint: These results follow from the relations

$$X^{-1}BX = X^{-1}AX - bI, \quad X^{-1}CX = cX^{-1}AX.$$

13. Suppose that a matrix A has the property that there is a nonsingular matrix Y such that

$$B = Y^{-1}AY$$

is a symmetric matrix. Conclude that A is diagonalizable. *Hint:* Choose Z so that $D = Z^{-1}BZ$ is a diagonal matrix. Set $X = YZ$ and conclude that $D = X^{-1}AX$.

14. Let A, B, and X be $n \times n$-matrices. Suppose that X is nonsingular. Set

$$D = X^{-1}AX, \qquad E = X^{-1}BX.$$

(a) Show that

$$DE = X^{-1}ABX, \qquad ED = X^{-1}BAX,$$

$$AB = XDEX^{-1}, \qquad BA = XEDX^{-1}.$$

(b) Conclude that $DE = ED$ if and only if $AB = BA$. That is, A and B commute if and only if D and E commute.

(c) $D^k = X^{-1}A^kX$, $A^k = XD^kX^{-1}$ for every integer k.

Hint: Part (c) can be established by induction. For if $D^k = X^{-1}A^kX$ for some integer k, then by (a) with $E = D^k$ and $B = A^k$, we have $D^{k+1} = X^{-1}A^{k+1}X$.

15. Continuing with Exercise 14, suppose that D and E are diagonal matrices. That is, suppose that A and B are diagonalizable by X. Then:

(a) ED is a diagonal matrix. Moreover, $ED = DE$ and $AB = BA$.

(b) AB and BA are diagonalizable by X.

(c) The main diagonal entries of DE are the eigenvalues of $AB = BA$.

(d) $rA + sB$ is diagonalizable by X for all numbers r and s. Verify that the main diagonal entries of $rD + sE$ are the eigenvalues of $rA + sB$.

16. Let a, b, and c be positive numbers. Form the matrices

$$Y = \begin{bmatrix} a & b & c \\ a & 0 & -2c \\ a & -b & c \end{bmatrix}, \quad Y^T = \begin{bmatrix} a & a & a \\ b & 0 & -b \\ c & -2c & c \end{bmatrix}, \quad E = \begin{bmatrix} a\sqrt{3} & 0 & 0 \\ 0 & b\sqrt{2} & 0 \\ 0 & 0 & c\sqrt{6} \end{bmatrix}.$$

Show that:

(a) The columns of Y are mutually orthogonal.

(b) The lengths of the columns of Y are the main diagonal entries of E.

(c) $Y^TY = E^2$.

(d) The matrix $X = YE^{-1}$ is an orthogonal matrix.

(e) The matrix $A = YY^T$ is a symmetric matrix with the property that $X^TAX = E^2$.

(f) Conclude that the eigenvalues of $A = YY^T$ are $3a^2$, $2b^2$, and $6c^2$, the square of the lengths of the columns of Y.

17. Let A be a symmetric matrix and select an orthogonal matrix X such that $X^TAX = D$ is a diagonal matrix.

(a) Show that the columns of $Y = AX = XD$ are mutually orthogonal.

(b) Show that $YY^T = A^2$ and $X^TA^2X = D^2$.

18. Show that the rows of the matrix

$$Y = \begin{bmatrix} 0 & 1 & -1 \\ 1 & 0 & 0 \\ 0 & 1 & 1 \end{bmatrix}$$

are mutually orthogonal. Show further that the columns of Y are also mutually orthogonal. However, Y is not an orthogonal matrix. Why not? Can the columns of Y be scaled so as to form an orthogonal matrix? Do so.

19. Show that the matrix

$$Y = \begin{bmatrix} \cos\theta & \sin\theta \\ -\sin\theta & \cos\theta \end{bmatrix}$$

is an orthogonal matrix. Show that Y^T is also orthogonal, and calculate Y^TY.

20. Let A be an $n \times n$-matrix. If A is diagonalizable by an orthogonal matrix Y, show that A must be symmetric.

21. Let X be an (real) orthogonal matrix. Select vectors $\mathbf{x}$ and $\mathbf{y}$ and set

$$\mathbf{u} = X\mathbf{x}, \qquad \mathbf{v} = X\mathbf{y}.$$

Conclude that:

(a) $\mathbf{u}^T\mathbf{v} = \mathbf{x}^T\mathbf{y}$.

(b) The angle between $\mathbf{u}$ and $\mathbf{v}$ is the same as the angle between $\mathbf{x}$ and $\mathbf{y}$.

(c) The vectors $\mathbf{u}$ and $\mathbf{x}$ have the same length.

(d) The vectors $\mathbf{v}$ and $\mathbf{y}$ have the same length.

22. Show that the eigenvalues of a real orthogonal matrix X lie on the unit circle in the xy-plane. Hence the real eigenvalues must be 1 or -1. *Hint:* If $X\mathbf{z} = \lambda\mathbf{z}$, then

$$\overline{(X\mathbf{z})^T} = \overline{(\lambda\mathbf{z})^T}$$

and

$$\overline{\mathbf{z}}^T\overline{X}^T = \overline{\lambda}\overline{\mathbf{z}}^T.$$

But X is real, so that

$$\overline{\mathbf{z}}^TX^T = \overline{\lambda}\overline{\mathbf{z}}^T.$$

Hence $\overline{\mathbf{z}}^T\mathbf{z} = \overline{\mathbf{z}}^TI\mathbf{z} = \overline{\mathbf{z}}^TX^TX\mathbf{z} = (\overline{\lambda}\mathbf{z}^T)(\lambda\mathbf{z}) = \overline{\lambda}\lambda\mathbf{z}^T\mathbf{z}$, so that $\overline{\lambda}\lambda = 1$.

23. Consider the matrices

$$N = \begin{bmatrix} 0 & a & c \\ 0 & 0 & b \\ 0 & 0 & 0 \end{bmatrix}, \qquad N^2 = \begin{bmatrix} 0 & 0 & ab \\ 0 & 0 & 0 \\ 0 & 0 & 0 \end{bmatrix}, \qquad N^3 = \begin{bmatrix} 0 & 0 & 0 \\ 0 & 0 & 0 \\ 0 & 0 & 0 \end{bmatrix},$$

where a, b, and c are not all zero. The matrix N has the property that $N^3 = 0$.

(a) Verify that N is not diagonalizable.

(b) Verify that for each number λ, the matrix

$$A = \lambda I + N = \begin{bmatrix} \lambda & a & c \\ 0 & \lambda & b \\ 0 & 0 & \lambda \end{bmatrix}$$

is not diagonalizable.

Hint: Suppose that N is diagonalizable. Then there is a nonsingular matrix X such that

$$D = X^{-1}NX \quad \text{and hence that} \quad D^3 = X^{-1}N^3X = 0$$

since $N^3 = 0$. It follows that $D = 0$ and $N = XDX^{-1} = 0$, which is not the case. If $A = \lambda I + N$ were diagonalizable, then $H = A - \lambda I$ would be diagonalizable, by virtue of the result given in Exercise 12. Hence A is not diagonalizable.

24. Recall that an $n \times n$-matrix N is nilpotent if $N^m = 0$ for some integer m.

(a) Show that the eigenvalues of a nilpotent matrix N are all zero. *Hint:* Use the fact that $N\mathbf{x} = \lambda\mathbf{x}$ implies that $N^m\mathbf{x} = \lambda^m\mathbf{x}$.

(b) Conclude that $(-\lambda)^n$ is the characteristic polynomial of a $n \times n$ nilpotent matrix N.

(c) Show conversely that if the eigenvalues of a matrix N are all zero, then N is nilpotent. Find the characteristic polynomial for N and use the Cayley–Hamilton theorem.

25. Show that:

(a) The transpose of a nilpotent matrix is nilpotent.

(b) An upper triangular matrix whose main diagonal entries are all zero is a nilpotent matrix.

(c) A lower triangular matrix whose main diagonal entries are all zero is a nilpotent matrix.

(d) $N = 0$ is nilpotent.

26. Show that:
(a) The only diagonalizable nilpotent matrix N is the matrix $N = 0$.

(b) If N is a nonzero nilpotent matrix and λ is any number, $A = \lambda I + N$ is not diagonalizable.

Hint: Parts (a) and (b) can be established by the argument given in the Hint for Exercise 23.

27. Suppose that all of the eigenvalues of an $n \times n$-matrix A are equal to a number a.

(a) Show that $(a - \lambda)^n$ is the characteristic polynomial for A.

(b) Show that $A = aI + N$, where N is nilpotent. *Hint:* Use the Cayley–Hamilton theorem.

28. Consider the matrices

$$A = \begin{bmatrix} a & b & c & d \\ e & a & b & c \\ f & e & a & b \\ g & f & e & a \end{bmatrix}, \qquad R = \begin{bmatrix} 0 & 0 & 0 & 1 \\ 0 & 0 & 1 & 0 \\ 0 & 1 & 0 & 0 \\ 1 & 0 & 0 & 0 \end{bmatrix}.$$

(a) Show that $S = AR$ is symmetric.

(b) Show that $R^2 = I$ and hence that $R^{-1} = R$.

(c) Show that $AR = RA^T$ and hence that $A^T = RAR$.

29. Let a and b be real numbers with b not zero. Consider the real and complex matrices

$$A = \begin{bmatrix} a & b \\ -b & a \end{bmatrix}, \qquad B = \begin{bmatrix} a + bi & 0 \\ 0 & a - bi \end{bmatrix} \qquad (i^2 = -1).$$

(a) Show that the characteristic polynomial for A is

$$p(\lambda) = (\lambda - a)^2 + b^2.$$

(b) Conclude that the eigenvalues of A are $\lambda_1 = a + bi$ and $\lambda_2 = a - bi$.

(c) Verify that the vectors $\mathbf{x}_1 = (1, i)$ and $\mathbf{x}_2 = (i, 1)$ satisfy the relations

$$A\mathbf{x}_1 = \lambda_1\mathbf{x}_1, \qquad A\mathbf{x}_2 = \lambda_2\mathbf{x}_2.$$

(d) Conclude that $\mathbf{x}_1$ and $\mathbf{x}_2$ are eigenvectors of A corresponding to the eigenvalues λ_1 and λ_2, respectively.

(e) Construct the matrices

$$X = [\mathbf{x}_1 \quad \mathbf{x}_2] = \begin{bmatrix} 1 & i \\ i & 1 \end{bmatrix}, \qquad X^{-1} = \frac{1}{2}\begin{bmatrix} i & -i \\ -1 & 1 \end{bmatrix}.$$

(f) Verify that $B = X^{-1}AX$. Since B is a diagonal matrix, we see again that its main diagonal entries $a + bi$ and $a - bi$ are the eigenvalues of A.

(g) Set $Y = (1/\sqrt{2})X$. Verify that $Y^{-1} = Y^*$ the conjugate transpose of Y. Verify that $B = Y^*AY$. A matrix Y whose conjugate transpose is its inverse is called a **unitary matrix**. A real unitary matrix is an orthogonal matrix.

Supplementary Proofs

Verify the following proofs, supplying additional arguments if necessary.

■ **P1.** Let $\mathbf{x}_1, \mathbf{x}_2, \ldots, \mathbf{x}_q$ be a set of q mutually orthogonal n-dimensional unit vectors.

1. Show that they are linearly independent.

2. Show that if $q < n$, there exists a unit vector $\mathbf{y}$ orthogonal to each of these vectors.

3. Show that if $q < n$, there exists a set of $m = n - q$ mutually orthogonal unit vectors $y_1, y_2, \ldots, y_m$ such that the vectors $x_1, \ldots, x_q, y_1, \ldots, y_m$ are mutually orthogonal unit vectors.

To prove property 2, let X be the $n \times q$-matrix having the vectors $x_1, x_2, \ldots, x_q$ as its column vectors. Choose y to be a nonzero solution of the homogeneous linear equation $X^T y = 0$. Scale y so that $y^T y = 1$. The vector y is the desired vector.

Property 3 follows by repeated applications of Property 2.   ▬

▬ **P2.** Let $x_1, x_2, \ldots, x_q$ be q mutually orthogonal eigenvectors of a real symmetric $n \times n$-matrix A. If $q < n$, there exists an eigenvector y of A orthogonal to the vectors $x_1, x_2, \ldots, x_q$.

To prove this, we can suppose that the eigenvectors $x_1, x_2, \ldots, x_q$ have been scaled so that $x_i^T x_i = 1$ for $i = 1, 2, \ldots, q$. These vectors are the columns of an $n \times q$-matrix X such that

$$X^T X = I, \qquad AX = XD,$$

where I is the $q \times q$-identity matrix and D is a $q \times q$ diagonal matrix whose main diagonal entries are eigenvalues of A. Suppose that $q < n$. Select $m = n - q$ nonzero mutually orthogonal vectors $y_1, y_2, \ldots, y_m$ which are orthogonal to the eigenvectors $x_1, x_2, \ldots, x_m$. Normalize these vectors so that $y_j^T y_j = 1$ $(j = 1, 2, \ldots, m)$. The y-vectors are the column vectors of an $n \times m$-matrix Y having the following properties:

$$Y^T Y = J, \quad Y^T X = 0, \quad X^T Y = 0, \quad Y^T AX = Y^T XD = 0, \quad X^T AY = 0,$$

where J is the $m \times m$-identity matrix. The $n \times n$-matrix

$$Z = [X \qquad Y]$$

is nonsingular because its columns are mutually orthogonal nonzero vectors and so are linearly independent. The matrix $C = Y^T AY$ is symmetric. Let u be an eigenvector of C with μ as the corresponding eigenvalue of C. We shall show that the vector $y = Yu$ is an eigenvector of A. To do so, set $w = Ay - \mu y = AYu - \mu Yu$. Then

$$
\begin{aligned}
X^T w &= X^T AYu - \mu X^T Yu = 0, & w^T X &= 0, \\
Y^T w &= Y^T AYu - \mu Y^T Yu = Cu - \mu u = 0, & w^T Y &= 0.
\end{aligned}
$$

We have, accordingly,

$$w^T Z = w^T [X \qquad Y] = [w^T X \qquad w^T Y] = [0 \qquad 0] = 0.$$

Because Z is nonsingular, this implies that $w = Ay - \mu y = 0$. It follows that y is an eigenvector of A orthogonal to the eigenvectors $x_1, x_2, \ldots, x_q$, as was to be proved.   ▬

4.4 SIMILARITY OF MATRICES

In the preceding section we were sometimes able to restructure a matrix A into a diagonal matrix D by finding a nonsingular matrix X such that

$$X^{-1}AX = D.$$

We were able to construct this nonsingular matrix X by knowing about the eigenvectors of A. We can now generalize this idea as follows:

> A square matrix B is said to be *similar* to a given square matrix A if there is a nonsingular matrix P such that
> $$B = P^{-1}AP.$$

Notice that here, in this more general case, B is not necessarily a diagonal matrix. An equivalent definition that is often more useful since it avoids having to calculate P^{-1} is the following:

> A square matrix B is said to be *similar* to a given matrix A if there is a nonsingular matrix P such that
> $$AP = PB.$$

■ **EXAMPLE 1**

If

$$A = \begin{bmatrix} 4 & -2 \\ 1 & 3 \end{bmatrix}, \qquad B = \begin{bmatrix} -7 & -16 \\ 7 & 14 \end{bmatrix},$$

we can verify that B is similar to A if we can find a nonsingular matrix P such that

$$B = P^{-1}AP \quad \text{or equivalently if} \quad AP = PB.$$

In this example, such a P does exist. It turns out that

$$P = \begin{bmatrix} 1 & 1 \\ 2 & 3 \end{bmatrix}, \qquad P^{-1} = \begin{bmatrix} 3 & -1 \\ -2 & 1 \end{bmatrix}, \qquad \text{and}$$

$$P^{-1}AP = \begin{bmatrix} 3 & -1 \\ -2 & 1 \end{bmatrix}\begin{bmatrix} 4 & -2 \\ 1 & 3 \end{bmatrix}\begin{bmatrix} 1 & 1 \\ 2 & 3 \end{bmatrix} = \begin{bmatrix} -7 & -16 \\ 7 & 14 \end{bmatrix} = B.$$

Equivalently, it is easily verified that $AP = PB$. Finding the nonsingular matrix P is in general no easy task. In this example, we actually knew it in advance.

■ **EXAMPLE 2** ───
Consider the matrices

$$A = \begin{bmatrix} 5 & 4 \\ 4 & 5 \end{bmatrix}, \qquad B = \begin{bmatrix} 9 & 0 \\ 0 & 1 \end{bmatrix}.$$

In asking whether B is similar to A, we once again seek a nonsingular matrix P such that

$$P^{-1}AP = B.$$

Here we observe that A is symmetric, hence diagonalizable. Furthermore, B is a diagonal matrix made up of the eigenvalues of A. Hence we may construct a nonsingular matrix X whose columns are eigenvectors of A and such that

$$X^{-1}AX = B.$$

This X plays the same role as P in our definition of similarity, so we see that B is similar to A.

───

Example 2 illustrates that the diagonalization of a matrix A into a matrix D is a special case of similarity. With this new terminology, we can now say that a matrix A is *diagonalizable* if and only if it is *similar* to a diagonal matrix D.

Returning to the general case, we now examine some of the properties of this new concept of similarity.

We first observe that:

┌───┐
│ **1.** If B is similar to A, then A is similar to B. │
└───┘

For if B is similar to A, there is a nonsingular matrix P such that $B = P^{-1}AP$. Multiplying this equation on the left by P and on the right by P^{-1} gives

$$PBP^{-1} = P(P^{-1}AP)P^{-1} = A.$$

If we rename P^{-1} and call it Q (i.e., $P^{-1} = Q$), we get

$$Q^{-1}BQ = A,$$

so that A is similar to B. Rule 1 is referred to as the **symmetric law** of similarity. As a second observation, we have that:

┌───┐
│ **2.** A is similar to itself.   │
└───┘

This follows by choosing $P = I$. We get

$$P^{-1}AP = I^{-1}AI = A.$$

This is known as the *reflexive law* of similarity. Finally,

> **3.** If A is similar to B and B is similar to C, then A is similar to C.

We have $A = P^{-1}BP$, for some nonsingular P, $B = Q^{-1}CQ$, for some nonsingular Q. Substitution yields

$$A = P^{-1}BP = P^{-1}(Q^{-1}CQ)P = (P^{-1}Q^{-1})C(QP) = (QP)^{-1}C(QP).$$

Since Q and P are nonsingular, so is the product QP. We see that A is similar to C. This rule is referred to as the *transitive law* of similarity.

In the preceding sections we saw that if two matrices A and B were related by the equation $B = P^{-1}AP$, they shared many of the same properties. We can now rephrase some of these results in terms of similarity as follows:

PROPERTY 1. Let A and B be similar $n \times n$-matrices. Then:

(a) They have the same characteristic polynomial $p(\lambda)$.

(b) They have the same eigenvalues with the same multiplicities.

(c) $\det A = \det B$.

(d) They have the same trace.

Properties (b), (c), and (d) follow from property (a). The eigenvalues of A and B are the zeros of their common characteristic polynomial $p(\lambda)$ and hence are the same and have the same multiplicities. We also have $p(0) = \det A = \det B$. The traces of A and B are equal to the coefficient of $(-\lambda)^{n-1}$ in $p(\lambda)$ and so are equal.

Property 1 is often more useful in showing that two matrices are *not* similar. As an example, consider the following:

■ **EXAMPLE 3** ————————————————————————————

Suppose that A and B are 3×3-matrices. If 1, 1, and 2 are the eigenvalues of A and 1, 2, and 2 are the eigenvalues of B, then A and B have 1 and 2 as eigenvalues of different multiplicities. It follows that A and B are not similar in this case.

■ **EXAMPLE 4** ───────────────────────────────

Consider the matrices

$$A = \begin{bmatrix} 2 & 0 \\ 0 & 2 \end{bmatrix} \quad \text{and} \quad B = \begin{bmatrix} 4 & 0 \\ 0 & 1 \end{bmatrix}.$$

Here both matrices have the same determinant 4; however, the eigenvalues 2 and 2 of A are different from the eigenvalues 4 and 1 of B, so that A and B are not similar.

───────────────────────────────

The following example illustrates the fact that in each of the cases of Property 1, the converse is false.

■ **EXAMPLE 5** ───────────────────────────────

If

$$A = \begin{bmatrix} 0 & 1 & 0 & 0 \\ 0 & 0 & 1 & 0 \\ 0 & 0 & 0 & 0 \\ 0 & 0 & 0 & 0 \end{bmatrix}, \quad B = \begin{bmatrix} 0 & 1 & 0 & 0 \\ 0 & 0 & 0 & 0 \\ 0 & 0 & 0 & 1 \\ 0 & 0 & 0 & 0 \end{bmatrix},$$

it can readily be shown that (a), (b), (c), and (d) hold. However, A and B are not similar. Why?

───────────────────────────────

In addition to Property 1 similar matrices have the following

---

PROPERTY 2. Let A and B be similar matrices. Then:

(e) sA is similar to sB for every scalar s.

(f) $A - mI$ is similar to $B - mI$ for every scalar m. Moreover, if $A = mI$, then $B = mI = A$.

(g) A^k is similar to B^k for every integer k.

(h) A^T is similar to B^T.

(i) If A is invertible, so is B. Moreover, A^{-1} is similar to B^{-1}.

To establish this result, choose a nonsingular matrix P such that $B = P^{-1}AP$. Then

$$sB = P^{-1}(sA)P, \quad B - mI = P^{-1}AP - mI = P^{-1}(A - mI)P,$$

$$B^2 = P^{-1}APP^{-1}AP = P^{-1}A^2P, \quad B^3 = P^{-1}A^2PP^{-1}AP = P^{-1}A^3P, \quad \ldots,$$

$$B^k = P^{-1}A^{k-1}PP^{-1}AP = P^{-1}A^kP.$$

This establishes statements (e), (f), and (g). Next, because

$$B^T = P^T A^T (P^{-1})^T = P^T A^T (P^T)^{-1} = Q^{-1} A^T Q, \qquad Q = (P^T)^{-1},$$

it follows that A^T is similar to B^T, as stated in (h). Finally, if A is invertible, the inverse of B is given by the formula

$$B^{-1} = (P^{-1} A P)^{-1} = P^{-1} A^{-1} P.$$

We have seen that a diagonalizable matrix A is one that is similar to a diagonal matrix D. We can now once again rephrase in terms of similarity some of the key properties of the preceding section. We have:

PROPERTY 3.
   (a) An $n \times n$-matrix A is similar to a diagonal matrix D if and only if it has n linearly independent eigenvectors.
   (b) If A is similar to a diagonal matrix D, the main diagonal entries of D are the eigenvalues of A. Furthermore, if the n eigenvalues of A are distinct, the matrix A is similar to a diagonal matrix D.

According to the result in Property 3(a), the matrix A in the set

$$A = \begin{bmatrix} 2 & 1 \\ 0 & 2 \end{bmatrix}, \qquad D = \begin{bmatrix} 2 & 0 \\ 0 & 2 \end{bmatrix}$$

cannot be similar to D because A does not have two linearly independent eigenvectors. (The reader should verify this.) However, the main diagonal entries of D are the eigenvalues of A. This example illustrates that there are matrices which are not similar to diagonal matrices.

We now motivate the next result by looking at several examples.

■ **EXAMPLE 6.** ───────────────────────────────────────

Consider the matrices

$$A = \begin{bmatrix} a & 0 \\ 0 & b \end{bmatrix}, \qquad B = \begin{bmatrix} b & 0 \\ 0 & a \end{bmatrix}.$$

Observe that A and B are both diagonal matrices and that the main diagonal entries of B are just a permutation of the main diagonal entries of A. If P is the elementary (hence nonsingular) matrix

$$P = \begin{bmatrix} 0 & 1 \\ 1 & 0 \end{bmatrix},$$

then

$$AP = \begin{bmatrix} a & 0 \\ 0 & b \end{bmatrix}\begin{bmatrix} 0 & 1 \\ 1 & 0 \end{bmatrix} = \begin{bmatrix} 0 & a \\ b & 0 \end{bmatrix} = \begin{bmatrix} 0 & 1 \\ 1 & 0 \end{bmatrix}\begin{bmatrix} b & 0 \\ 0 & a \end{bmatrix} = PB.$$

Since $AP = PB$ and P is nonsingular, it follows that $P^{-1}AP = B$ (i.e., A and B are similar matrices). Moreover, if C is any 2×2-diagonal matrix

$$C = \begin{bmatrix} c & 0 \\ 0 & d \end{bmatrix},$$

then C is similar to A if and only if $C = A$ or $C = B$. This follows because if C is similar to A, then by Property 1, its eigenvalues c and d are the eigenvalues a and b of A in some order.

■ **EXAMPLE 7** _____

Consider the 3×3-diagonal matrices

$$A = \begin{bmatrix} 1 & 0 & 0 \\ 0 & 2 & 0 \\ 0 & 0 & 1 \end{bmatrix}, \qquad B = \begin{bmatrix} 2 & 0 & 0 \\ 0 & 1 & 0 \\ 0 & 0 & 1 \end{bmatrix}.$$

Once again, observe that the main diagonal entries of B are just a permutation of the main diagonal entries of A. If P is the elementary (hence nonsingular) matrix

$$P = P_{12} = \begin{bmatrix} 0 & 1 & 0 \\ 1 & 0 & 0 \\ 0 & 0 & 1 \end{bmatrix},$$

it follows that

$$AP = \begin{bmatrix} 1 & 0 & 0 \\ 0 & 2 & 0 \\ 0 & 0 & 1 \end{bmatrix}\begin{bmatrix} 0 & 1 & 0 \\ 1 & 0 & 0 \\ 0 & 0 & 1 \end{bmatrix} = \begin{bmatrix} 0 & 1 & 0 \\ 2 & 0 & 0 \\ 0 & 0 & 1 \end{bmatrix},$$

$$PB = \begin{bmatrix} 0 & 1 & 0 \\ 1 & 0 & 0 \\ 0 & 0 & 1 \end{bmatrix}\begin{bmatrix} 2 & 0 & 0 \\ 0 & 1 & 0 \\ 0 & 0 & 1 \end{bmatrix} = \begin{bmatrix} 0 & 1 & 0 \\ 2 & 0 & 0 \\ 0 & 0 & 1 \end{bmatrix},$$

so that $AP = PB$. Hence $P^{-1}AP = B$, and A and B are similar matrices.

PROPERTY 4. A diagonal matrix E is similar to a diagonal matrix D if and only if its main diagonal entries are a permutation of those of D.

For a proof of Property 4, see the Supplementary Proofs at the end of the section.

In view of Properties 3 and 4, we have

> **PROPERTY 5.** Two diagonalizable matrices A and B are similar if and only if they have the same eigenvalues with the same multiplicities.

To prove this, let D be a diagonal matrix similar to A and let E be a diagonal matrix similar to B. The matrices A and B are similar if and only if D and E are similar. The matrix E is similar to D if and only if its main diagonal entries, the eigenvalues of B, are a permutation of the main diagonal entries of D, that is, of the eigenvalues of A. In other words, A and B are similar if and only if they have the same eigenvalues with the same multiplicities.

Since symmetric matrices are diagonalizable, we have

> **PROPERTY 6.** Two symmetric matrices A and B are similar if and only if they have the same eigenvalues with the same multiplicities.

■ **EXAMPLE 8**

As a final example, suppose that

$$A = \begin{bmatrix} 5 & 8 \\ 2 & 5 \end{bmatrix}, \qquad B = A^T = \begin{bmatrix} 5 & 2 \\ 8 & 5 \end{bmatrix}.$$

In this case, the eigenvalues of A are the same as the eigenvalues of B. They are $\lambda_1 = 9$ and $\lambda_2 = 1$. Since the eigenvalues are distinct, both A and B are diagonalizable. From Property 5 it follows that A and $B = A^T$ are similar.

This example, in fact, illustrates the more general property.

> **PROPERTY 7.** Let A be an $n \times n$-matrix. Then A is similar to its transpose A^T.

By Property 5, this holds when A is diagonalizable. The proof of this result in the general case is complicated and will not be given here.

We have seen that an $n \times n$-matrix A is similar to a diagonal matrix D when A has n linearly independent eigenvectors. This condition was satisfied if all of the eigenvalues of A are distinct or if A is symmetric. We have also seen examples of matrices that were not similar to a diagonal matrix. This raises the question as to whether nondiagonalizable matrices can be made similar to matrices that are "almost diagonalizable." In the two-dimensional case, if

$$A = \begin{bmatrix} a & b \\ c & d \end{bmatrix}$$

where a, b, c, and d are real entries, it turns out that A is similar to a matrix of one of the following three types:

$$D = \begin{bmatrix} r & 0 \\ 0 & s \end{bmatrix}, \quad E = \begin{bmatrix} r & 1 \\ 0 & r \end{bmatrix}, \quad F = \begin{bmatrix} r & s \\ -s & r \end{bmatrix}.$$

These types are determined by the eigenvalues and eigenvectors of A. These matrices are called **canonical 2 × 2 matrices**. In the complex case, a matrix of type F is similar to a complex diagonal matrix D.

Turning to 3 × 3-matrices, we remark that a real 3 × 3-matrix A is similar to a matrix of one of the following four types:

$$D = \begin{bmatrix} r & 0 & 0 \\ 0 & s & 0 \\ 0 & 0 & t \end{bmatrix}, \quad E = \begin{bmatrix} r & 1 & 0 \\ 0 & r & 0 \\ 0 & 0 & t \end{bmatrix},$$

$$F = \begin{bmatrix} r & 1 & 0 \\ 0 & r & 1 \\ 0 & 0 & r \end{bmatrix}, \quad G = \begin{bmatrix} r & s & 0 \\ -s & r & 0 \\ 0 & 0 & t \end{bmatrix}.$$

Matrices of these types are called canonical real 3 × 3 matrices. Accordingly, we have four cases:

1. A is similar to a diagonal matrix D when A has three linearly independent (real) eigenvectors. This result was established in Property 3.

2. A is similar to a matrix of type E when A has at least a double eigenvalue and has two linearly independent eigenvectors with the property that every other eigenvector is a linear combination of these two.

3. A is similar to a matrix of type F when A has a triple eigenvalue and its eigenvectors are multiples of a single eigenvector.

4. A is similar to a matrix of type G when A has complex eigenvalues. In the complex case G is similar to a complex diagonal matrix D.

We shall not pause to establish these results here.

In the $n \times n$ case a matrix A is similar, by a real or complex matrix X, to a special bidiagonal matrix J that we call a Jordan matrix. A 5 × 5 bidiagonal matrix is of the form

$$
B = \begin{bmatrix}
\lambda_1 & b_1 & 0 & 0 & 0 \\
0 & \lambda_2 & b_2 & 0 & 0 \\
0 & 0 & \lambda_3 & b_3 & 0 \\
0 & 0 & 0 & \lambda_4 & b_4 \\
0 & 0 & 0 & 0 & \lambda_5
\end{bmatrix}.
$$

The diagonal just above the main diagonal is called the **super diagonal**. In the $n \times n$ case a matrix B is a **bidiagonal matrix** if its entries are zero except possibly on its main diagonal and on its super diagonal, as indicated above in the 5×5 case. Its main diagonal entries $\lambda_1, \lambda_2, \ldots, \lambda_n$ are its eigenvalues. For the present we denote the super-diagonal entries by $b_1, b_2, \ldots, b_{n-1}$.

A bidiagonal matrix B will be called a **canonical bidiagonal matrix** or a **Jordan matrix** if it has the following additional properties:

1. The super-diagonal entries $b_1, b_2, \ldots, b_{n-1}$ are either 0 or 1.
2. If s successive super-diagonal entries

$$
b_k, b_{k+1}, \ldots, b_{k+s-1}
$$

   are 1's, then the $s + 1$ main diagonal entries

$$
\lambda_k, \lambda_{k+1}, \ldots, \lambda_{k+s}
$$

are equal. A diagonal matrix is a Jordan matrix whose super-diagonal entries are zero. The zero matrix is also a Jordan matrix.

The 5×5 matrices

$$
\begin{bmatrix}
3 & 1 & 0 & 0 & 0 \\
0 & 3 & 1 & 0 & 0 \\
0 & 0 & 3 & 0 & 0 \\
0 & 0 & 0 & 2 & 1 \\
0 & 0 & 0 & 0 & 2
\end{bmatrix}, \qquad
\begin{bmatrix}
4 & 1 & 0 & 0 & 0 \\
0 & 4 & 1 & 0 & 0 \\
0 & 0 & 4 & 1 & 0 \\
0 & 0 & 0 & 4 & 0 \\
0 & 0 & 0 & 0 & 5
\end{bmatrix}
$$

are Jordan matrices. So are the matrices

$$
\begin{bmatrix}
3 & 1 & 0 & 0 & 0 \\
0 & 3 & 0 & 0 & 0 \\
0 & 0 & 2 & 1 & 0 \\
0 & 0 & 0 & 2 & 0 \\
0 & 0 & 0 & 0 & 2
\end{bmatrix}, \qquad
\begin{bmatrix}
4 & 1 & 0 & 0 & 0 \\
0 & 4 & 1 & 0 & 0 \\
0 & 0 & 4 & 1 & 0 \\
0 & 0 & 0 & 4 & 1 \\
0 & 0 & 0 & 0 & 4
\end{bmatrix}.
$$

Notice that a Jordan matrix has diagonal blocks that are Jordan matrices with 1's as its super-diagonal entries. Observe further that its main diagonal entries could all be the same.

We state without proof the following property.

PROPERTY 8. The following similarity relations hold:

(a) A real matrix A, whose eigenvalues are real, is similar by a real matrix X to a Jordan matrix J.

(b) A real matrix A having some complex eigenvalues is similar by a complex matrix X to a Jordan matrix J.

(c) A complex matrix A is similar to a Jordan matrix J.

Exercises

1. Find the eigenvalues and eigenvectors of the following 2×2 matrices. In each case find the canonical matrix of type D, E, or F which is similar to the given matrix.

$$U = \begin{bmatrix} 5 & 1 \\ -1 & 7 \end{bmatrix}, \quad V = \begin{bmatrix} 5 & 1 \\ 1 & 5 \end{bmatrix}, \quad W = \begin{bmatrix} 4 & 2 \\ -2 & 8 \end{bmatrix}, \quad X = \begin{bmatrix} 10 & 3 \\ 0 & -5 \end{bmatrix}.$$

Are any of these matrices similar?

2. Show that the following matrices P and Q are similar. Also show that R and S are similar.

$$P = \begin{bmatrix} 0 & 1 \\ 0 & 0 \end{bmatrix}, \quad Q = \begin{bmatrix} 0 & 4 \\ 0 & 0 \end{bmatrix}, \quad R = \begin{bmatrix} 5 & 1 \\ 0 & 5 \end{bmatrix}, \quad S = \begin{bmatrix} 5 & 4 \\ 0 & 5 \end{bmatrix}.$$

3. Construct a real 2×2 matrix having the complex number $2 + 3i$ as an eigenvalue.

4. Construct two distinct symmetric 2×2-matrices having 2 and 1 as their eigenvalues. Are they similar?

5. Show that the matrices

$$A = \begin{bmatrix} 2 & 0 \\ 0 & 2 \end{bmatrix}, \quad B = \begin{bmatrix} 2 & 1 \\ 0 & 2 \end{bmatrix}$$

are not similar.

6. Suppose that the eigenvalues of A are distinct and λ is not zero. Show that A and $A - \lambda I$ are not similar.

7. Let A and B have the same eigenvalues. Suppose that these eigenvalues are distinct. Show that A and B are similar.

8. Show that the matrices

$$A = \begin{bmatrix} 2 & 0 \\ 0 & 3 \end{bmatrix}, \quad B = \begin{bmatrix} 2 & 5 \\ 0 & 3 \end{bmatrix}, \quad C = \begin{bmatrix} 3 & 0 \\ 7 & 2 \end{bmatrix}, \quad D = \begin{bmatrix} 4 & -1 \\ 2 & 1 \end{bmatrix}$$

are similar matrices.

9. Consider the matrices

$$A = \begin{bmatrix} 2 & 0 \\ 0 & 3 \end{bmatrix}, \quad B = \begin{bmatrix} 4 & 5 \\ 0 & 6 \end{bmatrix}, \quad C = \begin{bmatrix} 9 & 0 \\ 7 & 6 \end{bmatrix}, \quad D = \begin{bmatrix} 8 & -2 \\ 4 & 2 \end{bmatrix}.$$

Show that B and D are similar to $2A$ and that C is similar to $3A$. Why are B and D similar? *Hint:* Use Property 2(e).

10. Find the eigenvalues and eigenvectors of the following 3×3 matrices.

$$K = \begin{bmatrix} 1 & 2 & 3 \\ 0 & -1 & 2 \\ 0 & 0 & 0 \end{bmatrix}, \quad L = \begin{bmatrix} 2 & 0 & 1 \\ 0 & 1 & 0 \\ 0 & 0 & 2 \end{bmatrix}, \quad M = \begin{bmatrix} 1 & 0 & 0 \\ 1 & 1 & 0 \\ 0 & 1 & 1 \end{bmatrix}, \quad N = \begin{bmatrix} 1 & 1 & 1 \\ -1 & 1 & 1 \\ 0 & 0 & 1 \end{bmatrix}.$$

Obtain a canonical matrix similar to each of these matrices.

11. Construct three nonsimilar matrices having 2, 2, and 2 as eigenvalues.

12. Prove that A is diagonalizable if and only if A is similar to a symmetric matrix.

13. Consider the matrices

$$A = \begin{bmatrix} 11 & 7 & -9 \\ 7 & 11 & -9 \\ -9 & -9 & 27 \end{bmatrix}, \quad B = \begin{bmatrix} 3 & 1 & -1 \\ 1 & 3 & -1 \\ -1 & -1 & 5 \end{bmatrix},$$

$$X = \begin{bmatrix} 1/\sqrt{6} & 1/\sqrt{3} & -\sqrt{2} \\ 1/\sqrt{6} & 1/\sqrt{3} & 1/\sqrt{2} \\ -2/\sqrt{6} & 1/\sqrt{3} & 0 \end{bmatrix}.$$

(a) Show that

$$D = X^T A X = \begin{bmatrix} 36 & 0 & 0 \\ 0 & 9 & 0 \\ 0 & 0 & 4 \end{bmatrix}, \quad E = X^T B X = \begin{bmatrix} 6 & 0 & 0 \\ 0 & 3 & 0 \\ 0 & 0 & 2 \end{bmatrix}.$$

(b) Conclude that the columns of X are unit eigenvectors of both A and B. Observe that the eigenvalues 6, 3, and 2 of B are square roots of the eigenvalues 36, 9, and 4 of A.

(c) Verify that $B^2 = A$ and that $E^2 = D$.

14. Let P, Q, Y be $m \times m$-matrices and R, S, Z be $n \times n$-matrices. Suppose that Y and Z are nonsingular. Construct the block diagonal matrices.

$$A = \begin{bmatrix} P & 0 \\ 0 & R \end{bmatrix}, \quad B = \begin{bmatrix} Q & 0 \\ 0 & S \end{bmatrix}, \quad C = \begin{bmatrix} R & 0 \\ 0 & P \end{bmatrix}, \quad X = \begin{bmatrix} Y & 0 \\ 0 & Z \end{bmatrix}.$$

(a) Show that if Q is similar to P by Y and S is similar to R by Z, then B is similar to A by X.

(b) Show that there is a permutation matrix P such that $P^T A P = C$. Conclude that A and C are similar.

15. Let A, B, and P be matrices connected by the relation

$$B = P^{-1} A P.$$

Show that if **y** is an eigenvector of B, then $\mathbf{x} = P\mathbf{y}$ is an eigenvector of A. Conclude that if $\mathbf{y}_1, \ldots, \mathbf{y}_k$ is a maximal set of linearly independent eigenvectors for B, then $\mathbf{x}_1 = P\mathbf{y}_1, \ldots, \mathbf{x}_k = P\mathbf{y}_k$ form a maximal set of linearly independent eigenvectors for A. It will be convenient to call k the *number* of eigenvectors of B and of A. Hence if two matrices are similar, they have the same number of eigenvectors.

16. What are the number of eigenvectors of each of the following matrices?

$$A = \begin{bmatrix} 0 & 1 & 0 & 0 \\ 0 & 0 & 1 & 0 \\ 0 & 0 & 0 & 1 \\ 0 & 0 & 0 & 0 \end{bmatrix}, \quad B = \begin{bmatrix} 0 & 1 & 0 & 0 \\ 0 & 0 & 1 & 0 \\ 0 & 0 & 0 & 0 \\ 0 & 0 & 0 & 0 \end{bmatrix},$$

$$C = \begin{bmatrix} 0 & 1 & 0 & 1 \\ 0 & 0 & 0 & 0 \\ 0 & 0 & 0 & 0 \\ 0 & 0 & 0 & 0 \end{bmatrix}, \quad D = \begin{bmatrix} 0 & 1 & 0 & 0 \\ 0 & 0 & 0 & 0 \\ 0 & 0 & 0 & 1 \\ 0 & 0 & 0 & 0 \end{bmatrix}.$$

Show further that no two of these matrices are similar.

17. Let P_{23} be the elementary permutation matrix obtained from I_4 by interchanging the second and third columns of I_4. The matrix P_{23} is its own inverse. Conclude that the matrix D described in Exercise 16 is similar to the matrix

$$E = P_{23}DP_{23} = \begin{bmatrix} 0 & 0 & 1 & 0 \\ 0 & 0 & 0 & 1 \\ 0 & 0 & 0 & 0 \\ 0 & 0 & 0 & 0 \end{bmatrix}.$$

18. By the order of a nilpotent matrix N is meant the least integer k such that $N^k = 0$. Observe that if $k = 1$, then $N = 0$. Verify that:

(a) If M is similar to a nilpotent matrix N, then M is nilpotent and has the same order as N.

(b) A nilpotent matrix is singular.

(c) No nonsingular matrix A is similar to a nilpotent matrix.

(d) If N is nilpotent, the matrix $A = cI + N$ is nonsingular unless $c = 0$.

(e) A nonzero nilpotent matrix N is similar to none of its powers $N^2, N^3, \ldots$.

19. Select three nonzero numbers p, q, and r. Construct matrices of the form

$$A = \begin{bmatrix} a & p & 0 & 0 \\ 0 & b & q & 0 \\ 0 & 0 & c & r \\ 0 & 0 & 0 & d \end{bmatrix}, \quad B = \begin{bmatrix} a & 1 & 0 & 0 \\ 0 & b & 1 & 0 \\ 0 & 0 & c & 1 \\ 0 & 0 & 0 & d \end{bmatrix}, \quad X = \begin{bmatrix} 1 & 0 & 0 & 0 \\ 0 & p & 0 & 0 \\ 0 & 0 & pq & 0 \\ 0 & 0 & 0 & pqr \end{bmatrix}.$$

Verify that $XAX^{-1} = B$. Conclude that A is similar to B.

20. Let D be a diagonal matrix whose main diagonal entries $\lambda_1, \lambda_2, \ldots, \lambda_n$ are nonnegative. The diagonal matrix whose main diagonal entries are the nonnegative square roots

$$\mu_1 = \sqrt{\lambda_1}, \quad \mu_2 = \sqrt{\lambda_2}, \quad \ldots, \quad \mu_n = \sqrt{\lambda_n}$$

of $\lambda_1, \lambda_2, \ldots, \lambda_n$ is called the *square root* of D and is denoted by $D^{1/2}$. Verify that

$$(D^{1/2})^2 = D.$$

Show that in Exercise 13, we have $E = D^{1/2}$.

21. Let A be a symmetric matrix. Select an orthogonal matrix X such that the matrix

$$D = X^T A X$$

is a diagonal matrix. The main diagonal entries, $\lambda_1, \lambda_2, \ldots, \lambda_n$ of D are eigenvalues of A.

(a) Show that under the transformation

$$\mathbf{x} = X\mathbf{y}$$

we have

$$\mathbf{x}^T A \mathbf{x} = \mathbf{y}^T D \mathbf{y} = \lambda_1 y_1^2 + \lambda_2 y_2^2 + \cdots + \lambda_n y_n^2,$$

where $y_1, y_2, \ldots, y_n$ are the components of $\mathbf{y}$.

(b) Suppose that the eigenvalues $\lambda_1, \lambda_2, \ldots, \lambda_n$ of A are nonnegative. Then, by Exercise 20, the matrix

$$D = X^T A X = \text{diag}\,(\lambda_1, \lambda_2, \ldots, \lambda_n)$$

has a square root $D^{1/2}$. Show that the matrix

$$A^{1/2} = X D^{1/2} X^T \quad \text{has the property that} \quad (A^{1/2})^2 = A.$$

The matrix $A^{1/2}$ is called the *square root* of A.

(c) Show that in Exercise 13, we have $B = A^{1/2}$.

22. A symmetric matrix A is said to be *positive definite* if the inequality

$$\mathbf{x}^T A \mathbf{x} > 0$$

holds unless $\mathbf{x} = \mathbf{0}$. Show that:

(a) A symmetric matrix A is positive definite if and only if its eigenvalues $\lambda_1, \lambda_2, \ldots, \lambda_n$ are all positive. *Hint:* Use the result in part (a) of Exercise 21.

(b) A positive definite matrix is nonsingular and so is invertible.

(c) The inverse of a positive definite symmetric matrix is a positive definite symmetric matrix.

23. Show that if Y is a nonsingular $n \times n$-matrix, the symmetric matrices YY^T and $Y^T Y$ are positive definite.

24. Let A be a diagonalizable square matrix. Select a nonsingular matrix Y such that

$$D = Y^{-1} A Y$$

is a diagonal matrix. We have $AY = YD$. The matrix $P = YY^T$ is a positive definite symmetric matrix.

(a) Show that $Q = AP = AYY^T = YDY^T$ is symmetric.

(b) Conclude that $A = QP^{-1}$ is the product of two symmetric matrices, one of which is positive definite.

(c) Conclude that because $Q = Q^T$, we have $AP = PA^T$. This tells us there exists a positive definite matrix P such that $AP = PA^T$.

(d) Conclude that the zeros of det $(Q - \lambda P)$ are the eigenvalues of A.

25. Let Q and R be symmetric matrices. Suppose that R is positive definite. Show that $A = QR$ and $B = RQ = RAR^{-1}$ are diagonalizable. *Hint:* Set $S = R^{1/2}$. Then $RS^{-1} = S$. Moreover, the matrix $F = SAS^{-1} = SQRS^{-1} = SQS$ is symmetric. Conclude that A is similar to the symmetric matrix F and so is diagonalizable. Because B is similar to A, it too is diagonalizable.

26. From the results given in Exercises 24 and 25 conclude that a matrix A is diagonalizable if and only if it is the product of two symmetric matrices, one of which is positive definite.

Supplementary Proof

▬ **P1.** Proof of Property 4:

In the general case, if D is a diagonal matrix and P_{jk} is an elementary matrix of the first kind, P_{jk} is its own inverse. Moreover, the matrix $P_{jk}DP_{jk}$ is similar to D and is obtained from D by interchanging the jth and kth main diagonal entries of D. Since every permutation P is a product $P = P_1P_2 \cdots P_k$ of elementary matrices of the first kind, its inverse is of the form $P^{-1} = P_k \cdots P_2P_1$, in reverse order. The matrix

$$E = P^{-1}DP = P_k \cdots P_2P_1DP_1P_2 \cdots P_k$$

is a diagonal matrix similar to D. The main diagonal entries of E are a permutation of the main diagonal entries of D. In this manner it is seen that if E is a diagonal matrix whose main diagonal entries are a permutation of those of D, then E is similar to D. Conversely, if a diagonal matrix E is similar to D its eigenvalues are the eigenvalues of D with the same multiplicities. Since the main diagonal entries of E are its eigenvalues, it follows that these entries are a permutation of the main diagonal entries of D, the eigenvalues of D. ▬

4.5   GERSCHGORIN INTERVALS AND DISKS

Sometimes, in applications involving a square matrix A, all that is needed is an approximate location of its eigenvalues. In stability problems, for example, we wish to know if the eigenvalues of A are negative or, if complex, if the real parts of the eigenvalues are negative. Information of this type can sometimes (but not always) be obtained by the use of Gerschgorin intervals or disks. Consider, for example, the matrices

$$D = \begin{bmatrix} 2 & 0 & 0 \\ 0 & -5 & 0 \\ 0 & 0 & 9 \end{bmatrix}, \quad E = \begin{bmatrix} 2 & 1 & 0 \\ 1 & -5 & -1 \\ 1 & -2 & 9 \end{bmatrix}, \quad F = \begin{bmatrix} 2 & -7 & 4 \\ 7 & -5 & -8 \\ 3 & -3 & 9 \end{bmatrix}.$$

In the case of the matrix D we see, by inspection, that its eigenvalues are 2, -5, and 9. For the matrix E let r_1, r_2, and r_3 be the sums of the absolute values of the off-diagonal entries in its three rows, respectively. We have

$$r_1 = 1 + 0 = 1, \quad r_2 = 1 + |-1| = 2, \quad r_3 = 1 + |-2| = 3.$$

Its main diagonal entries are 2, -5, and 9. According to the result described below, an eigenvalue λ of E must lie on one of the intervals

$$|2 - \lambda| \le r_1 = 1, \quad |-5 - \lambda| \le r_2 = 2, \quad |9 - \lambda| \le r_3 = 3.$$

These intervals can be written in the form

$$1 \le \lambda \le 3, \quad -7 \le \lambda \le -3, \quad 6 \le \lambda \le 12.$$

In fact, since these intervals do not overlap, each interval contains one eigenvalue of E. We conclude that the matrix E has two positive eigenvalues and one negative eigenvalue.

We next apply this procedure to the matrix F. Let

$$r_1 = |-7| + 4 = 11, \quad r_2 = 7 + |-8| = 15, \quad r_3 = 3 + |-3| = 6$$

be the sums of the absolute values of the off-diagonal entries in each of the three rows of F. The main diagonal entries of F are again 2, -5, and 9. A real eigenvalue λ of F must lie on one of the intervals

$$|2 - \lambda| \le r_1 = 11, \quad |-5 - \lambda| \le r_2 = 15, \quad |9 - \lambda| \le r_3 = 6.$$

Writing these intervals in the form

$$-9 \le \lambda \le 13, \quad -20 \le \lambda \le 10, \quad 3 \le \lambda \le 15,$$

we see that they overlap. At the moment all that we can say about the real eigenvalues of F is that they lie somewhere on the union of these intervals and so cannot exceed 15 or be less than -20.

To justify the arguments just made, consider the general 3×3-matrix

$$A = \begin{bmatrix} a_{11} & a_{12} & a_{13} \\ a_{21} & a_{22} & a_{23} \\ a_{31} & a_{32} & a_{33} \end{bmatrix}.$$

Let λ be an eigenvalue of A and let $\mathbf{x} = (x_1, x_2, x_3)$ be a corresponding eigenvector of A. We can suppose that this eigenvector has been scaled so that one of its components is 1 and the absolute values of each of the other two components do not exceed 1. Suppose for definiteness that $x_2 = 1$ and that $|x_1| \leq 1$, $|x_3| \leq 1$. The second equation in the eigenvector equation $(A - \lambda I)\mathbf{x} = \mathbf{0}$ is the equation

$$a_{21}x_1 + (a_{22} - \lambda)x_2 + a_{23}x_{23} = 0.$$

Since $x_2 = 1$ and $|x_1| \leq 1$, $|x_3| \leq 1$, we conclude from this equation that

$$|a_{22} - \lambda| = |-a_{21}x_1 - a_{23}x_3| \leq |a_{21}| + |a_{23}|.$$

The eigenvalue λ therefore must lie on the interval

$$|a_{22} - \lambda| \leq r_2 = |a_{21}| + |a_{23}|.$$

This justifies the arguments made with regard to the matrices E and F described above.

The arguments just made for the case $n = 3$ can be extended to the $n \times n$-case. This gives the following:

GERSCHGORIN'S THEOREM   Let r_i be the sum of the absolute values of the off-diagonal entries in the ith row of a real $n \times n$-matrix

$$A = [a_{ij}], \qquad i, j = 1, \ldots, n.$$

Then

(a) Each real eigenvalue λ of A lies on one of the Gerschgorin intervals

$$|a_{ii} - \lambda| \leq r_i, \qquad i = 1, \ldots, n$$

and so is in the union of these intervals.

(b) Suppose that the eigenvalues of A are all real. Suppose further that the union $\mathbf{U}_k$ of k of these intervals has no point in common with the union $\mathbf{V}_k$ of the $n - k$ remaining intervals. Then there are k eigenvalues in $\mathbf{U}_k$ and $n - k$ eigenvalues in $\mathbf{V}_k$.

As remarked earlier, conclusion (a) can be obtained by arguments like that made above for the case $n = 3$. Conclusion (b) follows from conclusion (b) in the next theorem.

■ **EXAMPLE 1** _____

Let us apply conclusion (b) to the 3×3-matrix $A = E$ described at the beginning of this section. It can be shown that the eigenvalues of E are real. Let $\mathbf{U}_2$ be the union of the first two Gerschgorin intervals

$$|2 - \lambda| \leq 1, \qquad |-5 - \lambda| \leq 2$$

for E. Then $\mathbf{V}_2$ is the Gerschgorin interval $|9 - \lambda| \le 3$. Clearly, $\mathbf{U}_2$ and $\mathbf{V}_2$ have no points in common. It follows from conclusion (b) that $\mathbf{U}_2$ contains two eigenvalues of E and that the Gerschgorin interval $\mathbf{V}_2$ contains a single eigenvalue of E. A similar argument will show that each of the Gerschgorin intervals for E contains a single eigenvalue of E.

Gerschgorin theorem 1 is deficient in that it does not take into account complex eigenvalues of a real matrix A. We now give an extension of this result that takes complex eigenvalues into account. In fact this extension, given in Gerschgorin theorem 2 below, applies to complex matrices as well as real matrices. Gerschgorin theorem I is an immediate consequence of Gerschgorin theorem II.

To make this extension so as to include complex eigenvalues of a real or complex matrix A, recall that if $z = u + iv$ is a complex number with u and v real, then z can be identified with the point (u, v) in the uv-plane. Moreover,

$$|z| = (u^2 + v^2)^{1/2}.$$

If $a = \alpha + i\beta$ is a fixed complex number corresponding to the point (α, β) and r is a positive number, the inequality

$$|a - z| \le r$$

is equivalent to the inequality

$$(\alpha - u)^2 + (\beta - v)^2 \le r^2.$$

The set of points (u, v) satisfying this inequality forms a disk of radius r with (α, β) as its center. Thus, in terms of complex numbers, the inequality

$$|a - z| \le r$$

represents a disk of radius r with a as its center.

We now state

GERSCHGORIN'S THEOREM II. Let r_i be the sum of the absolute values of the off-diagonal entries in the ith row of an $n \times n$-matrix

$$A = [a_{ij}], \qquad i, j = 1, \ldots, n.$$

Then

(a) Each eigenvalue λ of A lies in one of the Gerschgorin disks

$$|a_{ii} - \lambda| \le r_i, \qquad i = 1, \ldots, n$$

and so is in the union of these disks.

(b) Suppose that the union $\mathbf{U}_k$ of k of these disks has no point in common with the union $\mathbf{V}_k$ of the $n - k$ remaining disks. Then there are k eigenvalues in $\mathbf{U}_k$ and $n - k$ eigenvalues in $\mathbf{V}_k$.

The proof of conclusion (a) is like that of conclusion (a) in Gerschgorin theorem I, taking into account the geometrical interpretation of complex numbers.

Conclusion (b) can be established by a continuity argument. We shall sketch such a continuity argument. It proceeds as follows. Write A in the form

$$A = D + H,$$

where D is the diagonal matrix whose main diagonal entries are those of A. For t on the interval $0 \le t \le 1$, set

$$A(t) = D + tH.$$

Then $A(0) = D$ and $A(1) = A$. Let $\mathbf{U}_k$ and $\mathbf{V}_k$ be unions of Gerschgorin disks as described in conclusion (b). The centers of these Gerschgorin disks are the eigenvalues of $A(0) = D$. It follows that $A(0) = D$ has k eigenvalues in $\mathbf{U}_k$ and $n - k$ eigenvalues in $\mathbf{V}_k$. The eigenvalues of $A(t)$ are continuous functions of t. As t varies from 0 to 1, the k eigenvalues of $A(t)$ which were in $\mathbf{U}_k$ initially will remain in $\mathbf{U}_k$. Similarly, the $n - k$ eigenvalues that were in $\mathbf{V}_k$ initially will remain in $\mathbf{V}_k$. It follows that $\mathbf{U}_k$ contains k of the eigenvalues of $A(1) = A$ and that $\mathbf{V}_k$ contains $n - k$ of the eigenvalues of A, as stated in conclusion (b).

COROLLARY. Suppose that no two of the Gerschgorin disks for an $n \times n$-matrix A have a point in common. Then each of the n Gerschgorin disks contains a single eigenvalue. If A is real, its n eigenvalues are real.

For let $\mathbf{U}_1$ be one of these Gerschgorin disks and let $\mathbf{V}_1$ be the union of the remaining disks. Then $\mathbf{U}_1$ and $\mathbf{V}_1$ have no points in common. By conclusion (b), $\mathbf{U}_1$ contains a single eigenvalue. It follows that each disk contains a single eigenvalue. When A is real, these disks are centered on the real axis, so that no two eigenvalues can be complex conjugates of each other. This means that the eigenvalues of A are real, when its Gerschgorin disks are disjoint.

■ **EXAMPLE 2** ─────────────────────────────────

The two Gerschgorin disks for the matrix

$$A = \begin{bmatrix} -3 & 2 \\ -2 & -3 \end{bmatrix}$$

coincide. This disk is defined by the equation

$$|-3 - \lambda| \le 2$$

and is centered at the point $-3 + 0i$, that is, the point $(-3, 0)$. The real part u of the point $\lambda = u + iv$ in this disk is negative. Because the eigenvalues of A are in this disk, they have negative real parts. In fact, $\lambda_1 = -3 + 2i$ and $\lambda_2 = -3 - 2i$ are the eigenvalues of A. The matrix

$$B = \begin{bmatrix} -3 & 2 \\ 2 & -3 \end{bmatrix}$$

has the same Gerschgorin disk, so its eigenvalues also have negative real parts. They are the negative real numbers $\lambda_1 = -5$ and $\lambda_2 = -1$.

■ **EXAMPLE 3** _____

The matrix

$$C = \begin{bmatrix} 2 & 1 \\ 2 & 8 \end{bmatrix}$$

has two nonintersecting Gerschgorin disks. They are defined by the equations

$$|2 - \lambda| \le 1, \qquad |8 - \lambda| \le 2.$$

Since these disks have no points in common, each disk must contain an eigenvalue of C. No point in the first disk can be conjugate to a point in the second disk. Because C is real, its eigenvalues are either real or else a pair of conjugate complex numbers. Consequently, the eigenvalues of C are real. Find them.

■ **EXAMPLE 4** _____

The Gerschgorin disks of the matrix

$$G = \begin{bmatrix} i & 1 \\ 1 & -i \end{bmatrix}$$

are defined by the inequalities

$$|i - \lambda| \le 1, \qquad |-i - \lambda| \le 1.$$

They are centered at the points $0 + i$ and $0 - i$ and have the point $\lambda = 0$ in common. It turns out that $\lambda = 0$ is an eigenvalue of G of multiplicity 2. Here is a complex matrix whose eigenvalues are real.

A second set of Gerschgorin disks for a matrix A can be obtained by defining the radius r_i of the ith disk to be the sum of the absolute values of the off-diagonal entries in the ith column of A. This follows because the columns of A are the rows of its transpose A^T and the eigenvalues of A^T are the eigenvalues of A. Thus the Gerschgorin disks for A^T using rows gives the Gerschgorin disks of A using columns.

Exercises

1. Find the Gerschgorin disks for each of the following matrices. In each case, what conclusions can you draw with regard to the location of its eigenvalue?

$$A = \begin{bmatrix} 5 & 1 \\ 2 & 9 \end{bmatrix}, \qquad B = \begin{bmatrix} 5 & 4 \\ 4 & -5 \end{bmatrix}, \qquad C = \begin{bmatrix} 5 & 4 \\ 4 & 5 \end{bmatrix}, \quad D = \begin{bmatrix} 1 & 1 \\ -1 & 1 \end{bmatrix},$$

$$E = \begin{bmatrix} 3 & -1 & 1 \\ -1 & 5 & -1 \\ 1 & -1 & 3 \end{bmatrix}, \quad F = \begin{bmatrix} 3 & 0 & 1 \\ -1 & 7 & 1 \\ 0 & 1 & 0 \end{bmatrix}, \quad G = \begin{bmatrix} 9 & -2 & 1 \\ 0 & -5 & 3 \\ 6 & 5 & 21 \end{bmatrix}.$$

2. Repeat Exercise 1 with each matrix replaced by its transpose. What additional information do you obtain?

3. Matrices of the form

$$H = \begin{bmatrix} 2 & -1 & 0 & 0 \\ -1 & 2 & -1 & 0 \\ 0 & -1 & 2 & -1 \\ 0 & 0 & -1 & 2 \end{bmatrix}, \quad K = \begin{bmatrix} 4 & -1 & 0 & 0 \\ -1 & 4 & -1 & 0 \\ 0 & -1 & 4 & -1 \\ 0 & 0 & -1 & 4 \end{bmatrix}$$

are encountered in the study of numerical solutions of ordinary and partial equations. Since they are symmetric, they have real eigenvalues. Find the Gerschgorin intervals for these matrices.

4. Use Gerschgorin disks to show that the following matrices have real and distinct eigenvalues.

$$L = \begin{bmatrix} 1 & 1 & 1 \\ 1 & 5 & 0 \\ 1 & 0 & 9 \end{bmatrix}, \quad M = \begin{bmatrix} 2 & 1 & 1 \\ 0 & 5 & 0 \\ 1 & 0 & 9 \end{bmatrix}, \quad N = \begin{bmatrix} 5 & 2 & 0 \\ 1 & 1 & 0 \\ 0 & 1 & -3 \end{bmatrix}.$$

5. Show that -2 and 4 are the eigenvalues of the matrix

$$\begin{bmatrix} 2 & 8 \\ 1 & 0 \end{bmatrix}.$$

The Gerschgorin disk $|\lambda - 2| \leq 8$ contains both eigenvalues. But the Gerschgorin disk $|\lambda| \leq 1$ contains no eigenvalue. Conclude that a Gerschgorin disk for a matrix need not contain an eigenvalue.

6. Show that $1 + i$ and $1 - i$ are the eigenvalues of the matrix

$$\begin{bmatrix} -2 & -2 \\ 1 & 0 \end{bmatrix}.$$

Show that the Gerschgorin disk $|\lambda + 2| \leq 2$ contains these two eigenvalues. This illustrates the fact that for a real matrix A, a Gerschgorin disk that contains a complex eigenvalue $\lambda_1 = a + bi$ of A also contains its complex conjugate $\lambda_2 = a - bi$. Prove this.

7. Let d_i be the ith main diagonal entry of an $n \times n$-matrix A. Let r_i be the sum of the absolute values of the off-diagonal entries in the ith row.

(a) Show that if for a fixed index i, the inequality

$$|d_i - d_j| > r_i + r_j$$

for all j distinct from i, there is one and only one eigenvalue λ of A satisfying the inequality

$$|\lambda - d_i| < r_i.$$

If A is real, so is λ.

(b) If the hypothesis in part (a) holds for every index i, the eigenvalues of A are distinct. If A is real, its eigenvalues are real. Moreover, for each i, there is a unique eigenvalue λ_i of A satisfying the inequality

$$|\lambda - d_i| \leq r_i.$$

Hint: This result follows from (b) in the Corollary of Gerschgorin theorem II with $k = 1$.

8. Apply the result described in Exercise 6 to the matrices described in Exercise 4.

9. Consider the matrices

$$P = \begin{bmatrix} 3 & 1 & 1 \\ 1 & 7 & 0 \\ 1 & 0 & 11 \end{bmatrix}, \qquad Q = \begin{bmatrix} 3 & 1 & 1 \\ -1 & -5 & 0 \\ 1 & 0 & 5 \end{bmatrix}, \qquad R = \begin{bmatrix} 5 & 2 & 0 \\ 1 & -2 & 0 \\ 0 & 1 & -5 \end{bmatrix}.$$

(a) Show that the eigenvalues of P are positive.

(b) Show that Q has two positive eigenvalues and one negative eigenvalue.

(c) Show that R has one positive eigenvalue and two negative eigenvalues.

10. For an upper or lower triangular matrix, the diagonal entries are its eigenvalues. Consequently, show that each of its Gerschgorin disks automatically contain an eigenvalue: its center.

4.6   APPLICATIONS INVOLVING CALCULUS

One of the basic differential equations we encounter in calculus is the equation

$$\frac{dy}{dt} = ay,$$

where a is a constant. It has many applications. For example, in the study of populations, this equation is sometimes used to describe the rate of growth of a population y. In this event our equation states that the time rate of change of y is proportional to y. Its solution is

$$y = ce^{at},$$

where c is the value of y at $t = 0$. When $a > 0$, y is growing. When $a < 0$, y is decaying. In mathematics, our differential equation can be viewed to be the defining equation for the exponential function e^{at}.

A more general growth equation is the differential equation

$$\frac{d\mathbf{y}}{dt} = A\mathbf{y}.$$

Here A is an $n \times n$-matrix, $\mathbf{y}$ is an n-dimensional vector, and $d\mathbf{y}/dt$ is an n-dimensional vector whose components are the derivatives of the components of $\mathbf{y}$. It too appears in applications in many fields.

A fundamental property of this differential equation is the following linearity property:

PROPERTY 1. If $\mathbf{y} = \mathbf{u}$ and $\mathbf{y} = \mathbf{v}$ are solutions of

$$\frac{d\mathbf{y}}{dt} = A\mathbf{y}$$

so is $\mathbf{y} = a\mathbf{u} + b\mathbf{v}$ for all scalars a and b. A solution $\mathbf{y}$ of this equation is uniquely determined by its value at $t = 0$.

This follows because when $\mathbf{y} = a\mathbf{u} + b\mathbf{v}$,

$$\frac{d\mathbf{y}}{dt} = a\frac{d\mathbf{u}}{dt} + b\frac{d\mathbf{v}}{dt} = aA\mathbf{u} + bA\mathbf{v} = A(a\mathbf{u} + b\mathbf{v}) = A\mathbf{y}.$$

The proof of the second statement can be found in a textbook on differential equations.

Looking at the one-dimensional case, one might suspect that in the n-dimensional case, the equation

$$\frac{d\mathbf{y}}{dt} = A\mathbf{y}$$

has a solution of the form

$$\mathbf{y} = \mathbf{x}e^{mt},$$

where $\mathbf{x}$ is the value of $\mathbf{y}$ at $t = 0$. To test this we make the computations

$$A\mathbf{y} - \frac{d\mathbf{y}}{dt} = A\mathbf{x}e^{mt} - \mathbf{x}me^{mt} = (A\mathbf{x} - m\mathbf{x})e^{mt}.$$

This quantity is zero if and only if $A\mathbf{x} = m\mathbf{x}$, that is, if and only if $\mathbf{x}$ is an eigenvector of A and $\lambda = m$ is the corresponding eigenvalue of A.

This gives the following:

PROPERTY 2.   If **x** is an eigenvector of A and $\lambda = m$ is the corresponding eigenvalue of A, then

$$\mathbf{y} = \mathbf{x}e^{mt}$$

is a solution of the homogeneous linear differential equation

$$\frac{d\mathbf{y}}{dt} = A\mathbf{y}.$$

For each t, **y** is an eigenvector of A. Accordingly, we call a solution **y** of this type an ***eigenvector solution*** of this equation.

Of course, only special solutions of A are eigenvector solutions. This follows because linear combinations of eigenvector solutions in general are not eigenvector solutions.

■ **EXAMPLE 1** _____

Consider the case in which the matrix A and two of its eigenvectors $\mathbf{x}_1$ and $\mathbf{x}_2$ are

$$A = \begin{bmatrix} 5 & 2 \\ 2 & 2 \end{bmatrix}, \qquad \mathbf{x}_1 = \begin{bmatrix} 1 \\ -2 \end{bmatrix}, \qquad \mathbf{x}_2 = \begin{bmatrix} 2 \\ 1 \end{bmatrix}.$$

The corresponding eigenvalues are $\lambda_1 = 1$ and $\lambda_2 = 6$, as one easily verifies. According to Property 2 the differential equation

$$\frac{d\mathbf{y}}{dt} = A\mathbf{y}$$

has as eigenvector solutions

$$\mathbf{y}_1 = \mathbf{x}_1 e^t = \begin{bmatrix} 1 \\ -2 \end{bmatrix} e^t = \begin{bmatrix} e^t \\ -2e^t \end{bmatrix}, \qquad \mathbf{y}_2 = \mathbf{x}_2 e^{6t} = \begin{bmatrix} 2 \\ 1 \end{bmatrix} e^{6t} = \begin{bmatrix} 2e^{6t} \\ e^{6t} \end{bmatrix}.$$

To verify that $\mathbf{y}_2$ is a solution, we use the eigenvector relation $A\mathbf{x}_2 = 6\mathbf{x}_2$ together with the computations

$$\frac{d\mathbf{y}_2}{dt} = \frac{d}{dt}(\mathbf{x}_2 e^{6t}) = \mathbf{x}_2 \frac{d}{dt} e^{6t} = \mathbf{x}_2 6 e^{6t} = A\mathbf{x}_2 e^{6t} = A\mathbf{y}_2.$$

Of course, this computation could have been carried out componentwise as follows:

$$\frac{d\mathbf{y}_2}{dt} = \frac{d}{dt}\begin{bmatrix} 2e^{6t} \\ e^{6t} \end{bmatrix} = \begin{bmatrix} 12e^{6t} \\ 6e^{6t} \end{bmatrix} = 6\begin{bmatrix} 2e^{6t} \\ e^{6t} \end{bmatrix} = 6\mathbf{y}_2,$$

$$A\mathbf{y}_2 = \begin{bmatrix} 5 & 2 \\ 2 & 2 \end{bmatrix}\begin{bmatrix} 2e^{6t} \\ e^{6t} \end{bmatrix} = \begin{bmatrix} 12e^{6t} \\ 6e^{6t} \end{bmatrix} = 6\begin{bmatrix} 2e^{6t} \\ e^{6t} \end{bmatrix} = 6\mathbf{y}_2 = \frac{d\mathbf{y}_2}{dt}.$$

Observe that in making these computations we have verified that for fixed t, $\mathbf{y}_2$ is an eigenvector of A corresponding to the eigenvalue $\lambda_2 = 6$. A similar computation for $\mathbf{y}_1$ tells us that $\mathbf{y}_1$ is an eigenvector solution of our differential equation. To find the solution $\mathbf{y}$ that has the value, say $\mathbf{x} = (10, -5)$ at $t = 0$, we solve the equation

$$\mathbf{x} = c_1\mathbf{x}_1 + c_2\mathbf{x}_2$$

for c_1 and c_2. This equation is equivalent to the pair of equations

$$c_1 + 2c_2 = 10, \qquad -2c_1 + c_2 = -5.$$

Solving, we find that $c_1 = 4$ and $c_2 = 3$. The function

$$\mathbf{y} = 4\mathbf{y}_1 + 3\mathbf{y}_2 = 4\mathbf{x}_1e^t + 3\mathbf{x}_2e^{6t} = \begin{bmatrix} 4e^t + 6e^{6t} \\ -8e^t + 3e^{6t} \end{bmatrix}$$

is the desired solution. This solution is not an eigenvector solution.

This example illustrates the following general property.

PROPERTY 3. Let A be a diagonalizable $n \times n$-matrix. There exist n eigenvector solutions $\mathbf{y}_1, \mathbf{y}_2, \ldots, \mathbf{y}_n$ of the differential equation

$$\frac{d\mathbf{y}}{dt} = A\mathbf{y}$$

having the following two properties:

(a) They are linearly independent.
(b) Every solution $\mathbf{y}$ of this equation is expressible as a linear combination

$$\mathbf{y} = c_1\mathbf{y}_1 + c_2\mathbf{y}_2 + \cdots + c_n\mathbf{y}_n$$

of these n solutions.

Because A is diagonalizable, it has n linearly independent eigenvectors $\mathbf{x}_1, \mathbf{x}_2, \ldots, \mathbf{x}_n$ and a set of corresponding eigenvalues $\lambda_1, \lambda_2, \ldots, \lambda_n$. According to Property 2, each eigenvector $\mathbf{x}_i$ determines an eigenvector solution $\mathbf{y}_i = \mathbf{x}_ie^{mt}$ with $m = \lambda_i$. The solutions $\mathbf{y}_1, \ldots, \mathbf{y}_n$ are linearly independent because the vectors $\mathbf{x}_1, \ldots, \mathbf{x}_n$ are. Given a vector $\mathbf{x}$, select constants $c_1, c_2, \ldots, c_n$ such that

$$\mathbf{x} = c_1\mathbf{x}_1 + c_2\mathbf{x}_2 + \cdots + c_n\mathbf{x}_n.$$

Then, by Property 1,

$$\mathbf{y} = c_1\mathbf{y}_1 + c_2\mathbf{y}_2 + \cdots + c_n\mathbf{y}_n$$

is a solution of $d\mathbf{y}/dt = A\mathbf{y}$ having $\mathbf{y} = \mathbf{x}$ when $t = 0$. This establishes Property 3.

As a further result we have:

PROPERTY 4. Let P be a nonsingular $n \times n$-matrix. Under the change of variables $\mathbf{y} = P\mathbf{z}$, the equation

$$\frac{d\mathbf{y}}{dt} = A\mathbf{y}$$

becomes

$$\frac{d\mathbf{z}}{dt} = B\mathbf{z}, \qquad \text{where} \qquad B = P^{-1}AP.$$

This follows because $\mathbf{z} = P^{-1}\mathbf{y}$ so that

$$\frac{d\mathbf{z}}{dt} = P^{-1}\frac{d\mathbf{y}}{dt} = P^{-1}A\mathbf{y} = P^{-1}AP\mathbf{z} = B\mathbf{z}.$$

When A is diagonalizable, we can choose P so that the similar matrix $B = P^{-1}AP$ is a diagonal matrix. In this case it is a simple matter to solve the equation $d\mathbf{z}/dt = B\mathbf{z}$, as illustrated by the following example.

■ **EXAMPLE 2** ─────────────────────────────────

Consider the matrices

$$A = \begin{bmatrix} 5 & 2 \\ 2 & 2 \end{bmatrix}, \qquad P = \begin{bmatrix} 1 & 2 \\ -2 & 1 \end{bmatrix},$$

$$P^{-1} = \begin{bmatrix} 1/5 & -2/5 \\ 2/5 & 1/5 \end{bmatrix}, \quad B = P^{-1}AP = \begin{bmatrix} 1 & 0 \\ 0 & 6 \end{bmatrix}.$$

Under the transformation $\mathbf{y} = P\mathbf{z}$ with $\mathbf{z} = (u, v)$, the equation $d\mathbf{y}/dt = A\mathbf{y}$ becomes

$$\frac{d\mathbf{z}}{dt} = \begin{bmatrix} du/dt \\ dv/dt \end{bmatrix} = B\mathbf{z} = \begin{bmatrix} 1 & 0 \\ 0 & 6 \end{bmatrix}\begin{bmatrix} u \\ v \end{bmatrix} = \begin{bmatrix} u \\ 6v \end{bmatrix}.$$

This equation is therefore equivalent to the scalar equations

$$\frac{du}{dt} = u, \qquad \frac{dv}{dt} = 6v,$$

whose general solutions are

$$u = ae^t, \qquad v = be^{6t}.$$

The general solution of $dy/dt = Ay$ is therefore

$$\mathbf{y} = P\mathbf{z} = \begin{bmatrix} 1 & 2 \\ -2 & 1 \end{bmatrix} \begin{bmatrix} ae^t \\ be^{6t} \end{bmatrix} = \begin{bmatrix} ae^t + 2be^{6t} \\ -2ae^t + be^{6t} \end{bmatrix},$$

so that

$$\mathbf{y} = a \begin{bmatrix} e^t \\ -2e^t \end{bmatrix} + b \begin{bmatrix} 2e^{6t} \\ e^{6t} \end{bmatrix} = a\mathbf{y}_1 + b\mathbf{y}_2.$$

It should be noted that in this case $\mathbf{y}_1$ and $\mathbf{y}_2$ are the eigenvector solutions of $dy/dt = Ay$ obtained in Example 1.

The following property of the equation $dy/dt = Ay$ is sometimes useful when the matrix A is not diagonalizable.

> **PROPERTY 5.** If $\mathbf{y}$ is a solution of
>
> $$\frac{d\mathbf{y}}{dt} = A\mathbf{y},$$
>
> then $\mathbf{z} = e^{-mt}\mathbf{y}$ satisfies the equation
>
> $$\frac{d\mathbf{z}}{dt} = (A - mI)\mathbf{z}.$$

This follows because

$$\frac{d\mathbf{z}}{dt} = \frac{d}{dt}\mathbf{y}e^{-mt} = \frac{d\mathbf{y}}{dt}e^{-mt} - m\mathbf{y}e^{-mt} = (A - mI)\mathbf{y}e^{-mt} = (A - mI)\mathbf{z}.$$

An application of this property is given in the following example.

In the examples given above, the matrix A was diagonalizable. We now consider a case when A is not diagonalizable.

■ **EXAMPLE 3** ————————————————————————————————

Consider the matrix

$$A = \begin{bmatrix} m & 1 & 0 \\ 0 & m & 1 \\ 0 & 0 & m \end{bmatrix} = mI + N, \qquad N = \begin{bmatrix} 0 & 1 & 0 \\ 0 & 0 & 1 \\ 0 & 0 & 0 \end{bmatrix}.$$

In this event we use the substitution $\mathbf{z} = e^{-mt}\mathbf{y}$ in the differential equation $dy/dt = Ay$. By Property 5 we have

$$\frac{d\mathbf{z}}{dt} = (A - mI)\mathbf{z} = N\mathbf{z}.$$

Let u, v, and w be the components of $\mathbf{z}$. Then

$$\frac{d\mathbf{z}}{dt} = \begin{bmatrix} du/dt \\ dv/dt \\ dw/dt \end{bmatrix} = N\mathbf{z} = \begin{bmatrix} 0 & 1 & 0 \\ 0 & 0 & 1 \\ 0 & 0 & 0 \end{bmatrix} \begin{bmatrix} u \\ v \\ w \end{bmatrix} = \begin{bmatrix} v \\ w \\ 0 \end{bmatrix}.$$

This gives us the scalar equations

$$\frac{du}{dt} = v, \qquad \frac{dv}{dt} = w, \qquad \frac{dw}{dt} = 0.$$

The general solutions of these equations are

$$w = a, \qquad v = at + b, \qquad u = \tfrac{1}{2}at^2 + bt + c,$$

where a, b, and c are arbitrary constants. The general solution $\mathbf{y} = \mathbf{z}e^{mt}$ of our equation $d\mathbf{y}/dt = A\mathbf{y}$ is therefore

$$\mathbf{y} = \begin{bmatrix} \tfrac{1}{2}at^2 + bt + c \\ at + b \\ a \end{bmatrix} e^{mt} = a \begin{bmatrix} \tfrac{1}{2}t^2 e^{mt} \\ te^{mt} \\ e^{mt} \end{bmatrix} + b \begin{bmatrix} te^{mt} \\ e^{mt} \\ 0 \end{bmatrix} + c \begin{bmatrix} e^{mt} \\ 0 \\ 0 \end{bmatrix}.$$

The last vector in the sum is an eigenvector solution. The other two are not.

So far we tacitly assumed that the eigenvalues are real. However, our results are valid when they are complex. Suppose that $m = a + bi$ is a complex eigenvalue of a real matrix A. Here a and b are real and $i^2 = -1$. A corresponding eigenvector $\mathbf{x}$ is of the form $\mathbf{x} = \mathbf{u} + i\mathbf{v}$, where $\mathbf{u}$ and $\mathbf{v}$ are real. Then, by Property 2, the vector

$$\mathbf{y} = \mathbf{x}e^{mt} = (\mathbf{u} + i\mathbf{v})e^{(a+ib)t}$$

is a solution of $d\mathbf{y}/dt = A\mathbf{y}$. In this expression

$$e^{(a+ib)t} = e^{at}e^{ibt} = e^{at}(\cos bt + i \sin bt)$$

by virtue of the Euler formula

$$e^{ibt} = \cos bt + i \sin bt.$$

From these relations we see that

$$\mathbf{y} = (\mathbf{u} + i\mathbf{v})e^{at}(\cos bt + i \sin bt) = \mathbf{y}_1 + i\mathbf{y}_2,$$

where

$$\mathbf{y}_1 = e^{at}(\mathbf{u} \cos bt - \mathbf{v} \sin bt), \qquad \mathbf{y}_2 = e^{at}(\mathbf{u} \sin bt + \mathbf{v} \cos bt).$$

We have

$$\frac{d\mathbf{y}}{dt} = \frac{d\mathbf{y}_1}{dt} + i\frac{d\mathbf{y}_2}{dt} = A\mathbf{y} = A\mathbf{y}_1 + iA\mathbf{y}_2.$$

Equating real and imaginary parts, we find that

$$\frac{d\mathbf{y}_1}{dt} = A\mathbf{y}_1, \qquad \frac{d\mathbf{y}_2}{dt} = A\mathbf{y}_2.$$

This establishes the following:

> **PROPERTY 6.** Let $m = a + ib$ be a complex eigenvalue of a real matrix A and let $\mathbf{x} = \mathbf{u} + i\mathbf{v}$ be a corresponding eigenvector. Then the vector functions $\mathbf{y}_1$ and $\mathbf{y}_2$ described above are real solutions of $d\mathbf{y}/dt = A\mathbf{y}$.

■ **EXAMPLE 4**

Consider the differential equation

$$\frac{d^2w}{dt} - \frac{4dw}{dt} + 13w = 0.$$

This differential equation can be put in matrix form by setting

$$\mathbf{y} = \begin{bmatrix} w \\ dw/dt \end{bmatrix}, \qquad A = \begin{bmatrix} 0 & 1 \\ -13 & 4 \end{bmatrix}.$$

We have

$$\frac{d\mathbf{y}}{dt} = \begin{bmatrix} dw/dt \\ d^2w/dt^2 \end{bmatrix} = \begin{bmatrix} dw/dt \\ -13w + 4\,dw/dt \end{bmatrix} = \begin{bmatrix} 0 & 1 \\ -13 & 4 \end{bmatrix}\begin{bmatrix} w \\ dw/dt \end{bmatrix} = A\mathbf{y}.$$

The complex number $m = 2 + 3i$ is an eigenvalue of A with

$$\mathbf{x} = \begin{bmatrix} 1 \\ 2 + 3i \end{bmatrix} = \begin{bmatrix} 1 \\ 2 \end{bmatrix} + \begin{bmatrix} 0 \\ 3 \end{bmatrix}i = \mathbf{u} + \mathbf{v}i$$

as a corresponding eigenvector. By Property 6, the equation $d\mathbf{y}/dt = A\mathbf{y}$ has the solutions

$$\mathbf{y}_1 = \begin{bmatrix} e^{2t}\cos 3t \\ e^{2t}(2\cos 3t - 3\sin 3t) \end{bmatrix}, \qquad \mathbf{y}_2 = \begin{bmatrix} e^{2t}\sin 3t \\ e^{2t}(2\sin 3t + 3\cos 3t) \end{bmatrix}.$$

It follows that

$$w_1 = e^{2t}\cos 3t, \qquad w_2 = e^{2t}\sin 3t$$

are solutions of our original differential equation. The general solution is

$$w = e^{2t}(a\cos 3t + b\sin 3t).$$

Exercises

1. Let A be one of the matrices

$$A_1 = \begin{bmatrix} 2 & 0 \\ 0 & 3 \end{bmatrix}, \qquad A_2 = \begin{bmatrix} 5 & 8 \\ 2 & 5 \end{bmatrix}, \qquad A_3 = \begin{bmatrix} 2 & 1 \\ 0 & 2 \end{bmatrix}, \qquad A_4 = \begin{bmatrix} 0 & 0 \\ 0 & 0 \end{bmatrix}.$$

   In each case find the solution of $dy/dt = Ay$.

2. Let A be one of the matrices

$$A_5 = \begin{bmatrix} 2 & 3 & -2 \\ 0 & 5 & 8 \\ 0 & 2 & 5 \end{bmatrix}, \qquad A_6 = \begin{bmatrix} 2 & 1 & 0 \\ 0 & 2 & 0 \\ 0 & 0 & 3 \end{bmatrix}, \qquad A_7 = \begin{bmatrix} 2 & 1 & 0 \\ 0 & 2 & 1 \\ 0 & 0 & 2 \end{bmatrix}.$$

   In each case find the solutions of $dy/dt = Ay$.

3. Let N be a nilpotent matrix. Choose q such that $N^{q+1} = 0$. Let Z be the function

$$Z = I + Nt + \frac{N^2 t^2}{2!} + \cdots + \frac{N^q t^q}{q!}.$$

   Show that Z satisfies the equation

$$\frac{dZ}{dt} = NZ.$$

   Apply this result to the matrix $N = A_7 - 2I$, where A_7 is the matrix in Exercise 2.

4. Let e^{At} be the matrix defined by the convergent power series

$$e^{At} = I + At + \frac{A^2 t^2}{2!} + \cdots + \frac{A^k t^k}{k!} + \cdots.$$

   Show that $Y = e^{At}$ satisfies the equation $dY/dt = AY$. Show further that $\mathbf{y} = e^{At}\mathbf{x}$ is the solution of $dy/dt = Ay$ having $\mathbf{y} = \mathbf{x}$ at $t = 0$.

5. Let $\mathbf{y}$ be a solution of $dy/dt = Ay$ and let $\mathbf{z}$ be a solution of $dz/dt = -A^T\mathbf{z}$. Show that

$$\frac{d}{dt}(\mathbf{z}^T\mathbf{y}) = \left(\frac{d\mathbf{z}}{dt}\right)^T \mathbf{y} + \mathbf{z}^T \frac{d\mathbf{y}}{dt} = (-A^T\mathbf{z})^T\mathbf{y} + \mathbf{z}^T A\mathbf{y} = 0.$$

   Conclude that $\mathbf{z}^T\mathbf{y} = \text{constant} = \mathbf{y}^T\mathbf{z}$.

6. Verify the result given in Exercise 5 by using solutions of $dy/dt = Ay$ and $dz/dt = -A^T\mathbf{z}$ for one of the matrices, say A_2, given in Exercise 1.

7. Give an alternative proof of the result in Exercise 5 using the result given in Exercise 4. The proof proceeds as follows. Choose vectors $\mathbf{u}$ and $\mathbf{v}$. Then:

   (a) $\mathbf{y} = e^{At}\mathbf{u}$ solves $dy/dt = Ay$ with $\mathbf{y} = \mathbf{u}$ at $t = 0$.

   (b) $\mathbf{z}^T = \mathbf{v}^T e^{-At}$ is the transpose of the solution $\mathbf{z}$ of the equation $dz/dt = -A^T\mathbf{z}$ with $\mathbf{z} = \mathbf{v}$ at $t = 0$.

   (c) $\mathbf{z}^T\mathbf{y} = \mathbf{v}^T e^{-At} e^{At} \mathbf{u} = \mathbf{v}^T\mathbf{u}$ for all values of t.

8. Let Y be a matrix whose columns $\mathbf{y}_1, \mathbf{y}_2, \ldots, \mathbf{y}_n$ are solutions of $dy/dt = Ay$.

The matrix Y is a matrix solution of $dY/dt = AY$. Show that:

(a) $Y = e^{At}$ is a solution of $dY/dt = AY$ having $Y = I$ at $t = 0$.

(b) If Y solves $dY/dt = AY$, so does $Y_c = YC$ for every constant matrix C.

(c) Let $A = A_2$, where A_2 is the 2×2-matrix given in Exercise 1. Find a solution Y of $dY/dt = AY$ for this case.

9. Let Y and Z be solutions of $dY/dt = AY$ and $dZ/dt = -A^TZ$, respectively.

(a) Show that $C = Z^TY$ is a constant matrix, that is $\dfrac{dC}{dt} = 0$.

(b) Show that the solutions Y and Z can be chosen so that $Z^TY = I$. Then Z^T is the inverse of Y for all values of t.

(c) Illustrate this result using one of the matrices given in Exercise 1.

5

Linear Spaces and Subspaces

5.1 LINEAR SPACES AND SUBSPACES

In the preceding pages we have been dealing with many different types of matrices, including elementary matrices, nonsingular and singular matrices, symmetric matrices, diagonal matrices, orthogonal matrices, and others. Each of these matrices had special properties, depending on which one was chosen. In this chapter we are concerned with certain common properties that are shared by all matrices of a particular class. We will see that if a class of matrices all satisfy a certain list of properties, that class of matrices will be called a *linear space*. In fact, if the properties listed are satisfied by any collection of elements, whether or not they are matrices, the collection will be called a linear space. We have already been dealing with linear spaces without saying so. For example, the collection of matrices of the same size form a linear space. So does the space $\mathbf{R}$ and the spaces $\mathbf{R}^2$, $\mathbf{R}^3, \ldots ,$ and $\mathbf{R}^n$, as described in Section 1.6.

Briefly, a *linear space* is a collection of elements on which it makes sense to add any two of the elements and scalar multiply any element with the consequence of getting back an element of the original collection. For example, if our original collection of elements is the set of all 3×3 matrices with the usual definitions of addition and scalar multiplication of matrices, it follows immediately that the sum of any two 3×3 matrices is again a 3×3 matrix. Also, if we multiply any 3×3 matrix by a scalar, we get back a new 3×3 matrix. The collection of all 3×3 matrices is a linear space.

More precisely, a linear space $\mathbf{V}$ is a set of elements $\mathbf{u}, \mathbf{v}, \mathbf{w}, \ldots ,$ and a set of scalars (real or complex numbers), $a, b, c, \ldots$ such that the sum $\mathbf{u} + \mathbf{v}$ of two elements $\mathbf{u}$ and $\mathbf{v}$ in $\mathbf{V}$ is in $\mathbf{V}$ and such that the product $a\mathbf{u}$ of an element $\mathbf{u}$ in $\mathbf{V}$ by a scalar a is in $\mathbf{V}$. The following basic properties of addition and scalar multiplication must also be satisfied:

1. $\mathbf{u} + \mathbf{v} = \mathbf{v} + \mathbf{u}, \mathbf{u} + (\mathbf{v} + \mathbf{w}) = (\mathbf{u} + \mathbf{v}) + \mathbf{w}$.
2. There is an element $\mathbf{0}$ such that $\mathbf{u} + \mathbf{0} = \mathbf{u}$ for every element $\mathbf{u}$.
3. For each element $\mathbf{u}$, there is an element $-\mathbf{u}$ such that $\mathbf{u} + (-\mathbf{u}) = \mathbf{0}$.
4. $a\mathbf{u} = \mathbf{u}a, \quad a(b\mathbf{u}) = (ab)\mathbf{u} = ab\mathbf{u}$.
5. $(a + b)\mathbf{u} = a\mathbf{u} + b\mathbf{u}, \quad a(\mathbf{u} + \mathbf{v}) = a\mathbf{u} + a\mathbf{v}$.
6. $1\mathbf{u} = \mathbf{u}, 0\mathbf{u} = \mathbf{0}, \quad (-1)\mathbf{u} = -\mathbf{u}$.

In almost all of our examples, we will use real numbers as the scalars unless otherwise specified.

It follows from the above that if $\mathbf{u}$ and $\mathbf{v}$ are in $\mathbf{V}$, every linear combination *a***u** + *b***v** of **u** and **v** is in **V**. Also, the difference **u** − **v** of two elements **u** and **v** in **V** can be interpreted as the sum **u** + (−**v**).

The reader can easily verify that properties 1 through 6 are satisfied by the set of all 3 × 3 matrices as defined above. Here the elements **u**, **v**, **w**, . . . are 3 × 3 matrices and the element **0** is the 3 × 3 zero matrix. If **u** is any 3 × 3 matrix, −**u** is the 3 × 3 matrix whose entries are the negative of the corresponding entries of **u**.

A linear space **V** is also called a ***vector space***. It is for this reason that the elements of **V** are frequently called *vectors* or *points*. In the example above, where the vector space **V** consists of the class of all 3 × 3 matrices, the vectors are 3 × 3 matrices. (The word *vector* is therefore used in a much broader context when dealing with linear spaces.) As in Euclidean spaces, the element **0** in **V** is considered to be the origin. Sometimes we denote the elements of our linear space by capital letters, such as *A*, *B*, *C*, *X*, *Y*, *Z*, . . . rather than by boldface letters. We even use lowercase letters when it is appropriate to do so.

We now consider several examples of linear spaces. In many cases we leave it as an exercise to verify that they are indeed linear spaces. It will not be necessary to verify Properties 1 to 6 for addition and scalar multiplication because, in the linear spaces that we consider, these properties will automatically be a consequence of the properties of addition and scalar multiplication of real (or complex) numbers. (Even so, it would be worthwhile for the reader to verify these properties mentally and to identify in each example the **0** element and for each element **u**, the corresponding element −**u**.) It will be necessary to verify that:

> **1.** The sum **u** + **v** of two elements **u** and **v** in the space is again in the space.
>
> **2.** The product *a***u** of an element **u** in the space by a scalar *a* is again in the space.

Properties 1 and 2 are often referred to as *closure under addition* and *scalar multiplication*, respectively.

■ **EXAMPLE 1** _____

The spaces **R**, **R**^{2}, **R**^{3}, . . . described in Chapter 1 are linear spaces.

■ **EXAMPLE 2** _____

There are many linear spaces that are of interest other than the spaces **R**^{n}. One such linear space is the class **M** of all *n* × *n*-matrices (*n* fixed). This class is a linear space because, if we add two *n* × *n*-matrices, we get an $n \times n$-matrix. Similarly, if we multiply an $n \times n$-matrix by a scalar, the matrix we get is an $n \times n$-matrix.

■ **EXAMPLE 3** _____

The class of all $m \times n$-matrices with m and n fixed is also a linear space. We shall encounter these matrices in our study of linear transformations.

■ **EXAMPLE 4** _____

The set of solutions $\mathbf{x}$ of a homogeneous linear matrix equation $A\mathbf{x} = \mathbf{0}$ is a linear space. This follows because if $A\mathbf{x} = \mathbf{0}$ and $A\mathbf{y} = \mathbf{0}$, then

$$A(\mathbf{x} + \mathbf{y}) = A\mathbf{x} + A\mathbf{y} = \mathbf{0} + \mathbf{0} = \mathbf{0}, \qquad A(c\mathbf{x}) = c(A\mathbf{x}) = \mathbf{0}.$$

We call this linear space the **null space** of the matrix A. It is also called the **kernel** of A. There are other important linear spaces associated with A which we describe later.

■ **EXAMPLE 5** _____

The set $\mathbf{P}_2$ of all polynomials

$$f(t) = a + bt + ct^2$$

of degree ≤ 2 (including the zero polynomial) is a linear space. It is comprised of all linear combinations of the elementary polynomials 1, t, t^2. Similarly, the set $\mathbf{P}_k$ of all polynomials

$$f(t) = a_0 + a_1 t + a_2 t^2 + \cdots + a_k t^k$$

of degree $\le k$ (including the zero polynomial) is a linear space. It is comprised of all linear combinations of the elementary polynomials 1, t, $t^2, \ldots, t^k$. These elementary polynomials generate the class $\mathbf{P}_k$. The class of polynomials of fixed degree k, say $k = 2$, is not a linear space because it does not contain the zero polynomial. Also, it is easy to construct two polynomials of degree k whose sum is of degree less than k.

■ **EXAMPLE 6** _____

The set $\mathbf{C}$ of all matrices $C = f(A)$ determined by all polynomials

$$f(A) = a_0 I + a_1 A + a_2 A^2 + \cdots + a_k A^k$$

of a fixed $n \times n$-matrix A is a linear space. By the use of the Cayley–Hamilton theorem it can be shown that A^n is a linear combination of the n matrices

$$I, \quad A, \quad A^2, \ldots, \quad A^{n-1}.$$

From this fact we can conclude that the same is true for all powers of A. We can then infer that every matrix B in $\mathbf{C}$ is also a linear combination of the matrices $I, A, A^2, \ldots, A^{n-1}$. We say that these matrices generate $\mathbf{C}$. In Chapter 6 we shall encounter other important linear spaces associated with matrices.

In describing various linear spaces, it is often desirable to have at hand the concept of a linear subspace.

A *linear subspace* $\mathbf{W}$ of a linear space $\mathbf{V}$ is a subset of $\mathbf{V}$ which is itself a linear space with the same operations of addition and scalar multiplication as in $\mathbf{V}$.

Thus if $\mathbf{u}$ and $\mathbf{v}$ are in $\mathbf{W}$, so is $\mathbf{u} + \mathbf{v}$. Also, if $\mathbf{u}$ is in $\mathbf{W}$, so is $a\mathbf{u}$ for any scalar a. As an example, consider the linear space $\mathbf{V}$ of all 3×3 matrices. Let $\mathbf{W}$ be the set of all 3×3 diagonal matrices. Certainly, $\mathbf{W}$ is a subset of $\mathbf{V}$ since $\mathbf{V}$ contains all 3×3 matrices. Furthermore, $\mathbf{W}$ is itself a linear space since the sum of any two 3×3 diagonal matrices is again a 3×3 diagonal matrix and any scalar multiple of a 3×3 diagonal matrix is again a 3×3 diagonal matrix under the usual operations of addition and scalar multiplication for 3×3 matrices. Properties 1 to 6 are all easily verified for $\mathbf{W}$. It follows that the set $\mathbf{W}$ of all 3×3 diagonal matrices is a linear subspace of the linear space $\mathbf{V}$ of all 3×3 matrices.

Once again, because of the linear spaces that we choose, in verifying whether a certain set $\mathbf{W}$ is a linear subspace of a given linear space $\mathbf{V}$, it will not be necessary to check Properties 1 to 6. They will hold automatically because of what we know about real (or complex) numbers. To verify that $\mathbf{W}$ is a linear subspace of a linear space $\mathbf{V}$, it will only be necessary to show that:

1. The set $\mathbf{W}$ is a subset of the linear space $\mathbf{V}$.
2. The set $\mathbf{W}$ is closed under the same addition and scalar multiplication as in $\mathbf{V}$:
   (a) The sum $\mathbf{u} + \mathbf{v}$ of two elements $\mathbf{u}$ and $\mathbf{v}$ in $\mathbf{W}$ is again in $\mathbf{W}$.
   (b) The product $a\mathbf{u}$ of an element $\mathbf{u}$ in $\mathbf{W}$ by a scalar a is again in $\mathbf{W}$.

We proceed to consider several examples of linear subspaces of some given linear spaces. The reader should verify that each example satisfies the criteria above.

■ **EXAMPLE 7** _____

The linear space **M** of all $n \times n$ matrices (n fixed) presented in Example
2 has many important linear subspaces. The class **S** of symmetric $n \times$
n matrices is a linear subspace of **M**, for **S** is certainly a subset of **M**
and the sum of two symmetric $n \times n$ matrices is again a symmetric n
$\times n$ matrix, as is the product of a symmetric $n \times n$ matrix by a scalar
once again a $n \times n$ symmetric matrix.

   The class **U** of all upper triangular $n \times n$ matrices is also a linear
subspace of **M**. To verify this, for the case, say $n = 3$, recall that an
upper triangular matrix is a matrix of the form

$$
\mathbf{U} = \begin{bmatrix} a & d & f \\ 0 & b & e \\ 0 & 0 & c \end{bmatrix}.
$$

Clearly, **U** is a subset of **M**. Furthermore, no nonzero entries below the
main diagonal can be introduced by addition and scalar multiplication of
upper triangular matrices. Similarly, the class **L** of all lower triangular
$n \times n$ matrices is a linear subspace of **M**. The class **D** of diagonal $n \times$
n matrices is a linear subspace of **M**. Observe that **D** is also a linear
subspace of **U**, **L**, and **S**.

   The class of all nonsingular $n \times n$ matrices, although obviously being
a very important subclass of **M**, is not a linear subspace of **M**. While
this class is certainly a subset of **M**, it is not closed under addition nor
under scalar multiplication. The problem is that the zero matrix is not a
nonsingular matrix. (The reader should think of two nonsingular matrices
whose sum is the zero matrix and also a multiple of a nonsingular matrix
that yields the zero matrix.)

■ **EXAMPLE 8** _____

Recall that $\mathbf{R}^2$ is the set of all number pairs (x, y) and that $\mathbf{R}^3$ is all triples
(x, y, z) of numbers. The set **S** of all triples $(x, y, 0)$, whose last
component is 0, forms a linear subspace of $\mathbf{R}^3$ that "behaves" like $\mathbf{R}^2$.
It is not the same as $\mathbf{R}^2$ because it is comprised of triples of numbers
and not pairs of numbers. In other words, $\mathbf{R}^2$ is not a subspace of $\mathbf{R}^3$,
although $\mathbf{R}^3$ has many subspaces that "behave" like $\mathbf{R}^2$. For example,
the set **T** of triples $(x, 0, z)$, whose second component is 0 is also a
linear subspace that "behaves" like $\mathbf{R}^2$. Even the set of points of the
form $(x, 1, z)$ "behaves somewhat" like $\mathbf{R}^2$ but is not a linear space,
because it does not contain the origin. Geometrically, it is a plane parallel
to the xz-plane **T**, a plane that is not a linear subspace of $\mathbf{R}^3$.

■ **EXAMPLE 9**

Consider the class **T** of all $n \times n$ tridiagonal matrices. For example, if $n = 5$, these are the 5×5 matrices, which look like

$$
\begin{bmatrix}
a_1 & b_1 & 0 & 0 & 0 \\
c_1 & a_2 & b_2 & 0 & 0 \\
0 & c_2 & a_3 & b_3 & 0 \\
0 & 0 & c_3 & a_4 & b_4 \\
0 & 0 & 0 & c_4 & a_5
\end{bmatrix}.
$$

The tridiagonal matrices have zeros everywhere above and below the three diagonals consisting of the main diagonal and the diagonals that are just above and below the main diagonal. Of course, some or all of the a's, b's, or c's may also be 0.

It is easy to verify that the $n \times n$ tridiagonal matrices **T** form a linear space. The reader should verify that the set **D** of all $n \times n$ diagonal matrices are a linear subspace of **T**.

In the next three examples we present some relationships between geometry, linear spaces, and subspaces. We also illustrate the concept of the sum of linear subspaces.

■ **EXAMPLE 10**

The set **L** of vectors (points) $\mathbf{x} = (x, y)$ in $\mathbf{R}^2$ that satisfy the equation

$$2x + 3y = 0$$

is a linear subspace of $\mathbf{R}^2$. Notice that this is a line through the origin. It is also the set of all vectors $\mathbf{x} = (x, y)$ orthogonal to the vector $\mathbf{n} = (2, 3)$. Its equation can therefore be written in the compact form

$$\mathbf{n}^T\mathbf{x} = 0.$$

The vector $\mathbf{n}$ is a normal to the line **L**. The vector $\mathbf{v} = (3, -2)$ is in **L**. So is every scalar multiple $\mathbf{x} = \mathbf{v}t$ of $\mathbf{v}$. This gives us the parametric representation

$$\mathbf{x} = \mathbf{v}t$$

of a line **L** through the origin. It states that the multiples of $\mathbf{v}$ form a line **L**. It should be noted that the set of points satisfying the equation

$$2x + 3y = 1$$

is another line **T**. But it is not a linear space, because it does not contain the origin. It contains the point $\mathbf{x}_0 = (2, -1)$. If $\mathbf{x}$ is in **T**, the point

$x - x_0$ is in **L** and hence is a multiple vt of **v**. This gives us the parametric representation

$$x = x_0 + vt$$

of the line **T** parallel to **L**. This equation states that

(a point in **T**) $= x_0 +$ (a point in **L**).

This suggests that we can represent **T** as a translate

$$T = x_0 + L$$

of the linear subspace **L**. The theory of "translates" plays an important role in the theory of linear spaces. However, we shall make limited use of this concept in this book.

Suppose that, in $\mathbf{R}^2$, we have two distinct lines $\mathbf{L}_1$ and $\mathbf{L}_2$ through the origin. They are linear subspaces of $\mathbf{R}^2$ and can be represented parametrically by the equations

$$x = v_1 t \quad \text{and} \quad x = v_2 t,$$

where v_1 is in $\mathbf{L}_1$ and v_2 is in $\mathbf{L}_2$. For definiteness, we suppose that $v_1 = (3, -2)$ and $v_2 = (2, 3)$. Observe that any point **x** in $\mathbf{R}^2$ is the sum

$$x = x_1 + x_2$$

of a point $x = v_1 t_1$ in $\mathbf{L}_1$ and a point $x_2 = v_2 t_2$ in $\mathbf{L}_2$. This gives us the parametric representation

$$x = v_1 t_1 + v_2 t_2$$

of the points **x** in $\mathbf{R}^2$. Equivalently, we have the equation

(a point in $\mathbf{R}^2$) $=$ (a point in $\mathbf{L}_1$) $+$ (a point in $\mathbf{L}_2$).

We rewrite this equation in the form

$$\mathbf{R}^2 = \mathbf{L}_1 + \mathbf{L}_2,$$

signifying that a point **x** in $\mathbf{R}^2$ is the sum of a point $x_1 = v_1 t_1$ in $\mathbf{L}_1$ and a point $x_2 = v_2 t_2$ in $\mathbf{L}_2$. Observe that, for definiteness, we chose v_1 and v_2 to be the column vectors of the matrix

$$V = [v_1 \quad v_2] = \begin{bmatrix} 3 & 2 \\ -2 & 3 \end{bmatrix}.$$

Our parametric representation of $\mathbf{R}^2$ therefore can be put in the form

$$x = v_1 t_1 + v_2 t_2 = [v_1 \quad v_2] \begin{bmatrix} t_1 \\ t_2 \end{bmatrix} = Vt, \qquad t = \begin{bmatrix} t_1 \\ t_2 \end{bmatrix}.$$

The equation $x = Vt$ is therefore a parametric representation of $\mathbf{R}^2$ for every choice of the nonsingular 2×2-matrix V. Later in the book we view the equation $\mathbf{x} = V\mathbf{t}$ to be a transformation of coordinates.

■ **EXAMPLE 11** —————————————————————————————

Let us take a look at the linear space $\mathbf{R}^3$. The set of vectors (points) $\mathbf{x} = (x, y, z)$ in $\mathbf{R}^3$ satisfying the equation

$$2x + 3y - z = 0$$

is a linear subspace $\mathbf{P}$ of $\mathbf{R}^3$. This is a two-dimensional plane through the origin. It is also the set of all vectors $\mathbf{x}$ orthogonal to the vector $\mathbf{n} = (2, 3, -1)$. Accordingly, this equation can be put in the form

$$\mathbf{n}^T\mathbf{x} = 0.$$

The vector $\mathbf{n}$ is a *normal* to $\mathbf{P}$. Observe that the set of points (x, y, z) satisfying the equation

$$2x + 3y - z = 1$$

is also a plane that we label $\mathbf{T}$. $\mathbf{T}$ is not a linear space, because it does not contain the origin. It is parallel to $\mathbf{P}$. The point $\mathbf{x}_0 = (1, 1, 4)$ is in $\mathbf{T}$. If $\mathbf{x}$ is in $\mathbf{T}$, the point $\mathbf{x} - \mathbf{x}_0$ is in $\mathbf{P}$. Consequently,

$$\mathbf{x} = \mathbf{x}_0 + (\mathbf{x} - \mathbf{x}_0) = \mathbf{x}_0 + (\text{a point in } \mathbf{P}).$$

We restate this fact by writing

$$\mathbf{T} = \mathbf{x}_0 + \mathbf{P}.$$

Thus the plane $\mathbf{T}$ in $\mathbf{R}^3$ is a translate of a plane $\mathbf{P}$ through the origin and so is a translate of a linear subspace $\mathbf{P}$ of $\mathbf{R}^3$.

Let $\mathbf{Q}$ be a second plane through the origin defined by the equation

$$x - 2y + z = 0.$$

It too is a linear subspace of $\mathbf{R}^3$. The plane $\mathbf{Q}$ intersects $\mathbf{P}$ in a line $\mathbf{L}$. The linear subspace $\mathbf{L}$ is accordingly the intersection of two linear subspaces $\mathbf{P}$ and $\mathbf{Q}$ of $\mathbf{R}^3$. The line $\mathbf{L}$ is comprised of all points (x, y, z) satisfying the equations

$$2x + 3y - z = 0$$
$$x - 2y + z = 0.$$

These equations can be written in the matrix form $A\mathbf{x} = \mathbf{0}$ in the usual manner. The matrix A is a 2×3-matrix whose rows are linearly independent. When A is such a matrix, the homogeneous linear equation $A\mathbf{x} = \mathbf{0}$ represents a line. What does it represent when the rows of A are linearly dependent?

Return to the two-dimensional linear subspace **P** of $\mathbf{R}^3$ of points $\mathbf{x} = (x, y, z)$ satisfying the equation

$$2x + 3y - z = 0.$$

The points (vectors) $\mathbf{v}_1 = (3, -2, 0)$ and $\mathbf{v}_2 = (1, 1, 5)$ are in **P**. So is every linear combination

$$\mathbf{x} = \mathbf{v}_1 t_1 + \mathbf{v}_2 t_2.$$

In fact, every point $\mathbf{x}$ in **P** can be represented in this manner. This equation, with t_1 and t_2 as parameters, is a parametric representation of **P**. When we view the vectors $\mathbf{v}_1$ and $\mathbf{v}_2$ to be the column vectors of the 3×2-matrix

$$V = [\mathbf{v}_1 \quad \mathbf{v}_2] = \begin{bmatrix} 3 & 1 \\ -2 & 1 \\ 0 & 5 \end{bmatrix},$$

we see that the parametric equation for **P** can be written as a matrix equation

$$\mathbf{x} = \mathbf{v}_1 t_1 + \mathbf{v}_2 t_2 = [\mathbf{v}_1 \quad \mathbf{v}_2] \begin{bmatrix} t_1 \\ t_2 \end{bmatrix} = V\mathbf{t}, \qquad \mathbf{t} = \begin{bmatrix} t_1 \\ t_2 \end{bmatrix}.$$

The matrix equation $\mathbf{x} = V\mathbf{t}$ is therefore a parametric representation of the linear space **P**, a plane in $\mathbf{R}^3$ passing through the origin. This equation can be given another interpretation. Observe that the parameter $\mathbf{t}$ is a point in $\mathbf{R}^2$. The equation $\mathbf{x} = V\mathbf{t}$ "transforms" $\mathbf{R}^2$ into **P**, a linear subspace of $\mathbf{R}^3$. In this interpretation the equation $\mathbf{x} = V\mathbf{t}$ represents a linear transformation. We study linear transformations in Chapter 6.

■ **EXAMPLE 11 (continued)** ────────────────────────────────

We continue with the study of the linear subspace **P** of $\mathbf{R}^3$, a plane in $\mathbf{R}^3$ defined, as described above, by the parametric equation

$$\mathbf{x} = \mathbf{v}_1 t_1 + \mathbf{v}_2 t_2,$$

where $\mathbf{v}_1 = (3, -2, 0)$ and $\mathbf{v}_2 = (1, 1, 5)$. This equation for **P** can be given an interpretation other than that given previously. Observe that

$$\mathbf{x} = \mathbf{v}_1 t$$

is the equation of a line $\mathbf{L}_1$, a linear subspace of **P** and of $\mathbf{R}^3$. Similarly, the equation

$$\mathbf{x} = \mathbf{v}_2 t$$

is the equation of a line $\mathbf{L}_2$, a second linear subspace of **P**. Our parametric equation for **P** therefore states that a point $\mathbf{x}$ in **P** is given by the formula

$$\mathbf{x} = (\text{a point in } \mathbf{L}_1) + (\text{a point in } \mathbf{L}_2).$$

It follows that our linear space **P** is the sum

$$\mathbf{P} = \mathbf{L}_1 + \mathbf{L}_2$$

of two linear spaces L_1 and L_2, which, in this case, are lines through the origin of R^3.

One final observation about our linear space R^3. The vector $v_3 = (2, 3, -1)$ is not in **P**. In fact, it is normal (perpendicular) to **P**. The equation

$$x = v_3 t$$

represents a third line L_3 through the origin. It too is a linear space. Every point **x** in R^3 is expressible as a linear combination

$$x = v_1 t_1 + v_2 t_2 + v_3 t_3$$

of v_1, v_2, and v_3. It follows that R^3 is the sum

$$R^3 = L_1 + L_2 + L_3$$

of the lines L_1, L_2, and L_3. It is also the sum

$$R^3 = P + L_3,$$

because $P = L_1 + L_2$. We also see that R^3 is the sum

$$R^3 = L_1 + Q,$$

where **Q** is the plane $Q = L_2 + L_3$. In fact, we have

$$R^3 = P + Q = L_1 + 2L_2 + L_3.$$

At first glance, this may appear to be incorrect because the term $2L_2$ is located where we should have the term L_2. However, it is correct because the sum $x + y$ of two vectors **x** and **y** in L_2 is in L_2, signifying that

$$2L_2 = L_2 + L_2 = L_2.$$

This illustrates a situation in which two of something is not more.

Let us now return to the study of some of the properties of subspaces of a general linear space **V**. Here are some properties of linear subspaces of a linear space **V** that we have used in our discussions: (i) If **W** is a linear subspace of **V** and **U** is a linear subspace of **W**, then **U** is a linear subspace of **V**. (ii) The set **O**, whose only element is the zero element **0**, is a linear subspace of every linear subspace of **V**. We consider **V** to be a linear subspace of itself. The same is true for subspaces; that is, a linear subspace is considered to be a linear subspace of itself. Finally, (iii) the intersection **T** of two linear subspaces **U** and **W** of **V** is a linear subspace of **U**, of **W**, and of **V**. The intersection **T** of **U** and **W** is, of course, the set of elements that are in both **U** and **W**. If **u** and **v** are in **T**, they are also in **U** and **W**. Their sum $u + v$ is in both **U** and **W** and so is in **T**. Similarly, if c is a scalar, then cu is in both **U** and **W** and so is in **T**. This verifies that **T** is indeed a linear space, a linear subspace of both **U** and **W**. It could happen that there is only one point in **T**, the point **u** = **0**. In this event **T** = **O** and we say that the linear subspaces **U** and **W** have only the **0** element in common.

Here are some properties of the intersection **T** of two linear subspaces **U** and **W** of a linear space **V**.

The intersection **T** of two linear subspaces **U** and **W** of a linear space **V** has the following properties:

1. **T** is a linear space.
2. **T** is a linear subspace of **U** and of **W**.
3. If **U** is a linear subspace of **W**, then **T** = **U**.
4. The zero element **0** is in **T**.

Property 1 was established above. Property 4 holds because **T** is a linear space. The proof of properties 2 and 3 will be left as an exercise.

In our examples we encountered special instances of the sum of linear subspaces in **V**. We shall now give a precise definition of this concept. Let **U** and **W** be linear subspaces of a linear space **V**. By the *sum*

$$\mathbf{S} = \mathbf{U} + \mathbf{W}$$

of **U** and **W** is meant the set **S** of all points **x** in **V** expressible in the form

$$\mathbf{x} = \mathbf{y} + \mathbf{z}$$

having **y** in **U** and **z** in **W**.

Here are some properties of a sum of this type.

The sum **S** of two linear subspaces **U** and **W** has the following properties:

1. **S** is a linear subspace of **V**.
2. If **U** and **W** have only the **0** element in common, then for each **x** in **S** there is a unique **y** in **U** and a unique **z** in **W** such that

   $$\mathbf{x} = \mathbf{y} + \mathbf{z}.$$

   In this event we say that **S** is a *direct sum* of **U** and **W**.
3. If **W** is a linear subspace of **U**, then **S** = **U**, so that **U** + **W** = **U**. In particular,

   $$\mathbf{U} + \mathbf{U} = \mathbf{U}.$$

To establish Property 1, let

$$\mathbf{x}_1 = \mathbf{y}_1 + \mathbf{z}_1 \quad \text{and} \quad \mathbf{x}_2 = \mathbf{y}_2 + \mathbf{z}_2$$

be two points in **S** with $\mathbf{y}_1$, $\mathbf{y}_2$ in **U** and $\mathbf{z}_1$, $\mathbf{z}_2$ in **W**. Then

$$\mathbf{x} = \mathbf{x}_1 + \mathbf{x}_2 = (\mathbf{y}_1 + \mathbf{y}_2) + (\mathbf{z}_1 + \mathbf{z}_2) = \mathbf{y} + \mathbf{z}$$

is the sum of the point $\mathbf{y} = \mathbf{y}_1 + \mathbf{y}_2$ in **U** and the point $\mathbf{z} = \mathbf{z}_1 + \mathbf{z}_2$ in **W**. Consequently, $\mathbf{x}$ is in **S**. Similarly, the point $c\mathbf{x}_1 = c\mathbf{y}_1 + c\mathbf{z}_1$ is in **S**. Hence **S** is a linear space, as was to be proved.

To establish Property 2 for sums, let

$$\mathbf{x} = \mathbf{y}_1 + \mathbf{z}_1 = \mathbf{y}_2 + \mathbf{z}_2$$

be two representations of an element $\mathbf{x}$ in **S** with $\mathbf{y}_1$, $\mathbf{y}_2$ in **U** and $\mathbf{z}_1$, $\mathbf{z}_2$ in **W**. Then we have the relation

$$\mathbf{y}_1 - \mathbf{y}_2 = \mathbf{z}_2 - \mathbf{z}_1.$$

The left member of this equation is in **U** and the right member is in **W**. When **U** and **W** have only the **0** element in common, each of these members must be the zero element **0**. In this event, we have $\mathbf{y}_1 = \mathbf{y}_2$ and $\mathbf{z}_1 = \mathbf{z}_2$, as stated in Property 2.

The proof of Property 3 is left as an exercise.

The sum

$$\mathbf{S} = \mathbf{U}_1 + \mathbf{U}_2 + \cdots + \mathbf{U}_q$$

of q linear subspaces $\mathbf{U}_1$, $\mathbf{U}_2$, ...,$\mathbf{U}_q$ of a linear space **V** is defined in the same manner. It is a direct sum if every vector $\mathbf{x}$ in **S** can be written *uniquely* as

$$\mathbf{x} = \mathbf{u}_1 + \mathbf{u}_2 + \cdots + \mathbf{u}_q,$$

where $\mathbf{u}_i \in \mathbf{U}_i$. Equivalently, it is a ***direct sum*** if these linear subspaces, pairwise, have nothing in common except the **0** element.

Exercises

1. Determine whether each of the following sets is a vector space or not. Explain.

   (a) $A = \{(x, y, z) \in \mathbf{R}^3 \mid x + y + z = 0\}$ (b) $B = \{(x, y, z) \in \mathbf{R}^3 \mid x + y + z = 1\}$

   (c) $C = \left\{ \begin{pmatrix} x & y \\ y & x \end{pmatrix} \mid x, y \in \mathbf{R} \right\}$ (d) $D = \left\{ \begin{pmatrix} x & y \\ y & 1 \end{pmatrix} \mid x, y \in \mathbf{R} \right\}$

   (e) $E = \{(x, y) \in \mathbf{R}^2 \mid x = 0 \; or \; y = 0\}$ (f) $F = \{(x, 0) \in \mathbf{R}^2 \mid x \in \mathbf{R}\}$

2. Here we are concerned with points and vectors in $\mathbf{R}^2$.

(a) Why is the set of multiples of a fixed vector $\mathbf{v} = (2, 3)$ a linear space? Why can this linear space, a line through the origin, be described by the equation

$$\mathbf{x} = \mathbf{v}t,$$

where t is an arbitrary number, called a parameter? Why is not the set of points, described by the equation

$$\mathbf{x} = \mathbf{v}t^2$$

with t as a parameter, a linear space? This set of points is the set of positive multiples of $\mathbf{v}$, which can also be described by the equation

$$\mathbf{x} = \mathbf{v}t \qquad (t \geq 0).$$

Is the set defined by the equation

$$\mathbf{x} = \mathbf{v}t^3$$

with t as a parameter a linear space? For fixed numbers a and b with $a < b$, what is the set of points described by the equation

$$\mathbf{x} = \mathbf{v}t \qquad (a \leq t \leq b)?$$

Here the inequalities in the parentheses tell us what values the parameter t can take.

(b) Let $\mathbf{u}$ be the vector $\mathbf{u} = (2, -3)$, show that the line described in (a) is the set of points $\mathbf{x}$ satisfying the equation

$$\mathbf{u}^T\mathbf{x} = 0.$$

This equation, with $\mathbf{u}$ fixed, therefore describes a linear space. The point $\mathbf{x}_0 = (1, -1)$ satisfies the equation

$$\mathbf{u}^T \mathbf{x} = 5.$$

Is the set of points $\mathbf{x}$, satisfying this equation, a linear space? Show that this set of points can also be described by the equation

$$\mathbf{x} = \mathbf{x}_0 + \mathbf{v}t.$$

where t is a parameter and $\mathbf{v}$ is the vector $\mathbf{v} = (2, 3)$ appearing in part (a). Sketch this line graphically.

(c) With $\mathbf{u}$ and $\mathbf{v}$ as in parts (a) and (b), show that every point (vector) $\mathbf{x}$ is expressible in the form

$$\mathbf{x} = s\mathbf{u} + t\mathbf{v}.$$

This equation, with s and t parameters, is therefore another way to describe the linear space $\mathbf{R}^2$. What configuration do you get when you restrict s and t to be nonnegative. Sketch it. What configuration do you get when you require the parameters s and t to satisfy the equation

$$s + t = 1?$$

Sketch it. Is it a linear space?

3. Here we are concerned with points and vectors in $\mathbf{R}^3$.

(a) With $\mathbf{v} = (2, 3, -1)$ and $\mathbf{u} = (1, 1, 5)$, are the statements and conclusions given in parts (a), (b) and (c) of Exercise 2 valid here also? If some of them are not valid, modify them so that they are valid.

(b) Does the set of points $\mathbf{x}$ satisfying the equations

$$\mathbf{u}^T\mathbf{x} = 0, \qquad \mathbf{v}^T\mathbf{x} = 0$$

form a linear space? Show that this set of points can be described parametrically by the equation

$$\mathbf{x} = \mathbf{w}t$$

for a suitably chosen vector $\mathbf{w}$. Is the vector $\mathbf{w}$ unique? Show that the vector $\mathbf{w}$ is orthogonal to the vectors $\mathbf{u}$ and $\mathbf{v}$.

(c) Show that every point $\mathbf{x}$ in $\mathbf{R}^3$ is expressible as a linear combination

$$\mathbf{x} = r\mathbf{u} + s\mathbf{v} + t\mathbf{w}$$

of the vectors $\mathbf{u}$, $\mathbf{v}$, $\mathbf{w}$. This equation can be viewed to be a parametric representation of $\mathbf{R}^3$ with r, s, and t as parameters. The terms $r\mathbf{u}$, $s\mathbf{v}$, $t\mathbf{w}$ in this equation represent points on lines $\mathbf{U}$, $\mathbf{V}$, $\mathbf{W}$ through the origin. Show that our equation is equivalent to the statement that $\mathbf{R}^3$ is the sum

$$\mathbf{R}^3 = \mathbf{U} + \mathbf{V} + \mathbf{W}$$

of three linear subspaces, which, in this case, are lines through the origin. Is this sum a ***direct*** sum?

4. (a) Show that the class of 2×2-matrices whose trace is zero is a linear space.

(b) Show that the class of 2×2-matrices whose determinant is zero is not a linear space.

(c) For a fixed nonzero vector $\mathbf{x}$, show that the class of 2×2-matrices A such that $A\mathbf{x} = 0$ is a linear space.

(d) Let λ be an eigenvalue of a 2×2-matrix A. Show that the set of corresponding eigenvectors of A is not a linear space because the point $\mathbf{x} = \mathbf{0}$ is missing. Show that if we adjoin this point, we obtain a linear space. This space is called an ***eigenspace*** of A.

5. Give an example of a sum of linear spaces that is not a direct sum.

6. Give an example of a linear space other than the ones we have described.

7. Show that the collection of all polynomials in a variable t is a linear space. Is every polynomial a linear combination of a finite number of polynomials?

8. Show that the class of functions $f(t)$ which are linear combinations of the functions

$$1, \quad \cos t, \quad \sin t, \quad \cos 2t, \quad \sin 2t$$

form a linear space $\mathbf{F}$. Let $\mathbf{C}$ be the subset of $\mathbf{F}$ which are linear combinations of the functions 1, $\cos t$ and $\cos 2t$. Is $\mathbf{C}$ a linear space? Let $\mathbf{S}$ be the linear subspace comprised of all linear combinations of $\sin t$ and $\sin 2t$. Is $\mathbf{F}$ the sum

$$\mathbf{F} = \mathbf{C} + \mathbf{S}$$

of these two subspaces? If so, is it a direct sum?

9. In calculus we learn that

   (a) the sum of two continuous functions is a continuous function.

   (b) the product of a continuous function by a scalar is a continuous function.

   Does this imply that the class of continous functions is a linear space? (There is a technicality here. We assume that our functions have the same domain.)

10. Do functions $f(x)$ that are differentiable on a given interval form a linear space? If so, why? Is it a linear subspace of the class of functions that are continuous on this interval?

11. In calculus, we learn that the sum of two convergent sequences of numbers is again a convergent sequence of numbers. Also, if we multiply each term of the sequence by the same number, the new sequence is also convergent. Can we conclude that the class of convergent sequences of numbers is a linear space?

12. Consider the class of infinite sequences

    $$\mathbf{a} = (a_0, a_1, a_2, \ldots)$$

    which have the property that only a finite number of the a_j's are different from zero. Show that this class is a linear space. Is there a connection between this space and the space of all polynomials of a single variable t?

5.2 GENERATORS AND BASES OF LINEAR SPACES

In $\mathbf{R}^3$, the xy-plane is the linear subspace comprised of all points of the form $(x, y, 0)$. Looking at the equation

$$(x, y, 0) = x(1, 0, 0) + y(0, 1, 0),$$

we see that the xy-plane is the set of all linear combinations of the fixed vectors $(1, 0, 0)$ and $(0, 1, 0)$. A similar situation is encountered in a general linear space $\mathbf{V}$. This section is devoted to the study of this phenomenon.

Let $\mathbf{V}$ be a linear space. There is a standard way to generate a linear subspace of $\mathbf{V}$. One need only take linear combinations of fixed elements in $\mathbf{V}$. For example, if $\mathbf{u}$, $\mathbf{v}$, and $\mathbf{w}$ are elements in $\mathbf{V}$, the set of all linear combinations

$$\mathbf{x} = a\mathbf{u} + b\mathbf{v} + c\mathbf{w}$$

of the elements $\mathbf{u}$, $\mathbf{v}$, and $\mathbf{w}$ is a linear subspace $\mathbf{U}$ of $\mathbf{V}$. The set $\mathbf{U}$ is a linear space because if we add two linear combinations of $\mathbf{u}$, $\mathbf{v}$, and $\mathbf{w}$, we get another linear combination of $\mathbf{u}$, $\mathbf{v}$, and $\mathbf{w}$. Also, if we multiply a linear combination of $\mathbf{u}$, $\mathbf{v}$, and $\mathbf{w}$ by a scalar, the element obtained is also a linear combination of $\mathbf{u}$, $\mathbf{v}$, and $\mathbf{w}$.

When a linear subspace $\mathbf{U}$ of $\mathbf{V}$ is obtained by taking all linear combinations of fixed elements $\mathbf{u}$, $\mathbf{v}$, and $\mathbf{w}$ of $\mathbf{V}$, we say that $\mathbf{U}$ is *generated* by the elements $\mathbf{u}$, $\mathbf{v}$, and $\mathbf{w}$ or, equivalently, we say that these elements *generate* U. The elements **u**, **v**, and **w** are called *generators* of U. The term *span* is also used. We say that the elements **u**, **v**, and **w** *span* U and that U is *spanned* by the elements **u**, **v**, and **w**. Moreover, U is the *span* of the elements **u**, **v**, and **w**.

In a more general case, we note that the set of all linear combinations of k elements $\mathbf{u}_1$, $\mathbf{u}_2$, . . ., $\mathbf{u}_k$ in **V** is a linear subspace U of **V**. These k elements *generate* U and U is *generated* by them. The elements $\mathbf{u}_1$, $\mathbf{u}_2$, . . . , $\mathbf{u}_k$ are called *generators* of U. Using the term *span*, we say that U is *spanned* by $\mathbf{u}_1$, $\mathbf{u}_2$, . . . , $\mathbf{u}_k$ and that these elements *span* U. We also say that U is the *span* of the elements $\mathbf{u}_1$, $\mathbf{u}_2$, . . . , $\mathbf{u}_k$.

Let us consider several examples.

■ **EXAMPLE 1** ───────────────────────────────────────

The two vectors $\mathbf{e}_1 = (1, 0)$ and $\mathbf{e}_2 = (0, 1)$ generate $\mathbf{R}^2$, for if $\mathbf{x} = (x, y)$ is any vector in $\mathbf{R}^2$, then $\mathbf{x} = (x, y) = x(1, 0) + y(0, 1) = x\mathbf{e}_1 + y\mathbf{e}_2$. The vectors $\mathbf{u} = (5, 2)$ and $\mathbf{v} = (2, 1)$ also generate $\mathbf{R}^2$ for writing any vector $\mathbf{x} = (x, y)$ in $\mathbf{R}^2$ as

$$(x, y) = a(5, 2) + b(2, 1)$$

gives

$$5a + 2b = x$$
$$2a + b = y.$$

Solving for a and b, we get

$$a = x - 2y \quad \text{and} \quad b = -2x + 5y.$$

Therefore,

$$(x, y) = (x - 2y)(5, 2) + (-2x + 5y)(2, 1)$$

[i.e., every vector in $\mathbf{R}^2$ can be written as a linear combination of the two vectors (5, 2) and (2, 1) in $\mathbf{R}^2$]. Therefore, these vectors generate $\mathbf{R}^2$. The vectors **u**, **v**, and $\mathbf{w} = (0, 0)$ also generate $\mathbf{R}^2$, but the vector **w** is not needed.

■ **EXAMPLE 2** ───────────────────────────────────────

The vectors

$$\mathbf{e}_1 = (1, 0, 0), \quad \mathbf{e}_2 = (0, 1, 0), \quad \mathbf{e}_3 = (0, 0, 1)$$

generate $\mathbf{R}^3$. For if $\mathbf{x} = (x, y, z)$ is any vector in $\mathbf{R}^3$, then

$$\mathbf{x} = (x, y, z) = x(1, 0, 0) + y(0, 1, 0) + z(0, 0, 1) = x\mathbf{e}_1 + y\mathbf{e}_2 + z\mathbf{e}_3.$$

The vectors

$$\mathbf{u} = (1, 0, 0), \qquad \mathbf{v} = (1, 1, 0), \qquad \mathbf{w} = (1, 1, 1), \qquad \mathbf{t} = (0, 1, 1)$$

also generate $\mathbf{R}^3$, even though we have more generators than are needed. To see this, let $\mathbf{x} = (x, y, z)$ be any vector in $\mathbf{R}^3$. We seek constants a, b, c, and d such that

$$\mathbf{x} = (x, y, z) = a(1, 0, 0) + b(1, 1, 0) + c(1, 1, 1) + d(0, 1, 1).$$

Writing this out yields the system of equations

$$a + b + c \qquad = x$$
$$b + c + d = y$$
$$c + d = z.$$

Setting $d = 0$ gives $a = x - y$, $b = y - z$, and $c = z$. We see that

$$x = (x, y, z) = (x - y)(1, 0, 0) + (y - z)(1, 1, 0) + z(1, 1, 1) + 0(0, 1, 1),$$

so that the vectors $\mathbf{u}$, $\mathbf{v}$, $\mathbf{w}$, and $\mathbf{t}$ generate $\mathbf{R}^3$. As can be seen, the vector $\mathbf{t}$ is not needed, so we can delete it and still have a set of generators $\mathbf{u}$, $\mathbf{v}$, and $\mathbf{w}$ of $\mathbf{R}^3$. The reader should verify that the vectors $\mathbf{v}$, $\mathbf{w}$, and $\mathbf{t}$ also generate $\mathbf{R}^3$.

The vectors $\mathbf{u}$, $\mathbf{w}$, and $\mathbf{t}$ do not generate $\mathbf{R}^3$. To see this, we argue as follows: Suppose that they do. Then for any vector $\mathbf{x} = (x, y, z)$ in $\mathbf{R}^3$, there exist constants a, b, and c such that

$$\mathbf{x} = (x, y, z) = a\mathbf{u} + b\mathbf{w} + c\mathbf{t} = a(1, 0, 0) + b(1, 1, 1) + c(0, 1, 1).$$

This gives the system of equations

$$a + b \qquad = x$$
$$b + c = y$$
$$b + c = z.$$

The last two equations imply that $y = z$. But this is impossible, since the vector $\mathbf{x} = (x, y, z)$ was an arbitrary vector in $\mathbf{R}^3$. Therefore, the vectors $\mathbf{u}$, $\mathbf{w}$, and $\mathbf{t}$ do not generate $\mathbf{R}^3$. They do, however, generate the linear subspace of $\mathbf{R}^3$ consisting of all 3-tuples (x, y, z) for which $y = z$.

■ **EXAMPLE 3** _____

The linear space $\mathbf{M}$ comprised of all 2×2-matrices is generated by the matrices

$$E_1 = \begin{bmatrix} 1 & 0 \\ 0 & 0 \end{bmatrix}, \qquad E_2 = \begin{bmatrix} 0 & 1 \\ 0 & 0 \end{bmatrix},$$

$$E_3 = \begin{bmatrix} 0 & 0 \\ 1 & 0 \end{bmatrix}, \qquad E_4 = \begin{bmatrix} 0 & 0 \\ 0 & 1 \end{bmatrix}.$$

This follows because

$$\begin{bmatrix} a & b \\ c & d \end{bmatrix} = aE_1 + bE_2 + cE_3 + dE_4.$$

Another set of generators is given by

$$E_1 + E_2, \qquad E_1 - E_2, \qquad E_3 - E_1 - E_2, \qquad E_1 + E_2 + E_3 + E_4,$$

although we would use these generators only in a very special situation.

■ **EXAMPLE 4** _____

The polynomials 1, t, and t^2 generate the linear space **P** comprised of all polynomials of the form

$$f(t) = a + bt + ct^2.$$

This follows immediately since any polynomial $f(t)$ in **P** can be written as

$$f(t) = a1 + bt + ct^2.$$

The polynomials $t - 1$, $1 + t$, and $t^2 - 1$ which are in **P** are another set of generators of **P**. To show that they generate **P**, we must produce constants α, β, γ so that any polynomial $f(t) = a + bt + ct^2$ in **P** can be written as

$$f(t) = a + bt + ct^2 = \alpha(t - 1) + \beta(1 + t) + \gamma(t^2 - 1).$$

Equating coefficients of like powers of t, we get

$$\gamma = c$$
$$\alpha + \beta \qquad = b$$
$$-\alpha + \beta - \gamma = a.$$

Solving this system gives

$$\alpha = \frac{1}{2}(-a + b - c), \qquad \beta = \frac{1}{2}(a + b + c), \qquad \gamma = c.$$

It follows that any polynomial $f(t) = a + bt + ct^2$ in **P** is generated by the polynomials $t - 1$, $1 + t$, and $t^2 - 1$ in **P**. The reader should verify that the polynomials 1, t, t^2, $t - 1$, $1 + t$, and $t^2 - 1$ also generate **P**, but in this case, we have more generators than are needed— so you can throw some away. Obviously, we can throw the last three away. Instead, we can throw away the first three. But we cannot throw away both t^2 and $t^2 - 1$ at the same time. Why not? However, we can keep both t^2 and $t^2 - 1$ if we also keep one of the polynomials t, $t - 1$, and $1 + t$.

In these examples we have illustrated the following.

DELETION PRINCIPLE FOR GENERATORS. Let $\mathbf{u}_1$, $\mathbf{u}_2$, . . . , $\mathbf{u}_k$ be k generators of a linear space **U**. If one of the generators is a linear combination of the others, it can be deleted giving us thereby $k - 1$ generators of **U**. In particular, if $\mathbf{u}_k$ is a linear combination of $\mathbf{u}_1$, $\mathbf{u}_2$, . . . , $\mathbf{u}_{k-1}$, then $\mathbf{u}_1$, $\mathbf{u}_2$, . . . , $\mathbf{u}_{k-1}$ generate **U**.

This principle states that if you have too many generators, you can throw one away. But if you throw one away, be sure that it is a linear combination of those that remain.

The proof of this principle in the general case is like that for the case $k = 3$. In this event **U** is comprised of all linear combinations

$$\mathbf{u} = a\mathbf{u}_1 + b\mathbf{u}_2 + c\mathbf{u}_3$$

of three elements $\mathbf{u}_1$, $\mathbf{u}_2$, and $\mathbf{u}_3$. Suppose that $\mathbf{u}_3$ is a linear combination

$$\mathbf{u}_3 = e\mathbf{u}_1 + f\mathbf{u}_2$$

of $\mathbf{u}_1$ and $\mathbf{u}_2$. Then

$$\mathbf{u} = a\mathbf{u}_1 + b\mathbf{u}_2 + c(e\mathbf{u}_1 + f\mathbf{u}_2) = (a + ce)\mathbf{u}_1 + (b + cf)\mathbf{u}_2$$

and so is a linear combination of $\mathbf{u}_1$ and $\mathbf{u}_2$. Consequently, **U** is generated by $\mathbf{u}_1$, and $\mathbf{u}_2$ as well as by $\mathbf{u}_1$, $\mathbf{u}_2$, and $\mathbf{u}_3$.

As an immediate consequence of the *deletion principle* we have the following important result.

PROPOSITION 1. Suppose that a linear subspace **U** possesses k generators $\mathbf{u}_1$, $\mathbf{u}_2$, . . . , $\mathbf{u}_k$ which are *linearly dependent* in the sense that one of them is a linear combination of the others. Then, by deletions, we can obtain a new set of r generators $\mathbf{v}_1$, $\mathbf{v}_2$, . . . , $\mathbf{v}_r$ which are linearly independent in the sense that no one of them is a linear combination of the others.

Here we observe that the concepts of linear dependence and linear independence are the same in a linear space as in the space $\mathbf{R}^m$, a space that was used implicitly when we first introduced these concepts. More formal definitions are the following.

We say that k elements $\mathbf{u}_1, \mathbf{u}_2, \ldots, \mathbf{u}_k$ in **V** are *linearly dependent* if the equation

$$a_1\mathbf{u}_1 + a_2\mathbf{u}_2 + \cdots + a_k\mathbf{u}_k = \mathbf{0}$$

has a solution $a_1, a_2, \ldots, a_k$, not all zero. If the only solution of this equation is the trivial solution $a_1 = a_2 = \cdots = a_k = 0$, the elements $\mathbf{u}_1, \mathbf{u}_2, \ldots, \mathbf{u}_k$ are *linearly independent*.

There are situations when it is natural to begin with more generators than are needed. One such arises when we consider the linear space **C** generated by the columns of a matrix A. This space is called the *column space* of A, a space that we study in Chapter 6.

■ **EXAMPLE 5**

Consider the matrices

$$A = \begin{bmatrix} 1 & -2 & 0 & 1 \\ 0 & 0 & 1 & 1 \\ 2 & -4 & 0 & 2 \end{bmatrix}, \quad B = \begin{bmatrix} 1 & 0 & 1 \\ 0 & 1 & 1 \\ 2 & 0 & 2 \end{bmatrix}, \quad C = \begin{bmatrix} 1 & 0 \\ 0 & 1 \\ 2 & 0 \end{bmatrix}.$$

In this case the matrix A has four columns generating **C**. The second column is a multiple of the first, so we can delete it as a generator. This gives the matrix B, whose columns also generate **C**. But the last column of B is the sum of the first two and so can be deleted. This gives us the matrix C whose columns generate **C**. No more deletions can be made because the columns of C are linearly independent. The columns of the matrices A, B, and C generate the same column space. There is no unique way of making these deletions. By deleting columns in another manner, following the same rules, we can obtain a second 3×2-matrix D whose columns are linearly independent and generate the column space **C** of A. For example, by deleting the first two columns of A, we obtain a matrix D of this type.

There is another principle for generators which we introduce by first looking at the following.

■ **EXAMPLE 4 (continued)**

We noted earlier that the polynomials $t - 1$, $1 + t$, and $t^2 - 1$ generated the polynomials

$$f(t) = a + bt + ct^2.$$

Observe that the polynomial t is the linear combination

$$t = \frac{1}{2}(t - 1) + \frac{1}{2}(1 + t) + 0(t^2 - 1)$$

of these generators. We have

$$t = \frac{1}{2}(t - 1) + \text{(others)}.$$

Our next principle states that because the coefficient of $t - 1$, being ½, is not zero, we can replace the generator $t - 1$ by t and thereby obtain a new set of generators, t, $1 + t$, and $t^2 - 1$. Obviously, we could not replace $t^2 - 1$ by t and obtain an allowable set of generators of our polynomials.

> **REPLACEMENT PRINCIPLE FOR GENERATORS.** Let $u_1, u_2, \ldots, u_k$ generate a linear subspace U of V. For a given index j, let v be an element in U expressible in the form
>
> $v = b u_j + \text{(a linear combination of the other generators)}.$
>
> If b is not zero, we can replace u_j by v and thereby obtain a new set of generators
>
> $u_1, \ldots, u_{j-1}, v, u_{j+1}, \ldots, u_k$
>
> of U. If the original generators are linearly independent, the new generators are linearly independent.

This principle states that if you wish to replace a generator by a new one, be sure that the new one is not a linear combination of the other generators.

    The proof of this replacement principle in the general case is like that for the case $k = 4$ and $j = 2$, which goes as follows. Suppose that u_1, u_2, u_3, and u_4 generate a subspace U of a vector space V. Then every u in U is a linear combination

$$u = a_1 u_1 + a_2 u_2 + a_3 u_3 + a_4 u_4$$

of the generators u_1, u_2, u_3, and u_4. Let v be an element of U expressible in the form

$$v = b u_2 + (c_1 u_1 + c_3 u_3 + c_4 u_4)$$

with b nonzero. Then

$$u_2 = \frac{1}{b}v - \frac{1}{b}(c_1 u_1 + c_3 u_3 + c_4 u_4).$$

Substituting this result in the formula for **u**, we find that

$$\mathbf{u} = \left(a_1 - \frac{a_2 c_1}{b}\right)\mathbf{u}_1 + \frac{a_2}{b}\mathbf{v} + \left(a_3 - \frac{a_2 c_3}{b}\right)\mathbf{u}_3 + \left(a_4 - \frac{a_2 c_4}{b}\right)\mathbf{u}_4.$$

It follows that $\mathbf{u}_1$, $\mathbf{v}$, $\mathbf{u}_3$, and $\mathbf{u}_4$ generate **U**. Moreover, if the original generators $\mathbf{u}_1$, $\mathbf{u}_2$, $\mathbf{u}_3$, and $\mathbf{u}_4$ are linearly independent, the element **v** could not be a linear combination of $\mathbf{u}_1$, $\mathbf{u}_3$, and $\mathbf{u}_4$. For if it were, the formula for **v** would tell us that the element $\mathbf{u}_2$ would also be a linear combination of $\mathbf{u}_1$, $\mathbf{u}_3$, and $\mathbf{u}_4$. This is not the case. From this fact we conclude that the generators $\mathbf{u}_1$, $\mathbf{v}$, $\mathbf{u}_3$, and $\mathbf{u}_4$ are linearly independent when the original generators are. This proves our replacement principle for a special case. The general case can be justified in the same manner.

We are now in a position to prove the following.

PROPOSITION 2. Let $\mathbf{u}_1$, $\mathbf{u}_2$, $\ldots$, $\mathbf{u}_n$ be n linearly independent elements of a linear space **V**. Let **U** be the linear subspace generated by these elements. Let $\mathbf{v}_1$, $\mathbf{v}_2$, $\ldots$, $\mathbf{v}_k$ be any k linearly independent elements in **U**. Then:

(a) $k \le n$.

(b) If $k = n$, then the elements $\mathbf{v}_1$, $\mathbf{v}_2$, $\ldots$, $\mathbf{v}_n$ generate **U**.

(c) If $k < n$, we can renumber the $\mathbf{u}_j$'s so that the elements $\mathbf{v}_1$, $\ldots$, $\mathbf{v}_k$, $\mathbf{u}_{k+1}$, $\ldots$, $\mathbf{u}_n$ generate **U**.

The proof in the general case is like that for the case $n = 4$. Since $\mathbf{v}_1$ is in **U**, it is a linear combination

$$\mathbf{v}_1 = a_1\mathbf{u}_1 + a_2\mathbf{u}_2 + a_3\mathbf{u}_3 + a_4\mathbf{u}_4$$

of the first set of generators of **U**. Since $\mathbf{v}_1$ is not zero, at least one of the a_j's is not zero. We can suppose that the $\mathbf{u}_j$'s have been renumbered so that a_1 is not zero. Then by our replacement principle, we can replace $\mathbf{u}_1$ by $\mathbf{v}_1$ giving us a second set of generators $\mathbf{v}_1$, $\mathbf{u}_2$, $\mathbf{u}_3$, and $\mathbf{u}_4$ of **U**. If $k = 1$, we are done. Otherwise, we can express $\mathbf{v}_2$ in the form

$$\mathbf{v}_2 = b_1\mathbf{v}_1 + b_2\mathbf{u}_2 + b_3\mathbf{u}_3 + b_4\mathbf{u}_4.$$

Since $\mathbf{v}_1$ and $\mathbf{v}_2$ are linearly independent, the numbers b_2, b_3, and b_4 are not all zero. After renumbering, we have b_2 different from zero. By our replacement principle, we can replace $\mathbf{u}_2$ by $\mathbf{v}_2$ in our second set of generators and thereby obtain a third set of generators $\mathbf{v}_1$, $\mathbf{v}_2$, $\mathbf{u}_3$, and $\mathbf{u}_4$. If $k = 2$, we are done. Otherwise, we can express $\mathbf{v}_3$ in the form

$$\mathbf{v}_3 = c_1\mathbf{v}_1 + c_2\mathbf{v}_2 + c_3\mathbf{u}_3 + c_4\mathbf{u}_4$$

with c_3 and c_4 not both zero. We can suppose that c_3 is not zero. Replacing $\mathbf{u}_3$ by $\mathbf{v}_3$ in our third set of generators, we obtain a fourth set of generators, $\mathbf{v}_1$, $\mathbf{v}_2$, $\mathbf{v}_3$, and $\mathbf{u}_4$. If $k = 3$, we are done. Otherwise, we can express $\mathbf{v}_4$ in the form

$$\mathbf{v}_4 = d_1\mathbf{v}_1 + d_2\mathbf{v}_2 + d_3\mathbf{v}_3 + d_4\mathbf{u}_4$$

with d_4 different from zero. Replacing $\mathbf{u}_4$ by $\mathbf{v}_4$, we find that the elements $\mathbf{v}_1$, $\mathbf{v}_2$, $\mathbf{v}_3$, and $\mathbf{v}_4$ generate U. We have $k = 4$, since otherwise the element $\mathbf{v}_5$ would be a linear combination of $\mathbf{v}_1$, $\mathbf{v}_2$, $\mathbf{v}_3$, and $\mathbf{v}_4$, contrary to our hypothesis. This proves our proposition for the case $n = 4$. The general case can be established in the same manner.

> A linear space **V** will be said to be *finite dimensional* if it possesses a finite set of generators $\mathbf{u}_1$, $\mathbf{u}_2$, $\ldots$, $\mathbf{u}_q$. Otherwise, it is *infinite dimensional*.

In view of Proposition 1,

> A finite-dimensional linear space **V** possesses a set of generators $\mathbf{v}_1$, $\mathbf{v}_2$, $\ldots$, $\mathbf{v}_n$ which are linearly independent. Such a set of generators is called a *basis* for **V**.

By Proposition 2,

> The number n of elements in a basis for **V** is the same for all bases. This number n is called the *dimension* of **V**.

> We see that in order for a set of vectors $\mathbf{v}_1$, $\mathbf{v}_2$, $\ldots$, $\mathbf{v}_n$ in a linear space **V** to be a *basis* for **V**, they must satisfy two conditions:
>
> 1. $\mathbf{v}_1$, $\mathbf{v}_2$, $\ldots$, $\mathbf{v}_n$ must generate **V**.
> 2. $\mathbf{v}_1$, $\mathbf{v}_2$, $\ldots$, $\mathbf{v}_n$ must be linearly independent.

■ **EXAMPLE 6** _____

The two vectors $\mathbf{e}_1 = (1, 0)$ and $\mathbf{e}_2 = (0, 1)$ form a basis for $\mathbf{R}^2$. As we saw in Example 1, they certainly generate $\mathbf{R}^2$. Furthermore, they are linearly independent, since they are the columns of the nonsingular 2×2 identity matrix

$$I = \begin{bmatrix} 1 & 0 \\ 0 & 1 \end{bmatrix}.$$

Since there are two vectors in a basis for $\mathbf{R}^2$, it is no surprise that the dimension of $\mathbf{R}^2$ is 2.

■ **EXAMPLE 7** _____

The vectors $\mathbf{u} = (5, 2)$ and $\mathbf{v} = (2, 1)$ also form a basis for $\mathbf{R}^2$. We have already seen in Example 1 that $\mathbf{u}$ and $\mathbf{v}$ generate $\mathbf{R}^2$. Since the 2×2-matrix

$$\begin{bmatrix} 5 & 2 \\ 2 & 1 \end{bmatrix}$$

whose columns are $\mathbf{u}$ and $\mathbf{v}$ has a nonzero determinant 1, and hence is nonsingular, it follows that $\mathbf{u}$ and $\mathbf{v}$ are linearly independent. (Of course, it is easy to see that they are linearly independent since one is not a multiple of the other.) Hence $\mathbf{u}$ and $\mathbf{v}$ are a basis for $\mathbf{R}^2$.

■ **EXAMPLE 8** _____

Any pair of linearly independent vectors in $\mathbf{R}^2$ are a basis for $\mathbf{R}^2$. So also are any two columns of a nonsingular 2×2-matrix. If a particular 2×2-matrix is diagonalizable, it has two linearly independent eigenvectors, which again form a basis for $\mathbf{R}^2$. The foregoing notions for $\mathbf{R}^2$ can be extended in like fashion to the spaces $\mathbf{R}^3$ and $\mathbf{R}^n$.

■ **EXAMPLE 9** _____

The polynomials $t - 1$, $1 + t$, and $t^2 - 1$ are a basis for the linear space $\mathbf{P}$ comprised of all polynomials of the form

$$f(t) = a + bt + ct^2.$$

We have already shown in Example 4 that they generate $\mathbf{P}$. It remains to show that they are linearly independent in $\mathbf{P}$. To this end, consider

$$c_1(t - 1) + c_2(1 + t) + c_3(t^2 - 1) = 0 = 0 + 0t + 0t^2 \text{ in } \mathbf{P}.$$

Equating coefficients of like powers of t gives

$$\begin{aligned} c_3 &= 0 \\ c_1 + c_2 \quad &= 0 \\ -c_1 + c_2 - c_3 &= 0. \end{aligned}$$

Solving this system, we get $c_1 = c_2 = c_3 = 0$. This shows that the polynomials $t - 1$, $1 + t$, and $t^2 - 1$ are linearly independent in **P**. Therefore, they are a basis for **P**. Any other basis for **P** will also have three polynomials. Of course, the polynomials 1, t, and t^2 are a basis for **P**.

Obviously, if **V** is finite dimensional, so are its linear subspaces. Moreover, the dimension k of a linear subspace **U** of **V**, which does not coincide with **V**, is less than the dimension n of **V**.

In view of property (c) in Proposition 2, we have the following:

> **EXTENSION PRINCIPLE FOR BASES.** Let **V** be an n-dimensional linear space and let $\mathbf{u}_1, \mathbf{u}_2, \ldots, \mathbf{u}_k$ form a basis for a linear subspace **U** of **V**. Then this basis for **U** can be extended by adjoining $n - k$ elements $\mathbf{v}_{k+1}, \ldots, \mathbf{v}_n$ to form a basis $\mathbf{v}_1, \mathbf{v}_2, \ldots, \mathbf{v}_n$ for **V** having $\mathbf{v}_1 = \mathbf{u}_1, \mathbf{v}_2 = \mathbf{u}_2, \ldots, \mathbf{v}_k = \mathbf{u}_k$.

For example, the space $\mathbf{R}^4$ has as a basis the column vectors

$$\mathbf{e}_1 = (1, 0, 0, 0), \qquad \mathbf{e}_2 = (0, 1, 0, 0), \qquad \mathbf{e}_3 = (0, 0, 1, 0),$$

$$\mathbf{e}_4 = (0, 0, 0, 1)$$

of the identity matrix I. The vectors

$$\mathbf{u} = (2, 1, 0, -1), \qquad \mathbf{v} = (2, 1, 1, 1)$$

are linearly independent and so are the generators of a linear subspace **U** of $\mathbf{R}^4$. According to our extension principle we can select vectors **w** and **z** such that the vectors **u**, **v**, **w**, and **z** generate $\mathbf{R}^4$ and are linearly independent. In fact, looking at Proposition 2, we can select **w** and **z** from the set $\mathbf{e}_1, \mathbf{e}_2, \mathbf{e}_3$, and $\mathbf{e}_4$. It can be shown that the choice $\mathbf{w} = \mathbf{e}_1$ and $\mathbf{z} = \mathbf{e}_4$ will do. We can interpret this result in terms of matrices. The vectors **u** and **v** are the column vectors of the matrix A shown below. The matrix B is obtained from A by adjoining the two columns $\mathbf{e}_1$ and $\mathbf{e}_4$.

$$A = \begin{bmatrix} 2 & 2 \\ 1 & 1 \\ 0 & 1 \\ -1 & 1 \end{bmatrix}, \qquad B = \begin{bmatrix} 2 & 2 & 1 & 0 \\ 1 & 1 & 0 & 0 \\ 0 & 1 & 0 & 0 \\ -1 & 1 & 0 & 1 \end{bmatrix}.$$

You might wonder how we selected the last two columns of B. We looked at A and selected the nonsingular 2×2-submatrix G of A whose entries are the boldfaced entries of A. We then selected columns of the 4×4 identity matrix I which had 0's in the rows corresponding to the rows of G.

Of course, usually there is no unique way to select the matrix G. In our case, other selections of G can be made, as shown in the following matrices:

$$C = \begin{bmatrix} 2 & 2 & 1 & 0 \\ 1 & 1 & 0 & 0 \\ 0 & 1 & 0 & 1 \\ -1 & 1 & 0 & 0 \end{bmatrix}, \quad D = \begin{bmatrix} 2 & 2 & 1 & 0 \\ 1 & 1 & 0 & 1 \\ 0 & 1 & 0 & 0 \\ -1 & 1 & 0 & 0 \end{bmatrix}.$$

The matrices B, C, and D are all nonsingular. Therefore, the columns are linearly independent. If we had adjoined the *last two columns* of the 4×4 identity matrix I to A, we would have obtained a singular matrix. (Why?) In that case, the columns would not have been linearly independent.

As an immediate consequence of the extension principle we have the following result.

Let **U** be a k-dimensional linear subspace of an n-dimensional linear space **V**. There is an $(n - k)$-dimensional subspace **W** of **V** such that **U** and **W** have nothing in common except the **0** element. Moreover, **V** is a direct sum of **U** and **W**, i.e.,

$$\mathbf{V} = \mathbf{U} + \mathbf{W}.$$

The vectors $\mathbf{v}_{k+1}, \ldots, \mathbf{v}_n$ described in our extension principle for bases generate an $(n - k)$-dimensional subspace **W** of this type.

Exercises

1. For the sets in Exercise 1, Section 5.1, which are subspaces, find a basis and write down the dimension.

2. By deletions, find a basis for the linear subspace of $\mathbf{R}^3$ generated by the vectors

   $$\mathbf{u} = (1, 0, 1), \quad \mathbf{v} = (0, 1, 1), \quad \mathbf{w} = (1, 1, 2), \quad \mathbf{z} = (2, 1, 3).$$

   Do the vectors

   $$\mathbf{x} = (1, -1, 0), \quad \mathbf{y} = (3, 1, 4)$$

   lie in this subspace? If so, do they form a basis for this subspace?

3. What is the dimension of the linear space comprised of all 2×2-matrices? Find a basis for this space. Find a two-dimensional linear subspace of this space.

4. Find a basis for the linear space comprised of all 2×3-matrices. What is its dimension?

5. Show that mn is the dimension of the class of all $m \times n$-matrices.

6. The dimension of the space of all 3×3-matrices is 9. Is there a linear subspace of dimension 5? If so, find one.

7. What is the dimension of the space of all 3 × 3 upper triangular matrices? What is the dimension of the space of all 3 × 3 upper triangular matrices whose main diagonal entries are zero? What is the dimension of the space of all 3 × 3 upper triangular matrices whose traces are zero? Construct a linear space of 3 × 3 upper triangular matrices whose dimension is 1.

8. Find a set of generators of the linear space **N** of all solutions **x** of the equation $A\mathbf{x} = \mathbf{0}$, where A is the 2 × 4-matrix

$$A = \begin{bmatrix} 1 & 0 & 1 & -1 \\ 1 & 1 & -1 & 3 \end{bmatrix}.$$

Find a basis for **N**. What is the dimension of **N**?

9. Let A be the matrix described in Exercise 8. Let **C** be the linear space generated by the columns of A.

   (a) Find a basis for **C**.

   (b) What is the dimension of **C**?

   (c) Find a submatrix B of A whose columns form a basis for **C**.

   (d) Find a basis for **C** whose vectors are not column vectors of A.

10. Construct a 3 × 5-matrix A whose column vectors generate a linear space **C** of dimension 2.

11. Construct three vectors **u**, **v**, **w** in $\mathbf{R}^5$ which generate a linear subspace **U** of $\mathbf{R}^5$ of dimension 3. Must these vectors be linearly independent? Determine a linear subspace **V** of $\mathbf{R}^5$ of dimension 2 such that

$$\mathbf{R}^5 = \mathbf{U} + \mathbf{V}.$$

Is this sum a direct sum? Let **x** and **y** be a basis for **V**. Are the vectors **u**, **v**, **w**, **x**, **y** linearly independent? If so, why?

5.3   BASIS OPERATORS AND CHANGES OF BASES

Let **V** be a finite-dimensional linear space. As was seen in the preceding section, we can select linearly independent elements $e_1, e_2, \ldots, e_n$ in **V** which generate **V**. Such a set of elements is called a *basis* for **V**. The number n of elements in a basis is the same no matter how the basis is chosen. We defined this number n to be the *dimension* of **V**.

Choosing a basis for **V** is the same operation that we perform when we choose coordinate axes in a Euclidean space, except that the concept of perpendicularity is not involved. Even in the two-dimensional Euclidean space, our analysis is sometimes simplified when we choose axes other than the usual *xy*-axes. In this section we describe relationships that exist between bases.

Once we have chosen a basis $\mathbf{e}_1, \mathbf{e}_2, \ldots, \mathbf{e}_n$ for a linear space $\mathbf{V}$, we have coordinatized $\mathbf{V}$. Consider, for the moment, the case $n = 3$. Viewing $\mathbf{V}$ geometrically as a set of points or vectors, a point $\mathbf{v}$ in $\mathbf{V}$ is expressible in the form

$$\mathbf{v} = x_1\mathbf{e}_1 + x_2\mathbf{e}_2 + x_3\mathbf{e}_3.$$

The numbers x_1, x_2, and x_3 are the coordinates of $\mathbf{v}$ relative to the basis $\mathbf{e}_1$, $\mathbf{e}_2$, and $\mathbf{e}_3$. The formula for $\mathbf{v}$ can be put in the form

$$\mathbf{v} = [\mathbf{e}_1 \quad \mathbf{e}_2 \quad \mathbf{e}_3]\begin{bmatrix} x_1 \\ x_2 \\ x_3 \end{bmatrix} = [\mathbf{e}_1 \quad \mathbf{e}_2 \quad \mathbf{e}_3]\mathbf{x}, \qquad \mathbf{x} = \begin{bmatrix} x_1 \\ x_2 \\ x_3 \end{bmatrix}.$$

This formula suggests that we introduce the basis operator

$$E = [\mathbf{e}_1 \quad \mathbf{e}_2 \quad \mathbf{e}_3],$$

a row matrix whose entries are the basis vectors of our basis for $\mathbf{V}$. We then have the equation

$$\mathbf{v} = E\mathbf{x} = x_1\mathbf{e}_1 + x_2\mathbf{e}_2 + x_3\mathbf{e}_3,$$

connecting the element $\mathbf{v}$ with its coordinate $\mathbf{x} = (x_1, x_2, x_3)$. The coordinate $\mathbf{x}$ is a member of $\mathbf{R}^3$. It follows that we have put an $\mathbf{R}^3$-structure on our three-dimensional linear space $\mathbf{V}$. The mapping $\mathbf{v} = E\mathbf{x}$ maps $\mathbf{R}^3$ into $\mathbf{V}$ in a one-to-one fashion and so has an inverse $\mathbf{x} = E^{-1}\mathbf{v}$. This inverse operator E^{-1} gives us the coordinate $\mathbf{x} = E^{-1}\mathbf{v}$ of an element $\mathbf{v}$.

Similarly, when $\mathbf{V}$ is n-dimensional and $\mathbf{e}_1, \mathbf{e}_2, \ldots, \mathbf{e}_n$ is a basis of $\mathbf{V}$, we can introduce the basis operator

$$E = [\mathbf{e}_1 \quad \mathbf{e}_2 \cdots \quad \mathbf{e}_n].$$

An element $\mathbf{v}$ in $\mathbf{V}$ is connected with its coordinate $\mathbf{x} = (x_1, x_2, \ldots, x_n)$ by the relation

$$\mathbf{v} = E\mathbf{x} = x_1\mathbf{e}_1 + x_2\mathbf{e}_2 + \cdots + x_n\mathbf{e}_n.$$

The basis operator E imposes an $\mathbf{R}^n$-structure on $\mathbf{V}$. The mapping, $\mathbf{v} = E\mathbf{x}$, maps $\mathbf{R}^n$ into $\mathbf{V}$ in a one-to-one fashion and so has an inverse, $\mathbf{x} = E^{-1}\mathbf{v}$ of $\mathbf{v}$.

■ **EXAMPLE 1** _____

Suppose that our linear space is R^3. In this case our basis can be any set of three linearly independent vectors in R^3. For example, we could choose the columns $\mathbf{e}_1$, $\mathbf{e}_2$, and $\mathbf{e}_3$ of the identity matrix

$$I = \begin{bmatrix} 1 & 0 & 0 \\ 0 & 1 & 0 \\ 0 & 0 & 1 \end{bmatrix}.$$

To see that I is a basis operator, consider the case in which $\mathbf{v} = (4, 2, 1)$. We can write $\mathbf{v}$ in the form

$$\mathbf{v} = (4, 2, 1) = (1, 0, 0)4 + (0, 1, 0)2 + (0, 0, 1)1.$$

Putting this in column vector form, we have

$$\mathbf{v} = \begin{bmatrix} 1 \\ 0 \\ 0 \end{bmatrix} 4 + \begin{bmatrix} 0 \\ 1 \\ 0 \end{bmatrix} 2 + \begin{bmatrix} 0 \\ 0 \\ 1 \end{bmatrix} 1 = \begin{bmatrix} 1 & 0 & 0 \\ 0 & 1 & 0 \\ 0 & 0 & 1 \end{bmatrix} \begin{bmatrix} 4 \\ 2 \\ 1 \end{bmatrix} = \begin{bmatrix} 4 \\ 2 \\ 1 \end{bmatrix}.$$

More generally, if $\mathbf{v} = (x, y, z)$ is any point in $\mathbf{R}^3$, then

$$\mathbf{v} = \begin{bmatrix} 1 \\ 0 \\ 0 \end{bmatrix} x + \begin{bmatrix} 0 \\ 1 \\ 0 \end{bmatrix} y + \begin{bmatrix} 0 \\ 0 \\ 1 \end{bmatrix} z = \begin{bmatrix} 1 & 0 & 0 \\ 0 & 1 & 0 \\ 0 & 0 & 1 \end{bmatrix} \begin{bmatrix} x \\ y \\ z \end{bmatrix} = \begin{bmatrix} x \\ y \\ z \end{bmatrix}.$$

The identity matrix I is therefore a basis operator for R^3. It is called the *natural basis operator*. In fact, any time the basis operator is the identity operator I, it follows that the coordinate of the vector $\mathbf{v}$ is itself the vector $\mathbf{v}$.

However, if we choose a different basis for R^3, say $(1, 0, 0)$, $(1, 1, 0)$, and $(1, 1, 1)$, and make these the columns of the nonsingular matrix

$$F = \begin{bmatrix} 1 & 1 & 1 \\ 0 & 1 & 1 \\ 0 & 0 & 1 \end{bmatrix},$$

this matrix F becomes our basis operator. The point

$$\mathbf{v} = (4, 2, 1)$$

now has coordinate (y_1, y_2, y_3) given by

$$\mathbf{v} = \begin{bmatrix} 4 \\ 2 \\ 1 \end{bmatrix} = F\mathbf{y} = \begin{bmatrix} 1 & 1 & 1 \\ 0 & 1 & 1 \\ 0 & 0 & 1 \end{bmatrix} \begin{bmatrix} y_2 \\ y_2 \\ y_3 \end{bmatrix}.$$

We can find this coordinate by first computing the inverse F^{-1} of the matrix F and then using the equation $\mathbf{y} = F^{-1}\mathbf{v}$. A simple calculation yields

$$F^{-1} = \begin{bmatrix} 1 & -1 & 0 \\ 0 & 1 & -1 \\ 0 & 0 & 1 \end{bmatrix}.$$

It follows that

$$\mathbf{y} = \begin{bmatrix} y_1 \\ y_2 \\ y_3 \end{bmatrix} = \begin{bmatrix} 1 & -1 & 0 \\ 0 & 1 & -1 \\ 0 & 0 & 1 \end{bmatrix} \begin{bmatrix} 4 \\ 2 \\ 1 \end{bmatrix} = \begin{bmatrix} 2 \\ 1 \\ 1 \end{bmatrix}.$$

The coordinate of the vector $\mathbf{v} = (4, 2, 1)$ relative to the basis operator F is $(2, 1, 1)$. In general, we see that if we denote by $\mathbf{v}$ the original coordinate of a point in R^3 relative to the basis operator $E = I$, the identity matrix, then the equation $\mathbf{v} = F\mathbf{y}$ imposes a new coordinate $\mathbf{y}$ on this point which is given by the equation $\mathbf{y} = F^{-1}\mathbf{v}$.

■ **EXAMPLE 2** _____

Let us consider a case in which a basis operator is not a matrix in the ordinary sense. In particular, let us consider the linear space $\mathbf{P}$ of all polynomials $\mathbf{p}$ of the form

$$\mathbf{p} = a + bt + ct^2.$$

This is a three-dimensional space having 1, t, and t^2 as a basis. The operator

$$E = [1 \quad t \quad t^2]$$

is the corresponding basis operator. We have

$$\mathbf{p} = E\mathbf{x} = [1 \quad t \quad t^2]\begin{bmatrix} a \\ b \\ c \end{bmatrix}, \qquad \mathbf{x} = E^{-1}\mathbf{p} = \begin{bmatrix} a \\ b \\ c \end{bmatrix}.$$

The vector $\mathbf{x} = \begin{bmatrix} a \\ b \\ c \end{bmatrix}$ is the coordinate of the polynomial $\mathbf{p}$ relative to the basis operator $[1 \quad t \quad t^2]$.

■ **EXAMPLE 3** _____

Let $\mathbf{M}$ be the linear space comprised of all 2×2-matrices

$$A = \begin{bmatrix} a & b \\ c & d \end{bmatrix}.$$

It has, as a basis, the matrices

$$E_1 = \begin{bmatrix} 1 & 0 \\ 0 & 0 \end{bmatrix}, \qquad E_2 = \begin{bmatrix} 0 & 1 \\ 0 & 0 \end{bmatrix},$$

$$E_3 = \begin{bmatrix} 0 & 0 \\ 1 & 0 \end{bmatrix}, \qquad E_4 = \begin{bmatrix} 0 & 0 \\ 0 & 1 \end{bmatrix}.$$

We have

$$A = aE_1 + bE_2 + cE_3 + dE_4.$$

Introducing the basis operator

$$E = [E_1 \quad E_2 \quad E_3 \quad E_4],$$

we obtain the relations

$$A = Ex = aE_1 + bE_2 + cE_3 + dE_4, \qquad x = E^{-1}A = (a, b, c, d).$$

This relation establishes a one-to-one correspondence between

$$A = \begin{bmatrix} a & b \\ c & d \end{bmatrix} \qquad \text{and} \qquad x = (a, b, c, d).$$

The vector $x = (a, b, c, d)$ gives the coordinates of the matrix A relative to the basis E_1, E_2, E_3, and E_4. This linear space is four-dimensional.

We now return to the consideration of a general linear space V of dimension n. We begin by considering the case $n = 3$. We refer to elements in V as "points" in V. We suppose that we have chosen a basis e_1, e_2, and e_3 for V together with its basis operator

$$E = [e_1 \quad e_2 \quad e_3].$$

Each point v in V is connected with its coordinate x in $\mathbf{R}^3$ relative to the basis e_1, e_2, and e_3 by the relations

$$v = Ex, \qquad x = E^{-1}v.$$

Let f_1, f_2, and f_3 be a second basis for V and let

$$F = [f_1 \quad f_2 \quad f_3]$$

be the corresponding basis operator for V. This basis operator establishes a one-to-one correspondence between a point v in V and its coordinate y in $\mathbf{R}^3$ relative to the basis f_1, f_2, and f_3 via the relations

$$v = Fy, \qquad y = F^{-1}v.$$

A point v in V accordingly has x-coordinates with respect to the e's and y-coordinates with respect to the f's. They are connected by the relations

$$v = Ex = Fy, \qquad y = F^{-1}Ex, \qquad x = E^{-1}Fy.$$

The operator

$$B = F^{-1}E = F^{-1}[e_1 \quad e_2 \quad e_3] = [F^{-1}e_1 \quad F^{-1}e_2 \quad F^{-1}e_3]$$

connecting the x- and y-coordinates, is a 3×3-matrix of real numbers, whose columns

$$b_1 = F^{-1}e_1, \qquad b_2 = F^{-1}e_2, \qquad b_3 = F^{-1}e_3$$

are the y-coordinates of e_1, e_2, and e_3, respectively. Observe that

$$e_1 = Fb_1, \qquad e_2 = Fb_2, \qquad e_3 = Fb_3$$

so that

$$E = [e_1 \quad e_2 \quad e_3] = [Fb_1 \quad Fb_2 \quad Fb_3] = F[b_1 \quad b_2 \quad b_3] = FB.$$

Alternatively, this follows because $E = FF^{-1}E = FB$. Similarly,

$$C = E^{-1}F = [E^{-1}\mathbf{f}_1 \quad E^{-1}\mathbf{f}_2 \quad E^{-1}\mathbf{f}_3] = [\mathbf{c}_1 \quad \mathbf{c}_2 \quad \mathbf{c}_3]$$

is a 3×3-matrix, whose columns are the x-coordinates of $\mathbf{f}_1$, $\mathbf{f}_2$, and $\mathbf{f}_3$, respectively. The matrices B and C are inverses of each other. The relations between E and F can also be computed as follows:

$$E = FF^{-1}E = FB, \qquad F = EE^{-1}F = EC.$$

Our results, extended to the n- dimensional case, can be summarized as follows.

Let E and F be basis operators for an n-dimensional linear space **V**. These basis operators determine x-coordinates and y-coordinates

$$\mathbf{x} = E^{-1}\mathbf{v}, \qquad \mathbf{y} = F^{-1}\mathbf{v}$$

of a point $\mathbf{v}$ in **V**. These coordinates are points in $\mathbf{R}^n$ and are connected by the relations

$$\mathbf{y} = B\mathbf{x}, \qquad \mathbf{x} = C\mathbf{y},$$

where B and C are the nonsingular $n \times n$-matrices

$$B = F^{-1}E, \qquad C = E^{-1}F, \qquad C = B^{-1}, \qquad B = C^{-1}.$$

We have $E = FB$ and $F = EC$. Conversely, if E is a basis operator for **V** and C is a nonsingular matrix, then $F = EC$ is another basis operator for **V**.

■ **EXAMPLE 4** _____

As in Example 2, we return to the study of the linear space **P** of polynomials

$$\mathbf{p} = a + bt + ct^2.$$

We have seen that the basis operator

$$E = [1 \quad t \quad t^2]$$

is a natural basis operator for **P**. This operator belongs, for $h = 0$, to a class of basis operators of the form

$$F = [1 \quad t - h \quad (t - h)^2],$$

one for each fixed value of h. It is easily verified that we have the relation

$$F = [1 \quad t - h \quad (t - h)^2] = [1 \quad t \quad t^2]\begin{bmatrix} 1 & -h & h^2 \\ 0 & 1 & -2h \\ 0 & 0 & 1 \end{bmatrix} = EC,$$

where

$$C = \begin{bmatrix} 1 & -h & h^2 \\ 0 & 1 & -2h \\ 0 & 0 & 1 \end{bmatrix}, \qquad B = C^{-1} = \begin{bmatrix} 1 & h & h^2 \\ 0 & 1 & 2h \\ 0 & 0 & 1 \end{bmatrix}.$$

The x- and y-coordinates of $\mathbf{p}$ are given by the relations

$$\mathbf{p} = F\mathbf{y}, \qquad \mathbf{p} = E\mathbf{x} = FB\mathbf{x}, \qquad \mathbf{y} = B\mathbf{x}.$$

Hence with $\mathbf{x} = (a, b, c)$ and $\mathbf{y} = (d, e, f)$,

$$\mathbf{y} = \begin{bmatrix} d \\ e \\ f \end{bmatrix} = \begin{bmatrix} 1 & h & h^2 \\ 0 & 1 & 2h \\ 0 & 0 & 1 \end{bmatrix} \begin{bmatrix} a \\ b \\ c \end{bmatrix} = \begin{bmatrix} a + bh + ch^2 \\ b + 2ch \\ c \end{bmatrix}.$$

Consequently,

$$d = a + bh + ch^2 = p(h), \qquad e = b + 2ch = p'(h), \qquad f = c = \tfrac{1}{2}p''(h),$$

where

$$p(t) = a + bt + ct^2, \qquad p'(t) = b + 2ct, \qquad p''(t) = 2c.$$

This gives the Taylor expansion

$$p(t) = p(h) + p'(h)t + \tfrac{1}{2}p''(h)(t - h)^2$$

of $p(t)$ about the point $t = h$. Taylor expansions play an important role in calculus. We have just seen that there is a connection between Taylor expansions for polynomials and change of basis in a corresponding linear space.

■ **EXAMPLE 5** _____

Let us explore an application of change of basis in analytic geometry. Using (u, v)-coordinates the equation

$$9u^2 - 8uv + 3v^2 = 11$$

represents an ellipse with its center at the origin. Let us verify this fact by showing that by a change of basis to a new xy-coordinate system this equation becomes the standard equation

$$x^2 + 11y^2 = 11$$

of an ellipse. To do so we observe that our original equation can be written in the form

$$\mathbf{u}^T A \mathbf{u} = 11$$

where

$$A = \begin{bmatrix} 9 & -4 \\ -4 & 3 \end{bmatrix}, \qquad \mathbf{u} = \begin{bmatrix} u \\ v \end{bmatrix}.$$

The points **u** belong to a linear space $\mathbf{U} = \mathbf{R}^2$ in which the natural basis is

$$\mathbf{e}_1 = (1, 0), \qquad \mathbf{e}_2 = (0, 1).$$

We shall transform this basis into a new basis comprised of *unit* eigenvectors $\mathbf{f}_1$ and $\mathbf{f}_2$ of A. Observe that A is a symmetric matrix. It is easily seen that $\lambda_1 = 1$ and $\lambda_2 = 11$ are the eigenvalues of A and that

$$\mathbf{f}_1 = (1/\sqrt{5}, 2/\sqrt{5}), \qquad \mathbf{f}_2 = (-2/\sqrt{5}, 1/\sqrt{5})$$

are corresponding unit eigenvectors. We transform our basis by setting

$$\mathbf{u} = \begin{bmatrix} u \\ v \end{bmatrix} = [\mathbf{f}_1 \quad \mathbf{f}_2] \begin{bmatrix} x \\ y \end{bmatrix} = \begin{bmatrix} 1/\sqrt{5} & -2/\sqrt{5} \\ 2/\sqrt{5} & 1/\sqrt{5} \end{bmatrix} \begin{bmatrix} x \\ y \end{bmatrix} = F\mathbf{x}$$

or equivalently, by setting

$$u = \frac{x}{\sqrt{5}} - \frac{2y}{\sqrt{5}}$$

$$v = \frac{2x}{\sqrt{5}} + \frac{y}{\sqrt{5}}.$$

Under this transformation we find that

$$9u^2 - 8uv + 3v^2 = x^2 + 11y^2,$$

giving us the standard equation

$$x^2 + 11y^2 = 11$$

for our ellipse. We do not need to carry out this computation, because we learned in Chapter 4 that an *orthogonal* matrix F, whose columns are eigenvectors of A, diagonalizes the matrix A. We have

$$\mathbf{u}^T A \mathbf{u} = \mathbf{x}^T F^T A F \mathbf{x} = \mathbf{x}^T D \mathbf{x} = x^2 + 11y^2,$$

because the main diagonal entries of the diagonal matrix D are the eigenvalues 1 and 11 of A. We used the fact that the transpose of an orthogonal matrix is its inverse. It should be noted that by choosing angle θ such that

$$\cos \theta = \frac{1}{\sqrt{5}}, \qquad \sin \theta = \frac{2}{\sqrt{5}},$$

our transformation takes the form

$$u = x \cos \theta - y \sin \theta, \qquad v = x \sin \theta + y \cos \theta.$$

From this equation we see that we have rotated our coordinate axes by an angle θ.

It is important to choose $\mathbf{f}_1$ and $\mathbf{f}_2$ to be unit eigenvectors. Otherwise, we would change the shape of the ellipse. For example, if we replaced $\mathbf{f}_1$ by $a\mathbf{f}_1$, with $a^2 = 11$, our equation would become

$$11x^2 + 11y^2 = 11 \quad \text{or} \quad x^2 + y^2 = 1.$$

In this event we would have transformed an ellipse into a circle.

Exercises

1. Find two bases for the linear space (plane) of points (x, y, z) satisfying the equation

$$5x - 2y + z = 0.$$

Determine a relation between them.

2. Find a basis for the linear space (line) of points (x, y) satisfying the equation

$$2x + y = 0.$$

How are other bases related to it?

3. Each of the equations

$$5x - 2y + z = 0$$

$$x + y - z = 0$$

determines a linear subspace of $\mathbf{R}^3$. Call them $\mathbf{P}$ and $\mathbf{Q}$. Find a basis for their intersection. Extend this basis to a basis for $\mathbf{P}$. Make another extension to obtain a basis for $\mathbf{Q}$. Can these basis vectors be used to obtain a basis for $\mathbf{R}^3$?

4. Consider the homogeneous linear equation $Ax = 0$, where A is the matrix

$$A = \begin{bmatrix} 1 & 2 & 0 & 1 & 3 \\ 0 & 0 & 1 & 1 & 1 \\ 1 & 2 & 1 & 2 & 4 \end{bmatrix}.$$

Find a basis for the linear space $\mathbf{N}$ of solutions $\mathbf{x}$ of $Ax = 0$. What is the dimension of $\mathbf{N}$? Find a second basis for $\mathbf{N}$. Find a basis for the linear space $\mathbf{R}$ generated by the rows of A. What is the dimension of $\mathbf{R}$? Can a basis for $\mathbf{N}$ be combined with a basis for $\mathbf{R}$ to obtain a basis for $\mathbf{R}^5$? Prove your answer.

5. Continuing with Exercise 4, find two bases for the linear space $\mathbf{C}$ generated by the columns of A. Find a relation between them. What is the dimension of $\mathbf{C}$? How is $\mathbf{C}$ related to the solutions $\mathbf{y}$ of $\mathbf{y}^T A = 0$?

6. Let E and F be bases operators of an n-dimensional linear space $\mathbf{V}$. The products $C = E^{-1}F$ and $B = F^{-1}E$ are nonsingular $n \times n$-matrices which are inverses of each other. Show that

$$\mathbf{v} = FE^{-1}\mathbf{u}$$

maps $\mathbf{V}$ into $\mathbf{V}$ in a one-to-one fashion. Show that the mapping

$$\mathbf{u} = EF^{-1}\mathbf{v}$$

is the inverse mapping. What is the mapping determined by the operator EE^{-1}?

5.4 ORTHOGONAL BASES AND THE GRAM–SCHMIDT PROCESS IN R^n

The concept of orthogonality in $\mathbf{R}^n$ was introduced in Section 1.6. Two vectors $\mathbf{x}$ and $\mathbf{y}$ were said to be orthogonal if the relation

$$\mathbf{x}^T\mathbf{y} = 0$$

held. The product $\mathbf{x}^T\mathbf{y}$ is called the *inner product* of $\mathbf{x}$ and $\mathbf{y}$. Various other notations, such as

$$\langle \mathbf{x}, \mathbf{y} \rangle, \quad (\mathbf{x}, \mathbf{y}), \quad \mathbf{x} \cdot \mathbf{y}, \quad \mathbf{x} * \mathbf{y}, \quad \langle \mathbf{x}|\mathbf{y} \rangle,$$

are also used for the inner product. In fact, it is more conventional to use one of these notations than the notation $\mathbf{x}^T\mathbf{y}$ that we used in Section 1.6. For this reason we adopt the notation $\langle \mathbf{x}, \mathbf{y} \rangle$ as our standard notation for an inner product. However, at times we shall revert to the notation $\mathbf{x}^T\mathbf{y}$ when it seems appropriate to do so. One of the advantages of using the notation $\langle \mathbf{x}, \mathbf{y} \rangle$ is that later it enables us to describe more general inner products without change of notation. Examples of more general inner products will be given at the end of this section. So, until otherwise specified, the *inner product* $\langle \mathbf{x}, \mathbf{y} \rangle$ is given by the formula

$$\langle \mathbf{x}, \mathbf{y} \rangle = \mathbf{x}^T\mathbf{y} = x_1y_1 + x_2y_2 + \cdots + x_ny_n.$$

In terms of the inner product, the *length* (or *norm*) $\|\mathbf{x}\|$ of a vector $\mathbf{x}$ is

$$\|\mathbf{x}\| = \langle \mathbf{x}, \mathbf{x} \rangle^{1/2} = (x_1^2 + x_2^2 + \cdots + x_n^2)^{1/2}.$$

A vector $\mathbf{x}$ is a *unit vector* if $\|\mathbf{x}\| = 1$. If $\mathbf{x}$ is any nonzero vector, then

$$\mathbf{u} = \frac{\mathbf{x}}{\|\mathbf{x}\|}$$

is a unit vector giving the direction of $\mathbf{x}$.

As remarked above, two vectors $\mathbf{x}$ and $\mathbf{y}$ are said to be *orthogonal* when the relation

$$\langle \mathbf{x}, \mathbf{y} \rangle = 0$$

holds. We saw in Section 1.6 that if θ is the angle between $\mathbf{x}$ and $\mathbf{y}$, then

$$\langle \mathbf{x}, \mathbf{y} \rangle = \|\mathbf{x}\| \, \|\mathbf{y}\| \cos \theta.$$

This equation can be used as the definition of the angle θ between $\mathbf{x}$ and $\mathbf{y}$ if one wishes to do so.

■ **EXAMPLE 1** _____

Let **U** be a linear subspace in $\mathbf{R}^3$ having as a basis the column vectors $\mathbf{g}_1$ and $\mathbf{g}_2$ of the matrix

$$G = [\mathbf{g}_1 \quad \mathbf{g}_2] = \begin{bmatrix} 1 & 1 \\ 0 & 2 \\ -1 & 1 \end{bmatrix}.$$

The matrix G is the corresponding basis operator. Computing the inner product of $\mathbf{g}_1$ and $\mathbf{g}_2$ mentally it is seen that $\langle \mathbf{g}_1, \mathbf{g}_2 \rangle = 0$. The basis vectors $\mathbf{g}_1$ and $\mathbf{g}_2$ are therefore orthogonal. Such a basis is called an *orthogonal basis* for **U**. It should be noted that because of this orthogonality relation, the matrix

$$G^T G = \begin{bmatrix} \mathbf{g}_1^T \\ \mathbf{g}_2^T \end{bmatrix} [\mathbf{g}_1 \quad \mathbf{g}_2] = \begin{bmatrix} \langle \mathbf{g}_1, \mathbf{g}_1 \rangle & \langle \mathbf{g}_1, \mathbf{g}_2 \rangle \\ \langle \mathbf{g}_2, \mathbf{g}_1 \rangle & \langle \mathbf{g}_2, \mathbf{g}_2 \rangle \end{bmatrix} = \begin{bmatrix} 2 & 0 \\ 0 & 6 \end{bmatrix}$$

is a diagonal matrix. Thus, in terms of the basis operator G, our basis is an orthogonal basis because $G^T G$ is a diagonal matrix.

_____

Consider a more general case in which **U** is a linear subspace of $\mathbf{R}^n$ of dimension q. Let $\mathbf{g}_1, \mathbf{g}_2, \ldots, \mathbf{g}_q$ be a basis for **U**. Such a basis is called an *orthogonal basis* for **U** if the basis vectors are mutually orthogonal, that is, if the relations

$$\langle \mathbf{g}_i, \mathbf{g}_j \rangle = 0 \quad \text{unless} \quad i = j$$

hold for $i, j = 1, \ldots, q$. An orthogonal basis whose basis vectors are unit vectors is called an *orthonormal basis*. In terms of the basis operator

$$G = [\mathbf{g}_1 \quad \mathbf{g}_2 \cdots \mathbf{g}_q]$$

our basis is an orthogonal basis if and only if the matrix

$$G^T G = [\langle \mathbf{g}_i, \mathbf{g}_j \rangle]$$

is a diagonal matrix. It is an orthonormal basis if and only if $G^T G = I$, the $q \times q$-identity matrix. It is clear that an orthogonal basis can be transformed into an orthonormal basis by dividing each of the vectors by its length.

The following result is useful.

PROPOSITION 1. If $\mathbf{g}_1, \mathbf{g}_2, \ldots, \mathbf{g}_q$ are q mutually orthogonal *nonzero* vectors, that is, if

$$\langle \mathbf{g}_i, \mathbf{g}_j \rangle = 0 \quad \text{whenever} \quad i \neq j; \quad i, j = 1, 2, \ldots, q,$$

then they are linearly independent.

In other words, if mutually orthogonal vectors are nonzero, they are automatically linearly independent.

The proof of this proposition in the general case is like that for the case $n = 3$. Suppose that $\mathbf{g}_1$, $\mathbf{g}_2$, and $\mathbf{g}_3$ are mutually orthogonal nonzero vectors. Consider the relation

$$b_1\mathbf{g}_1 + b_2\mathbf{g}_2 + b_3\mathbf{g}_3 = \mathbf{0}.$$

Forming the inner product with $\mathbf{g}_1$, we find that

$$b_1\langle \mathbf{g}_1, \mathbf{g}_1 \rangle + b_2\langle \mathbf{g}_1, \mathbf{g}_2 \rangle + b_3\langle \mathbf{g}_1, \mathbf{g}_3 \rangle = b_1\|\mathbf{g}_1\|^2 = 0.$$

Hence $b_1 = 0$. Similarly, by forming the inner products with $\mathbf{g}_2$ and $\mathbf{g}_3$, it is seen that $b_2 = 0$ and $b_3 = 0$ also. The vectors $\mathbf{g}_1$, $\mathbf{g}_2$, and $\mathbf{g}_3$ are therefore linearly independent.

■ **EXAMPLE 1 (continued)** ─────────────────────────────────

Consider again the two-dimensional subspace **U** in $\mathbf{R}^3$ having the vectors

$$\mathbf{g}_1 = (1, 0, -1), \qquad \mathbf{g}_2 = (1, 2, 1)$$

as a basis. As noted above, this basis is an orthogonal basis. Every vector **u** in **U** is expressible in the form

$$\mathbf{u} = z_1\mathbf{g}_1 + z_2\mathbf{g}_2.$$

Forming the inner product of **u** first with $\mathbf{g}_1$ and then with $\mathbf{g}_2$, we find that because $\mathbf{g}_1$ and $\mathbf{g}_2$ are orthogonal, we have

$$\langle \mathbf{g}_1, \mathbf{u} \rangle = z_1\langle \mathbf{g}_1, \mathbf{g}_1 \rangle + z_2\langle \mathbf{g}_1, \mathbf{g}_2 \rangle = z_1\|\mathbf{g}_1\|^2,$$

$$\langle \mathbf{g}_2, \mathbf{u} \rangle = z_1\langle \mathbf{g}_2, \mathbf{g}_1 \rangle + z_2\langle \mathbf{g}_2, \mathbf{g}_2 \rangle = z_2\|\mathbf{g}_2\|^2.$$

It follows that z-coordinates z_1 and z_2 of **u** with respect to $\mathbf{g}_1$ and $\mathbf{g}_2$ are given by the simple formulas

$$z_1 = \frac{\langle \mathbf{g}_1, \mathbf{u} \rangle}{\|\mathbf{g}_1\|^2}, \qquad z_2 = \frac{\langle \mathbf{g}_2, \mathbf{u} \rangle}{\|\mathbf{g}_2\|^2}.$$

In particular, when $\mathbf{u} = (1, 6, 5)$, then $z_1 = -4/2 = -2$, $z_2 = 18/6 = 3$.

Let **v** be any vector in $\mathbf{R}^3$. Set

$$y_1 = \frac{\langle \mathbf{g}_1, \mathbf{v} \rangle}{\|\mathbf{g}_1\|^2}, \qquad y_2 = \frac{\langle \mathbf{g}_2, \mathbf{v} \rangle}{\|\mathbf{g}_2\|^2}.$$

Then the vector

$$\mathbf{w} = \mathbf{v} - y_1\mathbf{g}_1 - y_2\mathbf{g}_2$$

is orthogonal to $\mathbf{g}_1$ and $\mathbf{g}_2$, as can be seen by the computations

$$\langle \mathbf{g}_1, \mathbf{w} \rangle = \langle \mathbf{g}_1, \mathbf{v} \rangle - y_1\|\mathbf{g}_1\|^2 - y_2\langle \mathbf{g}_1, \mathbf{g}_2 \rangle = 0,$$

$$\langle \mathbf{g}_2, \mathbf{w} \rangle = \langle \mathbf{g}_2, \mathbf{v} \rangle - y_1\langle \mathbf{g}_2, \mathbf{g}_1 \rangle - y_2\|\mathbf{g}_2\|^2 = 0.$$

When we have $\mathbf{v} = (3, 1, 1)$, then $y_1 = 1$ and $y_2 = 1$, so that

$$\mathbf{w} = \mathbf{v} - \mathbf{g}_1 - \mathbf{g}_2 = (3, 1, 1) - (1, 0, -1) - (1, 2, 1) = (1, -1, 1).$$

It is easily verified that $\mathbf{w}$ is orthogonal to $\mathbf{g}_1$ and $\mathbf{g}_2$.

This example illustrates the following result. It can be established by the same method.

PROPOSITION 2. Let $\mathbf{g}_1, \ldots, \mathbf{g}_q$ be an orthogonal basis for a q-dimensional linear subspace U of $\mathbf{R}^n$. Let

$$\mathbf{u} = z_1\mathbf{g}_1 + z_2\mathbf{g}_2 + \cdots + z_q\mathbf{g}_q$$

be a vector in U. The z-coordinates of $\mathbf{u}$ relative to this basis are given by the formulas

$$z_i = \frac{\langle \mathbf{g}_i, \mathbf{u} \rangle}{\|\mathbf{g}_i\|^2} \qquad (i = 1, \ldots, q).$$

Let $\mathbf{v}$ be a vector in $\mathbf{R}^n$. Set

$$y_i = \frac{\langle \mathbf{g}_i, \mathbf{v} \rangle}{\|\mathbf{g}_i\|^2} \qquad (i = 1, \ldots, q).$$

Then the vector

$$\mathbf{w} = \mathbf{v} - y_1\mathbf{g}_1 - y_2\mathbf{g}_2 - \cdots - y_q\mathbf{g}_q$$

is orthogonal to the basis vectors $\mathbf{g}_1, \mathbf{g}_2, \ldots, \mathbf{g}_q$ and hence to every vector $\mathbf{u}$ in U.

The proof of the last conclusion will be left as an exercise.

We have begun to see the importance of orthogonal bases in linear space theory. We now investigate a procedure for transforming a set of linearly independent vectors into a set of mutually orthogonal vectors. This will lead to a general procedure for finding an orthonormal basis for any linear space on which an inner product has been defined.

We will motivate this procedure by looking at a special case in two dimensions which will then be easily generalized. The idea will be very geometric in nature and is based on the idea of an orthogonal projection.

Suppose that $\mathbf{f}_1$ and $\mathbf{f}_2$ are any two vectors in the plane. Then we can orthogonally project $\mathbf{f}_2$ onto $\mathbf{f}_1$ provided that $\mathbf{f}_1$ is not the zero vector. This is illustrated in Figure 1. The orthogonal projection of $\mathbf{f}_2$ onto $\mathbf{f}_1$ is a vector that is either in the same or opposite direction of $\mathbf{f}_1$. It is therefore a multiple of $\mathbf{f}_1$. Denoting this projection of $\mathbf{f}_2$ onto $\mathbf{f}_1$ by $\mathbf{P}$, we have $\mathbf{P} = k\mathbf{f}_1$ where k is a constant. Let θ be the angle between $\mathbf{f}_1$ and $\mathbf{f}_2$, and assume $k > 0$. We

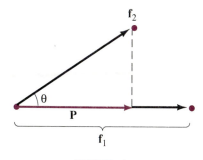

can now calculate the length of the orthogonal projection vector. We have

$$\|\mathbf{P}\| = \|k\mathbf{f}_1\| = \|\mathbf{f}_2\| \cos \theta = \frac{\|\mathbf{f}_1\| \, \|\mathbf{f}_2\| \cos \theta}{\|\mathbf{f}_1\|} = \frac{\langle \mathbf{f}_1, \mathbf{f}_2 \rangle}{\|\mathbf{f}_1\|},$$

so

$$\|k\mathbf{f}_1\| = k \, \|\mathbf{f}_1\| = \frac{\langle \mathbf{f}_1, \mathbf{f}_2 \rangle}{\|\mathbf{f}_1\|},$$

which implies that

$$k = \frac{\langle \mathbf{f}_1, \mathbf{f}_2 \rangle}{\|\mathbf{f}_1\|^2}.$$

It follows that $k = \langle \mathbf{f}_1, \mathbf{f}_2 \rangle / \langle \mathbf{f}_1, \mathbf{f}_1 \rangle$. Using this value of k, we see that

$$\mathbf{P} = \frac{\langle \mathbf{f}_1, \mathbf{f}_2 \rangle}{\langle \mathbf{f}_1, \mathbf{f}_1 \rangle} \mathbf{f}_1.$$

We leave it for the reader to verify that the same formula is true in the case when $k < 0$. It is helpful to draw a picture.

We can define the concept of an orthogonal projection in any linear space on which an inner product has been given. We have: For any two vectors $\mathbf{f}_1$ and $\mathbf{f}_2$ in a linear space on which an inner product is defined, if $\mathbf{f}_1 \neq 0$, the orthogonal projection of $\mathbf{f}_2$ onto $\mathbf{f}_1$ is given by

$$\mathbf{P} = \frac{\langle \mathbf{f}_1, \mathbf{f}_2 \rangle}{\langle \mathbf{f}_1, \mathbf{f}_1 \rangle} \mathbf{f}_1.$$

We return to the idea of an orthogonal projection in Chapter 6, where we pursue some of these ideas further.

For now, we have what we need in order to transform any set of linearly independent vectors into a set of mutually orthogonal ones, each of unit length. As an example, suppose that we are given a basis for $\mathbf{R}^2$. Call the basis vectors $\mathbf{f}_1$ and $\mathbf{f}_2$. These basis vectors $\mathbf{f}_1$ and $\mathbf{f}_2$ need not be orthogonal or of unit length. We wish to use these basis vectors $\mathbf{f}_1$ and $\mathbf{f}_2$ in order to construct a new basis for $\mathbf{R}^2$, which will consist of two vectors, say $\mathbf{g}_1$ and

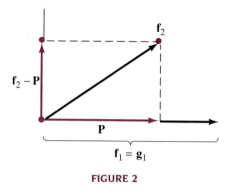

FIGURE 2

$\mathbf{g}_2$, which are mutually orthogonal and of unit length (i.e., an orthonormal basis). The procedure is as follows: Let $\mathbf{g}_1$ be one of the original vectors, say $\mathbf{f}_1$. We now seek a second vector $\mathbf{g}_2$ with the property that $\mathbf{g}_2$ is orthogonal to $\mathbf{g}_1$. Observe that the vector $\mathbf{f}_2 - \mathbf{P}$ is orthogonal to $\mathbf{g}_1 = \mathbf{f}_1$ (Figure 2). But from above, we know that

$$\mathbf{P} = \frac{\langle \mathbf{f}_1, \mathbf{f}_2 \rangle}{\langle \mathbf{f}_1, \mathbf{f}_1 \rangle} \mathbf{f}_1.$$

Since we have chosen $\mathbf{g}_1 = \mathbf{f}_1$, it follows that

$$\mathbf{P} = \frac{\langle \mathbf{f}_2, \mathbf{g}_1 \rangle}{\langle \mathbf{g}_1, \mathbf{g}_1 \rangle} \mathbf{g}_1.$$

We therefore choose

$$\mathbf{g}_2 = \mathbf{f}_2 - \mathbf{P} = \mathbf{f}_2 - \frac{\langle \mathbf{f}_2, \mathbf{g}_1 \rangle}{\langle \mathbf{g}_1, \mathbf{g}_1 \rangle} \mathbf{g}_1.$$

We have constructed two mutually orthogonal vectors $\mathbf{g}_1$ and $\mathbf{g}_2$ from our two original basis vectors $\mathbf{f}_1$ and $\mathbf{f}_2$. They are given by the formulas

$$\mathbf{g}_1 = \mathbf{f}_1 \quad \text{and} \quad \mathbf{g}_2 = \mathbf{f}_2 - \frac{\langle \mathbf{f}_2, \mathbf{g}_1 \rangle}{\langle \mathbf{g}_1, \mathbf{g}_1 \rangle} \mathbf{g}_1.$$

The vectors $\mathbf{g}_1$ and $\mathbf{g}_2$ can now be normalized to be of length 1 by dividing each of them by their respective lengths. The two new orthonormal vectors, from what we have learned earlier in this section, are linearly independent and hence form an orthonormal basis for $\mathbf{R}^2$. We illustrate the procedure with the following.

■ **EXAMPLE 2** _____

Suppose that in $\mathbf{R}^2$ we are given the basis vectors $\mathbf{f}_1 = (1, 2)$ and $\mathbf{f}_2 = (3, -1)$. As above, let $\mathbf{g}_1 = \mathbf{f}_1 = (1, 2)$. Then

$$g_2 = f_2 - \frac{\langle f_2, g_1 \rangle}{\langle g_1, g_1 \rangle} g_1$$

$$= (3, -1) - \frac{\langle (3, -1), (1, 2) \rangle}{\langle (1, 2), (1, 2) \rangle} (1, 2)$$

$$= (3, -1) - \tfrac{1}{5}(1, 2)$$

$$= (14/5, -7/5). \qquad (1.2)$$

Observe that the vectors g_1 and g_2 are orthogonal since their inner product $\langle g_1, g_2 \rangle = 0$. Dividing each of these vectors by their respective lengths yields

$$\frac{g_1}{\|g_1\|} = \frac{1}{\sqrt{5}}(1, 2) \quad \text{and} \quad \frac{g_2}{\|g_2\|} = \frac{5}{\sqrt{245}}(14/5, -7/5) = \frac{1}{\sqrt{245}}(14, -7).$$

These two vectors form an orthonormal basis for $\mathbf{R}^2$. We have constructed them from the given basis f_1 and f_2 for $\mathbf{R}^2$.

The method presented above generalizes and can be used to construct an orthonormal set of vectors from any given set of linearly independent vectors in a linear space on which an inner product has been defined. The method is known as a *Gram–Schmidt process*.

GRAM–SCHMIDT ORTHONORMALIZATION ALGORITHM. Let f_1, f_2, . . . , f_q be q linearly independent vectors in $\mathbf{R}^n$.

1. These vectors can be transformed into q mutually orthogonal vectors g_1, g_2, . . . , g_q by the following rules.

Set $g_1 = f_1$.

Set $g_2 = f_2 - \dfrac{\langle f_2, g_1 \rangle}{\langle g_1, g_1 \rangle} g_1$.

Set $g_3 = f_3 - \dfrac{\langle f_3, g_1 \rangle}{\langle g_1, g_1 \rangle} g_1 - \dfrac{\langle f_3, g_2 \rangle}{\langle g_2, g_2 \rangle} g_2$.

$$\vdots$$

Set $g_q = f_q - \dfrac{\langle f_q, g_1 \rangle}{\langle g_1, g_1 \rangle} g_1 - \dfrac{\langle f_q, g_2 \rangle}{\langle g_2, g_2 \rangle} g_2 - \cdots - \dfrac{\langle f_q, g_{q-1} \rangle}{\langle g_{q-1}, g_{q-1} \rangle} g_{q-1}$.

2. The vectors g_1, g_2, . . . , g_q can be normalized so that they are of length one by dividing each of them by their respective lengths. These q new vectors form an orthonormal set of vectors.

■ **EXAMPLE 3**

Consider the three linearly independent vectors

$$\mathbf{f}_1 = (1, 0, 1, 0), \qquad \mathbf{f}_2 = (2, 1, 2, 1), \qquad \mathbf{f}_3 = (-1, 2, -3, 4)$$

in $\mathbf{R}^4$. We seek to transform them into three mutually orthogonal vectors $\mathbf{g}_1$, $\mathbf{g}_2$, and $\mathbf{g}_3$. Using the algorithm above, we proceed as follows:

Set $\mathbf{g}_1 = \mathbf{f}_1 = (1, 0, 1, 0)$.

Set $\mathbf{g}_2 = \mathbf{f}_2 - \dfrac{\langle \mathbf{f}_2, \mathbf{g}_1 \rangle}{\langle \mathbf{g}_1, \mathbf{g}_1 \rangle} \mathbf{g}_1$

$$= (2, 1, 2, 1) - \frac{\langle (2, 1, 2, 1), (1, 0, 1, 0) \rangle}{\langle (1, 0, 1, 0), (1, 0, 1, 0) \rangle}(1, 0, 1, 0)$$

$$= (2, 1, 2, 1) - 2(1, 0, 1, 0)$$

$$= (0, 1, 0, 1).$$

Set $\mathbf{g}_3 = \mathbf{f}_3 - \dfrac{\langle \mathbf{f}_3, \mathbf{g}_1 \rangle}{\langle \mathbf{g}_1, \mathbf{g}_1 \rangle} \mathbf{g}_1 - \dfrac{\langle \mathbf{f}_3, \mathbf{g}_2 \rangle}{\langle \mathbf{g}_2, \mathbf{g}_2 \rangle} \mathbf{g}_2$

$$= (-1, 2, -3, 4) - \frac{\langle (-1, 2, -3, 4), (1, 0, 1, 0) \rangle}{\langle (1, 0, 1, 0), (1, 0, 1, 0) \rangle}(1, 0, 1, 0)$$

$$- \frac{\langle (-1, 2, -3, 4), (0, 1, 0, 1) \rangle}{\langle (0, 1, 0, 1), (0, 1, 0, 1) \rangle}(0, 1, 0, 1).$$

$$= (-1, 2, -3, 4) + 2(1, 0, 1, 0) - 3(0, 1, 0, 1)$$

$$= (1, -1, -1, 1).$$

Thus we have constructed a set of mutually orthogonal vectors

$$\mathbf{g}_1 = (1, 0, 1, 0), \qquad \mathbf{g}_2 = (0, 1, 0, 1), \qquad \mathbf{g}_3 = (1, -1, -1, 1)$$

from the vectors $\mathbf{f}_1$, $\mathbf{f}_2$, and $\mathbf{f}_3$. The set of vectors $\mathbf{f}_1$, $\mathbf{f}_2$, and $\mathbf{f}_3$ and the set of vectors $\mathbf{g}_1$, $\mathbf{g}_2$, and $\mathbf{g}_3$ generate the same subspace in $\mathbf{R}^4$.

By dividing each of the vectors $\mathbf{g}_1$, $\mathbf{g}_2$, and $\mathbf{g}_3$ by their respective lengths, we obtain the orthonormal set of vectors

$$\frac{\mathbf{g}_1}{\|\mathbf{g}_1\|} = \frac{1}{\sqrt{2}}(1, 0, 1, 0), \qquad \frac{\mathbf{g}_2}{\|\mathbf{g}_2\|} = \frac{1}{\sqrt{2}}(0, 1, 0, 1), \qquad \frac{\mathbf{g}_3}{\|\mathbf{g}_3\|} = \frac{1}{2}(1, -1, -1, 1).$$

As an immediate consequence of the Gram–Schmidt algorithm, we have:

Every linear subspace $\mathbf{U}$ of $\mathbf{R}^n$ possesses an orthogonal basis $\mathbf{g}_1$, $\mathbf{g}_2$, ... , $\mathbf{g}_q$. This basis can be extended to form an orthogonal basis $\mathbf{g}_1$, $\mathbf{g}_2$, ... , $\mathbf{g}_n$ of $\mathbf{R}^n$.

The proof of the extendibility of an orthogonal basis of **U** can be made in many ways. In the preceding section we showed that any basis of **U** can be extended to give us a basis for **R**^{n}. So our orthogonal basis $\mathbf{g}_1, \ldots, \mathbf{g}_q$ for **U** can be extended to give us a basis $\mathbf{g}_1, \ldots, \mathbf{g}_n$ for **R**^{n}. If the added vectors are not orthogonal, we can orthogonalize them by a Gram–Schmidt process. Because the first q vectors are already orthogonal, they are unaltered by the Gram–Schmidt process.

Let **U** be a linear subspace of **R**^{n}. A vector **w** is orthogonal to **U** if it is *orthogonal* to every vector **u** in **U**. It is easily verified that if a vector is orthogonal to the vectors in a basis for **U**, it is orthogonal to **U**. The set of all vectors **w** orthogonal to **U** is called the *orthogonal complement* of **U** and will be denoted by **U**$^\perp$. The reader should verify that **U**$^\perp$ is a linear subspace of **R**^{n}. As noted above, there is an orthogonal basis $\mathbf{g}_1, \mathbf{g}_2, \ldots, \mathbf{g}_n$ for **R**^{n} such that $\mathbf{g}_1, \mathbf{g}_2, \ldots, \mathbf{g}_q$ is a basis for **U**. The remaining basis vectors $\mathbf{g}_{q+1}, \ldots, \mathbf{g}_n$ form a basis for **U**$^\perp$. This gives us the following result.

> **PROPOSITION 3.** The orthogonal complement **U**$^\perp$ of a linear subspace **U** of **R**^{n} is a linear subspace of **R**^{n}. The space **R**^{n} is the direct sum
>
> $$\mathbf{R}^n = \mathbf{U} + \mathbf{U}^\perp$$
>
> of **U** and its orthogonal complement **U**$^\perp$. If q is the dimension of **U**, the dimension of **U**$^\perp$ is $n - q$.

■ **EXTENSION 1** _____

The purpose of introducing an inner product in **R**^{n} is to introduce a concept of length and angles. The concept of length and angles should not depend on the basis used. Suppose that in our original basis, the inner product of two vectors **u** and **v** is given by the formula

$$\mathbf{u}^T\mathbf{v}.$$

If we introduce a new basis with F as its basis operator, we have a relation of the form

$$\mathbf{u} = F\mathbf{x}, \qquad \mathbf{v} = F\mathbf{y}.$$

Then

$$\mathbf{u}^T\mathbf{v} = \mathbf{x}^T F^T F\mathbf{y} = \mathbf{x}^T A\mathbf{y} \qquad \text{where} \qquad A = F^T F.$$

If we are to preserve the concept of length and angle, we must take $\mathbf{x}^T A\mathbf{y}$ to be the formula for the inner product in terms of the new coordinates **x** and **y**. Thus, now,

$$\langle \mathbf{x}, \mathbf{y} \rangle = \mathbf{x}^T A\mathbf{y}.$$

The matrix A can be any **positive definite** symmetric matrix, that is, a symmetric matrix A having $\mathbf{x}^T A \mathbf{x} > 0$ for every nonzero vector $\mathbf{x}$. It is for this reason that we used the symbol $\langle \mathbf{x}, \mathbf{y} \rangle$ for the inner product. The results given above are valid for any inner product of this type.

A space $\mathbf{R}^n$ with an inner product is called an n-dimensional Euclidean space.

■ **EXTENSION 2** _____

This extension requires some knowledge of calculus. Let $\mathbf{P}$ be the class of all polynomials

$$\mathbf{p}: \qquad p(t) = a_0 + a_1 t + \cdots + a_n t^n$$

where n is any integer. If $\mathbf{p}$ and $\mathbf{q}$ are two polynomials, we define their inner product to be given by the formula

$$\langle \mathbf{p}, \mathbf{q} \rangle = \int_{-1}^{1} p(t) q(t) \, dt.$$

The square of the length of $\mathbf{p}$ is then

$$\|\mathbf{p}\|^2 = \langle \mathbf{p}, \mathbf{p} \rangle = \int_{-1}^{1} p(t)^2 \, dt.$$

When we begin with the polynomials $1, t, t^2, t^3, \ldots$ and apply the Gram–Schmidt process, we obtain a set of orthogonal polynomials. Except for a scale factor, the first five of these polynomials are

$$1, \quad t, \quad \frac{3t^2}{2} - \frac{1}{2}, \quad \frac{5t^3}{2} - \frac{3t}{2}, \quad \frac{35t^4}{8} - \frac{30t^2}{8} + \frac{3}{8}.$$

These polynomials are called Legendre polynomials and play a significant role in many physical problems. Other useful orthogonal polynomials are obtained by choosing a different inner product.

Exercises

1. Find the inner product between pairs of vectors in each of the following cases:

   (a) $\mathbf{f} = (1, 2, 3)$, $\quad \mathbf{g} = (-1, 0, 1)$, $\quad \mathbf{h} = (5, -2, -1)$

   (b) $\mathbf{x} = (1, 1, 1, 1)$, $\quad \mathbf{y} = (3, 1, 1, 3)$, $\quad \mathbf{z} = (2, -2, -4, 0)$

   (c) $\mathbf{u} = (1, 0, 1, 0, 1)$, $\quad \mathbf{v} = (1, -1, 1, 1, -1)$

2. In each of cases (a), (b), and (c) in Exercise 1, use the Gram–Schmidt process to transform the vectors into mutually orthogonal vectors. Then normalize them.

3. In the text we transformed the column vectors $\mathbf{f}_1$, $\mathbf{f}_2$, and $\mathbf{f}_3$ of the matrix F shown below into the columns $\mathbf{g}_1$, $\mathbf{g}_2$, and $\mathbf{g}_3$ of the matrix G by a Gram–Schmidt process.

$$F = \begin{bmatrix} 1 & 2 & -1 \\ 0 & 1 & 2 \\ 1 & 2 & -3 \\ 0 & 1 & 4 \end{bmatrix}, \qquad G = \begin{bmatrix} 1 & 0 & 1 \\ 0 & 1 & -1 \\ 1 & 0 & -1 \\ 0 & 1 & 1 \end{bmatrix}.$$

The relationship between these vectors given above can be rewritten in the form

$$\mathbf{f}_1 = \mathbf{g}_1, \qquad \mathbf{f}_2 = \mathbf{g}_1 b_{12} + \mathbf{g}_2, \qquad \mathbf{f}_3 = \mathbf{g}_1 b_{13} + \mathbf{g}_2 b_{23} + \mathbf{g}_3,$$

where $b_{12} = 2$, $b_{13} = -2$, and $b_{23} = 3$. Show that the relation $F = GB$ holds with

$$B = \begin{bmatrix} 1 & b_{12} & b_{13} \\ 0 & 1 & b_{23} \\ 0 & 0 & 1 \end{bmatrix} = \begin{bmatrix} 1 & 2 & -2 \\ 0 & 1 & 3 \\ 0 & 0 & 1 \end{bmatrix}, \qquad b_{ij} = \frac{\langle \mathbf{g}_i, \mathbf{f}_j \rangle}{\|\mathbf{g}_i\|^2}.$$

Observe that B is a normalized upper triangular matrix. [A normalized upper (or lower) triangular matrix is one whose main diagonal entries are all 1's.] Conclude that a relationship of this type exists whenever the columns of a matrix F are transformed into the columns of a matrix G by a Gram–Schmidt process, provided that the columns of F are linearly independent.

4. With F, G, and B as in Exercise 3, compute B^{-1}. Verify that $G = FB^{-1}$. Verify that B^{-1} is also a normalized upper triangular matrix.

5. Continuing with Exercises 3 and 4, compute the matrices $F^T F$ and $G^T G$. Show that they have the same determinant. This can be done by actual computation. An easier way is to use the fact that $F^T F = B^T G^T G B$ and $\det B = 1$. Why is this always the case when the columns of G are obtained from the columns of F by a Gram–Schmidt process?

6. The result given in Exercise 5 has a geometrical interpretation. The columns of F determine a parallelpiped and the columns of G determine a rectangular parallelpiped. Their volumes are given by the formula

$$\text{vol } F = [\det F^T F]^{1/2}, \qquad \text{vol } G = [\det G^T G]^{1/2}.$$

Verify this fact for a case when F is 3×2 and also when F is 3×3. What does vol F become when F is a column vector? Conclude that the Gram–Schmidt process preserves volumes.

7. Show that the real inner product $\langle \mathbf{x}, \mathbf{y} \rangle$ has the following properties.

(a) $\langle \mathbf{x}, \mathbf{x} \rangle > 0$ unless $\mathbf{x} = \mathbf{0}$, $\langle \mathbf{0}, \mathbf{0} \rangle = 0$.

(b) $\langle \mathbf{x}, \mathbf{y} \rangle = \langle \mathbf{y}, \mathbf{x} \rangle$.

(c) $\langle a\mathbf{x}, \mathbf{y} \rangle = \langle \mathbf{x}, a\mathbf{y} \rangle = a \langle \mathbf{x}, \mathbf{y} \rangle$ for every scalar a.

(d) $\langle \mathbf{x}, \mathbf{y} + \mathbf{z} \rangle = \langle \mathbf{x}, \mathbf{y} \rangle + \langle \mathbf{x}, \mathbf{z} \rangle$.

A linear space having associated with it a real-valued inner product $\langle \mathbf{x}, \mathbf{y} \rangle$ satisfying these properties is called a ***real inner product space***.

8. Let **M** be the linear space comprised of all $m \times n$-matrices $A, B, C, \ldots$. Here m and n are fixed. Show that we can make **M** an inner product space by setting

$$\langle A, B \rangle = \text{trace } A^T B.$$

Consider first the case $m = n = 2$.

Remark. In this section we used a basis operator F for a q-dimensional subspace $\mathbf{U}$ of $\mathbf{R}^n$. It is an $n \times q$-matrix. The matrix

$$G = F(F^T F)^{-1}$$

is another basis operator for $\mathbf{U}$. It has the property that

$$G^T F = I, \quad \text{the } q \times q\text{-identity matrix}.$$

This second basis is called the reciprocal basis of the basis determined by F. The matrix F maps $\mathbf{R}^q$ into $\mathbf{U}$. The inverse of this mapping is $F^{-1} = G^T$, mapping $\mathbf{U}$ into $\mathbf{R}^q$. In matrix theory, the matrix G^T is called the ***pseudoinverse*** of F. We will learn some of these ideas in Chapter 6.

6

Linear Transformations and Their Properties

6.1 LINEAR TRANSFORMATIONS AND LINEAR OPERATORS

In this chapter we are concerned with the notion of a linear transformation and a linear operator. We have already used these concepts in elementary mathematics when we studied linear functions. Recall that the equation $y = mx$, where m is a constant, represents a line through the origin. It transforms a number x into a number y by multiplying x by the fixed number m. This equation is an example of a *linear transformation*. The fixed number m is an *operator* that operates on x to obtain y. It has the following two properties: For every pair of numbers u and v,

1. $m(u + v) = mu + mv$.

2. $m(cu) = c(mu)$ for every number c.

Because of these properties, m is called a *linear operator*.

To see how the idea generalizes, suppose that we now consider the matrix equation $\mathbf{y} = A\mathbf{x}$, where A is a fixed $m \times n$-matrix, $\mathbf{x}$ is an $n \times 1$ column vector, and $\mathbf{y}$ is an $m \times 1$ column vector. Here the matrix A has taken the place of the constant m above, and the column vectors $\mathbf{x}$ and $\mathbf{y}$ have replaced the numbers x and y, respectively, in our original linear function. The matrix equation $\mathbf{y} = A\mathbf{x}$ transforms or maps $n \times 1$ column vectors $\mathbf{x}$ into $m \times 1$ column vectors $\mathbf{y}$. The equation represents a *linear transformation* from $\mathbf{R}^n$ into $\mathbf{R}^m$. Furthermore, the matrix A is an *operator*, operating on $\mathbf{x}$. It is a *linear operator* since it satisfies the same two properties as above; that is, for every pair of vectors $\mathbf{u}$ and $\mathbf{v}$,

1. $A(\mathbf{u} + \mathbf{v}) = A\mathbf{u} + A\mathbf{v}$.

2. $A(c\mathbf{u}) = c(A\mathbf{u})$ for every number c

The example above is but one of many linear transformations. In general, we have the following definition.

Let $\mathbf{U}$ and $\mathbf{V}$ be two linear spaces. A mapping

$$\mathbf{v} = T\mathbf{u}$$

of a point $\mathbf{u}$ in $\mathbf{U}$ into a point $\mathbf{v} = T\mathbf{u}$ in $\mathbf{V}$ will be said to be a linear transformation of $\mathbf{U}$ into $\mathbf{V}$ if it satisfies the following two properties:

1. $T(\mathbf{u}_1 + \mathbf{u}_2) = T\mathbf{u}_1 + T\mathbf{u}_2$ for all $\mathbf{u}_1$ and $\mathbf{u}_2$ in $\mathbf{U}$.

2. $T(a\mathbf{u}) = aT\mathbf{u}$ for all $\mathbf{u}$ in $\mathbf{U}$ and all scalars a.

The terminology "*T* preserves addition and scalar multiplication" is often used when referring to the fact that *T* satisfies properties 1 and 2, respectively.

The names *linear operator* and *linear mapping* are also used interchangeably with *linear transformation*.

■ **EXAMPLE 1** _____

Consider the transformation from $\mathbf{R}^3$ to $\mathbf{R}^2$ given by

$$T\begin{bmatrix} u_1 \\ u_2 \\ u_3 \end{bmatrix} = \begin{bmatrix} u_1 - u_2 \\ u_1 + u_3 \end{bmatrix} = \begin{bmatrix} v_1 \\ v_2 \end{bmatrix}.$$

The transformation *T* is linear, for if

$$\mathbf{u}_1 = \begin{bmatrix} u_1 \\ u_2 \\ u_3 \end{bmatrix} \quad \text{and} \quad \mathbf{u}_2 = \begin{bmatrix} w_1 \\ w_2 \\ w_3 \end{bmatrix}$$

are any two elements in $\mathbf{R}^3$, then

$$T(\mathbf{u}_1 + \mathbf{u}_2) = T\begin{bmatrix} u_1 + w_1 \\ u_2 + w_2 \\ u_3 + w_3 \end{bmatrix} = \begin{bmatrix} (u_1 + w_1) - (u_2 + w_2) \\ (u_1 + w_1) + (u_3 + w_3) \end{bmatrix}$$

$$= \begin{bmatrix} (u_1 - u_2) + (w_1 - w_2) \\ (u_1 + u_3) + (w_1 + w_3) \end{bmatrix} = \begin{bmatrix} u_1 - u_2 \\ u_1 + u_3 \end{bmatrix} + \begin{bmatrix} w_1 - w_2 \\ w_1 + w_3 \end{bmatrix}$$

$$= T(\mathbf{u}_1) + T(\mathbf{u}_2)$$

and

$$T(a\mathbf{u}) = T\begin{bmatrix} au_1 \\ au_2 \\ au_3 \end{bmatrix} = \begin{bmatrix} au_1 - au_2 \\ au_1 + au_3 \end{bmatrix} = a\begin{bmatrix} u_1 - u_2 \\ u_1 + u_3 \end{bmatrix} = aT(\mathbf{u})$$

for all $\mathbf{u}$ in $\mathbf{R}^3$ and all scalars *a*.

■ **EXAMPLE 2** _____

Let $\mathbf{V}$ be a two-dimensional linear space and let $E = [\mathbf{i} \quad \mathbf{j}]$ be a basis operator for $\mathbf{V}$. The operator E maps the points $\mathbf{u}_1 = (x, y)$ and $\mathbf{u}_2 = (u, v)$ in $\mathbf{R}^2$ into the points

$$E\mathbf{u}_1 = x\mathbf{i} + y\mathbf{j}, \qquad E\mathbf{u}_2 = u\mathbf{i} + v\mathbf{j}$$

in $\mathbf{V}$. We have $\mathbf{u}_1 + \mathbf{u}_2 = (x + u, y + v)$, so that

$$E(\mathbf{u}_1 + \mathbf{u}_2) = (x + u)\mathbf{i} + (y + v)\mathbf{j} = (x\mathbf{i} + y\mathbf{j}) + (u\mathbf{i} + v\mathbf{j}).$$

It follows that

$$E(\mathbf{u}_1 + \mathbf{u}_2) = E\mathbf{u}_1 + E\mathbf{u}_2.$$

In addition we have, for every scalar a, the relation

$$E(a\mathbf{u}) = ax\mathbf{i} + ay\mathbf{j} = a(x\mathbf{i} + y\mathbf{j}) = aE\mathbf{u}.$$

These two relations signify that the basis operator E is a linear transformation of $\mathbf{R}^2$ into $\mathbf{V}$ according to the definition.

In view of properties 1 and 2 of a linear mapping T, we have the relations

$$T(a_1\mathbf{u}_1 + a_2\mathbf{u}_2) = T(a_1\mathbf{u}_1) + T(a_2\mathbf{u}_2) = a_1 T\mathbf{u}_1 + a_2 T\mathbf{u}_2$$

$$T(a_1\mathbf{u}_1 + a_2\mathbf{u}_2 + a_3\mathbf{u}_3) = a_1 T\mathbf{u}_1 + a_2 T\mathbf{u}_2 + a_3 T\mathbf{u}_3,$$

and so on. In addition, by setting $a = 0$ in the relation $T(a\mathbf{u}) = T\mathbf{u}$, we find that $T\mathbf{0} = \mathbf{0}$. This gives us the following simple but important property of linear transformations.

> A linear transformation of $\mathbf{U}$ into $\mathbf{V}$ transforms the origin $\mathbf{0}$ of $\mathbf{U}$ into the origin $\mathbf{0}$ of $\mathbf{V}$. That is,
>
> $$T\mathbf{0} = \mathbf{0}.$$

This result tells us that the transformation T which sends a point (x, y) in $\mathbf{R}^2$ into the point $(x + 1, y)$ in $\mathbf{R}^2$ is not a linear transformation because it sends the origin $(0, 0)$ into the nonzero point $(1, 0)$. The transformation T of $\mathbf{R}^2$ into $\mathbf{R}^2$, which maps (x, y) into (x^2, y) maps $(0, 0)$ into $(0, 0)$, is not linear, however, because it does not have the additivity property (1) of a linear transformation.

Here are some examples of linear operators. We leave it as an exercise to verify the linearity of the operator.

1. The basis operator E described in Section 5.3 is a linear transformation. A special case was discussed at the beginning of this section. The general case can be verified in the same manner.

2. An $m \times n$-matrix A defines a linear transformation $\mathbf{y} = A\mathbf{x}$ that maps a point $\mathbf{x}$ in $\mathbf{R}^n$ into a point $\mathbf{y}$ in $\mathbf{R}^m$. Clearly,

$$A(\mathbf{x}_1 + \mathbf{x}_2) = A\mathbf{x}_1 + A\mathbf{x}_2 \quad \text{and} \quad A(a\mathbf{x}) = aA\mathbf{x}.$$

For example, the matrix

$$A = \begin{bmatrix} 1 & -1 & 0 \\ 2 & 3 & 4 \end{bmatrix}$$

as an operator maps via $\mathbf{y} = A\mathbf{x}$ the three-dimensional point $\mathbf{x} = (a, b, c)$ into the two-dimensional point $\mathbf{y} = (a - b, 2a + 3b + 4c)$, as can be seen by the computations

$$\mathbf{y} = A\mathbf{x} = \begin{bmatrix} 1 & -1 & 0 \\ 2 & 3 & 4 \end{bmatrix} \begin{bmatrix} a \\ b \\ c \end{bmatrix} = \begin{bmatrix} a - b \\ 2a + 3b + 4c \end{bmatrix}.$$

3. In the familiar xy-plane $\mathbf{R}^2$, let T be the transformation that maps the point $\mathbf{u} = (x, y)$ into the point $\mathbf{v} = (-x, y)$. This is a reflection of $\mathbf{u}$ about the y-axis. We have

$$\mathbf{v} = \begin{bmatrix} -x \\ y \end{bmatrix} = x \begin{bmatrix} -1 \\ 0 \end{bmatrix} + y \begin{bmatrix} 0 \\ 1 \end{bmatrix} = \begin{bmatrix} -1 & 0 \\ 0 & 1 \end{bmatrix} \begin{bmatrix} x \\ y \end{bmatrix} = A\mathbf{u},$$

where

$$A = \begin{bmatrix} -1 & 0 \\ 0 & 1 \end{bmatrix}.$$

The matrix A is a matrix representation of our reflection operator T. It is therefore a linear operator. This transformation is often described by the notation

$$T(x, y) = (-x, y).$$

4. Again in the xy-plane let T be the transformation that maps the vector $\mathbf{u} = (x, y)$ into the vector $\mathbf{v} = (y, -x)$. We have $T(x, y) = (y, -x)$. This is a counterclockwise rotation of $\mathbf{u}$ by 90°. We have

$$\mathbf{v} = \begin{bmatrix} y \\ -x \end{bmatrix} = x \begin{bmatrix} 0 \\ -1 \end{bmatrix} + y \begin{bmatrix} 1 \\ 0 \end{bmatrix} = \begin{bmatrix} 0 & 1 \\ -1 & 0 \end{bmatrix} \begin{bmatrix} x \\ y \end{bmatrix} = A\mathbf{u},$$

where now

$$A = \begin{bmatrix} 0 & 1 \\ -1 & 0 \end{bmatrix}$$

is a matrix representation of the rotation operator T.

5. The operator T of $\mathbf{R}^2$ into $\mathbf{R}^2$ which maps every point $\mathbf{u} = (x, y)$ into the origin $\mathbf{0} = (0, 0)$ is a linear transformation. It can be written in the form $\mathbf{v} = A\mathbf{u}$, where

$$A = \begin{bmatrix} 0 & 0 \\ 0 & 0 \end{bmatrix}.$$

It is the zero operator. For any linear spaces $\mathbf{U}$ and $\mathbf{V}$, the operator T which sends every point $\mathbf{u}$ in $\mathbf{U}$ into the origin $\mathbf{0}$ of $\mathbf{V}$ is a linear operator and is called the **zero operator**.

6. Let **u** be a *unit* (column) vector in $\mathbf{R}^n$. Because **u** is of unit length we have $\mathbf{u}^T\mathbf{u} = 1$. The matrix $P = \mathbf{u}\mathbf{u}^T$ is a projection operator that maps every vector **x** in $\mathbf{R}^n$ into a multiple $\mathbf{y} = P\mathbf{x} = \mathbf{u}\mathbf{u}^T\mathbf{x} = m\mathbf{u}$ of **u**, where $m = \mathbf{u}^T\mathbf{x}$. Thus $\mathbf{R}^n$ is projected on the line $\mathbf{y} = t\mathbf{x}$ through the origin. We have the relations

$$P^2 = \mathbf{u}\mathbf{u}^T\mathbf{u}\mathbf{u}^T = \mathbf{u}(1)\mathbf{u}^T = \mathbf{u}\mathbf{u}^T = P, \qquad P^T = (\mathbf{u}\mathbf{u}^T)^T = \mathbf{u}\mathbf{u}^T = P.$$

In a numerical case with $n = 3$ and $\mathbf{u} = (2/3, 1/3, 2/3)$ we have

$$P = \begin{bmatrix} 2/3 \\ 1/3 \\ 2/3 \end{bmatrix} \begin{bmatrix} 2/3 & 1/3 & 2/3 \end{bmatrix} = \begin{bmatrix} 4/9 & 2/9 & 4/9 \\ 2/9 & 1/9 & 2/9 \\ 4/9 & 2/9 & 4/9 \end{bmatrix}.$$

When $\mathbf{x} = (1, -1, 2)$ we have $m = \mathbf{u}^T\mathbf{x} = (1/3)(2 - 1 + 4) = 5/3$. In this event $\mathbf{y} = (5/3)\mathbf{u} = (10/9, 5/9, 10/9)$. What is $\mathbf{y} = P\mathbf{x}$ when $\mathbf{x} = (1, -1, 1)$? Verify that the operator $T = 2P$ is also a linear operator. So is $I - P$.

   Observe that $m = \mathbf{u}^T\mathbf{x}$, with **u** fixed, is a linear function of **x**. It maps vector $a\mathbf{x}$ into a number m. Here **u** can be an arbitrary vector and need not be a unit vector. The function $L(\mathbf{x}) = \mathbf{u}^T\mathbf{x}$ with **u** fixed is called a *linear form*. It is a linear operator that maps a vector into a scalar.

7. Let **u** and **v** be two orthogonal unit vectors in $\mathbf{R}^n$. They generate a 2-plane **W** through the origin. The matrices

$$P = \mathbf{u}\mathbf{u}^T + \mathbf{v}\mathbf{v}^T \qquad \text{and} \qquad Q = I - P$$

are linear operators. The operator P projects a vector **x** into a vector **y** in the 2-plane **W**. The operator Q drops a perpendicular **z** from the point **x** to the point **y** in **W**. We have $\mathbf{x} = \mathbf{y} + \mathbf{z}$.

8. Let **u** and **v** be unit (column) vectors in $\mathbf{R}^m$ and $\mathbf{R}^n$, respectively. The projection operator $P = \mathbf{u}\mathbf{v}^T$ maps a vector **x** into a multiple $\mathbf{y} = k\mathbf{u}$ of **u**. Here $k = \mathbf{v}^T\mathbf{x}$. Observe that $P\mathbf{v} = \mathbf{u}\mathbf{v}^T\mathbf{v} = \mathbf{u}$, and $P^T\mathbf{u} = \mathbf{v}$. The projection operator P is a linear operator.

9. Let **N** be the class of all $n \times n$-matrices $A, B, \ldots$. The operators S and D defined by the relations

$$SA = \tfrac{1}{2}(A + A^T), \qquad DA = \tfrac{1}{2}(A - A^T)$$

are linear operators on **N**. Note that $SA + DA = A$ and $SA - DA = A^T$. Here

$$S(A + B) = SA + SB, \qquad D(A + B) = DA + DBr,$$

$$S(aA) = aSA, \qquad\qquad D(aA) = aDA,$$

as one readily verifies.

10. Let **M** be the class of all $m \times n$-matrices A. Each matrix A defines a linear operator $\mathbf{v} = A\mathbf{u}$. When we change variables by means of the equations $\mathbf{u} = C\mathbf{x}$ and $\mathbf{v} = D\mathbf{y}$, where C and D are nonsingular, then

$$\mathbf{y} = D^{-1}\mathbf{v} = D^{-1}A\mathbf{u} = D^{-1}AC\mathbf{x}.$$

Accordingly, the operator A is transformed into the operator $D^{-1}AC$. This transformation of A is linear in A.

11. Let **P** be the class of polynomials

$$\mathbf{p} = a + bt + ct^2 + dt^3$$

of degree ≤ 3, including the zero polynomial. The derivative operator D is defined by the formula

$$D\mathbf{p} = b + 2ct + 3dt^2.$$

Verify that D is a linear operator on **P**. When we choose

$$E = \begin{bmatrix} 1 & t & t^2 & t^3 \end{bmatrix}$$

to be a basis operator for **P**, we have

$$\mathbf{p} = E\mathbf{x}, \qquad D\mathbf{p} = E\mathbf{y},$$

where **x** and **y** are the column vectors

$$\mathbf{x} = (a, b, c, d), \qquad \mathbf{y} = (b, 2c, 3d, 0).$$

Observe that

$$\mathbf{y} = \begin{bmatrix} b \\ 2c \\ 3d \\ 0 \end{bmatrix} = a\begin{bmatrix} 0 \\ 0 \\ 0 \\ 0 \end{bmatrix} + b\begin{bmatrix} 1 \\ 0 \\ 0 \\ 0 \end{bmatrix} + c\begin{bmatrix} 0 \\ 2 \\ 0 \\ 0 \end{bmatrix} + d\begin{bmatrix} 0 \\ 0 \\ 3 \\ 0 \end{bmatrix} = A\mathbf{x},$$

where

$$A = \begin{bmatrix} 0 & 1 & 0 & 0 \\ 0 & 0 & 2 & 0 \\ 0 & 0 & 0 & 3 \\ 0 & 0 & 0 & 0 \end{bmatrix}.$$

The matrix A is a matrix representation of the operator D. The derivative operator D can also be represented by the row matrix

$$F = \begin{bmatrix} 0 & 1 & 2t & 3t^2 \end{bmatrix} = EA.$$

This follows because if $\mathbf{x} = (a, b, c, d)$, then $E\mathbf{x} = \mathbf{u}$ and

$$F\mathbf{x} = \begin{bmatrix} 0 & 1 & 2t & 3t^2 \end{bmatrix} \begin{bmatrix} a \\ b \\ c \\ d \end{bmatrix} = b + 2ct + 3dt^2 = Du = DE\mathbf{x}.$$

We have $F = DE$ and $D = FE^{-1}$. Also, $F = EA$, so that $A = E^{-1}DE$.

12. Let **M** be the class of $m \times n$-matrices A and let **N** be the class of $n \times m$-matrices B. These linear spaces have the same dimension mn. If A is in **M**, its transpose A^T is in **N**. The operator T which maps a $m \times n$-matrix A into its transpose A^T is a linear operator. This follows because
$$(A_1 + A_2)^T = A_1^T + A_2^T, \qquad (cA)^T = cA^T.$$

Consider now two linear spaces **U** and **V** of dimensions n and m, respectively. Let
$$E = \begin{bmatrix} \mathbf{e}_1 & \mathbf{e}_2 & \cdots & \mathbf{e}_n \end{bmatrix}, \qquad G = \begin{bmatrix} \mathbf{g}_1 & \mathbf{g}_2 & \cdots & \mathbf{g}_m \end{bmatrix}$$
be bases operators for **U** and **V**, respectively. The operator E induces **x**-coordinates on **U** via the equation $\mathbf{u} = E\mathbf{x}$. Similarly, the operator G induces a **y**-coordinate system on **V** via the equation $\mathbf{v} = G\mathbf{y}$. Consider a linear transformation
$$\mathbf{v} = T\mathbf{u}$$
mapping a point **u** in **U** into a point **v** in **V**. The **x**-coordinate of **u** and the **y**-coordinate of **v** are connected by the relations
$$\mathbf{v} = G\mathbf{y} = T\mathbf{u} = TE\mathbf{x}, \qquad \mathbf{y} = G^{-1}TE\mathbf{x} = A\mathbf{x}, \qquad A = G^{-1}TE.$$
The operator A maps the **x**-coordinate of **u** into the **y**-coordinate of **v**. The operator A is an $m \times n$-matrix of real numbers whose columns
$$\mathbf{a}_1 = G^{-1}T\mathbf{e}_1, \quad \mathbf{a}_2 = G^{-1}T\mathbf{e}_2, \quad \ldots, \quad \mathbf{a}_n = G^{-1}T\mathbf{e}_n$$
are the **y**-coordinates of the points $T\mathbf{e}_1, T\mathbf{e}_2, \ldots, T\mathbf{e}_n$ in **V**. This gives the following result.

For a fixed linear transformation T of **U** into **V**, there is a unique $m \times n$-matrix
$$A = G^{-1}TE$$
for each choice of bases E on **U** and G on **V**. Such a matrix A is called a *matrix representation* of T.

Conversely, for each $m \times n$-matrix A, the corresponding linear operator T on **U** into **V** is given by the formula $T = GAE^{-1}$.

If T is a linear transformation from $\mathbf{R}^n$ to $\mathbf{R}^m$, often the bases that are chosen for $\mathbf{R}^n$ and $\mathbf{R}^m$ are the *standard bases* or so-called *natural bases*. In this case the basis operators E and G are both identity matrices. When this occurs, the matrix representation A of the linear transformation T is especially easy to compute since

$$A = G^{-1}TE = ITE = TE = TI = T.$$

This means that the columns of A are obtained by calculating what the linear transformation T does to each of the corresponding columns of the identity matrix I. We illustrate this with the following:

■ **EXAMPLE 3** ───

Consider the linear transformation from $\mathbf{R}^3$ to $\mathbf{R}^2$ given by

$$T\begin{bmatrix} u_1 \\ u_2 \\ u_3 \end{bmatrix} = \begin{bmatrix} u_1 - u_2 \\ u_1 + u_3 \end{bmatrix} = \begin{bmatrix} v_1 \\ v_2 \end{bmatrix}.$$

Let $\mathbf{e}_1 = (1, 0, 0)$, $\mathbf{e}_2 = (0, 1, 0)$, and $\mathbf{e}_3 = (0, 0, 1)$ be a basis for $\mathbf{R}^3$, with $\mathbf{g}_1 = (1, 0)$ and $\mathbf{g}_2 = (0, 1)$ a basis for $\mathbf{R}^2$. The basis operators for $\mathbf{R}^3$ and $\mathbf{R}^2$ are

$$E = \begin{bmatrix} 1 & 0 & 0 \\ 0 & 1 & 0 \\ 0 & 0 & 1 \end{bmatrix} \quad \text{and} \quad G = \begin{bmatrix} 1 & 0 \\ 0 & 1 \end{bmatrix},$$

respectively. Since $E = I$, the 3×3 identity matrix and $G = I$, the 2×2 identity matrix, it follows that the x-coordinate of $\mathbf{u}$ and the y-coordinate of $\mathbf{v}$ are connected by the relation $\mathbf{y} = A\mathbf{x}$, where $A = G^{-1}TE$. But $G = I$ implies that $G^{-1} = I$. Therefore, $A = TE$; that is, the operator A is the 2×3-matrix whose columns are $T(\mathbf{e}_1)$, $T(\mathbf{e}_2)$, and $T(\mathbf{e}_3)$. Computing yields

$$T(\mathbf{e}_1) = T\begin{bmatrix} 1 \\ 0 \\ 0 \end{bmatrix} = \begin{bmatrix} 1 - 0 \\ 1 + 0 \end{bmatrix} = \begin{bmatrix} 1 \\ 1 \end{bmatrix},$$

$$T(\mathbf{e}_2) = T\begin{bmatrix} 0 \\ 1 \\ 0 \end{bmatrix} = \begin{bmatrix} 0 - 1 \\ 0 + 0 \end{bmatrix} = \begin{bmatrix} -1 \\ 0 \end{bmatrix},$$

$$T(\mathbf{e}_3) = T\begin{bmatrix} 0 \\ 0 \\ 1 \end{bmatrix} = \begin{bmatrix} 0 - 0 \\ 0 + 1 \end{bmatrix} = \begin{bmatrix} 0 \\ 1 \end{bmatrix}.$$

The matrix representation A of the linear transformation T relative to the basis operators $E = I$ and $G = I$ is

$$A = \begin{bmatrix} 1 & -1 & 0 \\ 1 & 0 & 1 \end{bmatrix}.$$

Suppose now that the basis vectors in $\mathbf{R}^n$ are the natural basis vectors. Then the basis operator E is the identity matrix I. Suppose, however, that the basis vectors in $\mathbf{R}^m$ are vectors other than the natural basis vectors. Then the basis operator G will not be the identity matrix. The matrix representation A of the linear transformation T from $\mathbf{R}^n$ to $\mathbf{R}^m$ given by the formula

$$A = G^{-1}TE = G^{-1}TI$$

is calculated as follows:

> **1.** Obtain the matrix TI by calculating what T does to each of the natural basis column vectors of I and making these vectors the columns of TI.
>
> **2.** Compute G^{-1} and multiply it on the right by TI.

■ **EXAMPLE 3 (revisited)** ——————————————————————

Suppose that E is once again the natural basis operator for $\mathbf{R}^3$, so that E is the 3×3-identity matrix

$$E = \begin{bmatrix} 1 & 0 & 0 \\ 0 & 1 & 0 \\ 0 & 0 & 1 \end{bmatrix}.$$

Suppose that our new basis for $\mathbf{R}^2$ is $\mathbf{g}_1 = (1, 1)$ and $\mathbf{g}_2 = (2, 3)$, so that the basis operator for $\mathbf{R}^2$ is

$$G = \begin{bmatrix} 1 & 2 \\ 1 & 3 \end{bmatrix}.$$

From above,

$$TE = TI = \begin{bmatrix} 1 & -1 & 0 \\ 1 & 0 & 1 \end{bmatrix}.$$

A simple calculation gives

$$G^{-1} = \begin{bmatrix} 3 & -2 \\ -1 & 1 \end{bmatrix}.$$

Finally, the matrix representation A of the linear transformation T above, relative to these basis operators E and G, is given by the 2×3-matrix

$$A = G^{-1}TE = \begin{bmatrix} 3 & -2 \\ -1 & 1 \end{bmatrix} \begin{bmatrix} 1 & -1 & 0 \\ 1 & 0 & 1 \end{bmatrix} = \begin{bmatrix} 1 & -3 & -2 \\ 0 & 1 & 1 \end{bmatrix}.$$

If in the example above both basis operators E and G differ from the identity matrix, the following procedure should be followed in order to calculate the matrix representation A of the linear transformation T relative to the basis operators E and G.

---

1. Calculate what T does to each of the basis column vectors in the basis operator E.
2. Construct the matrix TE using the corresponding vectors obtained in step 1 as the columns of TE.
3. Calculate the inverse G^{-1} of G.
4. Compute the product $G^{-1}TE$ which is the matrix representation A of T.

■ **EXAMPLE 3 (again revisited)** ─────────────────────────

Suppose that we choose a new basis for $\mathbf{R}^3$ whose basis operator is

$$E = \begin{bmatrix} 1 & 0 & 1 \\ 0 & -1 & 1 \\ 0 & -2 & 0 \end{bmatrix}.$$

Suppose that the basis operator for $\mathbf{R}^2$ is

$$G = \begin{bmatrix} 1 & 2 \\ 1 & 3 \end{bmatrix}.$$

Using the definition of T above, we calculate

$$T\begin{bmatrix} 1 \\ 0 \\ 0 \end{bmatrix} = \begin{bmatrix} 1 \\ 1 \end{bmatrix}, \qquad T\begin{bmatrix} 0 \\ -1 \\ -2 \end{bmatrix} = \begin{bmatrix} 1 \\ -2 \end{bmatrix}, \qquad T\begin{bmatrix} 1 \\ 1 \\ 0 \end{bmatrix} = \begin{bmatrix} 0 \\ 1 \end{bmatrix}.$$

This gives the matrix

$$TE = \begin{bmatrix} 1 & 1 & 0 \\ 1 & -2 & 1 \end{bmatrix}.$$

From above, the matrix

$$G^{-1} = \begin{bmatrix} 3 & -2 \\ -1 & 1 \end{bmatrix}.$$

The matrix representation A of T relative to E and G is given by

$$A = G^{-1}TE = \begin{bmatrix} 3 & -2 \\ -1 & 1 \end{bmatrix}\begin{bmatrix} 1 & 1 & 0 \\ 1 & -2 & 1 \end{bmatrix} = \begin{bmatrix} 1 & 7 & -2 \\ 0 & -3 & 1 \end{bmatrix}.$$

In applications we usually have the choice of E and G at our disposal. Normally, they should be chosen so that the matrix $A = G^{-1}TE$ is as simple as possible for a given linear operator T. Recall that if C is a nonsingular $n \times n$-matrix and E is a basis operator for **U**, $F = EC$ is a second basis operator for **U**. Similarly, if D is a nonsingular $m \times m$-matrix and G is a basis operator for **V**, then $H = GD$ is a second basis operator for **V**. When E and G are our bases operators, the matrix

$$A = G^{-1}TE$$

is a matrix representation of T. When F and H are the bases operators for **U** and **V**, respectively, the matrix

$$B = H^{-1}TF = D^{-1}G^{-1}TEC = D^{-1}AC$$

is the matrix representation of T. Thus A and $B = D^{-1}AC$ are two matrix representations of the same operator T. This gives the following result.

> Let A be a $m \times n$-matrix representation of a linear transformation T of an n-dimensional linear space **U** into an m-dimensional linear space **V**. A second $m \times n$-matrix B is a matrix representation of T if and only if B is expressible in the form
>
> $$B = D^{-1}AC,$$
>
> where C and D are nonsingular matrices of dimensions n and m, respectively.

Of particular interest is the case when **V** = **U**. Then we are concerned with a linear mapping $T: \mathbf{U} \to \mathbf{U}$ of **U** into itself. If E is a basis operator for **U**, then $A = E^{-1}TE$ is the corresponding matrix representation of T. For a second basis operator $F = EC$ for **U** determined by a nonsingular matrix C, the corresponding matrix representation of T is

$$B = F^{-1}TF = C^{-1}E^{-1}TEC = C^{-1}AC.$$

The matrices A and B are therefore similar. Thus similar matrices represent the same linear operator T under different coordinate systems. Observe further that the operator $S = FE^{-1} = ECE^{-1}$ is a linear operator on **U** to **U** such that $F = SE$. The matrix C is the matrix representation of S relative to the coordinate system defined by E.

We restate this result as follows.

> Two matrices A and B represent the same linear transformation T of a linear space **U** into itself if and only if they are similar.

The operator T such that $Tu = u$ for every u in U is the identity operator on U. Its matrix representation is the identity matrix I for every basis operator of U.

Let T be a linear operator mapping U into U. We form polynomials in T in the usual manner. For example, if $p(\lambda)$ is the polynomial

$$p(\lambda) = a + b\lambda + c\lambda^2 + d\lambda^3,$$

then

$$p(T) = aI + bT + cT^2 + dT^3.$$

If $A = E^{-1}TE$ is a matrix representation of T determined by a basis operator E, then $A^k = E^{-1}T^kE$ for every integer k. It follows that

$$p(A) = E^{-1}p(T)E, \qquad p(T) = Ep(A)E^{-1}$$

for every polynomial $p(\lambda)$. The matrix $p(A)$ therefore is the corresponding matrix representation for the operator $p(T)$.

Exercises

Which of the following transformations are linear? If linear, find a matrix representation.

1. $T(x, y) = (x + y, x - y)$

2. $T(x, y) = (y + 1, x)$

3. $T(x, y) = (x, y)$

4. $T(x, y) = (y, 0)$

5. $T(x, y) = (x, x, x)$

6. $T(x, y) = (x + y, x - y, y - x)$

7. $T(x, y, z) = (2x + z, y - z, z)$

8. $T(x, y, z) = 2(z, y, x)$

9. $T(x, y, z) = (x, y)$

10. $T(x, y, z) = (2x - y, y - 3z)$

11. $T(x, y, z) = (0, 0)$

12. $T(x, y, z) = (y^2, x - z)$

13. Show that the linear transformation $y = Ax$ of $\mathbf{R}^2$ into itself defined by the matrix

$$A = \begin{bmatrix} \cos\theta & -\sin\theta \\ \sin\theta & \cos\theta \end{bmatrix}$$

rotates a vector x counterclockwise through an angle θ.

14. A linear transformation T of $\mathbf{R}^3$ into $\mathbf{R}^2$ has the values

$$T(1, 0, 0) = (0, 1), \qquad T(0, 1, 0) = (1, -1), \qquad T(0, 0, 1) = (2, 3).$$

Find a matrix representation A of this transformation. Find the value of $T(1, 1, 1)$.

15. A linear transformation T of $\mathbf{R}^3$ into $\mathbf{R}^2$ has the values

$$T(1, 1, 1) = (2, 2), \qquad T(0, 1, 1) = (0, 1), \qquad T(0, 0, 1) = (-1, 1).$$

Find a matrix representation A of this transformation. Find the value of $T(1, 2, 3)$.

16. Let **M** be the class of all 2 × 2-matrices. Which of the following operators on **M** are linear operators?

(a) TA = trace of A (b) TA = det A

(c) $TA = BAC$ (B and C are fixed)

17. Let $\mathbf{P}_n$ be the class of polynomials in x of degree n, including the zero polynomial. Consider the transformation

$$Tp(x) = xp(x)$$

mapping $\mathbf{P}_2$ into $\mathbf{P}_3$. Is it linear? Is every polynomial in $\mathbf{P}_3$ the image under T of a polynomial in $\mathbf{P}_2$?

6.2   THE NULL SPACE AND THE RANGE OF A LINEAR TRANSFORMATION

In this section we are concerned with a linear transformation

$$T\colon \mathbf{U} \to \mathbf{V}, \qquad \mathbf{v} = T\mathbf{u},$$

which maps a point **u** in a n-dimensional linear space **U** into a point **v** in a m-dimensional linear space **V**. We write

$$\mathbf{v} = T\mathbf{u}.$$

The space **U** is called the ***domain*** of T. Associated with the linear transformation T are two very important subsets. One of them is a subset of the domain **U** and the other is a subset of **V**. Both of these subsets will turn out to be linear subspaces. The first of these subsets is called the ***null space*** of T. We denote it by **N**. It is defined as follows:

> The ***null space*** **N** of the linear transformation T is the set of points **u** in **U** which get mapped under T into the origin **0** of **V** (i.e., the set of solutions **u** of $T\mathbf{u} = \mathbf{0}$).

We can now prove that

> The null space **N** of T is a linear subspace of **U**.

We can show this in the following way:

1. Let $\mathbf{u}_1$ and $\mathbf{u}_2$ be any two elements of **N**. Then $T\mathbf{u}_1 = \mathbf{0}$ and $T\mathbf{u}_2 = \mathbf{0}$. Since T is linear, it follows that

$$T(\mathbf{u}_1 + \mathbf{u}_2) = T\mathbf{u}_1 + T\mathbf{u}_2 = \mathbf{0} + \mathbf{0} = \mathbf{0}$$

(i.e., $\mathbf{u}_1 + \mathbf{u}_2$ is an element of **N**).

2. Similarly, if **u** is an element of **N**, then $T\mathbf{u} = \mathbf{0}$. Let a be any scalar. Then, because T is linear, it follows that

$$T(a\mathbf{u}) = a(T\mathbf{u}) = a\mathbf{0} = \mathbf{0}.$$

which implies that $a\mathbf{u}$ is an element of **N**.

The null space **N** of T is therefore a linear subspace of **U**.

The null space **N** of T is also called the **kernel** of T.

■ **EXAMPLE 1** _____

Consider the linear transformation T from $\mathbf{R}^3$ to $\mathbf{R}^2$ given by

$$T\begin{bmatrix} u_1 \\ u_2 \\ u_3 \end{bmatrix} = \begin{bmatrix} u_1 - u_2 \\ u_1 + u_2 \end{bmatrix} = \begin{bmatrix} v_1 \\ v_2 \end{bmatrix}.$$

The null space **N** of $\mathbf{R}^3$ is the set of all vectors

$\begin{bmatrix} u_1 \\ u_2 \\ u_3 \end{bmatrix}$ in $\mathbf{R}^3$ with the property that $T\begin{bmatrix} u_1 \\ u_2 \\ u_3 \end{bmatrix} = \begin{bmatrix} 0 \\ 0 \end{bmatrix}$,

that is,

the set of all $\begin{bmatrix} u_1 \\ u_2 \\ u_3 \end{bmatrix}$ such that $\begin{bmatrix} u_1 - u_2 \\ u_1 + u_2 \end{bmatrix} = \begin{bmatrix} 0 \\ 0 \end{bmatrix}$.

Solving, we get $u_1 = 0$ and $u_2 = 0$. We observe that there is no restriction on u_3. It follows that the null space **N** of T consists of all vectors in $\mathbf{R}^3$ of the form

$$\begin{bmatrix} 0 \\ 0 \\ a \end{bmatrix} \quad \text{where } a \text{ is any scalar.}$$

We can therefore write the null space **N** as

$$a\begin{bmatrix} 0 \\ 0 \\ 1 \end{bmatrix} \quad \text{for all scalars } a.$$

We see that **N** is generated by the vector $\begin{bmatrix} 0 \\ 0 \\ 1 \end{bmatrix}$. It is a basis for **N**. The null space **N** of T therefore has dimension 1.

In general, there is a name associated with the dimension of the null space **N** of a linear transformation *T*. We have the following definition:

The dimension *q* of the null space **N** is called the ***nullity*** of *T*.

In Example 1 we see that the nullity *q* of *T* is *q* = 1.

■ **EXAMPLE 2** _____

Consider the matrix operator

$$A = \begin{bmatrix} 1 & 2 & 0 & 4 \\ -1 & -2 & 0 & -4 \end{bmatrix},$$

which maps **R**^{4} into **R**^{2} via the equation **y** = *A***x**. We have

$$\begin{bmatrix} y_1 \\ y_2 \end{bmatrix} = \begin{bmatrix} 1 & 2 & 0 & 4 \\ -1 & -2 & 0 & -4 \end{bmatrix} \begin{bmatrix} x_1 \\ x_2 \\ x_3 \\ x_4 \end{bmatrix}.$$

The null space **N** of the linear transformation *T* associated with the matrix *A* is the set of all

$$\begin{bmatrix} x_1 \\ x_2 \\ x_3 \\ x_4 \end{bmatrix}$$

in **R**^{4} with the property that *T***x** = *A***x** = **0**. Setting *A***x** = **0** above gives

$$\begin{bmatrix} 1 & 2 & 0 & 4 \\ -1 & -2 & 0 & -4 \end{bmatrix} \begin{bmatrix} x_1 \\ x_2 \\ x_3 \\ x_4 \end{bmatrix} = \begin{bmatrix} 0 \\ 0 \end{bmatrix}.$$

To solve this system, we proceed to reduce the augmented matrix

$$\begin{bmatrix} 1 & 2 & 0 & 4 & \vdots & 0 \\ -1 & -2 & 0 & -4 & \vdots & 0 \end{bmatrix}$$

to row echelon form. We obtain the matrix

$$\begin{bmatrix} 1 & 2 & 0 & 4 & \vdots & 0 \\ 0 & 0 & 0 & 0 & \vdots & 0 \end{bmatrix}.$$

We conclude that

$$x_1 + 2x_2 + 0x_3 + 4x_4 = 0.$$

If we set
$$x_2 = a, \qquad x_3 = b, \qquad \text{and} \qquad x_4 = c,$$

where a, b, and c are any scalars, then
$$x_1 = -2a - 4c$$

and the null space **N** of T consists of all vectors in $\mathbf{R}^4$ of the form
$$\begin{bmatrix} -2a - 4c \\ a \\ b \\ c \end{bmatrix}.$$

We write this as
$$\begin{bmatrix} -2a - 4c \\ a \\ b \\ c \end{bmatrix} = a\begin{bmatrix} -2 \\ 1 \\ 0 \\ 0 \end{bmatrix} + b\begin{bmatrix} 0 \\ 0 \\ 1 \\ 0 \end{bmatrix} + c\begin{bmatrix} -4 \\ 0 \\ 0 \\ 1 \end{bmatrix}.$$

We see that the vectors $\mathbf{z}_1 = (-2, 1, 0, 0)$, $\mathbf{z}_2 = (0, 0, 1, 0)$, and $\mathbf{z}_3 = (-4, 0, 0, 1)$ generate **N**. Furthermore, they are linearly independent. Hence they form a basis for **N**. Therefore, the nullity q of **N** is $q = 3$.

■ **EXAMPLE 3** _____

Consider the case in which our operator is the matrix
$$A = \begin{bmatrix} 1 & 1 & 0 \\ 0 & 1 & 1 \\ 0 & 0 & 0 \end{bmatrix}.$$

The linear transformation $\mathbf{y} = A\mathbf{x}$ maps $\mathbf{R}^3$ into itself. The null space **N** of A consists of all vectors $\mathbf{x}$ in $\mathbf{R}^3$ for which
$$A\mathbf{x} = \begin{bmatrix} 1 & 1 & 0 \\ 0 & 1 & 1 \\ 0 & 0 & 0 \end{bmatrix}\begin{bmatrix} x_1 \\ x_2 \\ x_3 \end{bmatrix} = \begin{bmatrix} 0 \\ 0 \\ 0 \end{bmatrix} = \mathbf{0}.$$

By inspection we see that
$$x_1 = -x_2 \qquad \text{and} \qquad x_3 = -x_2.$$

Setting $x_2 = -a$, we have $x_1 = x_3 = a$. The null space **N** of A are all vectors in $\mathbf{R}^3$ of the form
$$\begin{bmatrix} a \\ -a \\ a \end{bmatrix} = a\begin{bmatrix} 1 \\ -1 \\ 1 \end{bmatrix} \qquad \text{for all real numbers } a.$$

It follows that the nullity q of **N** is $q = 1$.

Now let us turn to the second linear subspace associated with a linear operator T mapping $\mathbf{U}$ into $\mathbf{V}$. This is the range $\mathbf{R}$ of T. It is defined as follows:

> The set of points $\mathbf{v}$ in $\mathbf{V}$ which are images $\mathbf{v} = T\mathbf{u}$ of points $\mathbf{u}$ in $\mathbf{U}$ form a subset $\mathbf{R}$ of $\mathbf{V}$, called the *range* of T.

We can now prove that

> The range $\mathbf{R}$ of T is a linear subspace of $\mathbf{V}$.

We show this as follows:

1. Let $\mathbf{v}_1$ and $\mathbf{v}_2$ be any two points in $\mathbf{R}$. Then there exists points $\mathbf{u}_1$ and $\mathbf{u}_2$ in $\mathbf{U}$ such that $\mathbf{v}_1 = T\mathbf{u}_1$ and $\mathbf{v}_2 = T\mathbf{u}_2$. By the linearity of T and the fact that $\mathbf{U}$ is a linear space, we have

$$\mathbf{v}_1 + \mathbf{v}_2 = T\mathbf{u}_1 + T\mathbf{u}_2 = T(\mathbf{u}_1 + \mathbf{u}_2).$$

Consequently, $\mathbf{v}_1 + \mathbf{v}_2$ is the image of the point $\mathbf{u}_1 + \mathbf{u}_2$ in $\mathbf{U}$ and therefore is in the range $\mathbf{R}$ of T.

2. Similarly, if $\mathbf{v}$ is in $\mathbf{R}$, there is a point $\mathbf{u}$ in $\mathbf{U}$ such that $\mathbf{v} = T\mathbf{u}$. If a is any scalar, then because T is linear and $\mathbf{U}$ is a linear space, we have

$$a\mathbf{v} = a(T\mathbf{u}) = T(a\mathbf{u}).$$

It follows that $a\mathbf{v}$ is the image of the point $a\mathbf{u}$ in $\mathbf{U}$ and is therefore in the range $\mathbf{R}$ of T. The range $\mathbf{R}$ of T is therefore a linear subspace of $\mathbf{V}$.

When $\mathbf{U} = \mathbf{R}^n$ and $\mathbf{V} = \mathbf{R}^m$, our linear transformation takes the form

$$\mathbf{y} = \begin{bmatrix} y_1 \\ y_2 \\ y_3 \\ \cdot \\ \cdot \\ \cdot \\ y_m \end{bmatrix} = A\,\mathbf{x} = \begin{bmatrix} a_{11} & a_{12} & a_{13} & \cdots & a_{1n} \\ a_{21} & a_{22} & a_{23} & \cdots & a_{2n} \\ a_{31} & a_{32} & a_{33} & \cdots & a_{3n} \\ \cdot & \cdot & \cdot & \cdots & \cdot \\ \cdot & \cdot & \cdot & \cdots & \cdot \\ \cdot & \cdot & \cdot & \cdots & \cdot \\ a_{m1} & a_{m2} & a_{m3} & \cdots & a_{mn} \end{bmatrix} \begin{bmatrix} x_1 \\ x_2 \\ x_3 \\ \cdot \\ \cdot \\ \cdot \\ x_n \end{bmatrix}$$

$$= x_1\mathbf{a}_1 + x_2\mathbf{a}_2 + x_3\mathbf{a}_3 + \cdots + x_n\mathbf{a}_n,$$

where

$$\mathbf{a}_1 = \begin{bmatrix} a_{11} \\ a_{21} \\ a_{31} \\ \cdot \\ \cdot \\ \cdot \\ a_{m1} \end{bmatrix}, \quad \mathbf{a}_2 = \begin{bmatrix} a_{12} \\ a_{22} \\ a_{32} \\ \cdot \\ \cdot \\ \cdot \\ a_{m2} \end{bmatrix}, \quad \cdots, \quad \mathbf{a}_n = \begin{bmatrix} a_{1n} \\ a_{2n} \\ a_{3n} \\ \cdot \\ \cdot \\ \cdot \\ a_{mn} \end{bmatrix}$$

are the column vectors of A. The range of A therefore consists of all vectors $\mathbf{y}$ that are linear combinations of the columns of A. Hence the range of A is the column space of A. It follows that the dimension r of the range $\mathbf{R}$ of A is the column rank of A as defined in Chapter 3.

In general, there is a name associated with the dimension r of the range $\mathbf{R}$ of a linear transformation T. We have the following definition:

> The dimension r of the range $\mathbf{R}$ of a linear transformation T is called the *rank* of T.

From above we see that if $\mathbf{U} = \mathbf{R}^n$ and $\mathbf{V} = \mathbf{R}^m$, a linear transformation T is defined by $\mathbf{y} = T\mathbf{x} = A\mathbf{x}$, where A is an $m \times n$ matrix. The rank r of this linear transformation T is the column rank of A.

Consequently, our new definition of rank is equivalent to the definitions given in Chapter 3.

■ **EXAMPLE 2 (revisited)** ⸺⸺⸺⸺⸺⸺⸺⸺⸺⸺⸺⸺
Consider the matrix operator

$$A = \begin{bmatrix} 1 & 2 & 0 & 4 \\ -1 & -2 & 0 & -4 \end{bmatrix}.$$

This maps $\mathbf{R}^4$ into $\mathbf{R}^2$ via the equation $\mathbf{y} = A\mathbf{x}$. The column vectors

$$\mathbf{a} = \begin{bmatrix} 1 \\ -1 \end{bmatrix}, \qquad \mathbf{b} = \begin{bmatrix} 2 \\ -2 \end{bmatrix}, \qquad \mathbf{c} = \begin{bmatrix} 0 \\ 0 \end{bmatrix}, \qquad \mathbf{d} = \begin{bmatrix} 4 \\ -4 \end{bmatrix}$$

of A generate the range $\mathbf{R}$ of T. They are, however, not linearly independent. If we wish to obtain a basis for $\mathbf{R}$, we must extract from these vectors a maximal linearly independent set. By inspection, we observe that the column vectors $\mathbf{b}$, $\mathbf{c}$, and $\mathbf{d}$ are all multiples of the first column vector $\mathbf{a}$. We have

$$\mathbf{b} = 2\mathbf{a}, \qquad \mathbf{c} = 0\mathbf{a}, \quad \text{and} \quad \mathbf{d} = 4\mathbf{a}.$$

For any vector $\mathbf{x} = (x_1, x_2, x_3, x_4)$ in $\mathbf{R}^4$, the image

$$\mathbf{y} = A\mathbf{x} = x_1\mathbf{a} + x_2\mathbf{b} + x_3\mathbf{c} + x_4\mathbf{d} = (x_1 + 2x_2 + 4x_4)\mathbf{a}$$

of $\mathbf{x}$ is therefore a multiple of the column $\mathbf{a}$ of A. Hence the range $\mathbf{R}$ of A consists of all multiples $m\mathbf{a}$ of $\mathbf{a}$ and so is one-dimensional. It follows that the rank r of A is $r = 1$. It should be noted that the elements of the range $\mathbf{R}$ are also multiples of the second column $\mathbf{b}$ of A and multiples of the fourth column $\mathbf{d}$ of A.

⸺⸺⸺⸺⸺⸺⸺⸺⸺⸺⸺⸺⸺⸺⸺⸺⸺⸺⸺⸺⸺⸺⸺⸺⸺⸺⸺⸺

Before presenting some further examples for determining the range $\mathbf{R}$ of a linear transformation T, it should be mentioned that there is a very interesting result which relates the dimensions of the domain **U**, the null space **N**, and the range **R**.

We shall refer to this result as the *dimension proposition:*

Suppose that T is a linear transformation from the linear space **U** to the linear space **V**. If the dimension of **U** is n, if the dimension of the null space **N** is q, and if the dimension of the range **R** is r, then

$$q + r = n,$$

that is,

nullity of T + rank of T = dimension of **U**

In other words, the dimension of the null space of T added to the dimension of the range of T gives the dimension of the domain **U**.

Before proving this result we will look at some of its consequences. We begin by once again returning to

■ **EXAMPLE 2 (revisited again)** ————————————————————

Here $\mathbf{y} = A\mathbf{x}$, where A is the 2×4 matrix

$$A = \begin{bmatrix} 1 & 2 & 0 & 4 \\ -1 & -2 & 0 & -4 \end{bmatrix},$$

which maps $\mathbf{R}^4$ into $\mathbf{R}^2$. Suppose that we have not found the range **R**. When we first encountered this example, we found the null space **N** and saw that it could be generated by the three linearly independent vectors $(-2, 1, 0, 0)$, $(0, 0, 1, 0)$, and $(-4, 0, 0, 1)$. Consequently, the nullity q of T is $q = 3$. Since $\mathbf{U} = \mathbf{R}^4$, whose dimension is 4, it follows by the result above that the rank r of T is

$$r = n - q = 4 - 3 = 1.$$

This tells us in advance that the column space of the matrix A is one-dimensional. We therefore know that three of the columns of A will be multiples of one of the other columns, as of course we found out above. This reasoning can often be helpful in determining the range **R**.

■ **EXAMPLE 1 (revisited)** ————————————————————

Recall that T mapped $\mathbf{R}^3$ to $\mathbf{R}^2$ via the rule

$$T\begin{bmatrix} u_1 \\ u_2 \\ u_3 \end{bmatrix} = \begin{bmatrix} u_1 - u_2 \\ u_1 + u_2 \end{bmatrix} = \begin{bmatrix} v_1 \\ v_2 \end{bmatrix}.$$

We found the null space to be all those vectors in $\mathbf{R}^3$ that were multiples of the vector $(0, 0, 1)$, from which we concluded that the nullity q of T is $q = 1$. Since the domain $\mathbf{U}$ is $\mathbf{R}^3$, whose dimension is 3, it follows from the result above that the rank r of T is $r = n - q = 3 - 1 = 2$. This is useful since it tells us in advance that the range $\mathbf{R}$ of T is two-dimensional. In fact, since the range $\mathbf{R}$ is a subspace of $\mathbf{R}^2$, we can immediately conclude that the range $\mathbf{R}$ of T is $\mathbf{R}^2$ itself since $\mathbf{R}^2$ cannot have any subspaces of dimension 2 other than itself.

■ **EXAMPLE 3 (revisited)** ⎯⎯⎯⎯⎯⎯⎯⎯⎯⎯⎯⎯⎯⎯⎯⎯⎯⎯⎯⎯

Here the linear transformation $\mathbf{y} = A\mathbf{x}$ maps $\mathbf{R}^3$ into itself where the operator A is the 3×3 matrix

$$A = \begin{bmatrix} 1 & 1 & 0 \\ 0 & 1 & 1 \\ 0 & 0 & 0 \end{bmatrix}.$$

We found the null space $\mathbf{N}$ to be all vectors in $\mathbf{R}^3$ which were multiples of the vector $(1, -1, 1)$. Hence the nullity q of T is $q = 1$. Since $\mathbf{U} = \mathbf{R}^3$, $n = 3$ and $r = n - q = 3 - 1 = 2$. The range $\mathbf{R}$ therefore has dimension 2. This immediately tells us that the three columns of the matrix A are not linearly independent but that two of the columns of A will be a basis for the column space of A. It does not tell us which two we should choose. We observe, however, that the first column of A, which we denote by $\mathbf{a}_1$, is a linear combination of the second and third columns $\mathbf{a}_2$ and $\mathbf{a}_3$, respectively, of A, for

$$\mathbf{a}_1 = \begin{bmatrix} 1 \\ 0 \\ 0 \end{bmatrix} = \begin{bmatrix} 1 \\ 1 \\ 0 \end{bmatrix} - \begin{bmatrix} 0 \\ 1 \\ 0 \end{bmatrix} = \mathbf{a}_2 - \mathbf{a}_3.$$

One easily checks that $\mathbf{a}_2$ and $\mathbf{a}_3$ are linearly independent. It follows that the vectors $\mathbf{a}_2 = (1, 1, 0)$ and $\mathbf{a}_3 = (0, 1, 0)$ are a basis for the range $\mathbf{R}$. In this example the reader can verify that $\mathbf{a}_1$ and $\mathbf{a}_2$ as well as $\mathbf{a}_1$ and $\mathbf{a}_3$ also can serve as a basis for the range $\mathbf{R}$.

Observe that the columns $\mathbf{a}_1$, $\mathbf{a}_2$, and $\mathbf{a}_3$ of A are given by the relations

$$\mathbf{a}_1 = A\mathbf{e}_1, \qquad \mathbf{a}_2 = A\mathbf{e}_2, \qquad \mathbf{a}_3 = A\mathbf{e}_3,$$

where

$$\mathbf{e}_1 = (1, 0, 0), \qquad \mathbf{e}_2 = (0, 1, 0), \qquad \text{and} \qquad \mathbf{e}_3 = (0, 0, 1)$$

form the standard coordinate basis for $\mathbf{R}^3$.

This example illustrates the following result:

Let e_1, e_2, e_3, . . . , e_n be a basis for the n-dimensional linear space **U**. Then the range **R** of a linear operator T that maps **U** into **V** is generated by the vectors Te_1, Te_2, Te_3, . . . , Te_n in **V**. A basis for the range **R** can be selected from this set of n vectors by choosing from them a maximal linearly independent set.

To prove this, let $E = [e_1 \quad e_2 \quad e_3 \quad \cdots \quad e_n]$ be the corresponding basis operator which induces an x-coordinate system on **U**. Then an element **u** in **U** is expressible in the form

$$\mathbf{u} = E\mathbf{x} = x_1 e_1 + x_2 e_2 + x_3 e_3 + \cdots + x_n e_n.$$

It follows, by the linearity of T, that every element $\mathbf{v} = T\mathbf{u}$ in the range **R** of T is a linear combination

$$\mathbf{v} = T\mathbf{u} = TE\mathbf{x} = x_1 Te_1 + x_2 Te_2 + x_3 Te_3 + \cdots + x_n Te_n$$

of the elements Te_1, Te_2, Te_3, . . . , Te_n in **V**. These elements therefore generate the range **R** of T. By successively deleting the vectors that are linearly dependent on the vectors which precede them, we obtain a basis for the range **R**.

Observe that we used the operator

$$R = TE = [Te_1 \quad Te_2 \quad Te_3 \quad \cdots \quad Te_n],$$

which maps $\mathbf{R}^n$ into **V**. It follows that when R and E are known, we can determine T by the formula $T = RE^{-1}$. This means that if we have a basis e_1, e_2, e_3, . . . , e_n for **U**, and hence know the basis operator E, and if we know what T does to the basis vectors e_1, e_2, e_3, . . . , e_n, we can completely determine the linear transformation T. We illustrate this fact in the following:

■ **EXAMPLE 4** _____

Suppose that $\mathbf{U} = \mathbf{R}^3$ and $\mathbf{V} = \mathbf{R}^2$ and E is the matrix

$$E = \begin{bmatrix} 1 & 1 & 1 \\ 0 & 1 & 1 \\ 0 & 0 & 1 \end{bmatrix}.$$

The columns e_1, e_2, and e_3 of E form a basis for $\mathbf{R}^3$ so that E is the corresponding basis operator. A simple calculation gives

$$E^{-1} = \begin{bmatrix} 1 & -1 & 0 \\ 0 & 1 & -1 \\ 0 & 0 & 1 \end{bmatrix}.$$

Suppose that T is a linear operator which maps $\mathbf{R}^3$ into $\mathbf{R}^2$ with the property that

$$T\mathbf{e}_1 = \begin{bmatrix} 1 \\ 1 \end{bmatrix}, \qquad T\mathbf{e}_2 = \begin{bmatrix} -1 \\ 1 \end{bmatrix}, \qquad T\mathbf{e}_3 = \begin{bmatrix} 0 \\ 2 \end{bmatrix}.$$

We have

$$R = TE = [T\mathbf{e}_1 \quad T\mathbf{e}_2 \quad T\mathbf{e}_3] = \begin{bmatrix} 1 & -1 & 0 \\ 1 & 1 & 2 \end{bmatrix}.$$

Consequently,

$$T = RE^{-1} = \begin{bmatrix} 1 & -1 & 0 \\ 1 & 1 & 2 \end{bmatrix} \begin{bmatrix} 1 & -1 & 0 \\ 0 & 1 & -1 \\ 0 & 0 & 1 \end{bmatrix} = \begin{bmatrix} 1 & -2 & 1 \\ 1 & 0 & 1 \end{bmatrix}.$$

We have

$$\mathbf{y} = \begin{bmatrix} y_1 \\ y_2 \end{bmatrix} = T\mathbf{x} = \begin{bmatrix} 1 & -2 & 1 \\ 1 & 0 & 1 \end{bmatrix} \begin{bmatrix} x_1 \\ x_2 \\ x_3 \end{bmatrix} = \begin{bmatrix} x_1 - 2x_2 + x_3 \\ x_1 + x_3 \end{bmatrix}.$$

The linear transformation T, from $\mathbf{R}^3$ into $\mathbf{R}^2$, in terms of coordinates, is given by

$$y_1 = x_1 - 2x_2 + x_3$$
$$y_2 = x_1 + x_3.$$

Observe that the first and third columns of T are equal. Furthermore, the columns

$$\begin{bmatrix} 1 \\ 1 \end{bmatrix} \quad \text{and} \quad \begin{bmatrix} -2 \\ 0 \end{bmatrix}$$

are linearly independent. The rank r of T is therefore $r = 2$.

We conclude this section by proving the ***dimension proposition***. The proof is based on the following scheme:

1. Choose a basis for the null space $\mathbf{N}$ of T.
2. Extend the basis for $\mathbf{N}$ to a basis of the domain $\mathbf{U}$.
3. Show that the set of images of the basis elements of $\mathbf{U}$ which are not in the null space $\mathbf{N}$ are a basis for the range $\mathbf{R}$ of T (i.e., they are linearly independent and generate the range $\mathbf{R}$ of T).

We proceed as follows: Let $z_1, z_2, z_3, \ldots, z_q$ be a basis for the null space **N** of T. Adjoin $r = n - q$ linearly independent elements $\mathbf{f}_1, \mathbf{f}_2, \mathbf{f}_3, \ldots, \mathbf{f}_r$ to form a basis

$$\mathbf{f}_1, \quad \mathbf{f}_2, \quad \mathbf{f}_3, \quad \ldots, \quad \mathbf{f}_r, \quad z_1, \quad z_2, \quad z_3, \quad \ldots, \quad z_q$$

for **U**. (Here we have tacitly assumed that $0 < q < n$. When $q = n$, T is the zero operator. We exclude this case. When $q = 0$, there are no z's in the discussion given below. We leave the consideration of this case to the reader.)

The corresponding basis operator

$$F = [\mathbf{f}_1, \mathbf{f}_2, \mathbf{f}_3, \ldots, \mathbf{f}_r, \; z_1, z_2, z_3, \ldots, z_q]$$

when operated on (multiplied on the left) by the linear transformation T gives

$$TF = [T\mathbf{f}_1, T\mathbf{f}_2, T\mathbf{f}_3, \ldots, T\mathbf{f}_r, \quad Tz_1, Tz_2, Tz_3, \ldots, Tz_q]$$

$$= [\mathbf{v}_1, \mathbf{v}_2, \mathbf{v}_3, \ldots, \mathbf{v}_r, \quad \mathbf{0}, \mathbf{0}, \mathbf{0}, \ldots, \mathbf{0}],$$

where

$$\mathbf{v}_1 = T\mathbf{f}_1, \quad \mathbf{v}_2 = T\mathbf{f}_2, \quad \mathbf{v}_3 = T\mathbf{f}_3, \quad \ldots, \quad \mathbf{v}_r = T\mathbf{f}_r$$

and

$$Tz_1 = \mathbf{0}, \quad Tz_2 = \mathbf{0}, \quad Tz_3 = \mathbf{0}, \quad \ldots, \quad Tz_q = \mathbf{0}$$

since each of the z's are elements of the null space **N**. Setting $S = TF$ we have

$$S = TF = [\mathbf{v}_1, \mathbf{v}_2, \mathbf{v}_3, \ldots, \mathbf{v}_r, \mathbf{0}, \mathbf{0}, \mathbf{0}, \ldots, \mathbf{0}].$$

The operator S maps $\mathbf{R}^n$ into **V**. Its range is **R**, the range of T. We now show that the vectors

$$\mathbf{v}_1 = T\mathbf{f}_1, \quad \mathbf{v}_2 = T\mathbf{f}_2, \quad \mathbf{v}_3 = T\mathbf{f}_3, \quad \ldots, \quad \mathbf{v}_r = T\mathbf{f}_r$$

are linearly independent in the range **R** of T.

To do so, let $c_1, c_2, c_3, \ldots, c_r$ be any r constants and consider

$$c_1\mathbf{v}_1 + c_2\mathbf{v}_2 + c_3\mathbf{v}_3 + \cdots + c_r\mathbf{v}_r = c_1T\mathbf{f}_1 + c_2T\mathbf{f}_2 + c_3T\mathbf{f}_3 + \cdots + c_rT\mathbf{f}_r$$

$$= T(c_1\mathbf{f}_1 + c_2\mathbf{f}_2 + c_3\mathbf{f}_3 + \cdots + c_r\mathbf{f}_r) = \mathbf{0}.$$

We will now show that $c_1 = c_2 = c_3 = \cdots = c_r = 0$.

We see that the linear combination

$$c_1\mathbf{f}_1 + c_2\mathbf{f}_2 + c_3\mathbf{f}_3 + \cdots + c_r\mathbf{f}_r$$

is in the null space **N** of T. But that means that this linear combination can be written as some linear combination of the basis vectors on **N**,

$$c_1\mathbf{f}_1 + c_2\mathbf{f}_2 + c_3\mathbf{f}_3 + \cdots + c_r\mathbf{f}_r = a_1z_1 + a_2z_2 + a_3z_3 + \cdots + a_qz_q$$

for some constants $a_1, a_2, a_3, \ldots, a_q$. Moving everything to the left side of the equation yields

$$c_1\mathbf{f}_1 + c_2\mathbf{f}_2 + c_3\mathbf{f}_3 + \cdots + c_r\mathbf{f}_r - a_1\mathbf{z}_1 - a_2\mathbf{z}_2 - a_3\mathbf{z}_3 - \cdots - a_q\mathbf{z}_q = \mathbf{0}.$$

But the vectors

$$\mathbf{f}_1, \mathbf{f}_2, \mathbf{f}_3, \ldots, \mathbf{f}_r, \qquad \mathbf{z}_1, \mathbf{z}_2, \mathbf{z}_3, \ldots, \mathbf{z}_q$$

are a basis for $\mathbf{U}$ and hence are linearly independent. Therefore,

$$c_1 = c_2 = c_3 = \cdots = c_r = a_1 = a_2 = a_3 = \cdots = a_q = 0.$$

This shows that the vectors

$$\mathbf{v}_1 = T\mathbf{f}_1, \quad \mathbf{v}_2 = T\mathbf{f}_2, \quad \mathbf{v}_3 = T\mathbf{f}_3, \quad \ldots, \quad \mathbf{v}_r = T\mathbf{f}_r$$

are linearly independent in the range $\mathbf{R}$.

Finally, we show that the vectors

$$\mathbf{v}_1 = T\mathbf{f}_1, \quad \mathbf{v}_2 = T\mathbf{f}_2, \quad \mathbf{v}_3 = T\mathbf{f}_3, \quad \ldots, \quad \mathbf{v}_r = T\mathbf{f}_r$$

generate $\mathbf{R}$. To this end, let $\mathbf{v}$ be any element in the range $\mathbf{R}$. Then there is an element $\mathbf{u}$ in $\mathbf{U}$ such that $\mathbf{v} = T\mathbf{u}$. But $\mathbf{u}$ in $\mathbf{U}$ implies that $\mathbf{u} = F\mathbf{x}$ for some $\mathbf{x}$ in $\mathbf{R}^n$ since F is a basis operator for $\mathbf{U}$. It follows that

$$\mathbf{v} = T\mathbf{u} = TF\mathbf{x} = S\mathbf{x} = x_1\mathbf{v}_1 + x_2\mathbf{v}_2 + x_3\mathbf{v}_3 + \cdots + x_r\mathbf{v}_r$$

This shows that the vectors

$$\mathbf{v}_1 = T\mathbf{f}_1, \quad \mathbf{v}_2 = T\mathbf{f}_2, \quad \mathbf{v}_3 = T\mathbf{f}_3, \quad \cdots, \quad \mathbf{v}_r = T\mathbf{f}_r$$

generate $\mathbf{R}$. Since they are linearly independent, they form a basis for $\mathbf{R}$. The dimension of $\mathbf{R}$ is therefore r, so that the rank of T is

$$r = n - q$$

Equivalently,

$$q + r = n,$$

that is,

$$\text{nullity of } T + \text{rank of } T = \text{dimension of } \mathbf{U},$$

as was to be proved.

Exercises

1. For each of the following linear transformations, determine the null space by finding a basis. Use this information for finding the nullity.

(a) $T\begin{bmatrix} x \\ y \end{bmatrix} = \begin{bmatrix} x + 2y \\ -3x \end{bmatrix}$ (b) $T\begin{bmatrix} x \\ y \end{bmatrix} = \begin{bmatrix} x \\ 0 \end{bmatrix}$

(c) $T \begin{bmatrix} u_1 \\ u_2 \end{bmatrix} = \begin{bmatrix} u_1 + u_2 \\ u_1 - u_2 \end{bmatrix}$ (d) $T \begin{bmatrix} u_1 \\ u_2 \end{bmatrix} = \begin{bmatrix} 0 \\ 0 \end{bmatrix}$

2. Repeat Exercise 1 for the following linear transformations:

(a) $T \begin{bmatrix} x \\ y \\ z \end{bmatrix} = \begin{bmatrix} x + y \\ z \end{bmatrix}$ (b) $T \begin{bmatrix} x \\ y \\ z \end{bmatrix} = \begin{bmatrix} x + y \\ x + z \\ y + z \end{bmatrix}$

(c) $T \begin{bmatrix} u_1 \\ u_2 \\ u_3 \\ u_4 \\ u_5 \end{bmatrix} = \begin{bmatrix} u_1 - u_3 \\ u_2 \\ u_3 - u_5 \end{bmatrix}$ (d) $T \begin{bmatrix} x \\ y \\ z \end{bmatrix} = \begin{bmatrix} 2x + y \\ 0 \\ x - y \\ z \end{bmatrix}$

3. For each of the linear transformations in Exercise 1, determine the rank without first finding the range by using the proposition on dimension. Then find the range.

4. Repeat Exercise 3 for each of the linear transformations in Exercise 2.

5. Determine the linear transformation T by writing down what T does to a typical point in its domain.

(a) $T \begin{bmatrix} 1 \\ 0 \end{bmatrix} = \begin{bmatrix} 0 \\ 1 \end{bmatrix}$, $T \begin{bmatrix} 0 \\ 1 \end{bmatrix} = \begin{bmatrix} 1 \\ 0 \end{bmatrix}$

(b) $T \begin{bmatrix} 1 \\ 0 \end{bmatrix} = \begin{bmatrix} 1 \\ -1 \end{bmatrix}$, $T \begin{bmatrix} 0 \\ 1 \end{bmatrix} = \begin{bmatrix} 2 \\ 3 \end{bmatrix}$

(c) $T \begin{bmatrix} 1 \\ 0 \\ 0 \end{bmatrix} = \begin{bmatrix} 2 \\ -1 \end{bmatrix}$, $T \begin{bmatrix} 0 \\ 1 \\ 0 \end{bmatrix} = \begin{bmatrix} 1 \\ 1 \end{bmatrix}$, $T \begin{bmatrix} 0 \\ 0 \\ 1 \end{bmatrix} = \begin{bmatrix} 0 \\ 2 \end{bmatrix}$

(d) $T \begin{bmatrix} 1 \\ 0 \\ 0 \end{bmatrix} = \begin{bmatrix} 4 \\ 5 \end{bmatrix}$, $T \begin{bmatrix} 0 \\ 1 \\ 0 \end{bmatrix} = \begin{bmatrix} -1 \\ 0 \end{bmatrix}$, $T \begin{bmatrix} 0 \\ 0 \\ 1 \end{bmatrix} = \begin{bmatrix} 2 \\ -3 \end{bmatrix}$

(e) $T \begin{bmatrix} 1 \\ 0 \\ 0 \\ 0 \end{bmatrix} = \begin{bmatrix} 3 \\ 4 \\ 5 \end{bmatrix}$, $T \begin{bmatrix} 0 \\ 1 \\ 0 \\ 0 \end{bmatrix} = \begin{bmatrix} 1 \\ 1 \\ 0 \end{bmatrix}$, $T \begin{bmatrix} 0 \\ 0 \\ 1 \\ 0 \end{bmatrix} = \begin{bmatrix} 0 \\ 3 \\ 6 \end{bmatrix}$, $T \begin{bmatrix} 0 \\ 0 \\ 0 \\ 1 \end{bmatrix} = \begin{bmatrix} 9 \\ 4 \\ 1 \end{bmatrix}$

6. Write down the matrix A that corresponds to each of the linear transformations T in Exercise 5.

7. Determine a basis for the null space and the nullity of each of the linear transformations in Exercise 5.

8. Without first determining the range, write down the rank of each of the linear transformations in Exercise 5 by using Exercise 7 and the proposition on dimension. Then find the range.

9. Verify the results in Exercise 8 by calculating the rank of each of the matrices in Exercise 5 (see Chapter 3).

10. In Exercise 5e, calculate

$$T\begin{bmatrix} 6 \\ 9 \\ -5 \\ 1 \end{bmatrix}.$$

11. (a) Suppose that T is a linear transformation from $\mathbf{R}^3$ into itself and that the null space of T consists of only the $\mathbf{0}$ vector in $\mathbf{R}^3$. Prove that the inverse transformation T^{-1} exists. (One often refers to the null space as being trivial when it consists of only the $\mathbf{0}$ vector.)

    (b) If A is the matrix representation of T relative to the standard basis, what is the matrix representation of T^{-1}?

    (c) Let

$$T\begin{bmatrix} x \\ y \\ z \end{bmatrix} = \begin{bmatrix} 1 & 3 & 0 \\ 2 & 0 & 3 \\ 0 & 1 & 2 \end{bmatrix}\begin{bmatrix} x \\ y \\ z \end{bmatrix}$$

    be a linear transformation from $\mathbf{R}$ into itself. Show that the null space of T consists of only the $\mathbf{0}$ vector (i.e, the null space is trivial). Find T^{-1} and its matrix representation.

12. Prove that a linear transformation from $\mathbf{R}^3$ into itself with a trivial null space has range $\mathbf{R}^3$.

13. Generalize Exercise 12 to $\mathbf{R}^n$.

14. Suppose that T is a linear transformation from $\mathbf{R}^n$ into $\mathbf{R}^m$ with a trivial null space. What can you say about the range of T?

15. Suppose that T is a linear transformation from $\mathbf{R}^n$ into itself. Suppose that the null space of T is all of $\mathbf{R}^n$. What can you say about the range of T? What is T?

16. Suppose that T is a linear transformation from $\mathbf{R}^n$ into $\mathbf{R}^m$. Suppose that the null space of T is all of $\mathbf{R}^n$. What can you say about the range of T?

A Topic for Self-Study

A METHOD FOR FINDING A BASIS OPERATOR FOR THE RANGE OF A

This method is a variation of the Gram–Schmidt process. It produces an orthogonal basis $\mathbf{u}_1 \cdots \mathbf{u}_r$ for the range of A. This basis forms a basis operator $U = [\mathbf{u}_1 \cdots \mathbf{u}_r]$ for the range of A.

The following operations will be used. They do not alter the range of A.

1. Scale a column of A.

2. Delete a zero column.

3. Delete a column that is a multiple of another.

4. Orthogonalize one column to another.

The method of orthogonalization of a vector $\mathbf{v}$ to a vector $\mathbf{u}$ is the following:

1. Compute $\mathbf{w} = (\mathbf{u}^T\mathbf{u})\mathbf{v} - (\mathbf{u}^T\mathbf{v})\mathbf{u}$.

2. Replace $\mathbf{v}$ by $s\mathbf{w}$, where s is a convenient scale factor.

We illustrate the procedure by finding a basis $\mathbf{u}_1, \mathbf{u}_2, \ldots$ for the range of the matrix

$$A = \begin{bmatrix} 100 & 50 & 30 & 2 & 182 & 12 \\ 200 & -25 & 35 & 0 & 210 & 14 \\ 300 & 0 & 60 & 0 & 360 & 24 \\ 400 & 75 & 15 & 2 & 92 & 6 \end{bmatrix}.$$

Step 1: Scaling and a choice of $\mathbf{u}_1$. We begin by scaling A. We obtain the matrix

$$A_1 = \begin{bmatrix} 1 & 2 & 6 & 1 & 91 \\ 2 & -1 & 7 & 0 & 105 \\ 3 & 0 & 12 & 0 & 180 \\ 4 & 3 & 3 & 1 & 46 \end{bmatrix}.$$

Notice that we deleted the last column because it was identical to column 3.

We now choose $\mathbf{u}_1$ to be the vector

$$\mathbf{u}_1 = \begin{bmatrix} 1 \\ 0 \\ 0 \\ 1 \end{bmatrix}.$$

We delete $\mathbf{u}_1$ and orthogonalize the remaining columns to $\mathbf{u}_1$. We obtain, after rescaling, the matrix

$$A_2 = \begin{bmatrix} 1 & -1 & 3 & 9 \\ 4 & -2 & 14 & 42 \\ -6 & 0 & 24 & 72 \\ -1 & 1 & -3 & -9 \end{bmatrix}.$$

We illustrate the computation used by orthogonalizing column 5 of A_1 to $\mathbf{u}_1$ as follows. We have

$$\mathbf{v} = \begin{bmatrix} 91 \\ 105 \\ 180 \\ 46 \end{bmatrix}, \qquad \mathbf{u} = \mathbf{u}_1 = \begin{bmatrix} 1 \\ 0 \\ 0 \\ 1 \end{bmatrix},$$

$$\mathbf{u}^T\mathbf{u} = 2, \qquad \mathbf{u}^T\mathbf{v} = 137,$$

$$\mathbf{w} = (\mathbf{u}^T\mathbf{u})\mathbf{v} - (\mathbf{u}^T\mathbf{v})\mathbf{u} = 2\begin{bmatrix} 91 \\ 105 \\ 180 \\ 46 \end{bmatrix} - 137\begin{bmatrix} 1 \\ 0 \\ 0 \\ 1 \end{bmatrix} = \begin{bmatrix} 45 \\ 210 \\ 360 \\ -45 \end{bmatrix} = 5\begin{bmatrix} 9 \\ 42 \\ 72 \\ -9 \end{bmatrix}.$$

We replace $\mathbf{v}$ by $\mathbf{u}$ with $r = 1/5$. The other columns of A_2 are computed in the same manner.

Step 2: Choice of $\mathbf{u}_2$. We now select the second column of A_2 to be the vector $\mathbf{u}_2$. We have

$$\mathbf{u}_2 = \begin{bmatrix} -1 \\ -2 \\ 0 \\ 1 \end{bmatrix}.$$

We now delete $\mathbf{u}_2$ from A_2 and orthogonalize the remaining columns of A_2 to $\mathbf{u}_2$. We obtain the matrix A_3 shown below. Before exhibiting A_3 let us illustrate the computation by orthogonalizing the last column of A_2 to $\mathbf{u} = \mathbf{u}_2$. Denoting the last column by $\mathbf{v}$, we have

$$\mathbf{u}^T\mathbf{u} = 6, \qquad \mathbf{u}^T\mathbf{v} = -102,$$

$$\mathbf{w} = (\mathbf{u}^T\mathbf{u})\mathbf{v} - (\mathbf{u}^T\mathbf{v})\mathbf{u} = 6\begin{bmatrix} 9 \\ 42 \\ 72 \\ -9 \end{bmatrix} + 102\begin{bmatrix} -1 \\ -2 \\ 0 \\ 1 \end{bmatrix} = \begin{bmatrix} -48 \\ 48 \\ 432 \\ 48 \end{bmatrix} = 48\begin{bmatrix} -1 \\ 1 \\ 9 \\ 1 \end{bmatrix}.$$

Scaling $\mathbf{w}$ we obtain column 3 of the matrix

$$A_3 = \begin{bmatrix} -1 & -1 & -1 \\ 1 & 1 & 1 \\ 9 & 9 & 9 \\ 1 & 1 & 1 \end{bmatrix}.$$

The other columns of A_3 were obtained in the same manner. Deleting the last two columns, we get

$$\mathbf{u}_3 = \begin{bmatrix} -1 \\ 1 \\ 9 \\ 1 \end{bmatrix}.$$

The matrix

$$U = [\mathbf{u}_1 \quad \mathbf{u}_2 \quad \mathbf{u}_3] = \begin{bmatrix} 1 & -1 & -1 \\ 0 & -2 & 1 \\ 0 & 0 & 9 \\ 1 & 1 & 1 \end{bmatrix}$$

is a basis operator for the range of A.

6.3   APPLICATIONS TO MATRICES AND LINEAR EQUATIONS

In the preceding section we found that a linear operator T mapping a linear space $\mathbf{U}$ into a linear space $\mathbf{V}$ has associated with it a linear subspace $\mathbf{R}(T)$ of V, called its *range* and a linear subspace $\mathbf{N}(T)$ of U, called its *null space*. Here we denote the range of T by $\mathbf{R}(T)$ and the null space of T by $\mathbf{N}(T)$ to remind the reader that each of these subspaces depends on the choice of the linear transformation T. The rank $r(T)$ of T is the dimension of $\mathbf{R}(T)$ and its nullity $q(T)$ is the dimension of $\mathbf{N}(T)$.

In this section we study the case in which $\mathbf{U} = \mathbf{R}^n$ and $\mathbf{V} = \mathbf{R}^m$ and T is a fixed $m \times n$-matrix A. Some of the results we describe below are a restatement of results given earlier. The matrix A has associated with it various matrices such as A^T, $A^T A$, AA^T, and so on. Each of these has associated with it a range and a null space. We shall be concerned with the connections between these subspaces. Ranges will be denoted by the symbols $\mathbf{R}(A)$, $\mathbf{R}(A^T)$, ... Null spaces will be denoted by $\mathbf{N}(A)$, $\mathbf{N}(A^T)$, ... The symbols $r(A)$, $r(A^T)$, ... denote ranks, and $q(A)$, $q(A^T)$, ... denote nullities. In the preceding section and in Section 3.4, we established the following relations.

> **1.** $r(A) = r(A^T)$, $q(A) + r(A) = n$, $q(A^T) + r(A^T) = m$.

Recall that the orthogonal complement $\mathbf{S}^\perp$ of a set $\mathbf{S}$ in $\mathbf{R}^n$ is the set of all vectors $\mathbf{u}$ in $\mathbf{R}^n$ which are orthogonal to every vector $\mathbf{w}$ in $\mathbf{S}$. If $\mathbf{u}$ and $\mathbf{v}$ are orthogonal to $\mathbf{w}$, so also is every linear combination $a\mathbf{u} + b\mathbf{v}$. It follows that $\mathbf{S}^\perp$ is a linear subspace of $\mathbf{R}^n$. The orthogonal complement of $\mathbf{S}^\perp$ is the linear space generated by $\mathbf{S}$ and is equal to $\mathbf{S}$ when $\mathbf{S}$ is a linear space. When $\mathbf{S}$ is a linear space, $\mathbf{R}^n$ is the direct sum

$$\mathbf{R}^n = \mathbf{S} + \mathbf{S}^\perp$$

of $\mathbf{S}$ and its orthogonal complement. As a consequence, we have

> **2.** The sum of the dimensions of **S** and $\mathbf{S}^{\perp}$ equals n.

We have the following relations.

> **3.** $N(A)$ and $\mathbf{R}(A^T)$ are orthogonal complements in $\mathbf{R}^n$.
>
> **4.** $\mathbf{R}(A)$ and $N(A^T)$ are orthogonal complements in $\mathbf{R}^m$.
>
> **5.** $\mathbf{R}^n = N(A) + \mathbf{R}(A^T)$, a direct sum.
>
> **6.** $\mathbf{R}^m = \mathbf{R}(A) + N(A^T)$, a direct sum.

In order to prove relations 3 and 5, observe that the range $\mathbf{R}(A^T)$ of A^T is the column space of A^T and so is the row space of A. The equation $Ax = 0$ tells us that every vector $\mathbf{x}$ in the null space $N(A)$ of A is orthogonal to the rows of A and hence to every vector in the row space $\mathbf{R}(A^T)$ of A. It follows that $N(A)$ and $\mathbf{R}(A^T)$ are orthogonal complements of each other. Consequently, relations 3 and 5 hold. Similarly, interchanging the roles of A and A^T yields the relations 4 and 6.

We now turn to the symmetric operators A^TA and AA^T associated with A. The operator A^TA maps $\mathbf{R}^n$ into itself; similarly, the operator AA^T maps $\mathbf{R}^m$ into itself.

> **7.** $N(A^TA) = N(A)$, $\mathbf{R}(A^TA) = \mathbf{R}(A^T) = N(A)^{\perp}$.
>
> **8.** $N(AA^T) = N(A^T)$, $\mathbf{R}(AA^T) = \mathbf{R}(A) = N(A^T)^{\perp}$.
>
> **9.** $\mathbf{R}^n = N(A^TA) + \mathbf{R}(A^TA)$.
>
> **10.** $\mathbf{R}^m = N(AA^T) + \mathbf{R}(AA^T)$.

Clearly, $A^TAx = 0$ when $Ax = 0$. Conversely, when $A^TAx = 0$, the vector $\mathbf{y} = Ax$ has the property that $\mathbf{y}^T\mathbf{y} = \mathbf{x}^TA^TA\mathbf{x} = 0$, so that $\mathbf{y} = Ax = 0$. Consequently, $N(A^TA) = N(A)$. Using this result, we see that properties 7, 8, 9, and 10 follow from properties 3, 4, 5, and 6 applied to A^TA and AA^T. In particular, $\mathbf{R}(A^TA) = \mathbf{R}(A^T)$ because each has as its orthogonal complement in $\mathbf{R}^n$ the set $N(A^TA) = N(A)$.

Looking at the dimensions of these linear subspaces, we obtain the following result.

> **11.** The ranks of the operators A, A^T, A^TA, and AA^T are all equal.
>
> **12.** The nullities of A and A^TA are equal. So also are the nullities of A^T and AA^T.
>
> **13.** If $m = n$, the nullities of A, A^T, A^TA, and AA^T as well as their ranks are equal.

If $m = n$ and $A^T A = A A^T$, the matrix A is said to be *normal*. A normal matrix A maps $\mathbf{R}^n$ into $\mathbf{R}^n$. For a normal matrix A, we have the following special properties.

---

14. If $m = n$ and A is normal, then $A^T A = A A^T$, so that

$$N(A^T A) = N(A A^T) = N(A) = N(A^T)$$

$$R(A^T A) = R(A A^T) = R(A) = R(A^T).$$

A symmetric matrix is normal. So also is an orthogonal matrix and a constant multiple of an orthogonal matrix. For, if $A = cB$, where B is orthogonal, we have

$$A^T A = c^2 B^T B = c^2 I = c^2 B B^T = A A^T.$$

We now turn our attention to the study of a linear equation

$$A\mathbf{x} = \mathbf{h}.$$

Such an equation need not have a solution. This is the case when

$$A = \begin{bmatrix} 1 & -1 & 2 \\ 0 & 0 & 0 \end{bmatrix}, \qquad \mathbf{h} = \begin{bmatrix} 1 \\ 1 \end{bmatrix}.$$

We can determine when a system of linear equations, or equivalently a linear matrix equation $A\mathbf{x} = \mathbf{h}$ has a solution by the following solvability criteria.

SOLVABILITY CRITERIA. The equation

$$A\mathbf{x} = \mathbf{h}$$

is solvable for $\mathbf{x}$ if one of the following conditions is met. If one of these conditions holds, so also do the others.

   1. $\mathbf{h}$ is in the range $R(A)$ of A.
   2. A and $M = [A \quad \mathbf{h}]$ have the same range.
   3. A and $M = [A \quad \mathbf{h}]$ have the same rank.
   4. The reduced row echelon forms for A and $M = [A \quad \mathbf{h}]$ have the same number of nonzero rows.
   5. $\mathbf{h}$ is orthogonal to the null space $N(A^T)$ of A^T.
   6. $\mathbf{h}^T \mathbf{y} = 0$ whenever $A^T \mathbf{y} = \mathbf{0}$.

Condition 1 states that there is a vector $\mathbf{x}$ such that $\mathbf{h} = A\mathbf{x}$. Statement 2 is an alternative way of saying the $\mathbf{h}$ is in the range of A. Condition 3 holds if and only if **h** is in the range **R**(*A*) of *A*. Condition 4 is a restatement of condition 3 because the number of nonzero rows in the reduced row echelon form of a matrix is its rank. Condition 5 states that **h** is in $N(A)^\perp = R(A)$. Finally, condition 6 is an alternative way of stating condition 5.

Condition 4 is perhaps the simplest condition to apply. It has the advantage that finding the reduced row echelon matrix of the augmented matrix $M = [A \quad \mathbf{h}]$ is an effective first step in solving the equation $A\mathbf{x} = \mathbf{h}$ when the equation is solvable. So the procedure of finding the reduced row echelon matrix for M can be used to find a solution of $A\mathbf{x} = \mathbf{h}$, if it exists, and to determine the nonsolvability of $A\mathbf{x} = \mathbf{h}$ when it is not solvable. The technique used is described in Section 2.4.

■ **EXAMPLE 1** _____

Consider the system of equations encountered at the beginning of Section 2.4:

$$x - 2y + z = 1$$
$$2x - 4y - 3z = -8.$$

In matrix form, $A\mathbf{x} = \mathbf{h}$, where

$$A = \begin{bmatrix} 1 & -2 & 1 \\ 2 & -4 & -3 \end{bmatrix}, \qquad \mathbf{x} = \begin{bmatrix} x \\ y \\ z \end{bmatrix}, \qquad \mathbf{h} = \begin{bmatrix} 1 \\ -8 \end{bmatrix}.$$

We saw that the row reduced echelon form for the augmented matrix $[A \mid \mathbf{h}]$ was

$$[F \mid \mathbf{k}] = \begin{bmatrix} 1 & -2 & 0 & | & -1 \\ 0 & 0 & 1 & | & 2 \end{bmatrix}.$$

The number of nonzero rows of $[F \mid \mathbf{k}]$ is 2 and the number of nonzero rows of F is 2. By condition 4 the system has a solution.

In this simple example, it is easy to verify the other conditions. To verify condition 1, we see that **h** is in the range **R**(*A*) of *A* since

$$\mathbf{h} = \begin{bmatrix} 1 \\ -8 \end{bmatrix} = -1 \begin{bmatrix} 1 \\ 2 \end{bmatrix} + 0 \begin{bmatrix} -2 \\ -4 \end{bmatrix} + 2 \begin{bmatrix} 1 \\ -3 \end{bmatrix},$$

(i.e., **h** is a linear combination of the columns of *A*).

Finally, to verify condition 5, we show that **h** is orthogonal to the null space $N(A^T)$ of A^T. To do so, we first compute the null space of A^T. We have

$$A^T\mathbf{x} = \begin{bmatrix} 1 & 2 \\ -2 & -4 \\ 1 & -3 \end{bmatrix} \begin{bmatrix} x_1 \\ x_2 \end{bmatrix} = \begin{bmatrix} 0 \\ 0 \\ 0 \end{bmatrix},$$

that is,

$$x_1 + 2x_2 = 0$$
$$-2x_1 - 4x_2 = 0$$
$$x_1 - 3x_2 = 0.$$

A simple calculation yields $x_1 = x_2 = 0$.

The null space $N(A^T)$ consists of the **0** vector alone. It follows that

$$\mathbf{h} = \begin{bmatrix} 1 \\ -8 \end{bmatrix}$$

is orthogonal to $N(A^T)$.

■ **EXAMPLE 2** _____

Consider the system of equations encountered in Section 2.4, Example 4:

$$A\mathbf{x} = \begin{bmatrix} 1 & 3 & 0 & -2 \\ 0 & 0 & 1 & 0 \\ 3 & 9 & -1 & -6 \end{bmatrix} \begin{bmatrix} x \\ y \\ z \\ w \end{bmatrix} = \begin{bmatrix} 5 \\ 3 \\ 13 \end{bmatrix} = \mathbf{h}.$$

We saw that the row reduced echelon form for the augmented matrix $[A \mid \mathbf{h}]$ was

$$[F \mid \mathbf{k}] = \begin{bmatrix} 1 & 3 & 0 & -2 & \mid & 0 \\ 0 & 0 & 1 & 0 & \mid & 0 \\ 0 & 0 & 0 & 0 & \mid & 1 \end{bmatrix}.$$

The number of nonzero rows of the augmented matrix $[F \mid \mathbf{k}]$ is 3 and the number of nonzero rows of F is 2. The system has no solution.

The vector $\mathbf{h} = \begin{bmatrix} 5 \\ 3 \\ 13 \end{bmatrix}$ cannot be written as a linear combination of the columns of A, as the reader can easily verify, and hence $\mathbf{h}$ is not in the range $R(A)$ of A.

Furthermore, $\mathbf{h} = \begin{bmatrix} 5 \\ 3 \\ 13 \end{bmatrix}$ is not orthogonal to the null space $N(A^T)$ of A^T, for calculating $N(A^T)$, we have

$$\begin{bmatrix} 1 & 0 & 3 \\ 3 & 0 & 9 \\ 0 & 1 & -1 \\ -2 & 0 & -6 \end{bmatrix} \begin{bmatrix} x_1 \\ x_2 \\ x_3 \end{bmatrix} = \begin{bmatrix} 0 \\ 0 \\ 0 \\ 0 \end{bmatrix}.$$

We see that

$$x_1 \quad\ + 3x_3 = 0$$

$$3x_1 \quad\ + 9x_3 = 0$$

$$x_2 -\ x_3 = 0$$

$$-2x_1 \quad\ - 6x_3 = 0.$$

It follows that

$$x_1 = -3x_3$$

$$x_2 = \quad x_3.$$

Letting $x_3 = a$, we get

$$x_1 = -3a, \qquad x_2 = a, \qquad x_3 = a.$$

The null space $\mathbf{N}(A^T)$ of A^T is generated by the vector

$$\mathbf{x} = \begin{bmatrix} -3 \\ 1 \\ 1 \end{bmatrix}$$

and is one-dimensional. Since

$$\mathbf{h}^T\mathbf{x} = \begin{bmatrix} 5 & 3 & 13 \end{bmatrix} \begin{bmatrix} -3 \\ 1 \\ 1 \end{bmatrix} = -15 + 3 + 13 = 1 \neq 0,$$

we see that $\mathbf{h}$ is not orthogonal to $\mathbf{N}(A^T)$.

Finally, we restate and prove the main result concerning the solvability of $A\mathbf{x} = \mathbf{h}$ when this equation has at least one solution $\mathbf{x}_0$. Here, as before, A is an $m \times n$-matrix.

If $\mathbf{x}_0$ is a solution of $A\mathbf{x} = \mathbf{h}$, then:

1. If $r(A) = n$, $\mathbf{x}_0$ is the only solution.
2. If $r(A) < n$, every solution $\mathbf{x}$ of $A\mathbf{x} = \mathbf{h}$ is expressible in the form

$$\mathbf{x} = \mathbf{x}_0 + a_1\mathbf{z}_1 + \cdots + a_q\mathbf{z}_q,$$

where q is the nullity of A and $\mathbf{z}_1, \ldots, \mathbf{z}_q$ are q linearly independent solutions of $A\mathbf{z} = \mathbf{0}$.

If $A\mathbf{x} = \mathbf{h}$ and $A\mathbf{x}_0 = \mathbf{h}$, then, by subtraction, $\mathbf{z} = \mathbf{x} - \mathbf{x}_0$ is a solution of $A\mathbf{z} = \mathbf{0}$. When $r(A) = n$, then $q = 0$, so that $\mathbf{z} = \mathbf{0}$ and $\mathbf{x} = \mathbf{x}_0$. However, when $r(A) < n$, we have $q > 0$. In this case $\mathbf{z} = \mathbf{x} - \mathbf{x}_0$ is in the null space of A and so is a linear combination of vectors $\mathbf{z}_1, \ldots, \mathbf{z}_q$ forming a basis of this null space.

Exercises

1. Consider the matrix

$$A = \begin{bmatrix} 1 & 1 & 0 \\ 0 & 1 & 1 \end{bmatrix}.$$

   (a) Show that the rank of A is 2.

   (b) Conclude that the equation $A\mathbf{x} = \mathbf{h}$ is solvable for all choices of $\mathbf{h}$.

   (c) What is the range of A?

2. Show that if the rank of an $m \times n$-matrix A is m, then $A\mathbf{x} = \mathbf{h}$ is solvable for all choices of $\mathbf{h}$. What is the range of A?

3. In each of the following cases, determine the solvability of the equations $A\mathbf{x} = \mathbf{h}_1$ and $A\mathbf{x} = \mathbf{h}_2$.

   (a) $A = \begin{bmatrix} 1 & 0 \\ 1 & 0 \\ 0 & 1 \end{bmatrix}$, $\quad \mathbf{h}_1 = \begin{bmatrix} 1 \\ 2 \\ 1 \end{bmatrix}$, $\quad \mathbf{h}_2 = \begin{bmatrix} 1 \\ 0 \\ -1 \end{bmatrix}$

   (b) $A = \begin{bmatrix} 1 & 1 & 0 & 0 \\ 0 & 1 & 0 & 1 \\ 1 & 0 & 0 & -1 \end{bmatrix}$, $\quad \mathbf{h}_1 = \begin{bmatrix} 3 \\ 2 \\ 1 \end{bmatrix}$, $\quad \mathbf{h}_2 = \begin{bmatrix} 2 \\ 2 \\ 5 \end{bmatrix}$

   (c) $A = \begin{bmatrix} 1 & 0 & 1 \\ 1 & 1 & -1 \\ 1 & -1 & 1 \\ 1 & 0 & -1 \end{bmatrix}$, $\quad \mathbf{h}_1 = \begin{bmatrix} 1 \\ -1 \\ -1 \\ 1 \end{bmatrix}$, $\quad \mathbf{h}_2 = \begin{bmatrix} 0 \\ 1 \\ 1 \\ 2 \end{bmatrix}$

4. Consider the matrices

$$A = \begin{bmatrix} 2 & 2 & 1 & 7 \\ 1 & 1 & 2 & 8 \\ 1 & 1 & 0 & 2 \end{bmatrix}, \quad F = \begin{bmatrix} 1 & 1 & 0 & 2 \\ 0 & 0 & 1 & 3 \\ 0 & 0 & 0 & 0 \end{bmatrix}.$$

   In Section 2.4 we saw that the matrix F is the reduced row echelon matrix for the matrix A.

   (a) Conclude that the rank of A is 2 and that the nullity of A is 2.

   (b) Observe that columns 1 and 3 of F are the first two columns of the 3×3-identity matrix I_3. Conclude that columns 1 and 3 of A form a basis for the range $\mathbf{R}$ of A. These columns determine the matrix

$$R = \begin{bmatrix} 2 & 1 \\ 1 & 2 \\ 1 & 0 \end{bmatrix}$$

   as a basis operator for the range of $\mathbf{R}$.

   (c) Why are the transposes of the first two rows of F a basis for the range of A^T?

(d) By solving $F\mathbf{x} = \mathbf{0}$, show that the columns of

$$N = \begin{bmatrix} 1 & -2 \\ -1 & 0 \\ 0 & -3 \\ 0 & 1 \end{bmatrix}$$

form a basis for the null space of A.

(e) Solving $R^T\mathbf{y} = \mathbf{0}$, show that $\mathbf{y} = (-2, 1, 3)$ generates the null space of A^T.

5. Let A be an $m \times n$-matrix of rank r. Let F be its reduced row echelon form.

   (a) Show that the first r rows of F generate the row space of A (the range of A^T).

   (b) For $i \le r$, let (i, k_i) be the leading entry 1 in the ith row of F. Verify that the k_ith column of F is the ith column of the $m \times n$-identity matrix I_m.

   (c) Let $\mathbf{v}_i$ be the k_ith column of A. Show that the column vectors $\mathbf{v}_1, \mathbf{v}_2, \ldots, \mathbf{v}_r$ form a basis for the range $\mathbf{R}$ of A. They are the columns of a basis operator $V = [\mathbf{v}_1, \mathbf{v}_2, \ldots, \mathbf{v}_r]$ for $\mathbf{R}$.

   (d) Show, as explained in Section 2.4, that a basis for the null space $\mathbf{N}$ of A can be obtained by finding r linearly independent solutions of $F\mathbf{z} = \mathbf{0}$.

   (e) Show that a basis for the null space of A^T can be found by solving $V^T\mathbf{y} = \mathbf{0}$.

6. In Section 2.4 it was shown that the matrix

$$F = \begin{bmatrix} 1 & -1 & 2 & 0 & 2 & 0 \\ 0 & 0 & 0 & 1 & -1/2 & 0 \\ 0 & 0 & 0 & 0 & 0 & 1 \\ 0 & 0 & 0 & 0 & 0 & 0 \end{bmatrix}$$

is the reduced row echelon form of the matrix

$$A = \begin{bmatrix} -1 & 1 & -2 & 0 & -2 & 0 \\ -2 & 2 & -4 & 4 & -6 & -2 \\ 2 & -2 & 4 & -3 & 5 & 1 \\ 2 & -2 & 4 & 4 & 2 & 0 \end{bmatrix}.$$

It was also shown that the columns of the matrix

$$N = \begin{bmatrix} -2 & -2 & 1 \\ 0 & 0 & 1 \\ 0 & 1 & 0 \\ 1/2 & 0 & 0 \\ 1 & 0 & 0 \\ 0 & 0 & 0 \end{bmatrix}$$

form a basis for the null space of A.

(a) Use the results given in Exercise 5 to show that the columns of the submatrix

$$V = \begin{bmatrix} -1 & 0 & 0 \\ -2 & 4 & -2 \\ 2 & -3 & 1 \\ 2 & 4 & 0 \end{bmatrix}$$

form a basis for the range $\mathbf{R}$ of A.

(b) Looking at F, find a basis for the range of A^T. Verify that this basis determines the orthogonal complement of the null space of A.

(c) Find the null space of A^T by solving the equation $V^T\mathbf{y} = \mathbf{0}$.

7. Consider the equation $A\mathbf{x} = \mathbf{h}$, where A is an $m \times n$-matrix of rank r.

(a) Show that the equation

$$A^TA\mathbf{x} = A^T\mathbf{h}$$

is always solvable, no matter how $\mathbf{h}$ is chosen (*Hint:* Use the solvability criteria.) A solution $\mathbf{x}$ of $A^TA\mathbf{x} = A^T\mathbf{h}$ is called a *least square solution* of $A\mathbf{x} = \mathbf{h}$. These types of solutions are discussed in Section 6.6.

(b) Verify that if A has rank $r = n$, a least square solution of $A\mathbf{x} = \mathbf{h}$ is given by the formula

$$\mathbf{x}_0 = (A^TA)^{-1}A^T\mathbf{h}.$$

(c) Show that if $\mathbf{x}_0$ is a least square solution of $A\mathbf{x} = \mathbf{h}$ and $\mathbf{z}$ is a solution of $A\mathbf{x} = \mathbf{0}$, then $\mathbf{x} = \mathbf{x}_0 + \mathbf{z}$ is a least square solution of $A\mathbf{x} = \mathbf{h}$.

8. Apply the results given in Exercise 7 to find the least square solutions of the equations $A\mathbf{x} = \mathbf{h}$ found in Exercise 3.

9. Verify that the matrix

$$A = \begin{bmatrix} 0 & 0 \\ 1 & 0 \end{bmatrix}$$

has the property that

$$\mathbf{R}(A) = \mathbf{N}(A), \qquad \mathbf{R}(A^T) = \mathbf{N}(A^T).$$

Conclude that $\mathbf{N}(A^T)$ is the orthogonal complement of $\mathbf{N}(A)$. Construct a 4×4-matrix of rank 2 having these properties.

6.4   ORTHOGONAL PROJECTIONS

The purpose of this section is to introduce the concept of an orthogonal projection operator for a linear subspace $\mathbf{L}$ of $\mathbf{R}^n$. We begin with the case $n = 2$. Let $\mathbf{L}$ be a line in $\mathbf{R}^2$ passing through the origin O. In Figure 1 we represent $\mathbf{L}$ by a horizontal line. We view $\mathbf{L}$ to be a one-dimensional linear space comprised of multiples $\mathbf{v}t$ of a fixed vector $\mathbf{v}$, as indicated in the diagram. The line $\mathbf{L}^\perp$ is the orthogonal complement of $\mathbf{L}$. Select any vector $\mathbf{x}$ in $\mathbf{R}^2$ as shown in Figure 1. Drop a perpendicular from $\mathbf{x}$ to a point $\mathbf{y}$ in $\mathbf{L}$. The point $\mathbf{y}$ considered as a vector emanating from the origin is the orthogonal projection of the vector $\mathbf{x}$ on $\mathbf{L}$. Also drop a perpendicular from the point $\mathbf{x}$ to a point $\mathbf{z}$ in $\mathbf{L}^\perp$. The vector $\mathbf{z}$ is the orthogonal projection of the vector $\mathbf{x}$ on $\mathbf{L}^\perp$. We have

$$\mathbf{x} = \mathbf{y} + \mathbf{z}.$$

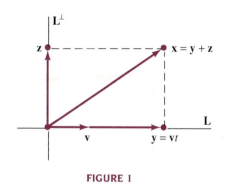

FIGURE 1

We propose to find formulas for **y** and **z** in terms of the vectors **x** and **v**. We first note that

$$\mathbf{y} = \mathbf{v}t, \qquad \mathbf{z} = \mathbf{x} - \mathbf{y}$$

for a suitable number t. Because **z** is orthogonal to **v** we have $\mathbf{v}^T\mathbf{z} = 0$. Consequently,

$$\mathbf{v}^T\mathbf{z} = \mathbf{v}^T\mathbf{x} - \mathbf{v}^T\mathbf{v}t = \mathbf{v}^T\mathbf{x} - kt = 0, \qquad k = \mathbf{v}^T\mathbf{v}.$$

Solving for t we find that

$$t = k^{-1}\mathbf{v}^T\mathbf{x}, \qquad \mathbf{y} = \mathbf{v}k^{-1}\mathbf{v}^T\mathbf{x} = k^{-1}\mathbf{v}\mathbf{v}^T\mathbf{x} = E\mathbf{x},$$

where E is the matrix

$$E = k^{-1}\mathbf{v}\mathbf{v}^T = \frac{\mathbf{v}\mathbf{v}^T}{\mathbf{v}^T\mathbf{v}}.$$

We call E the orthogonal projection operator for **L**. We call the complement

$$F = I - E$$

of E the orthogonal projection operator for $\mathbf{L}^\perp$. We do so because

$$F\mathbf{x} = (I - E)\mathbf{x} = \mathbf{x} - E\mathbf{x} = \mathbf{x} - \mathbf{y} = \mathbf{z}.$$

Hence

$$\mathbf{x} = \mathbf{y} + \mathbf{z} = E\mathbf{x} + F\mathbf{x}.$$

Let us look at the following.

■ EXAMPLE 1

Let **L** be the line generated by the vector $\mathbf{v} = (1, 1)$. Its orthogonal complement $\mathbf{L}^\perp$ is generated by $\mathbf{w} = (-1, 1)$. We have

$$k = \mathbf{v}^T\mathbf{v} = \begin{bmatrix} 1 & 1 \end{bmatrix}\begin{bmatrix} 1 \\ 1 \end{bmatrix} = 2, \qquad E = k^{-1}\mathbf{v}\mathbf{v}^T = 2^{-1}\begin{bmatrix} 1 \\ 1 \end{bmatrix}\begin{bmatrix} 1 & 1 \end{bmatrix},$$

so that

$$E = \frac{1}{2}\begin{bmatrix} 1 & 1 \\ 1 & 1 \end{bmatrix}, \qquad F = I - E = \frac{1}{2}\begin{bmatrix} 1 & -1 \\ -1 & 1 \end{bmatrix}.$$

Observe that the matrices E and F are symmetric. Moreover,

$$E^2 = E, \qquad F^2 = F, \qquad FE = EF = 0.$$

When $\mathbf{x} = (a, b)$,

$$\mathbf{x} = \begin{bmatrix} a \\ b \end{bmatrix}, \quad \mathbf{y} = E\mathbf{x} = \frac{1}{2}\begin{bmatrix} a + b \\ a + b \end{bmatrix} \text{ is in } \mathbf{L}, \quad \mathbf{z} = F\mathbf{x} = \frac{1}{2}\begin{bmatrix} a - b \\ b - a \end{bmatrix} \text{ is in } \mathbf{L}^{\perp}.$$

We have $\mathbf{x} = \mathbf{y} + \mathbf{z} = E\mathbf{x} + F\mathbf{x}$.

The discussion made above holds in the n-dimensional space $\mathbf{R}^n$ for a line $\mathbf{L}$ through the origin O. Such a line $\mathbf{L}$ is a one-dimensional linear subspace comprised of all multiples $\mathbf{v}t$ of a fixed vector $\mathbf{v}$. Its orthogonal complement $\mathbf{L}^{\perp}$ consists of all vectors $\mathbf{z}$ orthogonal to $\mathbf{v}$. Hence we have $\mathbf{v}^T\mathbf{z} = 0$ for all vectors $\mathbf{z}$ in $\mathbf{L}^{\perp}$. Figure 1 remains valid with the vertical line chosen to be the line in $\mathbf{L}^{\perp}$ generated by the orthogonal projection $\mathbf{z}$ of $\mathbf{x}$ on $\mathbf{L}^{\perp}$.

This gives the following:

PROPOSITION 1. Let $\mathbf{L}$ be the line in $\mathbf{R}^n$ generated by a vector $\mathbf{v}$. Let $\mathbf{L}^{\perp}$ be its orthogonal complement. Let E and F be the matrices defined by the relations

$$k = \mathbf{v}^T\mathbf{v}, \qquad E = k^{-1}\mathbf{v}\mathbf{v}^T, \qquad F = I - E.$$

The matrices E and F have the property that for each vector $\mathbf{x}$, the vector $\mathbf{y} = E\mathbf{x}$ is the orthogonal projection of $\mathbf{x}$ on $\mathbf{L}$ and the vector $\mathbf{z} = F\mathbf{x}$ is the orthogonal projection of $\mathbf{x}$ on $\mathbf{L}^{\perp}$. We have

$$\mathbf{x} = \mathbf{y} + \mathbf{z} = E\mathbf{x} + F\mathbf{x}.$$

In view of this result, E is called the *orthogonal projection operator* for $\mathbf{L}$ and F is called the *orthogonal projection operator* for $\mathbf{L}^{\perp}$.

Further properties of E and F are given below in Proposition 3.

Let us look at the following example in $\mathbf{R}^3$.

■ **EXAMPLE 2** ─────────────────────────────────────

Let $\mathbf{L}$ be the line generated by the vector $\mathbf{v} = (1, -1, 1)$. We have $k = \mathbf{v}^T\mathbf{v} = 3$,

$$E = k^{-1}\mathbf{v}\mathbf{v}^T = \frac{1}{3}\begin{bmatrix} 1 & -1 & 1 \\ -1 & 1 & -1 \\ 1 & -1 & 1 \end{bmatrix}, \qquad F = I - E = \frac{1}{3}\begin{bmatrix} 2 & 1 & -1 \\ 1 & 2 & 1 \\ -1 & 1 & 2 \end{bmatrix}.$$

The matrices E and F are symmetric and satisfy the relations

$$E^2 = E, \qquad F^2 = F, \qquad FE = EF = 0,$$

as one readily verifies. For $\mathbf{x} = (a, b, c)$,

$$\mathbf{y} = E\mathbf{x} = \frac{1}{3}\begin{bmatrix} a - b + c \\ -a + b - c \\ a - b + c \end{bmatrix}, \qquad \mathbf{z} = F\mathbf{x} = \frac{1}{3}\begin{bmatrix} 2a + b - c \\ a + 2b + c \\ -a + b + 2c \end{bmatrix}.$$

The vector $\mathbf{y} = E\mathbf{x}$ is the orthogonal projection of $\mathbf{x}$ on $\mathbf{L}$. The vector $\mathbf{z} = F\mathbf{x}$ is the orthogonal projection of $\mathbf{x}$ on the orthogonal complement $\mathbf{L}^{\perp}$ of $\mathbf{L}$. We have

$$\mathbf{x} = \mathbf{y} + \mathbf{z} = E\mathbf{x} + F\mathbf{x}.$$

Figure 1 is still valid for the case in which $\mathbf{L}$ is a k-dimensional linear subspace of $\mathbf{R}^n$ with $k < n$. In this case the horizontal line in Figure 1 is the line generated by the orthogonal projection $\mathbf{y}$ of $\mathbf{x}$ on $\mathbf{L}$. The vertical line is the line generated by the orthogonal projection $\mathbf{z}$ of $\mathbf{x}$ on $\mathbf{L}^{\perp}$. In this case the formulas for $\mathbf{y}$ and $\mathbf{z}$ are

$$\mathbf{y} = V\mathbf{t}, \qquad \mathbf{z} = \mathbf{x} - \mathbf{y}$$

where V is an $n \times k$-matrix whose columns form a basis for $\mathbf{L}$. The matrix V is of rank k. Because $\mathbf{z}$ is orthogonal to $\mathbf{L}$, it is orthogonal to the columns of V, the generators of $\mathbf{L}$. Hence $V^T\mathbf{z} = 0$ so that

$$V^T\mathbf{z} = V^T\mathbf{x} - V^T\mathbf{y} = V^T\mathbf{x} - V^TV\mathbf{t} = V^T\mathbf{x} - K\mathbf{t} = 0, \qquad K = V^TV.$$

The $k \times k$-matrix K is of rank k because V is of rank k. Hence K is nonsingular. Solving for $\mathbf{t}$, we find that

$$\mathbf{t} = K^{-1}V^T\mathbf{x}, \qquad \mathbf{y} = V\mathbf{t} = VK^{-1}V^T\mathbf{x}.$$

Setting

$$E = VK^{-1}V^T, \qquad F = I - E \qquad (K = V^TV)$$

we see that

$$\mathbf{y} = E\mathbf{x}, \qquad \mathbf{z} = \mathbf{x} - \mathbf{y} = \mathbf{x} - E\mathbf{x} = (I - E)\mathbf{x} = F\mathbf{x}.$$

The vector $\mathbf{y} = E\mathbf{x}$ is the orthogonal projection of $\mathbf{x}$ on $\mathbf{L}$. The vector $\mathbf{z} = F\mathbf{x}$ is the orthogonal projection of $\mathbf{x}$ on the orthogonal complement $\mathbf{L}^{\perp}$ of $\mathbf{L}$. We have

$$\mathbf{x} = \mathbf{y} + \mathbf{z} = E\mathbf{x} + F\mathbf{x}.$$

The matrices E and F are, respectively, the orthogonal projection operators for $\mathbf{L}$ and $\mathbf{L}^{\perp}$.

■ **EXAMPLE 3**

Let **L** be the two-dimensional linear subspace (a plane) in $\mathbf{R}^3$ having as a basis the columns of the matrix

$$V = \begin{bmatrix} 1 & 0 \\ 1 & 1 \\ 0 & 1 \end{bmatrix}.$$

We have

$$K = V^T V = \begin{bmatrix} 2 & 1 \\ 1 & 2 \end{bmatrix}, \qquad K^{-1} = \frac{1}{3}\begin{bmatrix} 2 & -1 \\ -1 & 2 \end{bmatrix}.$$

The orthogonal projection operator $E = VK^{-1}V^T$ for **L** is

$$E = \frac{1}{3}\begin{bmatrix} 1 & 0 \\ 1 & 1 \\ 0 & 1 \end{bmatrix}\begin{bmatrix} 2 & -1 \\ -1 & 2 \end{bmatrix}\begin{bmatrix} 1 & 1 & 0 \\ 0 & 1 & 1 \end{bmatrix} = \frac{1}{3}\begin{bmatrix} 2 & 1 & -1 \\ 1 & 2 & 1 \\ -1 & 1 & 2 \end{bmatrix}.$$

The orthogonal projection operation F for $\mathbf{L}^\perp$ is therefore

$$F = I - E = \frac{1}{3}\begin{bmatrix} 1 & -1 & 1 \\ -1 & 1 & -1 \\ 1 & -1 & 1 \end{bmatrix}.$$

(Explain the relationship of E and F to the matrices E and F obtained in Example 2.) As usual, every vector **x** is expressible in the form

$$\mathbf{x} = E\mathbf{x} + F\mathbf{x}$$

with $E\mathbf{x}$ in **L** and $F\mathbf{x}$ in $\mathbf{L}^\perp$.

The results given above lead us to the following:

PROPOSITION 2. Let **L** be a linear subspace of $\mathbf{R}^n$ of dimension $k < n$. Let $\mathbf{L}^\perp$ be its orthogonal complement. There exists a unique pair of $n \times n$-matrices E and F such that for every vector **x** in $\mathbf{R}^n$

$$\mathbf{y} = E\mathbf{x} \text{ is in } \mathbf{L}, \qquad \mathbf{z} = F\mathbf{x} \text{ is in } \mathbf{L}^\perp, \qquad \mathbf{x} = \mathbf{y} + \mathbf{z}.$$

These matrices can be constructed as follows: Choose an $n \times k$-matrix V whose columns form a basis for **L**. Set

$$K = V^T V, \qquad E = VK^{-1}V^T, \qquad F = I - E.$$

Further properties of the matrices E and F will be given in Proposition 3.

By construction, for every vector **x** in $\mathbf{R}^n$, the vector $\mathbf{y} = E\mathbf{x}$ is the orthogonal projection of **x** on **L**. The matrix E therefore projects $\mathbf{R}^n$ orthogonally onto **L** and so is called the ***orthogonal projection operator*** for **L**.

When **x** is in **L**, it is its own orthogonal projection. Accordingly, the operator E leaves **L** *invariant*. When **x** is in **L**, $\mathbf{z} = \mathbf{0}$ is the orthogonal projection of **x** on $\mathbf{L}^{\perp}$. We say therefore that E *annihilates* $\mathbf{L}^{\perp}$. Similarly, the matrix F projects $\mathbf{R}^n$ orthogonally onto $\mathbf{L}^{\perp}$ and is called the *orthogonal projection operator* for $\mathbf{L}^{\perp}$. The operator F leaves $\mathbf{L}^{\perp}$ invariant and annihilates **L**.

When $\mathbf{L} = \mathbf{R}^n$, our construction gives $E = I$ as its orthogonal projection operator and $F = 0$ as the orthogonal projection operator for the zero space **O**, the orthogonal complement of $\mathbf{R}^n$. Accordingly, we define I to be the orthogonal projection operator for $\mathbf{R}^n$ and the zero matrix 0 to be the orthogonal projection operator for the zero space **O**. The zero space **O** consists only of the vector $\mathbf{x} = \mathbf{0}$.

Returning to Proposition 2, we need to show that the matrices E and F are unique. Suppose that there is a second pair E_1 and F_1 having the properties described in Proposition 2. Then, for every vector **x**,

$$\mathbf{x} = E\mathbf{x} + F\mathbf{x} = E_1\mathbf{x} + F_1\mathbf{x}$$

with $E\mathbf{x}$ and $E_1\mathbf{x}$ in **L** and with $F\mathbf{x}$ and $F_1\mathbf{x}$ in $\mathbf{L}^{\perp}$. This gives us the equation

$$(E - E_1)\mathbf{x} = (F_1 - F)\mathbf{x}$$

whose left member is in **L** and whose right member is in $\mathbf{L}^{\perp}$. This is possible only if each member is **0**. Since **x** is arbitrary, this can happen only if $E_1 = E$ and $F_1 = F$.

PROPOSITION 3. The matrices E, F, and V described in Proposition 2 have the following properties:

(a) E and F are symmetric.
(b) $EV = V$; $FV = 0$.
(c) $I = F + E$, $E^2 = E$, $F^2 = F$; $FE = EF = 0$.
(d) $\mathbf{y} = E\mathbf{y}$ for every **y** in **L**; $\mathbf{z} = F\mathbf{z}$ for every **z** in $\mathbf{L}^{\perp}$.
(e) **y** is in **L** if and only if $F\mathbf{y} = \mathbf{0}$; **z** is in $\mathbf{L}^{\perp}$ if and only if $E\mathbf{z} = \mathbf{0}$.

The formulas for E and F tell us that they are symmetric. Property (b) follows because

$$EV = VK^{-1}V^TV = VK^{-1}K = V, \qquad FV = (I - E)V = V - V = 0.$$

To establish property (c) we observe that $I = F + E$. We make the following computations:

$$FE = FVK^{-1}V^T = 0, \qquad EF = (FE)^T = 0,$$

$$E^2 = (I - F)E = E - FE = E, \qquad F^2 = F(I - E) = F.$$

To establish properties (d) and (e), we use the relation

$$Fy = (I - E)y = y - Ey.$$

If $Fy = 0$, then $y = Ey$ is in **L**. Conversely, if **y** is in **L**, the right member of this equation is in **L** and the left member is in $\mathbf{L}^{\perp}$. This is possible only if these members are zero, that is, only if $Fy = 0$ and $y = Ey$. Similarly, **z** is in $L^{\perp}$ if and only if $Ez = 0$ or equivalently, if and only if $z = Fz$.

COROLLARY. Let E be a symmetric $n \times n$-matrix having the property that $E^2 = E$. Then E is the orthogonal projection for its range **L** in $\mathbf{R}^n$.

The proof of this result is left as an exercise.

Exercises

1. Let **L** be the line generated by the vector **v** for the case in which **v** is

    (a) (1, 2)   (b) (1, 0)   (c) (1, 0, 0)   (d) (1, 1, 1, 1)

    In each case find the orthogonal projection operators E and F for **L** and its orthogonal complement $\mathbf{L}^{\perp}$.

2. Let **L** be the linear subspace generated by columns of V for the case in which V is the matrix

    (a) $\begin{bmatrix} 1 & -1 \\ 0 & 1 \\ 1 & 0 \end{bmatrix}$ (b) $\begin{bmatrix} 0 & 1 \\ 1 & 0 \\ 0 & 1 \end{bmatrix}$ (c) $\begin{bmatrix} 1 & 0 \\ 0 & 1 \\ 1 & 0 \\ 0 & 1 \end{bmatrix}$

    In each case find the orthogonal projection operators E and F for **L** and for $\mathbf{L}^{\perp}$.

3. Show that for the line **L** generated by a unit vector **u**, its orthogonal projection operator E is $E = \mathbf{uu}^T$. Find E for the case in which

    (a) $\mathbf{u} = (3/5, 4/5)$ (b) $\mathbf{u} = (1/13)(3, -4, 12)$

4. Let $\mathbf{u}_1$, $\mathbf{u}_2$, and $\mathbf{u}_3$ be an orthonormal basis for a three-dimensional linear subspace **L** of $\mathbf{R}^n$. Set

    $$E_1 = \mathbf{u}_1\mathbf{u}_1^T, \qquad E_2 = \mathbf{u}_2\mathbf{u}_2^T, \qquad E_3 = \mathbf{u}_3\mathbf{u}_3^T.$$

    Show that the orthogonal projection operator E for **L** is

    $$E = E_1 + E_2 + E_3.$$

    Illustrate this by an example (a) when $n = 4$, (b) when $n = 3$. Generalize.

5. Continuing with Exercise 4, show that

$$E_1E_2 = 0, \quad E_2E_3 = 0, \quad E_3E_1 = 0, \quad E_1^2 = E_1, \quad E_2^2 = E_2, \quad E_3^2 = E_3.$$

Show that when $n = 3$, $E = I$ so that $I = E_1 + E_2 + E_3$.

6. Let $\mathbf{u}_1$, $\mathbf{u}_2$, and $\mathbf{u}_3$ be orthonormal eigenvectors of a symmetric 3×3-matrix A with λ_1, λ_2, and λ_3 as corresponding eigenvalues. Let E_1, E_2, and E_3 be defined as in Exercise 4. Show that

$$AE_1 = \lambda_1E_1, \quad AE_2 = \lambda_2E_2, \quad AE_3 = \lambda_3E_3.$$

$$A = \lambda_1E_1 + \lambda_2E_2 + \lambda_3E_3, \quad I = E_1 + E_2 + E_3.$$

6.5 ORTHOGONAL MATRICES AND ELEMENTARY ORTHOGONAL MATRICES

In Section 4.3 we saw how orthogonal matrices played a role when we wish to diagonalize a symmetric matrix. In this section we investigate some further applications of orthogonal matrices. But first let us recall the definition of an orthogonal matrix and describe some of its properties.

An $n \times n$-matrix X is said to be an **orthogonal matrix** if its transpose X^T is its inverse, that is, if

$$X^TX = XX^T = I.$$

An $n \times n$-matrix X is an orthogonal matrix if and only if its columns are orthonormal. Similarly, an $n \times n$-matrix X is an orthogonal matrix if and only if its rows are orthonormal.

Recall that $X^TX = I$ if and only if $XX^T = I$.

The defining relation $X^TX = I$ for an orthogonal matrix X is equivalent to the condition that the columns of X be orthonormal vectors. Similarly, the relation $XX^T = I$ is equivalent to the condition that the rows of X be orthonormal vectors. We often use these properties in constructing orthogonal matrices. For example, in the case $n = 2$, the matrices

$$R = \begin{bmatrix} \cos\theta & \sin\theta \\ -\sin\theta & \cos\theta \end{bmatrix}, \quad S = \begin{bmatrix} \cos\theta & \sin\theta \\ \sin\theta & -\cos\theta \end{bmatrix}$$

are orthogonal matrices because their columns are mutually orthogonal unit vectors. Observe that $\det R = 1$ and that $\det S = -1$. In fact, a 2×2 orthogonal matrix X is a R-matrix or an S-matrix according as $\det X = 1$ or $\det X = -1$.

PROPOSITION 1. Orthogonal matrices have the following properties:

(a) If X is an orthogonal matrix, so is its transpose.

(b) The product $Z = XY$ of two orthogonal matrices X and Y is an orthogonal matrix.

(c) If X and Y are orthogonal matrices, the matrices $U = XY^T$ and $V = Y^TX$ are orthogonal matrices such that

$$X = UY = YV.$$

Property (a) is immediate. Property (b) holds because

$$Z^TZ = Y^TX^TXY = Y^TY = I.$$

Property (c) holds because $X = (XY^T)Y = Y(Y^TX)$.

An orthogonal matrix, as an operator, preserves length, inner products, and angles. This follows from the result below.

PROPOSITION 2. Let X be an orthogonal matrix. Then:

(a) If $\mathbf{z} = X\mathbf{y}$, then $\mathbf{z}^T\mathbf{z} = \mathbf{y}^T\mathbf{y}$. That is, $\mathbf{y}$ and $\mathbf{z}$ have the same lengths.

(b) If $\mathbf{z}_1 = X\mathbf{y}_1$ and $\mathbf{z}_2 = X\mathbf{y}_2$, then $\mathbf{z}_1^T\mathbf{z}_2 = \mathbf{y}_1^T\mathbf{y}_2$.

Property (a) follows from property (b). Property (b) holds because

$$\mathbf{z}_1^T\mathbf{z}_2 = \mathbf{y}_1^TX^TX\mathbf{y}_2 = \mathbf{y}_1^TI\mathbf{y}_2 = \mathbf{y}_1^T\mathbf{y}_2.$$

If θ is the angle between $\mathbf{y}_1$ and $\mathbf{y}_2$, then

$$\cos\theta = \frac{\mathbf{y}_1^T\mathbf{y}_2}{\|\mathbf{y}_1\|\,\|\mathbf{y}_2\|} = \frac{\mathbf{z}_1^T\mathbf{z}_2}{\|\mathbf{z}_1\|\,\|\mathbf{z}_2\|}.$$

It follows that θ is also the angle between $\mathbf{z}_1$ and $\mathbf{z}_2$. Consequently, X preserves angles as well as inner products and lengths.

■ EXAMPLE 1 _____

Consider the matrix equation

$$\begin{bmatrix} 4/5 & -3/5 \\ 3/5 & 4/5 \end{bmatrix}\begin{bmatrix} 5 & -5 & 10 \\ 0 & 10 & 20 \end{bmatrix} = \begin{bmatrix} 4 & -10 & -4 \\ 3 & 5 & 22 \end{bmatrix}.$$

This equation is of the form $XA = B$, where X is an orthogonal matrix. Observe that the first columns of A and B have the same length. Their second columns are also of equal length. The same is true for their third columns. Conclude that the sum of the squares of the entries of B is equal to the sum of the squares of the entries of A. Observe also that the inner products of corresponding columns of A and B are equal.

This example illustrates the following result.

> **COROLLARY.** Multiplying a matrix A on the left by an orthogonal matrix X preserves lengths of columns. Multiplying A on the right by an orthogonal matrix Y preserves lengths of rows. The sum of the squares of the entries of A, of XA, of AY, and of XAY are the same.

Symmetric orthogonal matrices are connected with orthogonal projection operators as explained in the following

> **PROPOSITION 3.** Let S be a symmetric orthogonal matrix. Then the matrices
> $$E = \tfrac{1}{2}(I - S), \qquad F = \tfrac{1}{2}(I + S)$$
> have the following properties:
>
> **(a)** They are symmetric.
> **(b)** $E^2 = E$, $F^2 = F$, $EF = FE = 0$. Hence E and F are orthogonal projection operators.
> **(c)** $I = E + F$, $S = -E + F = I - 2E$.
> **(d)** $SE = -E$, $SF = F$.

Property (a) holds because S is symmetric. Since $S^2 = I$, we have
$$4EF = (I - S)(I + S) = I - S^2 = 0.$$
Hence $EF = FE = 0$. Obviously, $I = E + F$. It follows that
$$E = E(E + F) = E^2 + EF = E^2.$$
Similarly, $F = F^2$. Clearly, $S = F - E$. Hence
$$SE = (F - E)E = -E^2 = -E, \qquad SF = (F - E)F = F.$$

> **COROLLARY.** The range of E is the null space of F. The range of F is the null space of E. The ranges of E and F are orthogonal complements. For each vector $\mathbf{x}$, the vector $\mathbf{y} = S\mathbf{x}$ is the reflection of $\mathbf{x}$ across the range of F. A symmetric orthogonal matrix S is, accordingly, a ***reflection operator***.

The first three statements in the corollary are geometrical interpretations of properties (b) and (c) in proposition 3. By property (c) we have the relations
$$\mathbf{x} = F\mathbf{x} + E\mathbf{x}, \qquad S\mathbf{x} = F\mathbf{x} - E\mathbf{x}$$

for every vector $\mathbf{x}$. These relations are exhibited schematically in Figure 1. This figure tells us that the vector $S\mathbf{x}$ is the reflection of $\mathbf{x}$ across the range of F. A symmetric orthogonal matrix accordingly is a reflection operator.

Return to the case $n = 2$. As noted earlier, a 2×2 orthogonal matrix X is one of the following types:

$$R = \begin{bmatrix} \cos\theta & \sin\theta \\ -\sin\theta & \cos\theta \end{bmatrix}, \qquad S = \begin{bmatrix} \cos\theta & \sin\theta \\ \sin\theta & -\cos\theta \end{bmatrix}$$

according as $\det X = 1$ or $\det X = -1$. The matrix R is a rotation matrix. The matrix S, being symmetric, is a reflection matrix. Setting $\theta = 0$ in S, we obtain the matrix

$$S_0 = \begin{bmatrix} 1 & 0 \\ 0 & -1 \end{bmatrix}.$$

The matrix S_0 is a sign changer. We have

$$R = S_0 S, \qquad S = S_0 R,$$

as one readily verifies. Denoting the values of R and S by $R(\theta)$ and $S(\theta)$, respectively, we have the identity

$$S(\beta)S(\alpha) = R(\alpha - \beta).$$

In displayed form, this identity is

$$\begin{bmatrix} \cos\beta & \sin\beta \\ \sin\beta & -\cos\beta \end{bmatrix} \begin{bmatrix} \cos\alpha & \sin\alpha \\ \sin\alpha & -\cos\alpha \end{bmatrix} = \begin{bmatrix} \cos(\alpha - \beta) & \sin(\alpha - \beta) \\ -\sin(\alpha - \beta) & \cos(\alpha - \beta) \end{bmatrix}.$$

This result is obtained by the use of the formulas

$$\cos(\alpha - \beta) = \cos\alpha\cos\beta + \sin\alpha\sin\beta$$

$$\sin(\alpha - \beta) = \sin\alpha\cos\beta - \cos\alpha\sin\beta.$$

From the identity $S(\beta)S(\alpha) = R(\alpha - \beta)$, we see that the product of two S-matrices is an R-matrix. Because $n = 2$, this result can be obtained without

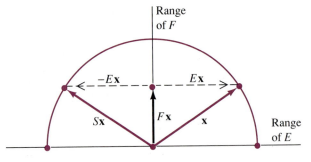

FIGURE I

the use of this identity. We need only to note that det $S(\beta)S(\alpha) = 1$. Hence $S(\beta)S(\alpha)$ is an *R*-matrix.

There is another way to describe the *S* matrix,

$$S = \begin{bmatrix} \cos 2\alpha & \sin 2\alpha \\ \sin 2\alpha & -\cos 2\alpha \end{bmatrix},$$

where, for convenience, we have set $\theta = 2\alpha$. According to proposition 3, the matrix *S*, being symmetric, is expressible in the form

$$S = I - 2E,$$

where *E* is the orthogonal projection operator

$$E = \tfrac{1}{2}(I - S).$$

Displaying this equation, we have

$$E = \frac{1}{2}\begin{bmatrix} 1 - \cos 2\alpha & -\sin 2\alpha \\ -\sin 2\alpha & 1 + \cos 2\alpha \end{bmatrix} = \begin{bmatrix} \sin^2\alpha & -\sin\alpha\cos\alpha \\ -\sin\alpha\cos\alpha & \cos^2\alpha \end{bmatrix}.$$

Hence *E* is given by the formula

$$E = \mathbf{u}\mathbf{u}^T \quad \text{with} \quad \mathbf{u} = (\sin\alpha, -\cos\alpha).$$

Since $SE = -E$, it follows that $S\mathbf{u} = -\mathbf{u}$. This result can be verified directly by computing $S\mathbf{u}$. It follows that the vector $\mathbf{u}$ is a (negative) eigenvector of *S*. Hence *S* is determined by its negative eigenvector $\mathbf{u}$. A multiple $\mathbf{v}$ of $\mathbf{u}$ is also a negative eigenvector. In terms of $\mathbf{v}$ we have

$$E = \frac{\mathbf{v}\mathbf{v}^T}{\mathbf{v}^T\mathbf{v}}.$$

The operator $S = I - 2E$ then takes the form

$$S = I - 2\frac{\mathbf{v}\mathbf{v}^T}{\mathbf{v}^T\mathbf{v}}.$$

In the *n*-dimensional case this formula gives us a special reflection operator

$$V = I - 2\frac{\mathbf{v}\mathbf{v}^T}{\mathbf{v}^T\mathbf{v}},$$

which we now denote by *V* instead of *S*, reserving the symbol *S* for a general symmetric operator. The operator *V* is uniquely determined by the vector $\mathbf{v}$. So we often call it a $\mathbf{v}$-operator or a $\mathbf{v}$-matrix. A second vector $\mathbf{w}$ determines the $\mathbf{w}$-operator

$$W = I - 2\frac{\mathbf{w}\mathbf{w}^T}{\mathbf{w}^T\mathbf{w}}.$$

Thus we have a **v**-operator V, **w**-operator W, **u**-operator U, and so on. A **v**-matrix is undefined for $\mathbf{v} = \mathbf{0}$. It will be convenient to set $V = I$ when $\mathbf{v} = \mathbf{0}$. We call an operator of this type an *elementary orthogonal operator* or an *elementary orthogonal matrix*. We also use the term an *elementary reflection operator* when **v** is nonzero.

This gives the following:

PROPOSITION 4. A nonzero vector **v** determines a unique elementary orthogonal matrix

$$V = I - 2\frac{\mathbf{v}\mathbf{v}^T}{\mathbf{v}^T\mathbf{v}},$$

which we call a **v**-*matrix*. It has the following properties:

(a) It is symmetric and therefore is its own inverse.
(b) For every vector **x**, the vector $\mathbf{y} = V\mathbf{x}$ is the vector

$$\mathbf{y} = \mathbf{x} - 2m\mathbf{v}, \qquad m = \frac{\mathbf{v}^T\mathbf{x}}{\mathbf{v}^T\mathbf{v}}.$$

The vector **y** is the reflection of **x** across the hyperplane

$$\mathbf{v}^T\mathbf{x} = 0.$$

It should be noted that $V\mathbf{x} = \mathbf{x}$ if and only if $\mathbf{v}^T\mathbf{x} = 0$. The hyperplane $\mathbf{v}^T\mathbf{x} = 0$ is therefore the positive eigenspace of V corresponding to the eigenvalue $\lambda = 1$. The operator V therefore reflects a vector **x** across its positive eigenspace. Because $V\mathbf{v} = -\mathbf{v}$, the vector **v** is an eigenvector of V corresponding to the eigenvalue $\lambda = -1$. The multiples of **v** form the negative eigenspace **L** of V. In fact, every vector **w** in **L** determines the same reflection operator V. This follows because if **w** is in **L**, then **w** is a multiple $\mathbf{w} = c\mathbf{v}$ of **v**, so that

$$W = I - 2\frac{\mathbf{w}\mathbf{w}^T}{\mathbf{w}^T\mathbf{w}} = I - 2\frac{(c\mathbf{v})(c\mathbf{v})^T}{(c\mathbf{v})^T(c\mathbf{v})} = I - 2\frac{\mathbf{v}\mathbf{v}^T}{\mathbf{v}^T\mathbf{v}} = V.$$

Notice further that because V is an orthogonal matrix, the image $\mathbf{y} = V\mathbf{x}$ has the same length as **x**. We have

$$\mathbf{y} = \mathbf{x} - 2m\mathbf{v}, \qquad m = \frac{\mathbf{v}^T\mathbf{x}}{\mathbf{v}^T\mathbf{v}}$$

so that $\mathbf{w} = \mathbf{x} - \mathbf{y}$ is a multiple of **v**. The **v**-matrix V is therefore also determined by the vector $\mathbf{w} = \mathbf{x} - \mathbf{y}$. This gives the following:

> **COROLLARY.** Let **x** and **y** be distinct vectors of the same length. The
> **w**-matrix W determined by $\mathbf{w} = \mathbf{x} - \mathbf{y}$ has the property that $\mathbf{y} = W\mathbf{x}$
> and $\mathbf{x} = W\mathbf{y}$.

The **w**-matrix W is the reflection operator

$$W = I - 2\frac{(\mathbf{x} - \mathbf{y})(\mathbf{x} - \mathbf{y})^T}{(\mathbf{x} - \mathbf{y})^T(\mathbf{x} - \mathbf{y})}.$$

■ **EXAMPLE 2**

Consider first the case in which $\mathbf{x} = (0, 1, 0)$ and $\mathbf{y} = (0, 0, 1)$. Then **v**
$= \mathbf{x} - \mathbf{y}$ is the vector $\mathbf{v} = (0, 1, -1)$. We have $\mathbf{v}^T\mathbf{v} = 2$ and

$$V = I - \mathbf{v}\mathbf{v}^T = I - \begin{bmatrix} 0 \\ 1 \\ -1 \end{bmatrix} [0 \quad 1 \quad -1] = \begin{bmatrix} 1 & 0 & 0 \\ 0 & 0 & 1 \\ 0 & 1 & 0 \end{bmatrix}.$$

This is the elementary permutation matrix P_{23} described in Chapter 2.
   Consider next the case in which $\mathbf{x} = (0, 1, 0)$ and $\mathbf{y} = (0, -1, 0)$. Then
$\mathbf{v} = \mathbf{x} - \mathbf{y} = (0, 2, 0)$, $\mathbf{v}^T\mathbf{v} = 4$, and

$$V = I - \frac{2}{4}\mathbf{v}\mathbf{v}^T = I - \frac{1}{2}\begin{bmatrix} 0 \\ 2 \\ 0 \end{bmatrix} [0 \quad 2 \quad 0] = \begin{bmatrix} 1 & 0 & 0 \\ 0 & -1 & 0 \\ 0 & 0 & 1 \end{bmatrix}.$$

In this event V is a sign changer.

■ **EXAMPLE 3**

Consider the vector $\mathbf{x} = (12, 0, 4, -3)$. The vectors

$\mathbf{y}_1 = (13, 0, 0, 0)$, $\mathbf{y}_2 = (12, 5, 0, 0)$, $\mathbf{y}_3 = (12, 0, 5, 0)$, $\mathbf{y}_4 = (12, 0, 4, 3)$

have the same length as **x**. Setting $\mathbf{v}_k = \mathbf{x} - \mathbf{y}_k$ we have $\mathbf{v}_1 = (-1, 0, 4, -3)$,
$\mathbf{v}_2 = (0, -5, 4, -3)$, $\mathbf{v}_3 = (0, 0, -1, -3)$, and $\mathbf{v}_4 = (0, 0, 0, -6)$. For
each k, the $\mathbf{v}_k$-matrix

$$V_k = I - 2\frac{\mathbf{v}_k\mathbf{v}_k^T}{\mathbf{v}_k^T\mathbf{v}_k}$$

transforms **x** into the vector $\mathbf{y}_k$ having the property that for $k < 4$, its last
$4 - k$ components are zero. Notice further that for $k > 1$ the first $k - 1$
components of $\mathbf{v}_k$ are zero. It follows that for each vector **z**, the first $k - 1$
components of **z** and $V_k\mathbf{z}$ are the same.
   The vectors

$\mathbf{z}_1 = (-13, 0, 0, 0)$, $\mathbf{z}_2 = (12, -5, 0, 0)$, $\mathbf{z}_3 = (12, 0, 5, 0)$, $\mathbf{z}_4 = (12, 0, 4, -3)$

also have the same length as **x**. Setting $\mathbf{w}_k = \mathbf{x} - \mathbf{z}_k$ we have

$$\mathbf{w}_1 = (25, 0, 4, -3), \qquad \mathbf{w}_2 = (0, 5, 4, -3),$$
$$\mathbf{w}_3 = (0, 0, -1, -3), \qquad \mathbf{w}_4 = (0, 0, 0, 0).$$

For each $k < 4$, the $\mathbf{w}_k$-matrix

$$W_k = I - 2\frac{\mathbf{w}_k \mathbf{w}_k^T}{\mathbf{w}_k^T \mathbf{w}_k}$$

transforms **x** into $\mathbf{x}_k$. For $k = 4$, we have $\mathbf{w}_4 = \mathbf{0}$, so that in this event $W_4 = I$.

These examples illustrate the result given in the following:

LEMMA 1. Let $\mathbf{x} = (x_1, x_2, \ldots, x_n)$ be a prescribed vector. Let $\mathbf{y}_k$ be the vector having the following properties:

(a) When $k < n$, its last $n - k$ components are zero.

(b) When $k > 1$, its first $k - 1$ components are those of **x**.

(c) Its kth component y_k is chosen to be $y_k = c_k$, where

$$c_k = (x_k^2 + x_{k+1}^2 + \cdots + x_n^2)^{1/2}.$$

Then $\mathbf{y}_k$ has the same length as **x** and its kth component is nonnegative. Set $\mathbf{v}_k = \mathbf{x} - \mathbf{y}_k$. Then the $\mathbf{v}_k$-matrix V_k transforms **x** into $\mathbf{y}_k$. Observe that $V_k = I$ when $c_k = x_k$, that is, when $\mathbf{v}_k = \mathbf{0}$. In addition, for every vector **z**, the first $k - 1$ components of **z** and $V_k\mathbf{z}$ coincide. Moreover, $V_k\mathbf{z} = \mathbf{z}$ when **z** is orthogonal to $\mathbf{v}_k$.

Notice in (c) that had we wished to do so, we could have chosen $y_k = -c_k$. Then the kth component would be nonpositive. For our purposes the choice $y_k = c_k$ is preferable. Then $y_k \geq 0$ with $y_k = 0$ when $c_k = 0$.

In what follows we shall refer to a $\mathbf{v}_k$-matrix V_k related to **x** as described in this lemma as *the $\mathbf{v}_k$-matrix V_k determined by* **x**.

As an application of V_k-matrices, consider the problem of reducing an $n \times n$-matrix A to an upper triangular matrix. In the case $n = 4$, we begin with the first of the following matrices:

$$A = \begin{bmatrix} a & \cdot & \cdot & \cdot \\ b & \cdot & \cdot & \cdot \\ c & \cdot & \cdot & \cdot \\ d & \cdot & \cdot & \cdot \end{bmatrix}, \qquad A_1 = V_1 A = \begin{bmatrix} e & f & g & h \\ 0 & i & \cdot & \cdot \\ 0 & j & \cdot & \cdot \\ 0 & k & \cdot & \cdot \end{bmatrix},$$

$$A_2 = V_2 A_1 = \begin{bmatrix} e & f & g & h \\ 0 & m & n & p \\ 0 & 0 & q & \cdot \\ 0 & 0 & r & \cdot \end{bmatrix}, \qquad A_3 = V_3 A_2 = \begin{bmatrix} e & f & g & h \\ 0 & m & n & p \\ 0 & 0 & s & t \\ 0 & 0 & 0 & u \end{bmatrix}.$$

We transform the matrix A into an upper triangular matrix U in the following four steps.

1. We choose V_1 to be the $\mathbf{v}_1$-matrix determined by the first column $\mathbf{x}_1$ of A. Then the product $A_1 = V_1 A$ has the form shown above with $e \geq 0$.

2. We next choose V_2 to be the $\mathbf{v}_2$-matrix determined by the second column $\mathbf{x}_2$ of A_1. Then $A_2 = V_2 A_1$ is of the form shown above with $m \geq 0$. Observe that the first row of A_1 is unaltered by this multiplication by V_2.

3. We now choose V_3 to be the $\mathbf{v}_3$-matrix determined by the third column $\mathbf{x}_3$ of A_2. The matrix $V_3 A_2$ is an upper triangular matrix whose first two rows coincide with the first two rows of A_2. The first three main diagonal entries of A_3 are nonnegative.

4. If the last main diagonal entry u of A_3 is nonnegative, we choose $U = A_3$. Equivalently, we set $U = V_4 A_3$, where $V_4 = I$. On the other hand, if $u < 0$, we select the $\mathbf{v}_4$-matrix V_4 determined by the last column $\mathbf{x}_4$ of A_3. The matrix $U = V_4 A_3$ differs from A_3 at most in the (4, 4)-entry. When this choice is made, the matrix U is an upper triangular matrix whose main diagonal entries are nonnegative.

Summarizing, we have

$$U = PA \qquad \text{with} \quad P = V_4 V_3 V_2 V_1.$$

The matrix P is the product of elementary orthogonal matrices and so is an orthogonal matrix.

The argument used for the case $n = 4$ is applicable to the general case. This gives the following result.

PROPOSITION 5. Let A be an $n \times n$-matrix. There are n elementary orthogonal matrices $V_1, V_2, \ldots, V_n$ such that

(a) The orthogonal matrix $P = V_n V_{n-1} \cdots V_2 V_1$ has the property that the matrix

$$U = PA$$

is an upper triangular matrix whose main diagonal entries are nonnegative.

(b) Hence $A = QU$ with $Q = P^T = V_1 V_2 \cdots V_n$.

(c) Each V_k is either the identity or else an elementary reflection matrix.

In describing Q we used the fact that the V's are symmetric. This proposition states that every $n \times n$-matrix A is the product of an orthogonal matrix Q and an upper triangular matrix U with nonnegative main diagonal entries. When A is an orthogonal matrix, the matrix $U = PA$ is also orthogonal. Since the main diagonal entries are positive, this is possible only if U is the identity I. It follows that $A = Q = V_1 V_2 \cdots V_n$. Disregarding the V's that are the identity, we have the following:

> **COROLLARY.** An orthogonal $n \times n$-matrix A is the product of q elementary reflection matrices, where $q \le n$.

The result described in Proposition 5 can be used to give us a numerically stable method for solving a linear equation

$$A\mathbf{x} = \mathbf{h}$$

where A is a nonsingular $n \times n$-matrix. We reduce A to an upper triangular matrix U by operating on this equation on the left successively by V_1-, $V_2- \cdots$ matrices. We obtain thereby the equation

$$U\mathbf{x} = \mathbf{k}.$$

Since U is upper triangular, this equation can be solved by backward substitution.

We have the following numerical example.

■ **EXAMPLE 4** ──

Consider the linear equation

$$A\mathbf{x} = \mathbf{h}, \quad \text{where } A = \begin{bmatrix} 2 & -2 & 12 & 5 \\ 1 & 8 & -4 & -3 \\ 0 & 3 & -2 & 0 \\ -2 & 2 & -5 & -1 \end{bmatrix}, \quad \mathbf{h} = \begin{bmatrix} 17 \\ 2 \\ 1 \\ -6 \end{bmatrix}.$$

We transform this equation to the form $U\mathbf{x} = \mathbf{k}$ by the use of $\mathbf{v}_k$-matrices. Here U is an upper triangular matrix. As in Gaussian elimination, we form the augmented matrix $M = [A \mid \mathbf{h}]$. We have

$$M = \left[\begin{array}{cccc|c} 2 & -2 & 12 & 5 & 17 \\ 1 & 8 & -4 & -3 & 2 \\ 0 & 3 & -2 & 0 & 1 \\ -2 & 2 & -5 & -1 & -6 \end{array} \right],$$

$$M_1 = \left[\begin{array}{cccc|c} 3 & 0 & 10 & 3 & 16 \\ 0 & 6 & -2 & -1 & 3 \\ 0 & 3 & -2 & 0 & 1 \\ 0 & 6 & -9 & -5 & -8 \end{array} \right].$$

The matrix M_1 is the matrix $M_1 = V_1 M$, where

$$V_1 = I - (\tfrac{1}{3})\mathbf{v}_1\mathbf{v}_1^T, \qquad \mathbf{v}_1^T = [-1 \quad 1 \quad 0 \quad -2]$$

is the $\mathbf{v}_1$-matrix determined by the first column $\mathbf{x}_1$ of M. We compute each column $\mathbf{y}$ of M_1 by the formula

$$\mathbf{y} = \mathbf{x} - \tfrac{1}{3}\mathbf{v}_1\mathbf{v}_1^T\mathbf{x},$$

where $\mathbf{x}$ is the corresponding column of M. For example, when $\mathbf{x}$ is the third column of M, we have $\mathbf{v}_1^T\mathbf{x} = -6$ and

$$\mathbf{y} = (12, -4, -2, -5) + 2(-1, 1, 0, -2) = (10, -2, -2, -9)$$

as the third column of M_1. We next construct the $\mathbf{v}_2$-matrix

$$V_2 = I - \tfrac{1}{27}\mathbf{v}_2\mathbf{v}_2^T, \qquad \mathbf{v}_2^T = [0 \quad -3 \quad 3 \quad 6]$$

determined by the second column $\mathbf{x}_2$ of M_1. Forming $M_2 = V_2 M_1$, we obtain the first of the matrices

$$M_2 = \begin{bmatrix} 3 & 0 & 10 & 3 & \vdots & 16 \\ 0 & 9 & -8 & -4 & \vdots & -3 \\ 0 & 0 & 4 & 3 & \vdots & 7 \\ 0 & 0 & 3 & 1 & \vdots & 4 \end{bmatrix}, \qquad M_3 = \begin{bmatrix} 3 & 0 & 10 & 3 & \vdots & 16 \\ 0 & 9 & -8 & -4 & \vdots & -3 \\ 0 & 0 & 5 & 3 & \vdots & 8 \\ 0 & 0 & 0 & 1 & \vdots & 1 \end{bmatrix}.$$

The matrix M_3 is the matrix $M_3 = V_3 M_2$, where V_3 is the V_3-matrix determined by the third column of M_2. The matrix $M_3 = V_3 M_2 = V_3 V_2 V_1 M$ is of the form $M_3 = [U \quad \vdots \quad \mathbf{k}]$, where U is an upper triangular matrix. Our original equation $A\mathbf{x} = \mathbf{h}$ is therefore equivalent to the equation $U\mathbf{x} = \mathbf{k}$, which in component form is

$$3x_1 + 0x_2 + 10x_3 + 3x_4 = 16$$
$$9x_2 - 8x_3 - 4x_4 = -3$$
$$5x_3 + 3x_4 = 8$$
$$x_4 = 1.$$

By backward substitution, we find that $x_4 = 1$, $x_3 = 1$, $x_2 = 1$, and $x_1 = 1$ give us the solution to our problem.

Exercises

Recall that a $\mathbf{v}$-matrix is defined by the formula

$$V = I - 2\frac{\mathbf{v}\mathbf{v}^T}{\mathbf{v}^T\mathbf{v}}.$$

1. Find the $\mathbf{v}$-matrix V for each of the following cases:

   (a) $\mathbf{v} = (1, 1)$ (b) $\mathbf{v} = (1, 0, 1)$ (c) $\mathbf{v} = (1, 1, 1, 1)$

2. Find the v-matrix V which transforms $\mathbf{x}$ into $\mathbf{y}$ in each of the following cases.

   (a) $\mathbf{x} = (2, 3)$, $\mathbf{y} = (3, 2)$ **(b)** $\mathbf{x} = (2, 1, 2)$, $\mathbf{y} = (0, 3, 0)$

   (c) $\mathbf{x} = (1, -1, 1, -1)$, $\mathbf{y} = (2, 0, 0, 0)$

3. The matrix A in the list

$$A = \begin{bmatrix} 3/5 & -4/5 \\ 4/5 & 3/5 \end{bmatrix}, \qquad A_1 = \begin{bmatrix} 1 & b \\ 0 & c \end{bmatrix}$$

   is an orthogonal matrix. According to step 1 in the proof of Proposition 5, the $\mathbf{v}_1$-matrix V_1 with $\mathbf{v}_1 = (-2/5, 4/5)$ transforms A into the matrix $A_1 = V_1 A$ of the form shown. Verify this. Why is A_1 an orthogonal matrix? Conclude, without computing, that $b = 0$ and $c = -1$.

4. Recall that two square matrices A and B are orthogonally similar if there is an orthogonal matrix P such that $B = P^T A P$.

   (a) Show that the matrices

$$A = \begin{bmatrix} 1 & 0 & 0 \\ 0 & 4/5 & 2/5 \\ 0 & 3/5 & -4/5 \end{bmatrix}, \quad U = \begin{bmatrix} 1 & 0 & 0 \\ 0 & -1 & 0 \\ 0 & 0 & 1 \end{bmatrix}, \quad V = \begin{bmatrix} 1 & 0 & 0 \\ 0 & 1 & 0 \\ 0 & 0 & -1 \end{bmatrix}$$

   are orthogonally similar. Why are they elementary reflection matrices?

   (b) Show that if U and V are elementary reflection matrices of the same size, they are orthogonally similar.

   (c) Show that for $n > 1$, the orthogonal matrix $V = -I$ is a reflection matrix that is not elementary.

   (d) Let $W = UV$ be the product of two elementary reflection matrices. Conclude that det $W = 1$ and hence that W is not an elementary reflection matrix.

5. Show that the matrices

$$B = \begin{bmatrix} 1 & 0 & 0 \\ 0 & 4/5 & -2/5 \\ 0 & 3/5 & 4/5 \end{bmatrix}, \quad R = \begin{bmatrix} 1 & 0 & 0 \\ 0 & 0 & -1 \\ 0 & 1 & 0 \end{bmatrix}, \quad S = \begin{bmatrix} 0 & 0 & -1 \\ 0 & 1 & 0 \\ 1 & 0 & 0 \end{bmatrix}$$

   are not elementary reflection matrices. However, each matrix is the product of two elementary reflection matrices in many ways. Find one way. These matrices are elementary rotation matrices.

6. Use the method described in Example 4 to solve the linear equation $A\mathbf{x} = \mathbf{h}$ in each of the following cases. In each case check your solution by substitution.

   (a) $A = \begin{bmatrix} 2 & -1 & 1 \\ 1 & 0 & -4 \\ 2 & 1 & 1 \end{bmatrix}$, $\mathbf{h} = \begin{bmatrix} 7 \\ -5 \\ 9 \end{bmatrix}$

   (b) $A = \begin{bmatrix} 1 & -1 & 2 \\ 1 & 0 & 3 \\ 1 & 0 & 4 \end{bmatrix}$, $\mathbf{h} = \begin{bmatrix} 2 \\ 4 \\ 6 \end{bmatrix}$

Notice that at least in case (b), it is simpler to use the Gaussian elimination procedure. This is because square roots are used in the present procedure. However, on a computer, square roots present no difficulties. For large systems, the method presented here is numerically stabler than the standard elimination method.

7. Consider the matrices

$$A = \begin{bmatrix} 7 & 2 & 1 & 2 & 0 \\ 2 & 60 & 60 & -63 & -54 \\ 1 & 60 & 45 & -67 & 18 \\ 2 & -63 & -67 & 64 & 72 \\ 0 & -54 & 18 & 72 & -45 \end{bmatrix}, \quad A_2 = \begin{bmatrix} 7 & 3 & 0 & 0 & 0 \\ 3 & 9 & 18 & 9 & 18 \\ 0 & 18 & 180 & 45 & -54 \\ 0 & 9 & 45 & 0 & -72 \\ 0 & 18 & -54 & -72 & -45 \end{bmatrix}.$$

As in Lemma 1, compute the v_2-matrix V_2 determined by the first column of A. Verify that

$$V_2 = \begin{bmatrix} 1 & 0 & 0 & 0 & 0 \\ 0 & 2/3 & 1/3 & 2/3 & 0 \\ 0 & 1/3 & 2/3 & -2/3 & 0 \\ 0 & 2/3 & -2/3 & -1/3 & 0 \\ 0 & 0 & 0 & 0 & 1 \end{bmatrix}.$$

Verify next that $A_2 = V_2 A V_2$.

8. With A_2 as in Exercise 7, construct the v_3-matrix V_3 determined by the second column of A_2 as described in Lemma 1. Verify that the matrix V_3 is the first of the matrices

$$V_3 = \begin{bmatrix} 1 & 0 & 0 & 0 & 0 \\ 0 & 1 & 0 & 0 & 0 \\ 0 & 0 & 2/3 & 1/3 & 2/3 \\ 0 & 0 & 1/3 & 2/3 & -2/3 \\ 0 & 0 & 2/3 & -2/3 & -1/3 \end{bmatrix}, \quad A_3 = \begin{bmatrix} 7 & 3 & 0 & 0 & 0 \\ 3 & 9 & 27 & 0 & 0 \\ 0 & 27 & 0 & 3 & 4 \\ 0 & 0 & 3 & 4 & 2 \\ 0 & 0 & 4 & 2 & 1 \end{bmatrix}.$$

Verify that $A_3 = V_3 A_2 V_3$.

9. With A_3 as in Exercise 8, construct the v_4-matrix V_4 determined by the third column as described in Lemma 1. Verify that V_4 is the first of the matrices

$$V_4 = \begin{bmatrix} 1 & 0 & 0 & 0 & 0 \\ 0 & 1 & 0 & 0 & 0 \\ 0 & 0 & 1 & 0 & 0 \\ 0 & 0 & 0 & 3/5 & 4/5 \\ 0 & 0 & 0 & 4/5 & -3/5 \end{bmatrix}, \quad A_4 = \begin{bmatrix} 7 & 3 & 0 & 0 & 0 \\ 3 & 9 & 27 & 0 & 0 \\ 0 & 27 & 0 & 5 & 0 \\ 0 & 0 & 5 & 4 & 2 \\ 0 & 0 & 0 & 2 & 1 \end{bmatrix}.$$

Verify that $A_4 = V_4 A_3 V_4$.

10. Observe that the matrix A_4 obtained in Exercise 9 is a tridiagonal matrix. It is related to the matrix A in Exercise 7 by the formula

$$A_4 = V_4 V_3 V_2 A V_2 V_3 V_4 = P^T A P \quad \text{with} \quad P = V_2 V_3 V_4.$$

Why is P an orthogonal matrix? The technique used in Exercises 7, 8, and 9 can be extended to the tridiagonalization of an $n \times n$ symmetric matrix A by the use of $n - 2$ elementary orthogonal matrices. The tridiagonal matrix T obtained in this manner is orthogonally similar to A and has the same eigenvalues as A. It is usually easier to find the eigenvalues of A by first transforming A into a tridiagonal matrix T and then finding the eigenvalues of T.

6.6   LEAST SQUARE SOLUTIONS

Let A be an $m \times n$-matrix. Earlier we found that the equation

$$A\mathbf{x} = \mathbf{h}$$

has a solution $\mathbf{x}$ if and only if $\mathbf{h}$ is in the range $\mathbf{R}$ of A. In Chapter 2 we transformed the equation $A\mathbf{x} = \mathbf{h}$ with $\mathbf{h}$ in $\mathbf{R}$ into its reduced row echelon form by the use of elementary row operations. This reduction is equivalent to multiplying the equation $A\mathbf{x} = \mathbf{h}$ on the left by a suitably chosen matrix B. We obtain thereby an equation of the form

$$BA\mathbf{x} = B\mathbf{h}.$$

In this process we changed the equation but not its solutions. The question arises as to whether or not we can effectively modify our procedure by first replacing our equation by another equation of the form

$$PA\mathbf{x} = P\mathbf{h}$$

before a row echelon reduction or before solving our equation by some other means. The new equation does not lose any solutions but may have extraneous solutions. The choice $P = A^T$ gives us the equation

$$A^TA\mathbf{x} = A^T\mathbf{h}.$$

which has many desirable properties, such as:

1. The matrix A^TA is a nonnegative symmetric matrix. In many important cases it is nonsingular and so is invertible.
2. The matrix A^TA has the same null space as A. That is, $A^TA\mathbf{z} = \mathbf{0}$ if and only if $A\mathbf{z} = \mathbf{0}$. Hence, if $\mathbf{x} = \mathbf{x}_0$ is a solution of $A^TA\mathbf{x} = A^T\mathbf{h}$, every solution is of the form $\mathbf{x} = \mathbf{x}_0 + \mathbf{z}$, where $\mathbf{z}$ is in the null space of A.
3. If $A^TA\mathbf{x} = A^T\mathbf{h}$ has more than one solution, there is a "shortest solution" $\mathbf{x}_0$. This solution is orthogonal to the null space of A. A method for constructing this shortest solution will be given in the next section.

4. Every solution of $A\mathbf{x} = \mathbf{h}$ is a solution of $A^TA\mathbf{x} = A^T\mathbf{h}$.

5. The equation $A^TA\mathbf{x} = A^T\mathbf{h}$ is always solvable because $A^T\mathbf{h}$ is in the range of A^TA.

6. As will be seen below, for every vector $\mathbf{h}$, the solution $\mathbf{x}_0$ of $A^TA\mathbf{x} = A^T\mathbf{h}$ has the following desirable property. The point $\mathbf{y}_0 = A\mathbf{x}_0$ is the point in the range $\mathbf{R}$ of A which is closest to the point $\mathbf{h}$. In many applications this result is all that is needed for the problem at hand.

Let us illustrate these properties by the following example.

■ **EXAMPLE 1** _____

Consider the case in which

$$A = \begin{bmatrix} 1 & 0 \\ 0 & 1 \\ -1 & 1 \end{bmatrix}, \quad \mathbf{h}_1 = \begin{bmatrix} 2 \\ 1 \\ -1 \end{bmatrix}, \quad \mathbf{h}_2 = \begin{bmatrix} 1 \\ 0 \\ 1 \end{bmatrix}.$$

Here we have two choices for $\mathbf{h}$. We have

$$A^TA = \begin{bmatrix} 2 & -1 \\ -1 & 2 \end{bmatrix}, \quad A^T\mathbf{h}_1 = \begin{bmatrix} 3 \\ 0 \end{bmatrix}, \quad A^T\mathbf{h}_2 = \begin{bmatrix} 0 \\ 1 \end{bmatrix}.$$

The matrix A^TA is nonsingular and has as its inverse

$$D = (A^TA)^{-1} = \frac{1}{3}\begin{bmatrix} 2 & 1 \\ 1 & 2 \end{bmatrix}.$$

When $\mathbf{h} = \mathbf{h}_1$, the solution of $A^TA\mathbf{x} = A^T\mathbf{h}_1$ is

$$\mathbf{x}_1 = DA^T\mathbf{h}_1 = \frac{1}{3}\begin{bmatrix} 2 & 1 \\ 1 & 2 \end{bmatrix}\begin{bmatrix} 3 \\ 0 \end{bmatrix} = \begin{bmatrix} 2 \\ 1 \end{bmatrix}.$$

In this case the vector $\mathbf{x}_1$ is also the solution of $A\mathbf{x} = \mathbf{h}_1$.
   In the case $\mathbf{h} = \mathbf{h}_2$, the solution of $A^TA\mathbf{x} = A^T\mathbf{h}_2$ is

$$\mathbf{x}_2 = DA^T\mathbf{h}_2 = \frac{1}{3}\begin{bmatrix} 2 & 1 \\ 1 & 2 \end{bmatrix}\begin{bmatrix} 0 \\ 1 \end{bmatrix} = \begin{bmatrix} 1/3 \\ 2/3 \end{bmatrix}.$$

In this case the equation $A\mathbf{x} = \mathbf{h}_2$ has no solution because $\mathbf{h}_2$ is not in the range $\mathbf{R}$ of A.
However the point

$$\mathbf{y}_2 = A\mathbf{x}_2 = \begin{bmatrix} 1 & 0 \\ 0 & 1 \\ -1 & 1 \end{bmatrix}\begin{bmatrix} 1/3 \\ 2/3 \end{bmatrix} = \begin{bmatrix} 1/3 \\ 2/3 \\ 1/3 \end{bmatrix}$$

has the property that the vector

$$\mathbf{r}_2 = \mathbf{h}_2 - \mathbf{y}_2 = \mathbf{h}_2 - A\mathbf{x}_2 = \begin{bmatrix} 1 \\ 0 \\ 1 \end{bmatrix} - \begin{bmatrix} 1/3 \\ 2/3 \\ 1/3 \end{bmatrix} = \begin{bmatrix} 2/3 \\ -2/3 \\ 2/3 \end{bmatrix}$$

is orthogonal to the columns of A and hence to the range **R** of A. The point $\mathbf{y}_2$ is therefore the point in **R** closest to the point $\mathbf{h}_2$.

In this example we chose a matrix A whose rank r is the same as the number n of columns of A. Later we discuss the case in which $r < n$.

The equation

$$A^T A\mathbf{x} = A^T\mathbf{h}$$

is called the *least square equation* associated with the equation $A\mathbf{x} = \mathbf{h}$. A solution $\mathbf{x}_0$ of the least square equation is called a *least square solution* of $A\mathbf{x} = \mathbf{h}$. The reason for this terminology is because a least square solution $\mathbf{x}_0$ of $A\mathbf{x} = \mathbf{h}$ yields a minimum to the quantity

$$\|\mathbf{h} - A\mathbf{x}\|^2 = (\mathbf{h} - A\mathbf{x})^T(\mathbf{h} - A\mathbf{x}).$$

This quantity is the square of the distance of the point $\mathbf{y} = A\mathbf{x}$ in the range **R** of A to the point $\mathbf{h}$, as stated in the following result.

PROPOSITION 1. Let $\mathbf{x}_0$ be a least square solution of $A\mathbf{x} = \mathbf{h}$. Then the inequality

$$\|\mathbf{h} - A\mathbf{x}\|^2 \geq \|\mathbf{h} - A\mathbf{x}_0\|^2$$

holds for all choices of **x**. That is, $\mathbf{x}_0$ minimizes the square of the length of the *residual* vector $\mathbf{r} = \mathbf{h} - A\mathbf{x}$.

This inequality states that the point $A\mathbf{x}_0$ is as close as you can get to the point $\mathbf{h}$ in $\mathbf{R}^m$.

To prove this result let **x** be a vector in $\mathbf{R}^n$. Introduce the vectors

$$\mathbf{y} = A\mathbf{x}, \qquad \mathbf{y}_0 = A\mathbf{x}_0, \qquad \mathbf{v} = A(\mathbf{x} - \mathbf{x}_0) = \mathbf{y} - \mathbf{y}_0,$$

$$\mathbf{r} = \mathbf{h} - A\mathbf{x}, \qquad \mathbf{r}_0 = \mathbf{h} - A\mathbf{x}_0.$$

Observe that if $\mathbf{r}_0 = \mathbf{0}$, then $\|\mathbf{r}\|^2 \geq 0 = \|\mathbf{r}_0\|^2$. Suppose therefore that $\mathbf{r}_0$ is nonzero. Considering the vectors $\mathbf{h}, \mathbf{y}, \mathbf{y}_0$ to be points in $\mathbf{R}^m$, these points are the vertices of a triangle, as shown in Figure 1. The vectors $\mathbf{r}_0, \mathbf{r}$, and **v** are the sides of this triangle. The vector **v** is orthogonal to $\mathbf{r}_0$. This follows from the computations

$$\mathbf{v}^T\mathbf{r}_0 = (\mathbf{x} - \mathbf{x}_0)^T A^T(\mathbf{h} - A\mathbf{x}_0) = (\mathbf{x} - \mathbf{x}_0)^T(A^T\mathbf{h} - A^T A\mathbf{x}_0) = 0.$$

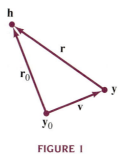

The last equality holds because $\mathbf{x}_0$ is a least square solution of $A\mathbf{x} = \mathbf{h}$. Our triangle is therefore a right triangle with $\mathbf{r}$ as its hypotenuse. It follows that

$$\|\mathbf{r}\|^2 = \|\mathbf{r}_0\|^2 + \|\mathbf{v}\|^2 \geq \|\mathbf{r}_0\|^2,$$

as was to be proved.

> **COROLLARY.** The vector $\mathbf{r}_0 = \mathbf{h} - A\mathbf{x}_0$ is orthogonal to the range **R** of A. The point $\mathbf{y}_0 = A\mathbf{x}_0$ is the point in **R** nearest to the point **h**.

In the next section we establish a formula

$$\mathbf{x} = X\mathbf{h}$$

for a least square solution of $A\mathbf{x} = \mathbf{h}$. It will be shown that the matrix X can be chosen so that its columns are orthogonal to the null space **N** of A. When this choice is made, the length of the least square solution $\mathbf{x} = X\mathbf{h}$ is as small as possible.

Let us look at a standard application of the ***principle of least squares***. In this application we have two related variables x and y. For example, x could be the height of a man and y could be his weight. In any event we seek to explore how these variables are related. To do so, we obtain a collection of sample points

$$(x_1, y_1), (x_2, y_2), \ldots, (x_n, y_n)$$

describing the relationship between x and y. We plot these points in the standard xy-plane. It may happen that the plotted points cluster about a line **L** as shown in Figure 2. In this event we seek to find a line **L** that in some sense best fits our data points. To describe a possible "best fit," let us consider the case in which we have only three points, as shown in Figure 3. We measure the vertical distances d_1, d_2, and d_3 of these points from a line **L**. We say that the line **L** best fits these points in the least square sense if the sum of squares

$$d_1^2 + d_2^2 + d_3^2$$

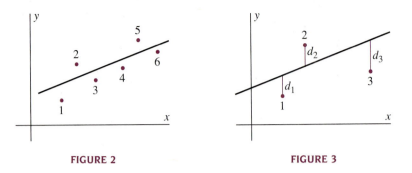

FIGURE 2   FIGURE 3

is as small as possible. When we have n points we make

$$d_1^2 + d_2^2 + \cdots + d_n^2$$

as small as possible.

Return to the case of three points. Let the line **L** be defined by the equation

$$y = mx + b.$$

The parameters m and b are to be determined. The vertical distance from the point (x_k, y_k) to the line **L** is given by the formula

$$d_k = y_k - (mx_k + b).$$

To determine the parameters m and b, we proceed in the manner used earlier in this section. We look at the equations

$$mx_1 + b = y_1$$
$$mx_2 + b = y_2 \quad \text{i.e.,} \quad \begin{bmatrix} x_1 & 1 \\ x_2 & 1 \\ x_3 & 1 \end{bmatrix} \begin{bmatrix} m \\ b \end{bmatrix} = \begin{bmatrix} y_1 \\ y_2 \\ y_3 \end{bmatrix}$$
$$mx_3 + b = y_3$$

first in component form and then in matrix form. We note that normally these equations do not have a solution for the unknown scalars m and b. But they have a least square solution as described earlier in this section. Using the matrix form $A\mathbf{z} = \mathbf{h}$ of these equations, we obtain the least square equation

$$A^T A\mathbf{z} = A^T\mathbf{h}$$

for $A\mathbf{z} = \mathbf{h}$. In our particular case with

$$A = \begin{bmatrix} x_1 & 1 \\ x_2 & 1 \\ x_3 & 1 \end{bmatrix}, \quad \mathbf{z} = \begin{bmatrix} m \\ b \end{bmatrix}, \quad \mathbf{h} = \begin{bmatrix} y_1 \\ y_2 \\ y_3 \end{bmatrix}$$

we find that the least square equation is

$$\begin{bmatrix} s & t \\ t & 3 \end{bmatrix} \begin{bmatrix} m \\ b \end{bmatrix} = \begin{bmatrix} f \\ g \end{bmatrix},$$

where

$$s = x_1^2 + x_2^2 + x_3^2, \qquad t = x_1 + x_2 + x_3,$$
$$f = x_1 y_1 + x_2 y_2 + x_3 y_3, \qquad g = y_1 + y_2 + y_3.$$

This equation has a unique solution for m and b. According to Proposition 1, this solution minimizes the square of the length of the vector

$$\mathbf{r} = \mathbf{h} - A\mathbf{z} = \begin{bmatrix} y_1 - mx_1 - b \\ y_2 - mx_2 - b \\ y_3 - mx_3 - b \end{bmatrix} = \begin{bmatrix} d_1 \\ d_2 \\ d_3 \end{bmatrix}.$$

Because the square of the length of $\mathbf{r}$ is

$$d_1^2 + d_2^2 + d_3^2,$$

the solution m and b gives us the line

$$y = mx + b$$

which is the "best fit" in the least square sense.

When there are n points, we proceed in the same manner. The least square equation is now

$$\begin{bmatrix} s & t \\ t & n \end{bmatrix} \begin{bmatrix} m \\ b \end{bmatrix} = \begin{bmatrix} f \\ g \end{bmatrix},$$

where

$$s = x_1^2 + x_2^2 + \cdots + x_n^2, \qquad t = x_1 + x_2 + \cdots + x_n,$$
$$f = x_1 y_1 + x_2 y_2 + \cdots + x_n y_n, \qquad g = y_1 + y_2 + \cdots + y_n.$$

Again the least square solution m and b gives us the desired line

$$y = mx + b,$$

which is the "best fit" of these n points in the least square sense.

As a numerical case, let us find the line **L** that best fits the five points

$$(1, 1), \quad (2, 1) \quad (3, 2), \quad (4, 4), \quad (5, 2)$$

in the least square sense. In this event

$$s = 1 + 4 + 9 + 16 + 25 = 55 \qquad t = 1 + 2 + 3 + 4 + 5 = 15$$
$$f = 1 + 2 + 6 + 16 + 10 = 35 \qquad g = 1 + 1 + 2 + 4 + 2 = 10.$$

Our parameters m and b are therefore given by the equations

$$55m + 15b = 35$$
$$15m + 5b = 10.$$

Hence $m = 1/2$ and $b = 1/2$. The line of best fit is

$$y = \frac{x}{2} + \frac{1}{2}.$$

In this case two of the points lie on the least square line. This is an unusual phenomenon. Usually, none of the points actually lie on the least square line.

Least square fits are not limited to lines. We have least square parabolas, least square cubics, and so on. However, we shall not pursue this topic further.

Exercises

1. Find a least square solution of $A\mathbf{x} = \mathbf{h}$ in each of the following cases for various choices of $\mathbf{h}$. Are the least square solutions unique? Which least square solutions, if any, also solve $A\mathbf{x} = \mathbf{h}$?

   (a) $A = \begin{bmatrix} 1 & 1 \\ 0 & 0 \end{bmatrix}$, $\qquad \mathbf{h}_1 = \begin{bmatrix} 1 \\ 1 \end{bmatrix}$, $\qquad \mathbf{h}_2 = \begin{bmatrix} 4 \\ 0 \end{bmatrix}$, $\qquad \mathbf{h}_3 = \begin{bmatrix} 0 \\ 1 \end{bmatrix}$

   (b) $A = \begin{bmatrix} 1 & 0 \\ 1 & 1 \\ 0 & 1 \end{bmatrix}$, $\qquad \mathbf{h}_1 = \begin{bmatrix} 1 \\ 0 \\ -1 \end{bmatrix}$, $\qquad \mathbf{h}_2 = \begin{bmatrix} 1 \\ 1 \\ 1 \end{bmatrix}$, $\qquad \mathbf{h}_3 = \begin{bmatrix} 1 \\ -1 \\ 1 \end{bmatrix}$

   (c) $A = \begin{bmatrix} 1 & 1 & 2 \\ 1 & -1 & 0 \\ 0 & 1 & 1 \end{bmatrix}$, $\qquad \mathbf{h}_1 = \begin{bmatrix} 1 \\ 1 \\ 1 \end{bmatrix}$, $\qquad \mathbf{h}_2 = \begin{bmatrix} 4 \\ 0 \\ 2 \end{bmatrix}$

2. Referring to Exercise 1, in each case, find the least square solution that is orthogonal to the null space of A.

3. Let $\mathbf{x}_1$ and $\mathbf{x}_2$ be least square solutions of $A\mathbf{x} = \mathbf{h}$ for $\mathbf{h} = \mathbf{h}_1$ and $\mathbf{h} = \mathbf{h}_2$, respectively. Show that for arbitrary numbers a and b, $\mathbf{x} = a\mathbf{x}_1 + b\mathbf{x}_2$ is a least square solution for $A\mathbf{x} = \mathbf{h}$ with $\mathbf{h} = a\mathbf{h}_1 + b\mathbf{h}_2$. Apply this result with $a = 2$ and $b = 3$ to the cases given in Exercise 1.

4. Find the least square line $y = mx + b$ in the xy-plane determined by the following points.

   (a) $(1, 0)$, $(2, 2)$, $(3, 2)$, $(4, 2)$, $(5, 5)$

   (b) $(1, 1)$, $(2, 2)$, $(2, 1)$, $(3, 2)$, $(3, 1)$

5. Consider five points
$$(x_1, y_1), \quad (x_2, y_2), \quad (x_3, y_3), \quad (x_4, y_4), \quad (x_5, y_5).$$

   To determine the parabola $y = ax^2 + bx + c$ of best fit in the least square sense, we construct the equation

$$\begin{bmatrix} x_1^2 & x_1 & 1 \\ x_2^2 & x_2 & 1 \\ x_3^2 & x_3 & 1 \\ x_4^2 & x_4 & 1 \\ x_5^2 & x_5 & 1 \end{bmatrix} \begin{bmatrix} a \\ b \\ c \end{bmatrix} = \begin{bmatrix} y_1 \\ y_2 \\ y_3 \\ y_4 \\ y_5 \end{bmatrix}.$$

Show that the least square equation for a, b, c is of the form

$$\begin{bmatrix} q & r & s \\ r & s & t \\ s & t & 5 \end{bmatrix} \begin{bmatrix} a \\ b \\ c \end{bmatrix} = \begin{bmatrix} e \\ f \\ g \end{bmatrix}.$$

Determine the entries q, r, s, t, e, f, and g. Show that the solution a, b, c of this least square equation determines the parabola

$$y = ax^2 + bx + c$$

of best fit in the least square sense for our five points. Apply this result to the five points

$$(1,0), \quad (2,1), \quad (3,2), \quad (4,4) \quad (5,7).$$

6.7 PSEUDOINVERSES AND RELATED TOPICS

In Chapter 1 we saw that every matrix A has a transpose A^T. In Chapter 2 we saw that every nonsingular square matrix A has an inverse A^{-1}. In this section we show that every matrix A has associated with it a generalized inverse, called its *pseudoinverse*. It has the property that the pseudoinverse of a nonsingular square matrix A is the inverse of A.

We begin by introducing the concepts of the left and right identities for a matrix A. We do so in the following:

PROPOSITION 1. Let r be the rank of an $m \times n$-matrix A. There exist unique symmetric matrices E and F of rank r such that

$$A = EA = AF.$$

We call E the left identity for A and F the right identity for A.

Before establishing this proposition let us look at the following examples.

■ **EXAMPLE 1** _____

Consider the matrix

$$A = \begin{bmatrix} 8 & 5 & 0 \\ 3 & 2 & 0 \\ 0 & 0 & 0 \\ 0 & 0 & 0 \end{bmatrix}.$$

It has rank 2. The 4×4-matrix

$$E = \begin{bmatrix} 1 & 0 & 0 & 0 \\ 0 & 1 & 0 & 0 \\ 0 & 0 & 0 & 0 \\ 0 & 0 & 0 & 0 \end{bmatrix}$$

is a symmetric matrix of rank 2. It has the property that $A = EA$ and so is the left identity for A. The 3×3-symmetric matrix

$$F = \begin{bmatrix} 1 & 0 & 0 \\ 0 & 1 & 0 \\ 0 & 0 & 0 \end{bmatrix}$$

has rank 2 and satisfies $AF = A$. It is the right identity for A.

■ **EXAMPLE 2** _____

The left and right identities of the matrix

$$A = \begin{bmatrix} 1 & 1 \\ 2 & 2 \end{bmatrix}$$

are the matrices

$$E = \frac{1}{5}\begin{bmatrix} 1 & 2 \\ 2 & 4 \end{bmatrix}, \qquad F = \frac{1}{2}\begin{bmatrix} 1 & 1 \\ 1 & 1 \end{bmatrix},$$

as can be seen by computing EA and AF.

■ **EXAMPLE 3** _____

Let A be a matrix of rank 1. There exist vectors $\mathbf{u}$ and $\mathbf{v}$ such that

$$A = \mathbf{u}\mathbf{v}^T.$$

The matrices

$$E = \frac{\mathbf{u}\mathbf{u}^T}{\mathbf{u}^T\mathbf{u}}, \qquad F = \frac{\mathbf{v}\mathbf{v}^T}{\mathbf{v}^T\mathbf{v}}$$

are its left and right identities. This follows because they are symmetric rank 1 matrices having the property that

$$EA = \frac{\mathbf{u}\mathbf{u}^T\mathbf{u}\mathbf{v}^T}{\mathbf{u}^T\mathbf{u}} = \mathbf{u}\mathbf{v}^T = A, \qquad AF = \frac{\mathbf{u}\mathbf{v}^T\mathbf{v}\mathbf{v}^T}{\mathbf{v}^T\mathbf{v}} = \mathbf{u}\mathbf{v}^T = A.$$

Hereafter, whenever a matrix is designated by A, it will be understood that E and F are its left and right identities, respectively.

We shall establish Proposition 1 by proving the following result.

The orthogonal projection operator E for the range $R(A)$ of A is the left identity for A. The orthogonal projection operator F for the range $R(A^T)$ of A^T is the right identity for A.

We shall prove the second statement. The first statement follows in the same manner.

The orthogonal projection operator F of $R(A^T)$ has the property that it leaves the vectors $\mathbf{y}$ in $R(A^T)$ invariant; that is, $F\mathbf{y}=\mathbf{y}$. Since the columns of A^T are in $R(A^T)$ we have $FA^T = A^T$. By transposition we find that

$$AF = A.$$

The matrix F has rank r. It remains to prove that no other symmetric matrix H of rank r has the property that

$$AH = A.$$

For, let H be any matrix of rank r having $AH = A$. We first show that H has the same null space as A. To do so, let $\mathbf{z}$ be a solution of $H\mathbf{z} = \mathbf{0}$.

$$A\mathbf{z} = AH\mathbf{z} = 0.$$

It follows that the null space $N(H)$ of H is a subspace of the null space $N(A)$. Since they have the same dimension, namely $m - r$, they must be identical. That is, $N(H) = N(A)$.

Consider next the matrix $Z = F - H$. We have

$$AZ = A(F - H) = AF - AH = A - A = 0.$$

But since $N(F) = N(H) = N(A)$, we also have $FZ = 0$, and $HZ = 0$. Hence

$$Z^2 = (F - H)Z = FZ - HZ = 0.$$

Since Z is symmetric it follows that $Z = 0$ and hence that $H = F$, as was to be proved.

COROLLARY 1. When A has rank n, $F = I_n$, the $n \times n$-identity matrix. When A has rank m, then $E = I_m$, the $m \times m$-identity matrix. When A is a nonsingular square $n \times n$-matrix, then $E = F = I_n = I$.

As a second corollary we have

COROLLARY 2. F is the left identity for A^T and E is the right identity for A^T.

For by transposition, the equation

$$A = EA = AF \qquad \text{becomes} \qquad A^T = FA^T = A^TE.$$

The main result in this section is given in the following:

PROPOSITION 2. Let r be the rank of A. Let E and F be, respectively, the left and right identities for A. There is a unique solution X of the equation

$$AX = E, \qquad XA = F,$$

which is of rank r. This solution X will be called the **pseudoinverse** of A. It will be denoted by A^{-1}. We have

$$AA^{-1} = E, \qquad A^{-1}A = F.$$

There is no standard notation for the pseudoinverse of A. We have chosen to use the notation A^{-1}. The notation A^{-1} for the pseudoinverse of A is appropriate in view of the following result.

The pseudoinverse A^{-1} of a nonsingular square matrix A is the inverse of A.

This follows because when A is a nonsingular square matrix, $E = F = I$, so that $X = A^{-1}$ is the solution of the equations

$$AX = I, \qquad XA = I.$$

We shall give a constructive proof of Proposition 2. To do so we select a basis operator U for the range $R(A)$ of A. For example, we can select the columns of U to be r linearly independent columns of A. The matrix U is an $m \times r$-matrix of rank r. According to the results described in Section 6.4, the matrix

$$E = UH^{-1}U^T \qquad \text{with} \quad H = U^TU$$

is the orthogonal projection for $R(A)$ and hence is the left identity for A.

We next select an $r \times n$-matrix V so that V^T is a basis operator for the range $R(A^T)$ of A^T. For example, we can choose the rows of V to be r linearly independent rows of A. By replacing U by V^T, we see that the matrix

$$F = V^TK^{-1}V \qquad \text{with} \quad K = VV^T$$

is the right identity for A.

Observe next that

$$EAF = AF = A.$$

Hence

$$A = EAF = UH^{-1}U^TAV^TK^{-1}V.$$

The $r \times r$-matrix

$$L = U^T A V^T$$

therefore has the property that

$$A = UH^{-1}LK^{-1}V.$$

Because A has rank r, the $r \times r$-matrix L has rank r. We shall show that

$$X = V^T L^{-1} U^T$$

has the properties described in Proposition 2. We have

$$AX = UH^{-1}LK^{-1}VV^T L^{-1} U^T = UH^{-1}LK^{-1}KL^{-1}U^T = UH^{-1}LL^{-1}U^T$$
$$= UH^{-1}U^T = E.$$

Similarly,

$$XA = V^T L^{-1} U^T UH^{-1}LK^{-1}V = V^T K^{-1} V = F.$$

It remains to show the uniqueness of this solution of $AX = E$ and $XA = F$. Let us postpone the proof of uniqueness and accept the fact that the solution is unique. We then have the following result.

PROPOSITION 3. Let the columns of U be a basis for the range of A. Select V such that the columns of V^T is a basis for the range of A^T. Set

$$L = U^T A V^T.$$

Then the pseudoinverse A^{-1} of A is given by the formula

$$A^{-1} = V^T L^{-1} U^T.$$

This formula for the pseudoinverse is called *MacDuffee's formula.* It is of the same type as the one originally given by E. H. Moore, one of the originators of the pseudoinverse.

Observe that it is not necessary to compute the left and right identities E and F for A in order to find A^{-1}. It should be noted, however, that F is the left identity for A^{-1} and E is its right identity. Combining this fact with the relations

$$A^{-1}A = F, \qquad AA^{-1} = E,$$

we see that A is the pseudoinverse of A^{-1}. Transposing, we have

$$A^T(A^{-1})^T = F, \qquad (A^{-1})^T A^T = E.$$

It follows that $(A^{-1})^T$ is the pseudoinverse of A^T. It will be convenient to introduce the special notation A^{-T} for the pseudoinverse of A^T. We record these results in the following:

COROLLARY. The pseudoinverse of A^{-1} is A. The pseudoinverse of A^T is $A^{-T} = (A^{-1})^T$.

Thus the pseudoinverse of the pseudoinverse of A is A. The pseudoinverse of the transpose of A is the transpose of the pseudoinverse of A.

■ **EXAMPLE 4**

Consider the matrices

$$A = \begin{bmatrix} 8 & 5 & 0 \\ 3 & 2 & 0 \\ 0 & 0 & 0 \\ 0 & 0 & 0 \end{bmatrix}, \quad U = \begin{bmatrix} 1 & 0 \\ 0 & 1 \\ 0 & 0 \\ 0 & 0 \end{bmatrix}, \quad V = \begin{bmatrix} 1 & 0 & 0 \\ 0 & 1 & 0 \end{bmatrix}.$$

The matrix U is a basis operator for the range of A and V^T is a basis operator for the range of A^T. We have

$$L = U^TAV^T = \begin{bmatrix} 8 & 5 \\ 3 & 2 \end{bmatrix}, \quad L^{-1} = \begin{bmatrix} 2 & -5 \\ -3 & 8 \end{bmatrix}.$$

It follows that

$$A^{-1} = V^TL^{-1}U^T = \begin{bmatrix} 2 & -5 & 0 & 0 \\ -3 & 8 & 0 & 0 \\ 0 & 0 & 0 & 0 \end{bmatrix}.$$

■ **EXAMPLE 5**

The matrices

$$A = \begin{bmatrix} 1 & 1 \\ 2 & 2 \end{bmatrix}, \quad U = \begin{bmatrix} 1 \\ 2 \end{bmatrix}, \quad V = [1 \quad 1]$$

have the property that U is a basis operator for the range of A and V^T is a basis operator for the range of A^T. In this case

$$L = U^TAV^T = 10.$$

Hence $L^{-1} = \dfrac{1}{10}$, so that

$$A^{-1} = V^TL^{-1}U^T = \frac{1}{10}V^TU^T = \frac{1}{10}\begin{bmatrix} 1 \\ 1 \end{bmatrix}[1 \quad 2] = \frac{1}{10}\begin{bmatrix} 1 & 2 \\ 1 & 2 \end{bmatrix} = \frac{1}{10}A^T.$$

■ **EXAMPLE 6**

Consider the case in which

$$A = \begin{bmatrix} 1 & 0 \\ 0 & 1 \\ 1 & 1 \end{bmatrix}, \quad U = A, \quad V = \begin{bmatrix} 1 & 0 \\ 0 & 1 \end{bmatrix}.$$

Here A is a basis operator for $R(A)$ and $V = V^T$ is a basis operator for $R(A^T)$. In this case

$$L = U^T A V^T = A^T A = \begin{bmatrix} 2 & 1 \\ 1 & 2 \end{bmatrix}, \qquad L^{-1} = \frac{1}{3} \begin{bmatrix} 2 & -1 \\ -1 & 2 \end{bmatrix}.$$

Hence

$$A^{-1} = V^T L^{-1} U^T = \frac{1}{3} \begin{bmatrix} 2 & -1 \\ -1 & 2 \end{bmatrix} \begin{bmatrix} 1 & 0 & 1 \\ 0 & 1 & 1 \end{bmatrix}$$

$$= \frac{1}{3} \begin{bmatrix} 2 & -1 & 1 \\ -1 & 2 & 1 \end{bmatrix}.$$

■ **EXAMPLE 7** ⸻⸻⸻⸻⸻⸻⸻⸻⸻⸻⸻⸻⸻⸻⸻

Let A be the matrix

$$A = \begin{bmatrix} 1 & 0 & 0 & 1 \\ 0 & 1 & 1 & 0 \end{bmatrix}.$$

In this case $V = A$ has the property that $V^T = A^T$ is a basis operator for the range of A^T. The matrix $U = I_2$ is a basis operator for the range of A. We have

$$L = U^T A V^T = A A^T = \begin{bmatrix} 2 & 0 \\ 0 & 2 \end{bmatrix}, \qquad L^{-1} = \begin{bmatrix} 1/2 & 0 \\ 0 & 1/2 \end{bmatrix}.$$

Hence

$$A^{-1} = V^T L^{-1} U^T = A^T (A A^T)^{-1} = \begin{bmatrix} 1/2 & 0 \\ 0 & 1/2 \\ 0 & 1/2 \\ 1/2 & 0 \end{bmatrix}.$$

Observe that in this example,

$$A^{-1} = \frac{1}{2} A^T.$$

■ **EXAMPLE 8** ⸻⸻⸻⸻⸻⸻⸻⸻⸻⸻⸻⸻⸻⸻⸻

Consider the matrices

$$A = \begin{bmatrix} 0 & 1 & 0 \\ 0 & 0 & 1 \\ 0 & 0 & 0 \end{bmatrix}, \qquad U = \begin{bmatrix} 1 & 0 \\ 0 & 1 \\ 0 & 0 \end{bmatrix}, \qquad V = \begin{bmatrix} 0 & 1 & 0 \\ 0 & 0 & 1 \end{bmatrix}.$$

The matrix U is a basis operator for the range of A and V^T is a basis operator for the range of A^T. We have

$$L = U^T A V^T = \begin{bmatrix} 1 & 0 \\ 0 & 1 \end{bmatrix}.$$

The pseudoinverse of A is therefore

$$A^{-1} = V^T L^{-1} U^T = V^T U^T = \begin{bmatrix} 0 & 0 & 0 \\ 1 & 0 & 0 \\ 0 & 1 & 0 \end{bmatrix} = A^T.$$

Note that here we also have $U^{-1} = U^T$ and $V^{-1} = V^T$.

■ **EXAMPLE 9** _____

Consider the zero matrices

$$A = \begin{bmatrix} 0 & 0 \\ 0 & 0 \\ 0 & 0 \end{bmatrix}, \qquad E = \begin{bmatrix} 0 & 0 & 0 \\ 0 & 0 & 0 \\ 0 & 0 & 0 \end{bmatrix}, \qquad F = \begin{bmatrix} 0 & 0 \\ 0 & 0 \end{bmatrix}.$$

The matrix E is the left identity for A and F is its right identity. The zero matrix A^T is the pseudoinverse of A. This illustrates the fact that the pseudoinverse of any zero matrix A is its transpose.

■ **EXAMPLE 10** _____

Let A be an orthogonal projection operator. It is a symmetric matrix with the property that $A^2 = A$. It follows that the left and right identity of A is A itself. Moreover, $A^{-1} = A$. Hence we have a situation in which

$$A = A^T = A^{-1} = A^{-T}.$$

Return to the study of an $m \times n$-matrix A. Consider first the case in which A has rank n. Then, as in Example 6, $U = A$ is a basis operator for the range of A. The matrix $V = V^T = I_n$, the $n \times n$-identity, is a basis operator for the range of A^T. We have

$$L = U^T A V^T = A^T A, \qquad L^{-1} = (A^T A)^{-1}.$$

It follows that

$$A^{-1} = V^T L^{-1} U^T = (A^T A)^{-1} A^T.$$

On the other hand, when A has rank m, we can select $U = I_m$, $V = A$. In this case,

$$L = A A^T, \qquad A^{-1} = A^T (A A^T)^{-1}.$$

This gives us the following:

PROPOSITION 4. Let A be an $m \times n$-matrix. If A has rank n, then $A^T A$ is nonsingular and

$$A^{-1} = (A^T A)^{-1} A^T.$$

On the other hand, if A has rank m, then AA^T is nonsingular and

$$A^{-1} = A^T (AA^T)^{-1}.$$

Observe that the matrices U and V appearing in Proposition 3 are, respectively, $m \times r$ and $r \times n$ and of rank r. It follows from the result just obtained that

$$U^{-1} = (U^T U)^{-1} U^T, \qquad V^{-1} = V^T (VV^T)^{-1}.$$

In Proposition 3 no special conditions were imposed on the choice of U and V. Suppose that having chosen U we choose $V = U^T A$, as is permissible (why?). Then

$$L = U^T A V^T = VV^T$$

and

$$A^{-1} = V^T L^{-1} U^T = V^T (VV^T)^{-1} U^T = V^{-1} U^T.$$

On the other hand, had we known V and chosen $U = AV^T$, we would have

$$L = U^T U, \qquad A^{-1} = V^T (U^T U)^{-1} U^T = V^T U^{-1},$$

as one readily verifies.

This gives the following:

PROPOSITION 5. Let A be an $m \times n$-matrix. Its pseudoinverse A^{-1} can be found by either of the following two methods.

(a) Choose a basis operator U for the range of A. Set $V = U^T A$. Then

$$A^{-1} = V^T (VV^T)^{-1} U^T = V^{-1} U^T.$$

(b) Choose V so that V^T is a basis operator for the range of A^T. Set $U = AV^T$. Then

$$A^{-1} = V^T (U^T U)^{-1} U^T = V^T U^{-1}.$$

By applying method (a) to each of the matrices

$$[A \quad 0], \quad [0 \quad A],$$

we obtain the following:

COROLLARY

$$[A \quad 0]^{-1} = \begin{bmatrix} A^{-1} \\ 0^T \end{bmatrix}, \qquad [0 \quad A]^{-1} = \begin{bmatrix} 0^T \\ A^{-1} \end{bmatrix}.$$

The details of the proof of this corollary will be left as an exercise. We apply methods (a) and (b) for finding A^{-1} in the following.

■ **EXAMPLE 11** ⎯⎯⎯⎯⎯⎯⎯⎯⎯⎯⎯⎯⎯⎯⎯⎯⎯⎯⎯⎯⎯⎯⎯⎯⎯⎯⎯⎯⎯

Consider the matrices

$$A = \begin{bmatrix} 1 & 0 & 3 \\ 0 & 2 & 0 \\ 1 & 2 & 3 \end{bmatrix}, \qquad U = \begin{bmatrix} 1 & 0 \\ 0 & 1 \\ 1 & 1 \end{bmatrix}.$$

The columns of U are scaled columns of A. They generate the range of A. Compute

$$V = U^T A = \begin{bmatrix} 2 & 2 & 6 \\ 1 & 4 & 3 \end{bmatrix}, \quad VV^T = \begin{bmatrix} 44 & 28 \\ 28 & 26 \end{bmatrix}, \quad (VV^T)^{-1} = \frac{1}{360} \begin{bmatrix} 26 & -28 \\ -28 & 44 \end{bmatrix}.$$

We have

$$V^{-1} = V^T(VV^T)^{-1} = \frac{1}{30} \begin{bmatrix} 2 & -1 \\ -5 & 10 \\ 6 & -3 \end{bmatrix}, \quad A^{-1} = V^{-1}U^T = \frac{1}{30} \begin{bmatrix} 2 & -1 & 1 \\ -5 & 10 & 5 \\ 6 & -3 & 3 \end{bmatrix}.$$

Using method (b) in Proposition 4 instead of method (a), we would proceed as follows. Choose

$$V = \begin{bmatrix} 2 & 0 & 6 \\ 0 & 10 & 0 \end{bmatrix},$$

whose rows are scaled rows of A. Set

$$U = AV^T = \begin{bmatrix} 20 & 0 \\ 0 & 20 \\ 20 & 20 \end{bmatrix} = 20 \begin{bmatrix} 1 & 0 \\ 0 & 1 \\ 1 & 1 \end{bmatrix}.$$

By Example 6 we have

$$\begin{bmatrix} 1 & 0 \\ 0 & 1 \\ 1 & 1 \end{bmatrix}^{-1} = \frac{1}{3} \begin{bmatrix} 2 & -1 & 1 \\ -1 & 2 & 1 \end{bmatrix}, \qquad \text{and}$$

$$A^{-1} = \frac{1}{60} \begin{bmatrix} 4 & -2 & 2 \\ -10 & 20 & 10 \\ 12 & -6 & 6 \end{bmatrix} = \frac{1}{30} \begin{bmatrix} 2 & -1 & 1 \\ -5 & 10 & 5 \\ 6 & -3 & 3 \end{bmatrix}.$$

PROPOSITION 6. Let B be an $n \times q$-matrix whose left identity is the right identity F of our $m \times n$-matrix A. The pseudoinverse C^{-1} of the product

$$C = AB$$

is given by the formula

$$C^{-1} = B^{-1}A^{-1}.$$

The matrices A, B, and C have the same rank r.

For let

$$X = B^{-1}A^{-1}.$$

We have

$$CX = ABB^{-1}A^{-1} = AFA^{-1} = AA^{-1} = E,$$

the left identity for A. Denoting the right identity for B by K, we have

$$XC = B^{-1}A^{-1}AB = B^{-1}FB = B^{-1}B = K.$$

Because $C = AB$, the rank s of C cannot exceed the rank r of A. From the equation $CX = E$, we see that $s \geq r$, the common rank of E and A. Hence C has rank r. Since A^{-1} has rank r, it follows from the relations $X = B^{-1}A^{-1}$ and $CX = E$, that X has rank r also. Consequently, X is the pseudoinverse of C. We leave it as an exercise to show that B has rank r.

■ **EXAMPLE 12**

Consider the matrix

$$A = \begin{bmatrix} 1 & 0 & 0 & 1 \\ 0 & 1 & 1 & 0 \\ 1 & 1 & 1 & 1 \end{bmatrix} = \begin{bmatrix} 1 & 0 \\ 0 & 1 \\ 1 & 1 \end{bmatrix} \begin{bmatrix} 1 & 0 & 0 & 1 \\ 0 & 1 & 1 & 0 \end{bmatrix} = UV.$$

In Examples 6 and 7 we found that

$$U^{-1} = \frac{1}{3} \begin{bmatrix} 2 & -1 & 1 \\ -1 & 2 & 1 \end{bmatrix}, \qquad V^{-1} = \frac{1}{2} \begin{bmatrix} 1 & 0 \\ 0 & 1 \\ 0 & 1 \\ 1 & 0 \end{bmatrix}.$$

The pseudoinverse of A is therefore

$$A^{-1} = V^{-1}U^{-1} = \frac{1}{6} \begin{bmatrix} 2 & -1 & 1 \\ -1 & 2 & 1 \\ -1 & 2 & 1 \\ 2 & -1 & 1 \end{bmatrix}.$$

COROLLARY. Let A be an $m \times n$-matrix. Then

$$(A^TA)^{-1} = A^{-1}A^{-T}, \qquad (AA^T)^{-1} = A^{-T}A^{-1},$$

$$A^{-1} = (A^TA)^{-1}A^T = A^T(AA^T)^{-1},$$

no matter what the rank of A is.

Because the left identity E for A is the right identity for A^T, we have

$$(A^TA)^{-1} = A^{-1}A^{-T}$$

by Proposition 6. Consequently,

$$(A^TA)^{-1}A^T = A^{-1}A^{-T}A^T = A^{-1}E = A^{-1}.$$

Similarly,

$$(AA^T)^{-1} = A^{-T}A^{-1}, \qquad A^T(AA^T)^{-1} = A^TA^{-T}A^{-1} = FA^{-1} = A^{-1}.$$

PROPOSITION 7. Let E and F be, respectively, the left and right identities for an $m \times n$-matrix A. Similarly, let H and K be, respectively, the left and right identities for a second $m \times n$-matrix B. Suppose further that

$$A^TB = 0, \qquad BA^T = 0.$$

Then the matrix

$$C = A + B$$

has the following two properties:

(a) $C^{-1} = A^{-1} + B^{-1}$.

(b) $E + H$ is the left identity for C and $F + K$ is the right identity for C.

We first note that by the last corollary,

$$A^{-1}B = (A^TA)^{-1}A^TB = 0, \qquad BA^{-1} = BA^T(AA^T)^{-1} = 0.$$

Similarly, because $B^TA = (A^TB)^T = 0$ and $BA^T = (AB^T)^T = 0$, we have, by the same argument,

$$B^{-1}A = 0, \qquad AB^{-1} = 0.$$

The matrix

$$X = A^{-1} + B^{-1}$$

therefore has the property that

$$CX = (A + B)(A^{-1} + B^{-1}) = AA^{-1} + BB^{-1} = E + H$$

$$XC = (A^{-1} + B^{-1})(A + B) = A^{-1}A + B^{-1}B = F + K.$$

It remains to show that $E + H$ and $F + K$ are the right and left identities for C. To this end observe that

$$EH = AA^{-1}BB^{-1} = A0B^{-1} = 0, \qquad HE = (EH)^T = 0.$$

Hence

$$(E + H)^2 = E^2 + EH + HE + H^2 = E^2 + H^2 = E + H.$$

The matrix $E + H$, being symmetric, is an orthogonal projection matrix. We have

$$EA = A, \qquad EB = EHB = 0, \qquad HA = HEA = 0, \qquad HB = B.$$

Hence

$$(E + H)C = (E + H)(A + B) = EA + EB + HA + HB = A + B = C.$$

The matrix $E + H$ is therefore the left identity for C. In the same manner it is seen that $F + K$ is the right identity for C. This establishes Proposition 7.

PROPOSITION 8. Let B be a matrix whose columns are orthogonal to the columns of A. Equivalently, suppose that

$$A^T B = 0.$$

Then the pseudoinverse of

$$C = [A \quad B] \qquad \text{is} \qquad C^{-1} = \begin{bmatrix} A^{-1} \\ B^{-1} \end{bmatrix}.$$

This follows because

$$C = [A \quad B] = [A \quad 0] + [0 \quad B].$$

The matrices $[A \quad 0]$ and $[0 \quad B]$ have the property that the columns of one are orthogonal to the columns of the other. Similarly, the rows of one are orthogonal to the rows of the other. By Proposition 7 and the corollary to Proposition 5, we find that

$$C^{-1} = [A \quad 0]^{-1} + [0 \quad B]^{-1} = \begin{bmatrix} A^{-1} \\ 0^T \end{bmatrix} + \begin{bmatrix} 0^T \\ B^{-1} \end{bmatrix} = \begin{bmatrix} A^{-1} \\ B^{-1} \end{bmatrix},$$

as was to be proved.

■ **EXAMPLE 13**

Consider the matrices

$$A = \begin{bmatrix} 1 & 0 \\ 0 & 1 \\ 1 & 1 \end{bmatrix}, \quad B = \begin{bmatrix} 1 \\ 1 \\ -1 \end{bmatrix}.$$

The pseudoinverse of A was found in Example 6. We have

$$A^{-1} = \frac{1}{3}\begin{bmatrix} 2 & -1 & 1 \\ -1 & 2 & 1 \end{bmatrix}, \quad B^{-1} = \frac{1}{3}[1 \quad 1 \quad -1].$$

The inverse of

$$C = [A \quad B] = \begin{bmatrix} 1 & 0 & 1 \\ 0 & 1 & 1 \\ 1 & 1 & -1 \end{bmatrix} \quad \text{is therefore} \quad C^{-1} = \frac{1}{3}\begin{bmatrix} 2 & -1 & 1 \\ -1 & 2 & 1 \\ 1 & 1 & -1 \end{bmatrix}.$$

Of course, since C is a nonsingular square matrix, its inverse could be obtained by standard methods for finding inverses.

As we said earlier, the proof of Proposition 2 will not be complete until we show that there is only one solution X of rank r of the equations

$$AX = E, \quad XA = F.$$

We do so now. Observe that the equation $XA = F$ tells us that the range $R(F)$ of F is in the range $R(X)$ of X. Since these two linear spaces have the same dimension, namely r, they are identical. That is, $R(X) = R(F)$. Because F is an orthogonal projection operator for these spaces, it follows that F is the left identity for X. So $FX = X$. The matrix X therefore solves the equations

$$AX = E, \quad X = FX.$$

These equations have a unique solution. For suppose they have a second solution Y. Then

$$AY = E, \quad Y = FY.$$

The matrix $Z = X - Y$ satisfies the equations

$$AZ = A(X - Y) = AX - AY = E - E = 0,$$
$$FZ = FX - FY = X - Y = Z.$$

The first equation states that the columns of Z are in the null space of A. Since A and F have the null space we have $FZ = Z = 0$. Hence $Y = X$, as was to be proved.

We have just seen that $X = A^{-1}$ is the only solution for the equations

$$AX = E, \qquad X = FX.$$

Setting $G = I_n - F$, we shall show that X solves these equations if and only if X solves the equations

$$A^T A X = A^T, \qquad GX = 0.$$

Because

$$GX = (I_n - F)X = X - FX$$

it follows that $GX = 0$ if and only if $X = FX$. If $AX = E$, then

$$A^T = A^T E = A^T A X.$$

On the other hand, if $A^T A X = A^T$, we have

$$E = A^{-T} A^T = A^{-T} A^T A X = AX.$$

The two systems of equations are therefore equivalent.

Noting that $G = I_n - F$ is the orthogonal projection operator for the null space of A, we have

---

PROPOSITION 9. The matrix $X = A^{-1}$ is the only solution of the equations

$$A^T A X = A^T, \qquad GX = 0$$

where G is the orthogonal projection operator for the null space of A.

As an immediate corollary we have:

COROLLARY 1. The transpose A^T of A is the pseudoinverse of A if and only if $A^T A A^T = A^T$ and hence if and only if $A A^T A = A$.

This follows from Proposition 9 with $X = A^T$. Here we use the fact that $GA^T = 0$ holds because $AG = 0$.

COROLLARY 2. There is a unique least square solution of

$$A\mathbf{x} = \mathbf{h}$$

orthogonal to the null space of A. This solution is given by the formula

$$\mathbf{x} = A^{-1}\mathbf{h}.$$

This follows from the relations

$$A^T A x = A^T A A^{-1} h, \qquad G x = G A^{-1} h = 0 h = 0.$$

In the following proposition we do not assume ahead of time that E and F are left and right identities.

PROPOSITION 10. The matrix $X = A^{-1}$ is the only matrix for which the matrices

$$E = AX, \qquad F = XA$$

have the following two properties:
   (a) They are symmetric.
   (b) $A = EA, \quad X = FX.$

The criterion given for A^{-1} in this proposition is often used as a definition for the pseudoinverse of A.

Clearly $X = A^{-1}$ has these properties with E and F as the left and right identities for A.

Let X by any matrix having the properties described in Proposition 10. Let r be the rank of A. From the equations $E = AX$ and $A = EA$ we see that E has rank r. It follows from Proposition 1 that E is the left identity for A. Similarly, from the relations $F = XA$ and $X = FX$, it follows that X and F have rank r and that F is the left identity of X.

It remains to show that F is the right identity for A. Because F is symmetric we have $F = A^T X^T$. The range of F is therefore contained in the range of A^T. Since they have the same dimension r, it follows that the two ranges are the same. Since F is an orthogonal projection operator, it is the left identity for A^T. Hence F is the right identity for A, as was to be proved.

Exercises

1. Show that the pseudoinverse of a row or column vector is its transpose divided by the square of its length. Thus, if U is a column vector and V is a row vector, we have

$$U^{-1} = \frac{U^T}{U^T U}, \qquad V^{-1} = V^{-1} = \frac{V^T}{VV^T}.$$

2. Use the result described in Exercise 1 to find the pseudoinverses of the vectors

$$U = \begin{bmatrix} 1 \\ 2 \\ 3 \end{bmatrix}, \quad V = [1 \quad 0 \quad 1 \quad 0], \quad W = \begin{bmatrix} 0 \\ 1 \\ 1 \\ 1 \\ 1 \end{bmatrix}, \quad Y = [1 \quad 1 \quad 1 \quad 1].$$

3. Recall that if A is a matrix of rank 1, it is the product

$$A = UV$$

of a column vector U and a row vector V. Show that

$$A^{-1} = \frac{A^T}{L}, \qquad L = (U^T U)(V V^T).$$

Show this in two ways. First by Proposition 3 and second by Proposition 8 together with the result given in Exercise 1.

4. Referring to Exercises 2 and 3, find the pseudoinverses of the following matrices.

$$A = UV, \quad B = WV, \quad C = V^T Y, \quad A^T, \quad B^T, \quad C^T$$

Hint: Use the fact that $A^{-T} = (A^{-1})^T$.

5. Choose $b \neq 0$. Show that the pseudoinverse of $B = bA$ is $B^{-1} = (1/b)A^{-1}$. Show that A and B have the same left and right identities.

6. Referring to Exercise 4, find the pseudoinverse of the matrices

$$2A, \quad 5B, \quad -7C.$$

7. Show that $A^T B = 0$ if and only if $B^T A = 0$. Construct two matrices A and B such that $A^T B = 0$.

8. Use the formulas given in Proposition 4 or Proposition 5 to find the pseudoinverses of the following matrices.

$$P = \begin{bmatrix} 1 & 1 \\ 2 & -1 \\ 1 & 1 \end{bmatrix}, \quad Q = \begin{bmatrix} 2 & 0 & 1 \\ 1 & 1 & 1 \end{bmatrix}, \quad R = \begin{bmatrix} 3 & 4 & -3 \\ 4 & -3 & 1 \\ 2 & -1 & -1 \\ 1 & 2 & 7 \end{bmatrix}.$$

9. With P and Q as in Exercise 8, compute the pseudoinverse

$$A = PQ$$

by the use of Proposition 6.

10. Suppose that $A^T B = 0$. Let r be the rank of A and s the rank of B. Show that the rank of the matrix

$$C = [A \quad B]$$

is $r + s$.

11. Consider the matrices

$$A = \begin{bmatrix} 2 & 1 \\ 2 & 1 \\ -2 & 1 \\ -2 & 1 \end{bmatrix}, \quad B = \begin{bmatrix} 3 & -1 \\ -3 & 1 \\ -3 & -1 \\ 3 & 1 \end{bmatrix}.$$

Verify that $A^T B = 0$. Compute A^{-1} and B^{-1}. Also compute C^{-1} for $C = [A \quad B]$. (Because A and B are of rank 2, C is of rank 4 and so is nonsingular.)

12. Show that if $AB^T = 0$, the pseudoinverse of

$$C = \begin{bmatrix} A \\ B \end{bmatrix} \quad \text{is} \quad C^{-1} = [A^{-1} \quad B^{-1}].$$

Why does this follow from Proposition 8?

13. Suppose that $A^T B = 0$ and $AB^T = 0$. Let r be the rank of A and s be the rank of B. Show that the rank of $C = A + B$ is $r + s$.

14. Let A be an $m \times n$-matrix of rank r. Let G be the orthogonal projection operator of the null space of A. It has rank $n - r$. For $b \neq 0$, let

$$P = A^T A + bG.$$

Show that

(a) $A^T AG = 0$, $GA^T A = 0$.

(b) P has rank $r + (n - r) = n$ and so is nonsingular.

(c) $P^{-1} = (A^T A)^{-1} + (1/b)G$.

(d) $PA^{-1} = A^T$, $A^{-1} = P^{-1}A^T = (A^T A)^{-1}A^T$.

15. Use the method described in Exercise 14 to find the pseudoinverse of

$$A = \begin{bmatrix} 1 & 1 \\ 2 & 2 \end{bmatrix}.$$

That is, show that

$$A^T A = \begin{bmatrix} 5 & 5 \\ 5 & 5 \end{bmatrix}, \qquad G = \frac{1}{2}\begin{bmatrix} 1 & -1 \\ -1 & 1 \end{bmatrix},$$

$$P = A^T A + 2G = \begin{bmatrix} 6 & 4 \\ 4 & 6 \end{bmatrix} = 2\begin{bmatrix} 3 & 2 \\ 2 & 3 \end{bmatrix}, \qquad P^{-1} = \frac{1}{10}\begin{bmatrix} 3 & -2 \\ -2 & 3 \end{bmatrix}.$$

Hence

$$A^{-1} = \frac{1}{10}\begin{bmatrix} 3 & -2 \\ -2 & 3 \end{bmatrix}\begin{bmatrix} 1 & 2 \\ 1 & 2 \end{bmatrix} = \frac{1}{10}\begin{bmatrix} 1 & 2 \\ 1 & 2 \end{bmatrix}.$$

16. Let $A = [P \quad Q \quad R]$ with $P^T Q = 0$, $P^T R = 0$, $Q^T R = 0$. Show that

$$A^{-1} = \begin{bmatrix} P^{-1} \\ Q^{-1} \\ R^{-1} \end{bmatrix}.$$

Generalize.

17. Let A be a matrix whose columns are mutually orthogonal. Let B be the matrix obtained from A by dividing each nonzero column by the square of its length. Show that $A^{-1} = B^T$. Also show that $B = A^{-T}$ and $B^{-1} = A^T$. Verify that if

$$A = \begin{bmatrix} 1 & 0 & 0 \\ 0 & 1 & 0 \\ 1 & 0 & 0 \end{bmatrix}, \quad \text{then} \quad B = \begin{bmatrix} 1/2 & 0 & 0 \\ 0 & 1 & 0 \\ 1/2 & 0 & 0 \end{bmatrix} \quad \text{and} \quad A^{-1} = \begin{bmatrix} 1/2 & 0 & 1/2 \\ 0 & 1 & 0 \\ 0 & 0 & 0 \end{bmatrix}.$$

18. Use the method described in Exercise 17 to find the pseudoinverses of

$$\begin{bmatrix} 2 & 1 \\ 0 & 1 \\ 2 & -1 \end{bmatrix}, \quad \begin{bmatrix} 1 & 0 & -1 \\ 0 & 1 & 0 \\ 1 & 0 & 1 \end{bmatrix}, \quad \begin{bmatrix} 2 & -1 \\ 1 & 1 \\ 1 & 1 \end{bmatrix}, \quad \begin{bmatrix} 2 & -1 & 0 \\ 1 & 1 & 1 \\ 1 & 1 & -1 \end{bmatrix}.$$

Which of these pseudoinverses are inverses?

19. Show that if the columns of A are orthonormal, A^T is its psudoinverse. Show also that if the rows of A are orthonormal, A^T is its pseudoinverse. Observe that A need not be square. In each case construct an example to illustrate this result.

20. With the help of Proposition 2, show that $X = A^{-1}$ is the only solution of the equations

$$A^T A X = A^T, \quad X^T X A = X^T.$$

Hint: Let G be the orthogonal projection operator for the null space of A. Then $AG = 0$ and by the second equation we have

$$X^T X A G = X^T G = 0 \quad \text{and} \quad GX = 0.$$

Hence our relations imply the relation

$$A^T A X = A^T, \quad GX = 0$$

given in Proposition 9. Conversely, by setting $G = I - F = I - XA$ in $X^T G = 0$, we see that $X^T X A = X^T$. It follows that the equations in this exercise are equivalent to the equations in Proposition 9.

21. Sometimes it is more convenient to use $B = X^T = A^{-T}$ instead of $X = A^{-1}$. (B is of the same size as A. It has the same range and same null space as A.) Show that in terms of $B = X^T$ the equations in Exercise 20 can be written in the form (use transposition)

$$A = BA^T A, \quad B = BB^T A.$$

Conclude that these equations imply that $B = A^{-T}$.

22. Continuing with Exercise 21, show that with $B = A^{-T}$

$$E = AB^T, \quad F = B^T A$$

are the left and right identities for A. Conclude further that

$$A = BA^T A = AB^T A = AA^T B, \quad B = AB^T B = BA^T B = BB^T A.$$

23. Consider the matrices

$$L = \begin{bmatrix} 1 & 0 & 0 & -1 \\ 0 & 1 & 1 & -1 \end{bmatrix}, \quad N = \begin{bmatrix} 3/5 & -1/5 & -1/5 & -2/5 \\ -1/5 & 2/5 & 2/5 & -1/5 \end{bmatrix}.$$

Verify that $N = L^{-T}$, the pseudoinverse of L^T. *Hint:* Use the result given in Exercise 21.

24. Consider the matrices

$$K = \begin{bmatrix} 1 & 0 & 1 & 1 \\ 0 & 1 & -1 & 0 \\ 1 & 1 & 0 & 1 \end{bmatrix}, \qquad M = \begin{bmatrix} 1/5 & -1/15 & 4/15 & 1/5 \\ 0 & 1/3 & -1/3 & 0 \\ 1/5 & 4/15 & -1/15 & 1/5 \end{bmatrix}.$$

(a) Verify that M is the pseudoinverse of K^T.

(b) Referring to Exercise 23, show that the rows of L are orthogonal to the rows of K and to the rows of M.

(c) Show that the rows of N are orthogonal to the rows of M and to the rows of K.

(d) Conclude that $KL^T = 0$, $MN^T = 0$, $KN^T = 0$, $ML^T = 0$.

(e) Set

$$A = \begin{bmatrix} K \\ L \end{bmatrix}, \qquad B = \begin{bmatrix} M \\ N \end{bmatrix}.$$

With the help of part (d), show that B^T is the pseudoinverse of A.

25. Let A, B, K, L, M, and N be the matrices described in Exercises 23 and 24. Verify that A and B are the matrices

$$A = \begin{bmatrix} 1 & 0 & 1 & 1 \\ 0 & 1 & -1 & 0 \\ 1 & 1 & 0 & 1 \\ 1 & 0 & 0 & -1 \\ 0 & 1 & 1 & -1 \end{bmatrix}, \qquad B = \begin{bmatrix} 1/5 & -1/15 & 4/15 & 1/5 \\ 0 & 1/3 & -1/3 & 0 \\ 1/5 & 4/15 & -1/15 & 1/5 \\ 3/5 & -1/5 & -1/5 & -2/5 \\ -1/5 & 2/5 & 2/5 & -1/5 \end{bmatrix}.$$

Show that A has rank 4 so that the columns of A are linearly independent. Use the formula described in Proposition 4 to obtain the pseudoinverse of A. The answer is B^T.

26. Let

$$A = \begin{bmatrix} K \\ L \end{bmatrix}, \qquad B = \begin{bmatrix} M \\ N \end{bmatrix}$$

be matrices of the same size with K and M of the same size. Suppose that $KL^T = 0$.

(a) Show that if B^T is the pseudoinverse of A, then M^T is the pseudoinverse of K and N^T is the pseudoinverse of L. *Hint:* Use Exercise 12.

(b) Show conversely that if M^T and N^T are pseudoinverses of K and L, respectively, then B^T is the pseudoinverse of A.

27. Let A be a matrix of the form

$$A = \begin{bmatrix} K & P \\ L & 0 \end{bmatrix},$$

where the columns of P form a basis for the null space of K^T and the rows of L

form a basis for the null space of K. Show that A is nonsingular and is therefore invertible. Show that

$$A^{-1} = \begin{bmatrix} K^{-1} & L^{-1} \\ P^{-1} & 0 \end{bmatrix}.$$

28. Let D be the first of the matrices

$$D = \begin{bmatrix} 2 & 0 & 0 & 0 & 0 \\ 0 & 4 & 0 & 0 & 0 \\ 0 & 0 & 6 & 0 & 0 \\ 0 & 0 & 0 & 0 & 0 \\ 0 & 0 & 0 & 0 & 0 \end{bmatrix}, \quad D^{-1} = \begin{bmatrix} 1/2 & 0 & 0 & 0 & 0 \\ 0 & 1/4 & 0 & 0 & 0 \\ 0 & 0 & 1/6 & 0 & 0 \\ 0 & 0 & 0 & 0 & 0 \\ 0 & 0 & 0 & 0 & 0 \end{bmatrix}.$$

(a) Verify that the second matrix D^{-1} is the pseudoinverse of D.

(b) Conclude that for any diagonal matrix D, its pseudoinverse is obtained from D by replacing each nonzero main diagonal entry by its reciprocal.

(c) When the main diagonal entries of a diagonal matrix D are nonnegative, the matrix D is said to be nonnegative. The matrix obtained from a nonnegative diagonal matrix D by replacing each main diagonal entry by its square root is called the **square root** of D and is denoted by $D^{1/2}$. We have

$$(D^{1/2})^2 = D.$$

Verify that for the matrix D appearing in part (a), we have

$$D^{1/2} = \begin{bmatrix} \sqrt{2} & 0 & 0 & 0 & 0 \\ 0 & \sqrt{4} & 0 & 0 & 0 \\ 0 & 0 & \sqrt{6} & 0 & 0 \\ 0 & 0 & 0 & 0 & 0 \\ 0 & 0 & 0 & 0 & 0 \end{bmatrix}.$$

29. Let B be a symmetric matrix. Let X be an orthogonal matrix such that

$$D = X^T B X$$

is a diagonal matrix.

(a) Show that the pseudoinverse of B is given by the formula

$$B^{-1} = XD^{-1}X^T.$$

(b) The matrix B is said to be nonnegative if D is nonnegative. When B is nonnegative, the matrix

$$B^{1/2} = XD^{1/2}X^T$$

is called the **square root** of B. Verify that

$$(B^{1/2})^2 = B.$$

(Note that the square root of a nonnegative symmetric matrix B was defined earlier in Exercise 21 of Section 4.4.)

30. Consider the matrices

$$A = \begin{bmatrix} 0 & 2 & 0 \\ 0 & 0 & -3 \\ 0 & 0 & 0 \\ 0 & 0 & 0 \end{bmatrix}, \qquad P = \begin{bmatrix} 0 & 0 & 0 \\ 0 & 2 & 0 \\ 0 & 0 & 3 \end{bmatrix}, \qquad R = \begin{bmatrix} 0 & 1 & 0 \\ 0 & 0 & -1 \\ 0 & 0 & 0 \\ 0 & 0 & 0 \end{bmatrix}.$$

Show that

(a) $P = (A^T A)^{1/2}$.

(b) $P^{-1} = \begin{bmatrix} 0 & 0 & 0 \\ 0 & 1/2 & 0 \\ 0 & 0 & 1/3 \end{bmatrix}$.

(c) $R = AP^{-1}$, $\quad R^{-1} = R^T$, $\quad R = RR^T R$.

(d) $A = RP = RA^T R$.

(e) Show that $Q = RPR^T$ is the square root of AA^T.

(f) Show that the values of λ for which

$$A - \lambda R = \begin{bmatrix} 0 & 2 - \lambda & 0 \\ 0 & 0 & \lambda - 3 \\ 0 & 0 & 0 \\ 0 & 0 & 0 \end{bmatrix}$$

is of rank less than the rank 2 of A are the nonzero eigenvalues of P.

31. Let A be an $m \times n$-matrix with E and F as its left and right identities. Set

$$P = (A^T A)^{1/2}, \qquad R = AP^{-1}.$$

Show that

(a) P has the same null space as A. Conclude that F is the left and right identities for P and P^{-1}.

(b) $R = ER = RF$, $\quad RR^T = E$, $\quad R^T = F$. Conclude that $R^{-1} = R^T$. Verify that $RR^T R = R$.

(c) $P = R^T A = A^T R$.

(d) $A = RP = RA^T R$.

(e) Set $Q = RPR^T$. Show that $Q = AR^T = RA^T$, $\quad Q^2 = AA^T$. Conclude that $Q = (A^T A)^{1/2}$.

(f) Let r be the rank of A. Conclude that P and R have rank r. Show that the rank of $A - \lambda P$ is less than r if and only if λ is a nonzero eigenvalue of P. These eigenvalues are called *singular values* of A. They are also called *principal values* of A. The largest of these is often taken to be the norm of A.

6.8   INVARIANT SUBSPACES AND CANONICAL FORMS

Let **U** be an n-dimensional (real) linear space and let T be a linear transformation that maps **U** into itself. Recall that for each basis $\mathbf{e}_1, \mathbf{e}_2, \ldots, \mathbf{e}_n$

or equivalently for each basis operator

$$E = [\mathbf{e}_1 \quad \mathbf{e}_2 \quad \cdots \quad \mathbf{e}_n]$$

there corresponds a unique matrix

$$A = E^{-1}TE.$$

This matrix is a **matrix representation** of T relative to this basis. The relationship between T and A can also be written in the form

$$TE = EA.$$

If F is a second basis operator for $\mathbf{U}$, the matrix

$$B = F^{-1}TF$$

is a second matrix representation of T. The matrices A and B are similar. This follows because if we form the matrix $P = E^{-1}F$, then $P^{-1} = F^{-1}E$ and

$$B = F^{-1}TF = F^{-1}EE^{-1}TEE^{-1}F = P^{-1}AP.$$

Conversely, if a matrix B is similar to A, there is a nonsingular matrix P such that $P^{-1}AP = B$. The basis operator $F = EP$ then has the property that

$$F^{-1}TF = P^{-1}E^{-1}TEP = P^{-1}AP = B.$$

We have accordingly established the following result.

Let A be a matrix representation of the operator T. A second matrix B is a matrix representation of T if and only if B is similar to A.

It should be noted that if T is the identity operator on $\mathbf{U}$, that is, if $T\mathbf{u} = \mathbf{u}$ for every $\mathbf{u}$ in $\mathbf{U}$, then T has a unique matrix representation, namely, the identity matrix I. Accordingly, we shall use the same symbol I for the identity operator on $\mathbf{U}$ as for the identity matrix.

If $T\mathbf{u} = \mathbf{0}$ for every $\mathbf{u}$ in $\mathbf{U}$, then T is the **null operator** $T = 0$. Its matrix representation is the zero matrix $A = 0$.

The concepts of singularity and nonsingularity hold for an operator T as well as for matrices. We say that T is **singular** if there is a nonzero vector $\mathbf{u}$ such that $T\mathbf{u} = \mathbf{0}$. Otherwise, T is said to be **nonsingular**. It is easily seen that T is singular if it has a singular matrix representation. It follows that T is nonsingular only if its matrix representations are nonsingular.

Return now to the study of a general operator T that maps $\mathbf{U}$ into itself. The problem at hand is to find the "simplest" matrix representation C of T, a representation that gives us at a glance some of the basic properties of T. Such a representation C is called a **canonical representation** of T or briefly a **canonical matrix**. A canonical representation of T is also a canonical representation of each of its matrix representations. An operator T can have more than one canonical representation, each of which emphasizes particular properties of T. For example, if T has a diagonal matrix D as a matrix representation, then D is a canonical representation of T. Any permutation of the main diagonal entries of D yields another canonical representation of T.

Let us consider in more detail the case in which T has a diagonal matrix D as a canonical representation. To simplify our presentation we consider the case in which $\mathbf{U}$ is three-dimensional, that is, the case $n = 3$. The general case proceeds in the same manner. Suppose therefore that T has the diagonal matrix

$$D = \operatorname{diag}(\lambda_1, \lambda_2, \lambda_3) = \begin{bmatrix} \lambda_1 & 0 & 0 \\ 0 & \lambda_2 & 0 \\ 0 & 0 & \lambda_3 \end{bmatrix}$$

as a matrix representation. Then there is a basis operator

$$E = [\mathbf{e}_1 \quad \mathbf{e}_2 \quad \mathbf{e}_3]$$

such that

$$TE = ED.$$

This relation can be rewritten in the form

$$TE = [T\mathbf{e}_1 \quad T\mathbf{e}_2 \quad T\mathbf{e}_3] = [\mathbf{e}_1 \quad \mathbf{e}_2 \quad \mathbf{e}_3] \begin{bmatrix} \lambda_1 & 0 & 0 \\ 0 & \lambda_2 & 0 \\ 0 & 0 & \lambda_3 \end{bmatrix} = [\lambda_1 \mathbf{e}_1 \quad \lambda_2 \mathbf{e}_2 \quad \lambda_3 \mathbf{e}_3].$$

This relation is equivalent to the eigenvalue relations

$$T\mathbf{e}_1 = \lambda_1 \mathbf{e}_1, \qquad T\mathbf{e}_2 = \lambda_2 \mathbf{e}_2, \qquad T\mathbf{e}_3 = \lambda_3 \mathbf{e}_3.$$

As in the case of matrices, a real number λ is an eigenvalue of T if there is a nonzero vector $\mathbf{e}$ in $\mathbf{U}$ such that

$$T\mathbf{e} = \lambda \mathbf{e}.$$

The vector $\mathbf{e}$ is a corresponding eigenvector of T. It follows that the diagonal entries $\lambda_1, \lambda_2, \lambda_3$ of D are eigenvalues of T and the basis vectors $\mathbf{e}_1, \mathbf{e}_2, \mathbf{e}_3$ are corresponding eigenvectors of T.

This illustrates the following result.

PROPOSITION 1. A linear transformation T has a diagonal matrix D as a canonical representation if and only if it has a basis comprised of n eigenvectors $\mathbf{e}_1, \mathbf{e}_2, \ldots, \mathbf{e}_n$ of T. The main diagonal entries $\lambda_1, \lambda_2, \ldots, \lambda_n$ of D are the corresponding eigenvalues of T.

This result can be given a geometrical interpretation as follows. An eigenvector $\mathbf{e}_k$ of T determines a line $\mathbf{L}_k$ comprised of all multiples $a\mathbf{e}_k$ of $\mathbf{e}_k$. We call such a line an *eigenvector line*. It is a one-dimensional linear subspace of $\mathbf{U}$. When T has n linearly independent eigenvectors $\mathbf{e}_1, \mathbf{e}_2, \ldots, \mathbf{e}_n$, these eigenvectors form a basis for $\mathbf{U}$ so that every vector $\mathbf{u}$ in $\mathbf{U}$ is expressible in the form

$$\mathbf{u} = a_1\mathbf{e}_1 + a_2\mathbf{e}_2 + \cdots + a_n\mathbf{e}_n.$$

Since $a_1\mathbf{e}_1$ is in $\mathbf{L}_1$, $a_2\mathbf{e}_2$ is in $\mathbf{L}_2$, and so on, it follows that our space $\mathbf{U}$ is the direct sum

$$\mathbf{U} = \mathbf{L}_1 + \mathbf{L}_2 + \cdots + \mathbf{L}_n$$

of these eigenvector lines. This gives the following result.

PROPOSITION 2. A linear transformation T on $\mathbf{U}$ into $\mathbf{U}$ has a diagonal matrix D as a matrix representation if and only if $\mathbf{U}$ is the direct sum

$$\mathbf{U} = \mathbf{L}_1 + \mathbf{L}_2 + \cdots + \mathbf{L}_n$$

of eigenvector lines.

An eigenvector line $\mathbf{L}$ has an additional property in that it is invariant under T. A linear subspace $\mathbf{V}$ of $\mathbf{U}$ is an *invariant subspace* under T if $T\mathbf{u}$ is in $\mathbf{V}$ whenever $\mathbf{u}$ is in $\mathbf{V}$. To see that an eigenvector line $\mathbf{L}$ is invariant under T, we note that $\mathbf{L}$ is comprised of all multiples $\mathbf{u} = a\mathbf{e}$ of an eigenvector $\mathbf{e}$ corresponding to an eigenvalue λ. The vector $T\mathbf{u} = T(a\mathbf{e}) = aT\mathbf{e} = a\lambda\mathbf{e}$ is also a multiple of $\mathbf{e}$ and so is in $\mathbf{L}$. Consequently, $\mathbf{L}$ is invariant under T, as was to be shown. Further examples of invariant subspaces will be given later.

Turn now to the general case in which T need not have n linearly independent eigenvectors. Let E be a basis operator and let

$$A = E^{-1}TE$$

be the corresponding matrix representation of T. Then

$$A^2 = E^{-1}TEE^{-1}TE = E^{-1}T^2E, \qquad A^3 = AA^2 = E^{-1}TEE^{-1}T^2E = E^{-1}T^3E,$$

and so, for each integer k,

$$A^k = E^{-1}T^kE.$$

Consequently, for a polynomial such as

$$p(\lambda) = a + b\lambda + c\lambda^2 + d\lambda^3,$$

we have

$$p(A) = aI + bA + cA^2 + dA^3 = E^{-1}(aI + bT + cT^2 + dT^3)E = E^{-1}p(T)E.$$

Similarly, for any polynomial $p(\lambda)$, we have

$$p(A) = E^{-1}p(T)E, \qquad p(T) = Ep(A)E^{-1}.$$

In other words, $p(A)$ is the matrix representation of the operator $p(T)$. Similarly, if B is a second matrix representation of T, then $p(B)$ is also a matrix representation of $p(T)$.

We have the following result:

> For any polynomial $p(\lambda)$, the nullspace of $p(T)$ is invariant under T.

This follows because

$$P(T)\mathbf{u} = \mathbf{0} \qquad \text{implies that} \qquad p(T)T\mathbf{u} = Tp(T)\mathbf{u} = \mathbf{0}.$$

As noted in Chapter 4, similar matrices have the same characteristic polynomial. Accordingly the matrix representations of T, being similar, have the same characteristic polynomial

$$\Delta(\lambda) = a_0 + a_1(-\lambda) + a_2(-\lambda)^2 + \cdots + a_{n-1}(-\lambda)^{n-1} + (-\lambda)^n.$$

By the Cayley–Hamilton theorem we have $\Delta(A) = 0$ for each matrix representation A of T. It follows that

$$\Delta(T) = E\Delta(A)E^{-1} = 0$$

also. We define $\Delta(\lambda)$ to be the ***characteristic polynomial*** for T. Its zeros, real or complex, are the eigenvalues of T as well as of each matrix representation A of T. For example, when $\Delta(\lambda)$, in factored form, is the polynomial

$$\Delta(\lambda) = (1 - \lambda)^4(2 - \lambda)^3(1 + \lambda^2)(13 - 4\lambda + \lambda^2)^2$$

then its distinct eigenvalues are $1, 2, i, -i, 2 + 3i$, and $2 - 3i$. The eigenvalue $\lambda = 1$ is of multiplicity 4, $\lambda = 2$ is of multiplicity 3, each of the pair of eigenvalues $\lambda = i$ or $-i$ is of multiplicity 1, while the pair of eigenvalues $2 + 3i$ and $2 - 3i$ are each of multiplicity 2. Observe that complex eigenvalues are determined by irreducible quadratic factors of $\Delta(\lambda)$. In the general case, the characteristic polynomial $\Delta(\lambda)$ is expressible as the product

$$\Delta(\lambda) = p_1(\lambda)p_2(\lambda) \cdots p_m(\lambda),$$

of polynomials, one for each of the real eigenvalues, and one for each pair of complex conjugate eigenvalues. These polynomials, which we call ***eigenfactors*** of $\Delta(\lambda)$, are one of the forms

$$p_j(\lambda) = (\lambda_j - \lambda)^{m_j}, \qquad p_k(\lambda) = (b_k^2 + (a_k - \lambda)^2)^{m_k}.$$

In the first case, λ_j is a real eigenvalue of multiplicity m_j. In the second case $b_k > 0$ and each of the eigenvalues $a_k + ib_k$ and $a_k - ib_k$ is of multiplicity m_k. The nullspace $\mathbf{C}_j$ of $p_j(T)$ will be called the ***eigenspace*** of T corresponding to the real eigenvalue λ_j. Similarly, the nullspace $\mathbf{C}_k$ of $p_k(T)$ will be called the *eigenspace* of T corresponding to the pair of complex eigenvalues $a_k + ib_k$ and $a_k - ib_k$.

The main result to be proved is given in the following:

PROPOSITION 3. The space $\mathbf{U}$ is the direct sum

$$\mathbf{U} = \mathbf{C}_1 + \mathbf{C}_2 + \cdots + \mathbf{C}_m$$

of the eigenspaces of T. Each of these eigenspaces is invariant under T. For $i = 1, \ldots, m$, let B_i be a matrix representation of T restricted to $\mathbf{C}_i$. The block diagonal matrix

$$B = \text{diag}(B_1, B_2, \ldots, B_m)$$

is a canonical matrix representation of T on $\mathbf{U}$. The characteristic polynomial for B_j is the eigenfactor $p_j(\lambda)$.

Other canonical matrices are obtained by specifying the nature of B_j.

Before establishing this result let us look at the following example. In this example the matrices used are already in canonical form.

■ **EXAMPLE 1** _____

Consider the case in which $\mathbf{U}$ is $\mathbf{R}^3$ and T is one of the matrices

$$A = \begin{bmatrix} 2 & 0 & 0 \\ 0 & 2 & 0 \\ 0 & 0 & 2 \end{bmatrix}, \qquad B = \begin{bmatrix} 2 & 1 & 0 \\ 0 & 2 & 0 \\ 0 & 0 & 2 \end{bmatrix}, \qquad C = \begin{bmatrix} 2 & 1 & 0 \\ 0 & 2 & 1 \\ 0 & 0 & 2 \end{bmatrix}.$$

In each case the characteristic polynomial is $\Delta(\lambda) = (2 - \lambda)^3$ and $\Delta(\lambda)$ has only one eigenfactor, namely, $(2 - \lambda)^3$. Also, in each case, the eigenspace of $\lambda = 2$ is the whole space $\mathbf{R}^3$. When $T = A$, every line through the origin is an eigenvector line. Also, every 2-plane through the origin is an invariant subspace under $T = A$.

When $T = B$, using the familiar *xyz*-coordinates, the *xz*-plane is invariant under T. Every line in this plane through the origin is an eigenvector line. The *xy*-plane is also invariant under $T = B$, but it contains only one eigenvector line, namely, the *x*-axis.

When $T = C$, the *x*-axis is the only eigenvector line under T. The *xy* plane is invariant under $T = C$ but it is not an eigenspace. Of course, the whole space $\mathbf{R}^3$ is invariant under T in each of these three cases.

On the other hand, when T is one of the first two of the matrices

$$F = \begin{bmatrix} 2 & 0 & 0 \\ 0 & 2 & 0 \\ 0 & 0 & 3 \end{bmatrix}, \qquad G = \begin{bmatrix} 2 & 1 & 0 \\ 0 & 2 & 0 \\ 0 & 0 & 3 \end{bmatrix}, \qquad H = \begin{bmatrix} 1 & 0 & 0 \\ 0 & 2 & 0 \\ 0 & 0 & 3 \end{bmatrix},$$

its characteristic polynomial $\Delta(\lambda) = (2 - \lambda)^2(3 - \lambda)$ has two eigenfactors, namely $p_1(\lambda) = (2 - \lambda)^2$ and $p_2(\lambda) = 3 - \lambda$. Computing the matrices

$$p_1(F) = (2I - F)^2, \qquad p_2(F) = 3I - F$$

as well as $p_1(G)$ and $p_2(G)$, we find that $p_1(G) = p_1(F)$ and that

$$p_1(F) = \begin{bmatrix} 0 & 0 & 0 \\ 0 & 0 & 0 \\ 0 & 0 & 1 \end{bmatrix}, \quad p_2(F) = \begin{bmatrix} 1 & 0 & 0 \\ 0 & 1 & 0 \\ 0 & 0 & 0 \end{bmatrix}, \quad p_2(G) = \begin{bmatrix} 1 & -1 & 0 \\ 0 & 1 & 0 \\ 0 & 0 & 0 \end{bmatrix}.$$

Looking at the last two matrices, we see that for F and for G, the eigenspace $\mathbf{C}_2$ corresponding to the eigenvalue $\lambda = 3$ is the z-axis. It is an eigenvector line. The eigenspace E_1 for the eigenvalue $\lambda = 2$ is the xy-plane. When $T = F$, every line in the xy-plane through the origin is an eigenvector line. When $T = G$, the x-axis is the only eigenvector line in the xy-plane. Observe that in each case $\mathbf{R}^3$ is the direct sum of the two eigenspaces $\mathbf{C}_1$ and $\mathbf{C}_2$.

When $T = H$, we have the situation considered earlier. The x-, y-, and z-axes are each eigenspaces corresponding to the eigenvalues 1, 2, and 3. Each eigenspace is also an eigenvector line. As noted in proposition 2, $\mathbf{R}^3$ is a direct sum of these three eigenspaces.

Finally, let us look at the case in which T is the 3×3-matrix

$$K = \begin{bmatrix} 2 & 3 & 0 \\ -3 & 2 & 0 \\ 0 & 0 & 5 \end{bmatrix}.$$

Its characteristic polynomial is

$$\Delta(\lambda) = (5 - \lambda)(9 + (2 - \lambda)^2).$$

The eigenfactors are accordingly

$$p_1(\lambda) = 5 - \lambda, \qquad p_2(\lambda) = 9 + (2 - \lambda)^2.$$

We have

$$p_1(K) = \begin{bmatrix} 3 & -3 & 0 \\ 3 & 3 & 0 \\ 0 & 0 & 0 \end{bmatrix}, \quad p_2(K) = \begin{bmatrix} 0 & 0 & 0 \\ 0 & 0 & 0 \\ 0 & 0 & 18 \end{bmatrix}.$$

The eigenspace $\mathbf{C}_1$, the nullspace of $p_1(K)$, is the z-axis. It is an eigenvector line. The nullspace of $p_2(K)$, the eigenspace $\mathbf{C}_2$, is the xy-plane. It contains no eigenvector lines. Again, $\mathbf{R}^3$ is the direct sum of $\mathbf{C}_1$ and $\mathbf{C}_2$.

Let us now turn to the proof of proposition 3. It is a consequence of the next proposition. In this proposition we express the characteristic polynomial $\Delta(\lambda)$ of T on $\mathbf{U}$ as the product

$$\Delta(\lambda) = p(\lambda)q(\lambda),$$

where $p(\lambda)$ and $q(\lambda)$ are each products of eigenfactors of $\Delta(\lambda)$. Then $p(\lambda)$ and $q(\lambda)$ have no zeros in common and so are relatively prime.

PROPOSITION 4. Let $p(\lambda)$ and $q(\lambda)$ be the relative prime factors of the characteristic polynomial $\Delta(\lambda)$ of T described above. Let **V** and **W** be the nullspaces of $p(T)$ and $q(T)$, respectively. Then:

(a) **V** and **W** are invariant subspaces of **U**.

(b) **V** is the range of $q(T)$ and **W** is the range of $p(T)$.

(c) **U** = **V** + **W**, a direct sum.

(d) The dimension of **V** equals the degree h of $p(\lambda)$. The dimension of **W** equals the degree k of $q(\lambda)$.

(e) Let B be a matrix representation of T restricted to **V** and let C be a matrix representation of T on **W**. Then the block diagonal matrix

$$A = \begin{bmatrix} B & 0 \\ 0 & C \end{bmatrix}$$

is a matrix representation of T on **U**.

(f) $p(\lambda)$ is the characteristic polynomial for B and $q(\lambda)$ is the characteristic polynomial for C.

This result is established in the Supplementary Proofs at the end of the section.

We are now in a position to show that proposition 3 is a consequence of Proposition 4. Let

$$\Delta(\lambda) = p_1(\lambda)p_2(\lambda) \cdots p_m(\lambda)$$

be the characteristic polynomial for T on **U**, expressed as the product of its eigenfactors. The proof in the general case is like that for the case $m = 3$. Applying Proposition 4 with

$$p(\lambda) = p_1(\lambda), \qquad q(\lambda) = p_2(\lambda)p_3(\lambda)$$

we find that **U** is the direct sum

$$\mathbf{U} = \mathbf{C}_1 + \mathbf{W}_1,$$

where $\mathbf{C}_1$ is an eigenspace, the nullspace of $p_1(\lambda)$, and $\mathbf{W}_1$ is the nullspace for $q(\lambda)$. Applying Proposition 4 again with $\mathbf{W}_1$ playing the role of **U** and with

$$p(\lambda) = p_2(\lambda), \qquad q(\lambda) = p_3(\lambda)$$

we find that $\mathbf{W}_1$ is the direct sum

$$\mathbf{W}_1 = \mathbf{C}_2 + \mathbf{C}_3,$$

where C_2 is the eigenspace for $p_2(\lambda)$ and C_3 is the eigenspace for $p_3(\lambda)$. It follows that U is the direct sum

$$U = C_1 + C_2 + C_3$$

of the eigenspaces determined by $p_1(\lambda), p_2(\lambda), p_3(\lambda)$, respectively. Moreover, in the same manner it is seen that T has a matrix representation

$$B = \text{diag } (B_1, B_2, B_3),$$

where for $j = 1, 2, 3$, B_j is a matrix representation of T restricted to C_j. In addition, $p_j(\lambda)$ is the characteristic polynomial for B_j.

As remarked earlier, we can replace B_j by any matrix similar to it. The matrix B_j either has one eigenvalue λ_j or else a pair of conjugate complex eigenvalues $a_j + ib_j$ and $a_j - ib_j$. It can be shown that in the first case B_j is similar to a block diagonal matrix

$$\text{diag } (J_1, J_2, \ldots, J_r)$$

called the *Jordan form*, where J_k is either the 1×1-matrix $[\lambda_j]$ or else is of the form $\lambda_j I + N$, where N has 1's on the superdiagonal and 0's elsewhere. For example, when J_k is a 3×3-matrix, it is the matrix

$$J_k = \begin{bmatrix} \lambda_j & 1 & 0 \\ 0 & \lambda_j & 1 \\ 0 & 0 & \lambda_j \end{bmatrix}.$$

The number of blocks is equal to the number of linearly independent eigenvectors B_j has in a maximal set. On the other hand when B_j has the complex eigenvalues $a_j + ib_j$ and $a_j - ib_j$, then J_k is either the 2×2-matrix

$$K_j = \begin{bmatrix} a_j & b_j \\ -b_j & a_j \end{bmatrix}$$

or else is of the form like that of the 6×6-case, which is

$$J_k = \begin{bmatrix} K_j & M & O \\ O & K_j & M \\ O & O & K_j \end{bmatrix}, \qquad M = \begin{bmatrix} 0 & 0 \\ 1 & 0 \end{bmatrix}, \qquad O = \begin{bmatrix} 0 & 0 \\ 0 & 0 \end{bmatrix}.$$

We shall not establish this result in this book.

Exercises

1. The whole space and the zero space are invariant subspaces of a matrix. Find the other invariant subspaces (if any):

   (a) Of the 2×2-matrices (use *xy*-coordinates):

$$A_1 = \begin{bmatrix} 2 & 0 \\ 0 & 3 \end{bmatrix}, \qquad A_2 = \begin{bmatrix} 2 & 0 \\ 0 & 2 \end{bmatrix}, \qquad A_3 = \begin{bmatrix} 2 & 0 \\ 0 & 2 \end{bmatrix}, \qquad A_4 = \begin{bmatrix} 2 & 1 \\ 0 & 2 \end{bmatrix},$$

$$A_5 = \begin{bmatrix} 2 & 1 \\ 0 & 3 \end{bmatrix}, \qquad A_6 = \begin{bmatrix} 4 & 3 \\ 3 & 4 \end{bmatrix}, \qquad A_7 = \begin{bmatrix} 4 & -3 \\ 3 & 4 \end{bmatrix}, \qquad A_8 = \begin{bmatrix} 0 & 0 \\ 1 & 3 \end{bmatrix}.$$

(b) Of the 3 × 3-matrices (use *x*-, *y*-, *z*-coordinates)

$$B_1 = \begin{bmatrix} 2 & 0 & 0 \\ 0 & 3 & 0 \\ 0 & 0 & 4 \end{bmatrix}, \qquad B_2 = \begin{bmatrix} 2 & 0 & 0 \\ 0 & 2 & 0 \\ 0 & 0 & 4 \end{bmatrix}, \qquad B_3 = \begin{bmatrix} 2 & 1 & 0 \\ 0 & 2 & 0 \\ 0 & 0 & 4 \end{bmatrix},$$

$$B_4 = \begin{bmatrix} 2 & 1 & 0 \\ 0 & 2 & 1 \\ 0 & 0 & 2 \end{bmatrix}, \qquad B_5 = \begin{bmatrix} 2 & 1 & 0 \\ 0 & 2 & 1 \\ 0 & 0 & 4 \end{bmatrix}, \qquad B_6 = \begin{bmatrix} 2 & 1 & 0 \\ 0 & 2 & 0 \\ 0 & 0 & 4 \end{bmatrix}.$$

2. The matrices

$$A = \begin{bmatrix} a & 0 & 0 & 0 \\ 0 & b & 0 & 0 \\ 0 & 0 & c & 0 \\ 0 & 0 & 0 & d \end{bmatrix}, \qquad B = \begin{bmatrix} a & 1 & 0 & 0 \\ 0 & a & 0 & 0 \\ 0 & 0 & c & 0 \\ 0 & 0 & 0 & d \end{bmatrix},$$

$$C = \begin{bmatrix} a & 1 & 0 & 0 \\ 0 & a & 0 & 0 \\ 0 & 0 & c & 1 \\ 0 & 0 & 0 & c \end{bmatrix}, \qquad D = \begin{bmatrix} a & 1 & 0 & 0 \\ 0 & a & 1 & 0 \\ 0 & 0 & a & 0 \\ 0 & 0 & 0 & d \end{bmatrix}, \qquad E = \begin{bmatrix} a & 1 & 0 & 0 \\ 0 & a & 1 & 0 \\ 0 & 0 & a & 1 \\ 0 & 0 & 0 & a \end{bmatrix}$$

are in Jordan form.

(a) By looking at their eigenvectors, show that no two of these matrices are similar.

(b) Suppose that the numbers a, b, c, and d are distinct. For each of these matrices find its eigenfactors and its eigenspaces.

(c) Discuss the case in which $a = b = c = d$. Show that in this case $(a - \lambda)^4$ is the only eigenfactor and the whole space is the only eigenspace of these matrices.

3. A square matrix A (and hence a linear transformation T mapping a linear space **U** into **U**) has associated with it a special polynomial $m(\lambda)$, called a ***minimum polynomial***. It is a polynomial of lowest degree such that $m(A) = 0$. It is unique apart from a nonzero scale factor.

(a) Let $p(\lambda)$ be a polynomial such that $p(A) = 0$. Show that $p(\lambda)$ is divisible by a minimum polynomial $m(\lambda)$. *Hint:* Use the division algorithm $p(\lambda) = q(\lambda)m(\lambda) + r(\lambda)$.

(b) Use this result to show that $m(\lambda)$ is unique, apart from a nonzero scale factor.

(c) Show that the zeros of $m(\lambda)$ are the eigenvalues of A.

(d) Show that if the eigenvalues of A are distinct, its characteristic polynomial $\Delta(\lambda)$ is a minimum polynomial for A.

4. Referring to Exercise 2, find a minimum polynomial for each of the matrices A, B, C, D, and E for the case in which $a = b = c = d = 2$.

5. Show that the polynomial $\lambda - c$ is a minimum polynomial for the matrix $A = cI$.

6. Let A be a diagonalizable matrix having exactly three distinct eigenvalues a, b, and c. Show that

$$m(\lambda) = (a - \lambda)(b - \lambda)(c - \lambda)$$

is a minimum polynomial for A.

7. Let A be an $n \times n$-matrix and let $\mathbf{x}_1$ be a nonzero vector in $\mathbf{R}^n$.

   (a) Show that there is a first integer k such that the vectors

   $$\mathbf{x}_1, A\mathbf{x}_1, \ldots, A^{k-1}\mathbf{x}_1, A^k\mathbf{x}_1$$

   are linearly dependent. Why is $k \leq n$? We call k the order of $\mathbf{x}_1$ relative to A.

   (b) Conclude that there is a polynomial $m_1(\lambda)$ of the form

   $$m_1(\lambda) = c_0 + c_1\lambda + \cdots + c_{k-1}\lambda^{k-1} + \lambda^k$$

   such that $m_1(A)\mathbf{x}_1 = \mathbf{0}$. Why is $m_1(\lambda)$ a polynomial of lowest degree such that $m_1(A)\mathbf{x}_1 = \mathbf{0}$?

   (c) Show that $m_1(\lambda)$ is a factor of a minimum polynomial of A. *Hint:* Use the division algorithm $m(\lambda) = q(\lambda)m_1(\lambda) + r(\lambda)$.

   (d) Let $\mathbf{X}_1$ be the linear subspace generated by the vectors

   $$\mathbf{x}_1, A\mathbf{x}_1, \ldots, A^{k-1}\mathbf{x}_1.$$

   The space $\mathbf{X}_1$ is called the *cyclic subspace* determined by $\mathbf{x}_1$. Show that $\mathbf{X}_1$ is comprised of all vectors $p(A)\mathbf{x}_1$, where $p(\lambda)$ is a polynomial. Show that $p(\lambda)$ can be restricted to be a polynomial of degree $<k$.

   (e) Find $\mathbf{X}_1$ and $m_1(\lambda)$ for the cases in which

   $$A = \begin{bmatrix} 0 & 1 & 0 & 0 \\ 0 & 0 & 1 & 0 \\ 0 & 0 & 0 & 0 \\ 0 & 0 & 0 & 0 \end{bmatrix}, \quad \mathbf{x}_1 = \begin{bmatrix} 1 \\ 1 \\ 0 \\ 0 \end{bmatrix} \quad or \quad \mathbf{x}_1 = \begin{bmatrix} 1 \\ 1 \\ 1 \\ 0 \end{bmatrix} \quad or \quad \mathbf{x}_1 = \begin{bmatrix} 0 \\ 1 \\ 0 \\ 1 \end{bmatrix}.$$

   Here we have three choices for $\mathbf{x}_1$.

   In the preceding pages it was shown that a square matrix is similar to a block diagonal matrix. Each block is a matrix with the property that it either has a single eigenvalue or else has a single pair of complex eigenvalues. In view of this result let us consider an $n \times n$-matrix A having a single eigenvalue λ_0. Its characteristic polynomial is $p(\lambda) = (\lambda_0 - \lambda)^n$. We have $p(A) = (-B)^n = 0$, where

   $$B = \lambda_0 I - A.$$

   The matrix B therefore has the property that $B^n = 0$. Such a matrix is called a *nilpotent matrix*. Thus the study of a matrix A with a single eigenvalue is equivalent to the study of a nilpotent matrix. We shall list some properties of nilpotent matrices in the form of exercises to be established by the reader.

As stated above, an $n \times n$-matrix B is **nilpotent** if there is an integer m such that $B^m = 0$. The smallest integer m such that $B^m = 0$ is called **the order of B** as a nilpotent matrix. In our exercises, B denotes a nilpotent $n \times n$-matrix and m is its order. Based on the discussion above, prove the following:

8. B^T is a nilpotent matrix of order m.

9. $\lambda = 0$ is the only eigenvalue of B. *Hint:* $B\mathbf{x} = \lambda\mathbf{x}$ implies that $B^m\mathbf{x} = \lambda^m\mathbf{x} = \mathbf{0}$.

10. $(-\lambda)^n$ is the characteristic polynomial for B. Hence $B^n = 0$ and $m \leq n$.

11. If B is symmetric, then $B = 0$. *Hint:* $B\mathbf{x} = \mathbf{0}$ for every $\mathbf{x}$. Why?

12. An upper triangular matrix with zero main diagonal entries is nilpotent because $(-\lambda)^n$ is its characteristic polynomial. For example, the matrix B given by

$$B = \begin{bmatrix} 0 & a & c \\ 0 & 0 & b \\ 0 & 0 & 0 \end{bmatrix},$$

is nilpotent, no matter how a, b, c are chosen, since

$$B^2 = \begin{bmatrix} 0 & 0 & ab \\ 0 & 0 & 0 \\ 0 & 0 & 0 \end{bmatrix}, \qquad B^3 = \begin{bmatrix} 0 & 0 & 0 \\ 0 & 0 & 0 \\ 0 & 0 & 0 \end{bmatrix}.$$

13. A lower triangular matrix with zero main diagonal entries is nilpotent.

14. Give an example of a nilpotent 4×4-matrix that is neither upper nor lower triangular.

15. If $m > 1$, there is a vector $\mathbf{x}$ such that $B^{m-1}\mathbf{x}$ is nonzero. *Hint:* Otherwise $B^{m-1} = 0$.

16. If $B^{k-1}\mathbf{x}$ is nonzero and $B^k\mathbf{x} = \mathbf{0}$, then $\mathbf{x}, B\mathbf{x}, \ldots, B^{k-1}\mathbf{x}$ are linearly independent. *Hint:* For $k = 3$, the relation $a\mathbf{x} + bB\mathbf{x} + cB^2\mathbf{x} = \mathbf{0}$ implies that $aB\mathbf{x} + bB^2\mathbf{x} = \mathbf{0}$, $aB^2\mathbf{x} = \mathbf{0}$. Hence $a = b = c = 0$.

17. If $m = n$ and $B^{n-1}\mathbf{x}$ is nonzero, then the $n \times n$-matrix

$$X = [B^{n-1}\mathbf{x} \quad B^{n-2}\mathbf{x} \quad \cdots \quad B\mathbf{x} \quad \mathbf{x}]$$

is nonsingular. Consider first the cases $n = 2$ and $n = 3$.

18. If $m = n = 2$, choose $\mathbf{x}$ such that $X = [B\mathbf{x} \quad \mathbf{x}]$ is nonsingular. Verify that

$$BX = B[B\mathbf{x} \quad \mathbf{x}] = [\mathbf{0} \quad B\mathbf{x}] = [B\mathbf{x} \quad \mathbf{x}]\begin{bmatrix} 0 & 1 \\ 0 & 0 \end{bmatrix} = XC.$$

Conclude that B is similar to

$$C = \begin{bmatrix} 0 & 1 \\ 0 & 0 \end{bmatrix}.$$

19. If $m = n = 3$, then B is similar to

$$C = \begin{bmatrix} 0 & 1 & 0 \\ 0 & 0 & 1 \\ 0 & 0 & 0 \end{bmatrix}.$$

Hint: Select $\mathbf{x}$ such that $X = [B^2\mathbf{x} \quad B\mathbf{x} \quad \mathbf{x}]$ is nonsingular. Then

$$BX = B[B^2\mathbf{x} \quad B\mathbf{x} \quad \mathbf{x}] = [0 \quad B^2\mathbf{x} \quad B\mathbf{x}] = [B^2\mathbf{x} \quad B\mathbf{x} \quad \mathbf{x}]C = XC.$$

20. If $m = 2$ and $n = 3$, choose $\mathbf{x}$ such that $B\mathbf{x}$ is not zero. Show that there is an eigenvector $\mathbf{y}$ of B such that the matrix $X = [B\mathbf{x} \quad \mathbf{x} \quad \mathbf{y}]$ is nonsingular. *Hint:* Choose $\mathbf{z}$ such that $[B\mathbf{x} \quad \mathbf{x} \quad \mathbf{z}]$ is nonsingular. If $B\mathbf{z} = \mathbf{0}$, choose $\mathbf{y} = \mathbf{z}$. Otherwise, $B\mathbf{z} = aB\mathbf{x} + b\mathbf{x}$. In this event choose $\mathbf{y} = \mathbf{z} - a\mathbf{x}$. Then

$$B\mathbf{y} = b\mathbf{x}, \qquad \mathbf{0} = B^2\mathbf{y} + bB\mathbf{x} = bB\mathbf{x}.$$

Hence $b = 0$ so that $B\mathbf{y} = \mathbf{0}$.

21. Continuing with Exercise 20, show that B is similar to

$$C = \begin{bmatrix} 0 & 1 & 0 \\ 0 & 0 & 0 \\ 0 & 0 & 0 \end{bmatrix}.$$

22. If $m = n = 4$, then B is similar to

$$\begin{bmatrix} 0 & 1 & 0 & 0 \\ 0 & 0 & 1 & 0 \\ 0 & 0 & 0 & 1 \\ 0 & 0 & 0 & 0 \end{bmatrix}.$$

Generalize to higher values of n.

Supplementary Proof

P1. Proof of Proposition 4.

The proof of Proposition 4 will be made in several steps. First note that **V** and **W** are invariant under T because they are nullspaces of the operators $p(T)$ and $q(T)$. Observe further that for every $\mathbf{u}$ in **U** and every pair of polynomials $f(\lambda)$ and $g(\lambda)$, the vectors

$$\mathbf{v} = g(T)q(T)\mathbf{u}, \qquad \mathbf{w} = f(T)p(T)\mathbf{u}$$

are in **V** and **W**, respectively, For then

$$p(T)\mathbf{v} = p(T)g(T)q(T)\mathbf{u} = g(T)\Delta(T)\mathbf{u} = \mathbf{0}$$

because $\Delta(T) = 0$. Similarly, $q(T)\mathbf{w} = \mathbf{0}$. Hence the range of $q(T)$ is in **V** and the range of $p(T)$ is in **W**.

In view of the fact that $p(\lambda)$ and $q(\lambda)$ are relative prime, we can select polynomials $f(\lambda)$ and $g(\lambda)$ such that

$$1 = g(\lambda)q(\lambda) + f(\lambda)p(\lambda).$$

It follows that

$$I = g(T)q(T) + f(T)p(T)$$

and hence that

$$\mathbf{u} = g(T)q(T)\mathbf{u} + f(T)p(T)\mathbf{u} = \mathbf{v} + \mathbf{w}$$

with $\mathbf{v} = g(T)q(T)\mathbf{u}$ in $\mathbf{V}$ and $\mathbf{w} = f(T)p(T)\mathbf{u}$ in $\mathbf{W}$. From this result we draw several conclusions. First, if $\mathbf{u}$ is in $\mathbf{V}$ and in $\mathbf{W}$, then $\mathbf{v} = \mathbf{w} = \mathbf{0}$ so that $\mathbf{u} = \mathbf{0}$. Consequently, $\mathbf{V}$ and $\mathbf{W}$ have only the zero vector in common. Looking at our equation we see that $\mathbf{U}$ is the direct sum

$$\mathbf{U} = \mathbf{V} + \mathbf{W}$$

of the nullspaces $\mathbf{V}$ and $\mathbf{W}$ of $p(T)$ and $q(T)$. Finally, a third look at this equation tells us that if $\mathbf{u}$ is in $\mathbf{V}$, then $\mathbf{v} = q(T)g(T)\mathbf{u}$ is in the range of $q(T)$, and it follows that $\mathbf{V}$ is the range of $q(T)$. Similarly, $\mathbf{W}$ is the range of $p(T)$. This establishes the first three conclusions in Proposition 4.

In order to establish the remaining conclusions, let r and s be the dimensions of $\mathbf{V}$ and $\mathbf{W}$, respectively. Let F be a basis operator for T on $\mathbf{V}$. Similarly, let G be a basis operator for T on $\mathbf{W}$. Then there is an $r \times r$-matrix B and a $s \times s$-matrix C such that

$$TF = FB, \qquad TG = GC.$$

Thus B is a matrix representation of T on $\mathbf{V}$ and C is a matrix representation of T on $\mathbf{W}$. The operator

$$E = [F \quad G]$$

is a basis operator for $\mathbf{U} = \mathbf{V} + \mathbf{W}$. We have

$$TE = T[F \quad G] = [TF \quad TG] = [FB \quad GC] = [F \quad G]\begin{bmatrix} B & 0 \\ 0 & C \end{bmatrix} = EA,$$

where A is the block diagonal matrix

$$A = \operatorname{diag}(B, C) = \begin{bmatrix} B & 0 \\ 0 & C \end{bmatrix}.$$

The matrix A is a matrix representation of T on $\mathbf{U}$, as stated in conclusion (5) in Proposition 4. In addition, the characteristic polynomials $\Delta(\lambda)$ for A, $\Delta_B(\lambda)$ for B, and $\Delta_C(\lambda)$ for C are connected by the relation

$$\Delta(\lambda) = \Delta_B(\lambda)\Delta_C(\lambda).$$

Because $\Delta(\lambda) = p(\lambda)q(\lambda)$ also, we have

$$\Delta(\lambda) = \Delta_B(\lambda)\Delta_C(\lambda) = p(\lambda)q(\lambda).$$

The proof of Proposition 4 will be complete when we have shown that

$$p(\lambda) = \Delta_B(\lambda), \qquad q(\lambda) = \Delta_C(\lambda).$$

This will imply that the dimension r of **V** is equal to the degree h of $p(\lambda)$. Also, the dimension s of **W** is equal to the degree k of $q(\lambda)$. To obtain this result, observe that our basis operator F of **V** has the property that

$$TF = FB, \qquad T^2F = TFB = FB^2, \quad \ldots, \quad T^kF = FB^k, \cdots.$$

Consequently,

$$g(T)F = Fg(B)$$

for every polynomial $g(\lambda)$. In particular,

$$p(T)F = Fp(B), \qquad q(T)F = Fq(B).$$

Since **V** is the nullspace for $p(T)$, we have $p(T)F = 0$ and hence $p(B) = 0$. In addition, the matrix $q(B)$ is nonsingular. For if there is a vector **y** such that $q(B)\mathbf{y} = \mathbf{0}$, then

$$q(T)F\mathbf{y} = Fq(B)\mathbf{y} = \mathbf{0}.$$

This implies that $F\mathbf{y}$ is in **W** as well as in **V**. This is possible only if $F\mathbf{y} = \mathbf{0}$ and hence only if $\mathbf{y} = \mathbf{0}$. It follows that $q(B)$ is nonsingular. Similarly, $q(C) = 0$ and $p(C)$ is nonsingular.

We shall show that $p(\lambda)$ and $\Delta_C(\lambda)$ are relatively prime. If this is not the case, they have a common factor $r(\lambda)$ of the form

$$r(\lambda) = \lambda_j - \lambda \quad \text{or} \quad r(\lambda) = b_k^2 + (a_k - \lambda)^2.$$

Since either λ_j or $a_k + ib_k$ is an eigenvalue of C, it follows that $r(C)$ is singular. Because $r(\lambda)$ is also a factor of $p(\lambda)$, this implies that $p(C)$ is singular, contrary to the fact that $p(C)$ is nonsingular. Consequently, $p(\lambda)$ and $\Delta_C(\lambda)$ are relatively prime. Similarly, $q(\lambda)$ and $\Delta_C(\lambda)$ are relatively prime. Combining these results with the relation

$$\Delta(\lambda) = \Delta_B(\lambda)\Delta_C(\lambda) = p(\lambda)q(\lambda),$$

we conclude that $p(\lambda) = \Delta_B(\lambda)$ and $q(\lambda) = \Delta_C(\lambda)$, as was to be proved. ■

7

Quadratic Forms and Geometric Interpretations

In the study of analytic geometry we learned that the equation

$$\frac{x^2}{3^2} + \frac{y^2}{2^2} = 1$$

or, equivalently, the equation

$$4x^2 + 9y^2 = 36$$

represents an ellipse whose axes coincide with the coordinate axes (Figure 1). The length of the semi-major axis is 3 and the length of the semi-minor axis is 2.

Similarly, the equation

$$\frac{x^2}{3^2} - \frac{y^2}{2^2} = 1$$

or, equivalently, the equation

$$4x^2 - 9y^2 = 36$$

represents a hyperbola whose axes coincide with the coordinate axes (Figure 2). The equation

$$xy = 4$$

also represents a hyperbola having the coordinate axes as its asymptotes. These equations are of the form

$$ax^2 + 2bxy + cy^2 = k.$$

We shall show that this general equation represents

1. An ellipse when $a > 0$, $ac - b^2 > 0$, $k > 0$.

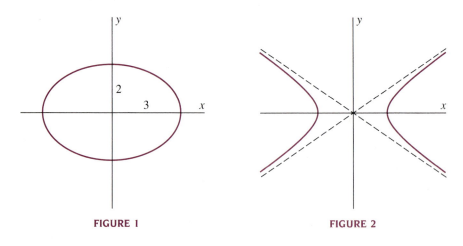

FIGURE 1   **FIGURE 2**

2. A hyperbola when $ac - b^2 < 0$.

We exclude the case $a = c > 0$, $b = 0$ in (1). This would give us a circle. If $k = 0$ in (2) we get a pair of straight lines. We exclude this case also. Consideration of the degenerate case in which $ac - bc^2 = 0$ will be omitted. Observe that each term in the expression

$$ax^2 + 2bxy + cy^2$$

is of degree 2 in x and y. To designate this fact we call this expression a *quadratic form* in x and y. Setting

$$A = \begin{bmatrix} a & b \\ b & c \end{bmatrix}, \qquad \mathbf{x} = \begin{bmatrix} x \\ y \end{bmatrix},$$

we see that A is symmetric and that

$$\mathbf{x}^T A \mathbf{x} = \begin{bmatrix} x & y \end{bmatrix} \begin{bmatrix} a & b \\ b & c \end{bmatrix} \begin{bmatrix} x \\ y \end{bmatrix} = ax^2 + 2bxy + cy^2$$

is a quadratic form. In general, if A is an $n \times n$-matrix and $\mathbf{x}$ is an n-dimensional column vector, the expression $\mathbf{x}^T A \mathbf{x}$ is a *quadratic form* in $\mathbf{x}$.

Return now to the two-dimensional case in which

$$A = \begin{bmatrix} a & b \\ b & c \end{bmatrix}, \qquad \mathbf{x} = \begin{bmatrix} x \\ y \end{bmatrix}$$

and our equation is

$$\mathbf{x}^T A \mathbf{x} = \begin{bmatrix} x & y \end{bmatrix} \begin{bmatrix} a & b \\ b & c \end{bmatrix} \begin{bmatrix} x \\ y \end{bmatrix} = ax^2 + 2bxy + cy^2 = k.$$

Observe that in the case in which $k = 1$ and

$$A = \begin{bmatrix} 1/9 & 0 \\ 0 & 1/4 \end{bmatrix},$$

this equation becomes the standard equation

$$\mathbf{x}^T A \mathbf{x} = \frac{x^2}{9} + \frac{y^2}{4} = 1$$

of an ellipse. Note that in this event the eigenvectors

$$\mathbf{x}_1 = \begin{bmatrix} 1 \\ 0 \end{bmatrix}, \qquad \mathbf{x}_2 = \begin{bmatrix} 0 \\ 1 \end{bmatrix}$$

determine the axes of the ellipse. This is also true in the general case. To find the eigenvectors of A, observe that the characteristic polynomial of A is

$$\det (A - \lambda I) = \lambda^2 - (a + c)\lambda + d,$$

where

$$d = \det A = ac - b^2.$$

By the use of the quadratic formula, it is seen that the solutions λ and μ of the characteristic equation

$$\det (A - \lambda I) = \lambda^2 - (a + c)\lambda + d = 0$$

of A are given by the formulas

$$\lambda = \frac{a + c - e}{2}, \qquad \mu = \frac{a + c + e}{2},$$

where

$$e = [(a + c)^2 - 4d]^{1/2} = [(a - c)^2 + 4b^2]^{1/2}.$$

These solutions are the eigenvalues of A. Note that $e > 0$ because the case $a = c$ and $b = 0$ is excluded. From these equations we see that

$$\mu = \lambda + e > \lambda, \qquad \lambda\mu = d = ac - b^2.$$

We can assume without loss of generality that $a \geq 0$ since this can be brought about by replacing A by $-A$, if necessary. When $d = ac - b^2 > 0$, we have $a > 0$, $c > 0$ and $0 < \lambda < \mu$. When $d < 0$, we have $\lambda < 0 < \mu$. To find the eigenvectors of A corresponding to the eigenvalues λ and μ, we use the eigenvector equations

$$(a - \lambda)x + y = 0, \qquad (a - \mu)x + y = 0,$$
$$x + (c - \lambda)y = 0, \qquad x + (c - \mu)y = 0.$$

Solving these equations, we find that the unit eigenvectors of A corresponding to λ and μ are, respectively,

$$\mathbf{x}_1 = \pm\frac{1}{f}\begin{bmatrix} -b \\ a - \lambda \end{bmatrix}, \qquad \mathbf{x}_2 = \pm\frac{1}{g}\begin{bmatrix} -b \\ a - \mu \end{bmatrix},$$

where

$$f = [b^2 + (a - \lambda)^2]^{1/2}, \qquad g = [b^2 + (a - \mu)^2]^{1/2}.$$

As was seen in Section 4.3, the eigenvectors $\mathbf{x}_1$ and $\mathbf{x}_2$ are orthogonal to each other. This result can also be obtained directly by verifying that $\mathbf{x}_1^T\mathbf{x}_2 = 0$. We have the relations

$$\mathbf{x}_1^T\mathbf{x}_1 = 1, \qquad \mathbf{x}_1^T\mathbf{x}_2 = \mathbf{x}_2^T\mathbf{x}_1 = 0, \qquad \mathbf{x}_2^T\mathbf{x}_2 = 1.$$

Combining these with the eigenvalue relations

$$A\mathbf{x}_1 = \lambda\mathbf{x}_1, \qquad A\mathbf{x}_2 = \mu\mathbf{x}_2,$$

we find that

$$\mathbf{x}_1^T A \mathbf{x}_1 = \lambda \mathbf{x}_1^T \mathbf{x}_1 = \lambda, \qquad \mathbf{x}_2^T A \mathbf{x}_2 = \mu \mathbf{x}_2^T \mathbf{x}_2 = \mu,$$

$$\mathbf{x}_2^T A \mathbf{x}_1 = \lambda \mathbf{x}_2^T \mathbf{x}_1 = 0, \qquad \mathbf{x}_1^T A \mathbf{x}_2 = 0.$$

Introducing the eigenvectors $\mathbf{x}_1$ and $\mathbf{x}_2$ as new basis vectors, the coordinate transformation

$$\mathbf{x} = u\mathbf{x}_1 + v\mathbf{x}_2$$

gives us new coordinates (u, v). Our quadratic form becomes

$$\mathbf{x}^T A \mathbf{x} = (u\mathbf{x}_1^T + v\mathbf{x}_2^T) A (u\mathbf{x}_1 + v\mathbf{x}_2) = \lambda u^2 + \mu v^2.$$

It follows that the equation

$$ax^2 + 2bxy + cy^2 = k$$

in xy-coordinates becomes

$$\lambda u^2 + \mu v^2 = k$$

in uv-coordinates. This equation can be rewritten in the form

$$\frac{\lambda}{k} u^2 + \frac{\mu}{k} v^2 = 1.$$

When $d = ac - b^2 > 0$, we have $0 < \lambda < \mu$. In this event, our equation represents an ellipse whose axes are in the directions of the eigenvectors $\mathbf{x}_1$ and $\mathbf{x}_2$ of our matrix A, as indicated in Figure 3. Its semi-major and semi-minor axes are $\sqrt{(k/\lambda)}$ and $\sqrt{(k/\mu)}$, respectively. On the other hand, when $d = ac - b^2 < 0$, we have $\lambda < 0 < \mu$. Then the equation

$$\lambda u^2 + \mu v^2 = k$$

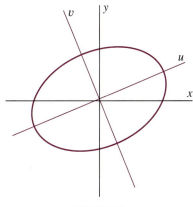

FIGURE 3

represents a hyperbola whose axes are in the directions of the eigenvectors $\mathbf{x}_1$ and $\mathbf{x}_2$.

In a numerical case in which our equation is

$$5x^2 - 4xy + 2y^2 = 6,$$

we have $a = 5$, $b = -2$, $c = 2$, and $k = 6$. In this event we have

$d = ac - b^2 = 6$, $\quad e = 5$, $\quad \lambda = 1$, $\quad \mu = 6$, $\quad r = 1$, $\quad f = 2\sqrt{5}$, and $g = \sqrt{5}$.

The unit eigenvectors of the associated matrix

$$A = \begin{bmatrix} 5 & -2 \\ -2 & 2 \end{bmatrix}$$

are

$$\mathbf{x}_1 = \begin{bmatrix} 1/\sqrt{5} \\ 2/\sqrt{5} \end{bmatrix}, \qquad \mathbf{x}_2 = \begin{bmatrix} 2/\sqrt{5} \\ -1\sqrt{5} \end{bmatrix}.$$

Our transformation of variables $\mathbf{x} = u\mathbf{x}_1 + v\mathbf{x}_2$ becomes

$$\begin{bmatrix} x \\ y \end{bmatrix} = u\begin{bmatrix} 1/\sqrt{5} \\ 2/\sqrt{5} \end{bmatrix} + v\begin{bmatrix} 2/\sqrt{5} \\ -1\sqrt{5} \end{bmatrix},$$

so that we have

$$x = (u + 2v)/\sqrt{5}, \qquad y = (2u - v)/\sqrt{5}.$$

Under this transformation the first of the following equations is transformed into the second:

$$5x^2 - 4xy + 2y^2 = 6, \qquad u^2 + 6v^2 = 6.$$

The second is the equation of an ellipse in standard form. Its major and minor semiaxes are $\sqrt{6}$ and 1.

In the three-dimensional case the equation

$$\frac{x^2}{4^2} + \frac{y^2}{3^2} + \frac{z^2}{2^2} = 1$$

is an equation (in standard form) of an ellipsoid whose axes are the coordinates axes, as shown in Figure 4. Its center is at the origin and the lengths of its semiaxes are 4, 3, and 2, respectively. The equation of the ellipsoid can be rewritten as

$$36x^2 + 64y^2 + 144z^2 = 576.$$

In matrix form this equation becomes

$$\mathbf{x}^T A \mathbf{x} = k,$$

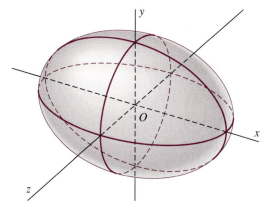

FIGURE 4

where $\quad A = \begin{bmatrix} 36 & 0 & 0 \\ 0 & 64 & 0 \\ 0 & 0 & 144 \end{bmatrix}, \quad \mathbf{x} = \begin{bmatrix} x \\ y \\ z \end{bmatrix}, \quad k = 576.$

The eigenvalues of A are

$$\lambda_1 = 36, \qquad \lambda_2 = 64, \qquad \lambda_3 = 144.$$

The unit coordinate vectors

$$\mathbf{x}_1 = \begin{bmatrix} 1 \\ 0 \\ 0 \end{bmatrix}, \qquad \mathbf{x}_2 = \begin{bmatrix} 0 \\ 1 \\ 0 \end{bmatrix}, \qquad \mathbf{x}_3 = \begin{bmatrix} 0 \\ 0 \\ 1 \end{bmatrix}$$

are the corresponding eigenvectors of A. The semiaxes of our ellipsoid are $4\mathbf{x}_1$, $3\mathbf{x}_2$, and $2\mathbf{x}_3$, respectively.

In a similar manner it is seen that the equations

$$4x^2 - 9y^2 + 16z^2 = 144 \quad \text{and} \quad 4x^2 - 9y^2 - 16z^2 = 144$$

are equations of hyperboloids of two different types, as shown schematically in Figure 5. The first is a hyperboloid of one sheet. The second is a hyperboloid of two sheets. Again their axes are the coordinate axes. Referring to Figure 2, we note that if we rotate the hyperbola about the y-axis, we obtain a hyperboloid of one sheet. (Here $c = a$.) On the other hand, if we rotate the hyperbola about the x-axis, we obtain a hyperboloid of the second kind. (Here $c = b$.) Of course, the equations

$$\frac{-x^2}{a^2} + \frac{y^2}{b^2} + \frac{z^2}{c^2} = 1, \qquad \frac{x^2}{a^2} + \frac{y^2}{b^2} - \frac{z^2}{c^2} = 1$$

are alternative standard equations of hyperboloids of two sheets.

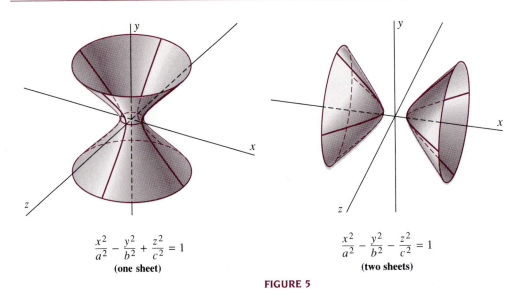

$$\frac{x^2}{a^2} - \frac{y^2}{b^2} + \frac{z^2}{c^2} = 1$$
(one sheet)

$$\frac{x^2}{a^2} - \frac{y^2}{b^2} - \frac{z^2}{c^2} = 1$$
(two sheets)

FIGURE 5

The equations for ellipsoids and hyperboloids given above are special cases of the general equation

$$ax^2 + by^2 + cz^2 + 2dxy + 2eyz + 2fxz = k,$$

which represents an ellipsoid or a hyperboloid except for degenerate cases. The left member of this equation is a quadratic form in the variables x, y, z. Again, this equation can be put in the matrix form

$$[x \quad y \quad z] \begin{bmatrix} a & d & f \\ d & b & e \\ f & e & c \end{bmatrix} \begin{bmatrix} x \\ y \\ z \end{bmatrix} = k.$$

This in turn can be written in the compact form

$$\mathbf{x}^T A \mathbf{x} = k,$$

where

$$A = \begin{bmatrix} a & d & f \\ d & b & e \\ f & e & c \end{bmatrix}, \qquad \mathbf{x} = \begin{bmatrix} x \\ y \\ z \end{bmatrix}.$$

This equation is the equation of an ellipsoid when $k > 0$ and $\mathbf{x}^T A \mathbf{x} > 0$ for all nonzero vectors $\mathbf{x}$. Its axes are in the direction of the eigenvectors of A. The proof is like that made in the general case which we consider next.

Turning now to the general case, let A be a symmetric $n \times n$-matrix. Then the function

$$\mathbf{x}^T A \mathbf{x}$$

is an *n*-dimensional **quadratic form.** As was seen in Section 4.3, the matrix *A* has a set of orthonormal eigenvectors $\mathbf{x}_1, \mathbf{x}_2, \ldots, \mathbf{x}_n$ corresponding to eigenvalues

$$\lambda_1 \leq \lambda_2 \leq \cdots \leq \lambda_n.$$

These vectors form the columns of an $n \times n$-matrix *X* whose transpose X^T is its inverse. That is,

$$X^T X = I.$$

Such a matrix is called an **orthogonal** matrix. Under the transformation

$$\mathbf{x} = X\mathbf{u}$$

our quadratic form becomes

$$\mathbf{x}^T A \mathbf{x} = \mathbf{u}^T X^T A X \mathbf{u} = \mathbf{u}^T D \mathbf{u},$$

where $D = X^T A X$ is the diagonal matrix whose main diagonal entries are the eigenvalues $\lambda_1, \lambda_2, \ldots, \lambda_n$ of *A*. We have, with $\mathbf{u} = (u_1, u_2, \ldots, u_n)$,

$$\mathbf{u}^T D \mathbf{u} = \lambda_1 u_1^2 + \lambda_2 u_2^2 + \cdots + \lambda_n u_n^2.$$

The equations

$$\mathbf{x}^T A \mathbf{x} = \mathrm{k}, \qquad \mathbf{u}^T D \mathbf{u} = \lambda_1 u_1^2 + \lambda_2 u_2^2 + \cdots + \lambda_n u_n^2 = k$$

are equations of the same "quadric" surface under different coordinate systems. Its center is at the origin. Its shape is determined by the eigenvalues of *A*. Its size is determined by the value of *k*. If the eigenvalues of *A* are positive, this surface is an $(n - 1)$-dimensional **ellipsoid** whose axes are the eigenvectors of *A*. Observe that when the eigenvalues of *A* are positive, we have

$$\mathbf{x}^T A \mathbf{x} = \mathbf{u}^T D \mathbf{u} = \lambda_1 u_1^2 + \lambda_2 u_2^2 + \cdots + \lambda_n u_n^2 > 0$$

unless $\mathbf{u} = \mathbf{0}$ or equivalently, unless $\mathbf{x} = \mathbf{0}$. A matrix *A* of this type is said to be **positive definite.** Thus ellipsoids are determined by positive definite matrices. They are also determined by **negative definite matrices,** that is, matrices *A* such that $-A$ is positive definite. The eigenvalues of a negative definite matrix are all negative. A matrix *A* is said to be **indefinite** if some of its eigenvalues are positive and others are negative. When *A* is indefinite, our quadric surface is called a **hyperboloid.** If *A* is singular, our quadric surface is said to be **degenerate.**

   Let us make another observation. For definiteness, we suppose that *A* is positive definite. Then the set of points **x** satisfying the equation

$$\mathbf{x}^T A \mathbf{x} = k_1,$$

with a fixed number k_1 is an ellipsoid $\mathbf{E}_1$ with its center at the origin. Let *t* be a fixed number and set $k_2 = t^2 k_1$. As indicated schematically in Figure 6, for each point $\mathbf{x}$ on the ellipsoid $\mathbf{E}_1$, the point $\mathbf{y} = t\mathbf{x}$ lies on the ellipsoid $\mathbf{E}_2$ defined by the equation

$$\mathbf{y}^T A \mathbf{y} = k_2,$$

as can be seen by the computations

$$\mathbf{y}^T A \mathbf{y} = t^2 \mathbf{x}^T A \mathbf{x} = t^2 k_1 = k_2.$$

Two ellipsoids $\mathbf{E}_1$ and $\mathbf{E}_2$ related in this manner are said to be *similar*. Thus the equation

$$\mathbf{x}^T A \mathbf{x} = k \qquad (k > 0)$$

defines a one-parameter family of similar ellipsoids.

We close this section by presenting the following.

CLASSIFICATION OF A REAL SYMMETRIC MATRIX A

1. A is *nonnegative* if the inequality $\mathbf{x}^T A \mathbf{x} \geq 0$ holds for all vectors $\mathbf{x}$. A is nonnegative if and only if its eigenvalues are nonnegative.

2. A is *positive definite* if $\mathbf{x}^T A \mathbf{x} > 0$ unless $\mathbf{x} = \mathbf{0}$. A is positive definite if and only if its eigenvalues are positive.

3. A is *nonpositive* if the inequality $\mathbf{x}^T A \mathbf{x} \leq 0$ holds for all vectors $\mathbf{x}$. A is nonpositive if and only if its eigenvalues are nonpositive.

4. A is *negative definite* if $\mathbf{x}^T A \mathbf{x} < 0$ unless $\mathbf{x} = \mathbf{0}$. A is negative definite if and only if its eigenvalues are negative.

5. Otherwise, A is *indefinite*. That is, A is indefinite if it has positive and negative eigenvalues.

A quadratic form $Q(\mathbf{x}) = \mathbf{x}^T A \mathbf{x}$ is nonnegative, positive definite, nonpositive, negative definite, indefinite according as the matrix A has these properties.

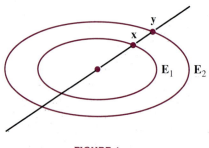

FIGURE 6

Exercises

1. Let A be one of the matrices

$$A_1 = \begin{bmatrix} 4 & 0 \\ 0 & 9 \end{bmatrix}, \qquad A_2 = \begin{bmatrix} 4 & 0 \\ 0 & -9 \end{bmatrix}, \qquad A_3 = \begin{bmatrix} 5 & 4 \\ 4 & 5 \end{bmatrix},$$

$$A_4 = \begin{bmatrix} -5 & 4 \\ 4 & -5 \end{bmatrix}, \qquad A_5 = \begin{bmatrix} 1 & 2 \\ 2 & 1 \end{bmatrix}, \qquad A_6 = \begin{bmatrix} 0 & 1 \\ 1 & 0 \end{bmatrix}.$$

In each case determine the curve defined by $\mathbf{x}^T A \mathbf{x} = k$.

(a) Is it an ellipse or a hyperbola?

(b) Determine the direction of its axes. That is, find the eigenvectors of A.

(c) Determine k so that the curve passes through the point $(2, 3)$.

(d) Is A positive definite, negative definite, or indefinite?

(e) Find the set of points for which $\mathbf{x}^T A \mathbf{x} = 0$.

(f) For the case $A = A_6$, sketch the curve $\mathbf{x}^T A \mathbf{x} = k$ when $k = 1$ and when $k = -1$. How are these curves related?

2. Let A be one of the matrices

$$A_7 = \begin{bmatrix} 1 & 0 & 0 \\ 0 & 4 & 0 \\ 0 & 0 & 9 \end{bmatrix}, \qquad A_8 = \begin{bmatrix} 1 & 0 & 0 \\ 0 & 4 & 0 \\ 0 & 0 & -9 \end{bmatrix}, \qquad A_9 = \begin{bmatrix} 3 & -1 & 1 \\ -1 & 3 & -1 \\ 1 & -1 & 3 \end{bmatrix},$$

$$A_{10} = \begin{bmatrix} 1 & 0 & 0 \\ 0 & -1 & 0 \\ 0 & 0 & -1 \end{bmatrix}, \qquad A_{11} = \begin{bmatrix} 0 & 0 & 1 \\ 0 & 1 & 0 \\ 1 & 0 & 0 \end{bmatrix}, \qquad A_{12} = \begin{bmatrix} 2 & 0 & 1 \\ 0 & 1 & 0 \\ 1 & 0 & 2 \end{bmatrix}.$$

In each case determine the surface defined by $\mathbf{x}^T A \mathbf{x} = k$.

(a) Which of these are ellipsoids?

(b) Determine the direction of its axes. (Find eigenvectors.)

(c) When $A = A_8$ the surface is an hyperboloid. Sketch the surface when $k = 36$ and also when $k = -36$. How do these hyperboloids differ?

(d) What happens when $k = 0$?

(e) Which A's are positive definite?

(f) When $A = A_{10}$, sketch the surface $\mathbf{x}^T A \mathbf{x} = 0$. This surface is a cone. In the theory of three-dimensional relativity, this cone is called the light cone.

3. Establish the identity

$$\mathbf{x}^T A \mathbf{x} - \mathbf{x}_1^T A \mathbf{x}_1 = 2\mathbf{x}_1^T A(\mathbf{x} - \mathbf{x}_1) + (\mathbf{x} - \mathbf{x}_1)^T A(\mathbf{x} - \mathbf{x}_1).$$

4. Let $\mathbf{x}_1$ be a point on the quadric surface $\mathbf{S}_k$ defined by the equation $\mathbf{x}^T A \mathbf{x} = k$. It can be shown that the (hyper)plane $\mathbf{T}$ of points $\mathbf{x}$ satisfying the equation

$$\mathbf{x}_1^T A \mathbf{x} = k \quad \text{or equivalently} \quad \mathbf{x}_1^T A(\mathbf{x} - \mathbf{x}_1) = 0$$

is the tangent space of $\mathbf{S}_k$ at $\mathbf{x}_1$. Show, by Exercise 3, that when A is positive definite, $\mathbf{x}_1$ is the only point of $\mathbf{T}$ that is on the ellipsoid $\mathbf{S}_k$.

5. Using the familiar xy-coordinates in $\mathbf{R}^2$, let (x_1, y_1) be a point on the conic section $\mathbf{C}$ defined by the equation

$$ax^2 + 2bxy + cy^2 = k.$$

Show, by Exercise 4, that the tangent to $\mathbf{C}$ at the point (x_1, y_1) is defined by the equation

$$ax_1x + b(x_1y + y_1x) + cy_1y = k.$$

Use this with $(x_1, y_1) = (2, 3)$ to find that tangents at $(2, 3)$ to each of the conic sections described in Exercise 1, part (c).

6. Let A be a symmetric matrix. Show that A is nonnegative if and only if it is expressible in the form $A = B^TB$. If B is nonsingular then A is positive definite. *Hint:* Use the fact that a nonnegative symmetric matrix A has a square root B which is also symmetric and nonnegative.

7. Show that if A is a positive definite symmetric matrix, then det $A > 0$. *Hint:* Use the fact that det A is the product of the eigenvalues of A.

8. Show that if A is a positive definite symmetric matrix, then (a) every principal minor B of A is positive definite and so has det $B > 0$, (b) the main diagonal entries of A are positive.

9. Show that if A is a nonsingular symmetric matrix, one of whose main diagonal entries is zero, then A is indefinite. *Hint:* Apply the result in part (b) of Exercise 8 to A and $-A$.

10. Show that if A is an indefinite symmetric matrix, there is a nonzero vector $\mathbf{x}$ such that $\mathbf{x}^TA\mathbf{x} = 0$. *Hint:* Suppose that $a = \mathbf{y}^TA\mathbf{y} > 0$ and $c = \mathbf{z}^TA\mathbf{z} < 0$. Set $b = \mathbf{y}^TA\mathbf{z}$. Since $ac < 0$, the vector $\mathbf{x} = t\mathbf{y} + \mathbf{z}$ has the property that

$$\mathbf{x}^TA\mathbf{x} = at^2 + 2bt + c = 0$$

for a suitable choice of t.

11. Suppose that A is a nonnegative but not a positive definite symmetric matrix. Show that there is a nonzero vector $\mathbf{x}$ such that $\mathbf{x}^TA\mathbf{x} = 0$. *Hint:* Choose $\mathbf{x}$ to be an eigenvector corresponding to the eigenvalue $\lambda = 0$ of A.

12. Let A be a symmetric matrix. Show that if $\mathbf{x}^TA\mathbf{x} = 0$ only when $\mathbf{x} = \mathbf{0}$, then A is either positive definite or else negative definite. *Hint:* Use the results given in Exercises 10 and 11.

13. Let (x_1, y_1, z_1) be a point on the hyperboloid

$$\frac{x^2}{a^2} + \frac{y^2}{b^2} - \frac{z^2}{c^2} = 1$$

of one sheet. Set

$$u = \frac{x_1}{a}, \qquad v = \frac{y_1}{b}, \qquad w = \frac{z_1}{c}$$

$$\alpha = uw - v, \qquad \beta = u + vw, \qquad \gamma = 1 + w^2.$$

(a) Verify that

$$u^2 + v^2 = 1 + w^2, \qquad u\alpha + v\beta = \gamma w, \qquad \alpha^2 + \beta^2 = \gamma^2.$$

(b) Show that, for each t, the point

$$x = a(u + t\alpha), \qquad y = b(v + t\beta), \qquad z = c(w + t\gamma)$$

lies on our hyperboloid. These points form a line on our hyperboloid. It passes through (x_1, y_1, z_1).

(c) Verify that a second such line on our hyperboloid is obtained when we use the alternative formulas

$$\alpha = uw + v, \qquad \beta = -u + vw, \qquad \gamma = 1 + w^2$$

for α, β, and γ.

(d) Verify that the tangent plane to our hyperboloid at the point (x_1, y_1, z_1) cuts the hyperboloid in the two lines described above.

(e) Discuss the case in which $(x_1, y_1, z_1) = (a, b, c)$.

7.2   RAYLEIGH QUOTIENTS

In the study of a symmetric matrix A, we frequently encounter the ratio

$$R(\mathbf{x}) = \frac{\mathbf{x}^T A \mathbf{x}}{\mathbf{x}^T \mathbf{x}}.$$

In dealing with this ratio we exclude the vector $\mathbf{x} = \mathbf{0}$. This ratio is called the **Rayleigh quotient** of A. Observe that if we set

$$\mathbf{y} = \frac{\mathbf{x}}{\|\mathbf{x}\|}, \qquad \|\mathbf{x}\| = (\mathbf{x}^T\mathbf{x})^{1/2},$$

then $\mathbf{y}$ is a unit vector in the direction of $\mathbf{x}$ and

$$\mathbf{y}^T A \mathbf{y} = \frac{\mathbf{x}^T A \mathbf{x}}{\|\mathbf{x}\|^2} = \frac{\mathbf{x}^T A \mathbf{x}}{\mathbf{x}^T \mathbf{x}} = R(\mathbf{x}).$$

It follows that the study of the Rayleigh quotient for A is equivalent to the study of the quadratic form

$$Q(\mathbf{y}) = \mathbf{y}^T A \mathbf{y}$$

for A on the class of all unit vectors $\mathbf{y}$. In addition, we have $R(\mathbf{y}) = Q(\mathbf{y})$ when $\mathbf{y}$ is a unit vector.

The Rayleigh quotient $R(\mathbf{x})$ of A gives us information about eigenvectors and eigenvalues of A. As we shall see presently, we have the following relations:

1. The maximum λ_{max} of $R(\mathbf{x})$ is the maximum eigenvalue of A. Moreover, a maximizer $\mathbf{x}_{max}$ of $R(\mathbf{x})$ is a corresponding eigenvector of A.

2. The minimum λ_{min} of $R(\mathbf{x})$ is the minimum eigenvalue of A and a minimizer $\mathbf{x}_{min}$ of $R(\mathbf{x})$ is a corresponding eigenvector of A.

According to these relations we can find the maximum and minimum eigenvalues of A by maximizing and minimizing $R(\mathbf{x})$. This can be done without finding the characteristic polynomial of A. Any algorithm for maximizing or minimizing $R(\mathbf{x})$ can be used. We shall not discuss such algorithms here.

One of the important properties of the Rayleigh quotient is the following:

PROPERTY 1. If $\mathbf{x}$ is an eigenvector of A, then $\lambda = R(\mathbf{x})$ is the corresponding eigenvalue of A.

This follows because when $A\mathbf{x} = \lambda\mathbf{x}$, we have

$$\mathbf{x}^T A\mathbf{x} = \mathbf{x}^T(\lambda\mathbf{x}) = \lambda\mathbf{x}^T\mathbf{x}, \qquad \lambda = \frac{\mathbf{x}^T A\mathbf{x}}{\mathbf{x}^T\mathbf{x}} = R(\mathbf{x}).$$

Associated with each vector $\mathbf{x}$ we have the vector

$$\mathbf{g} = A\mathbf{x} - R(\mathbf{x})\mathbf{x}.$$

We call this vector the *gradient* of $R(\mathbf{x})$ at $\mathbf{x}$. We do so because, apart from a nonessential scale factor, it is the gradient of $R(\mathbf{x})$ appearing in the theory of functions of n variables. Observe that in view of the relation

$$\mathbf{x}^T\mathbf{g} = \mathbf{x}^T A\mathbf{x} - R(\mathbf{x})\mathbf{x}^T\mathbf{x} = 0,$$

the vector $\mathbf{g}$ is orthogonal to $\mathbf{x}$. Moreover, if

$$\mathbf{g} = A\mathbf{x} - R(\mathbf{x})\mathbf{x} = \mathbf{0},$$

then $\mathbf{x}$ is an eigenvector of A with $\lambda = R(\mathbf{x})$ as the corresponding eigenvalue. Conversely, if $\mathbf{x}$ is an eigenvector, then $\mathbf{g} = \mathbf{0}$.

This gives us the following:

PROPERTY 2. The gradient

$$\mathbf{g} = A\mathbf{x} - R(\mathbf{x})\mathbf{x}$$

of $R(\mathbf{x})$ at $\mathbf{x}$ is orthogonal to $\mathbf{x}$. Moreover, $\mathbf{g} = \mathbf{0}$ if and only if $\mathbf{x}$ is an eigenvector of A.

As remarked above, the minimum value of the Rayleigh quotient $R(\mathbf{x})$ for A is the least eigenvalue of A. Similarly, the maximum value of $R(\mathbf{x})$ is the greatest eigenvalue of A.

■ **EXAMPLE 1**

$$A = \begin{bmatrix} 1 & 0 & 0 \\ 0 & 2 & 0 \\ 0 & 0 & 3 \end{bmatrix}, \qquad \mathbf{x} = \begin{bmatrix} x \\ y \\ z \end{bmatrix};$$

Here, the ratio $R(\mathbf{x})$ is

$$R(\mathbf{x}) = \frac{\mathbf{x}^T A \mathbf{x}}{\mathbf{x}^T \mathbf{x}} = \frac{x^2 + 2y^2 + 3z^2}{x^2 + y^2 + z^2}.$$

Clearly,

$$1 \leq \frac{\mathbf{x}^T A \mathbf{x}}{\mathbf{x}^T \mathbf{x}} = \frac{x^2 + 2y^2 + 3z^2}{x^2 + y^2 + z^2} \leq 3.$$

The Rayleigh quotient of A is therefore bounded from below by the least eigenvalue, 1, of A and is bounded from above by the largest eigenvalue, 3, of A.

To prove this result in the general case, let A be a symmetric $n \times n$-matrix whose n eigenvalues are

$$\lambda_1 \leq \lambda_2 \leq \cdots \leq \lambda_n$$

and whose corresponding *orthonormal* eigenvectors are $\mathbf{x}_1, \mathbf{x}_2, \ldots, \mathbf{x}_n$. These vectors are columns of an orthogonal matrix X. Because the transformation

$$\mathbf{x} = X\mathbf{u} = u_1\mathbf{x}_1 + u_2\mathbf{x}_2 + \cdots + u_n\mathbf{x}_n$$

is an orthogonal transformation, we have $X^T X = I$, so that

$$\mathbf{x}^T \mathbf{x} = \mathbf{u}^T X^T X \mathbf{u} = \mathbf{u}^T \mathbf{u} = u_1^2 + u_2^2 + \cdots + u_n^2.$$

Similarly, because $A\mathbf{x}_1 = \lambda_1\mathbf{x}_1, A\mathbf{x}_2 = \lambda_2\mathbf{x}_2, \ldots, A\mathbf{x}_n = \lambda_n\mathbf{x}_n$, we have

$$\mathbf{x}^T A \mathbf{x} = \lambda_1 u_1^2 + \lambda_2 u_2^2 + \cdots + \lambda_n u_n^2.$$

Using the inequalities $\lambda_1 \leq \lambda_2 \leq \cdots \leq \lambda_n$, we find that

$$\mathbf{x}^T A \mathbf{x} = \lambda_1 u_1^2 + \cdots + \lambda_n u_n^2 \geq \lambda_1(u_1^2 + \cdots + u_n^2) = \lambda_1 \mathbf{x}^T \mathbf{x},$$

$$\mathbf{x}^T A \mathbf{x} = \lambda_1 u_1^2 + \cdots + \lambda_n u_n^2 \leq \lambda_n(u_1^2 + \cdots + u_n^2) = \lambda_n \mathbf{x}^T \mathbf{x}.$$

Combining these inequalities and dividing by $\mathbf{x}^T\mathbf{x}$, we obtain the following:

PROPERTY 3. The Rayleigh quotient $R(\mathbf{x})$ for A satisfies the inequality

$$\lambda_{\min} = \lambda_1 \leq R(\mathbf{x}) = \frac{\mathbf{x}^T A \mathbf{x}}{\mathbf{x}^T \mathbf{x}} \leq \lambda_n = \lambda_{\max}$$

for every nonzero vector $\mathbf{x}$. If $\mathbf{x}$ is orthogonal to the first $j - 1$ eigenvectors $\mathbf{x}_1, \mathbf{x}_2, \ldots, \mathbf{x}_{j-1}$, then

$$\lambda_j \leq \frac{\mathbf{x}^T A \mathbf{x}}{\mathbf{x}^T \mathbf{x}} \leq \lambda_n.$$

If $\mathbf{x}$ is orthogonal to the last r eigenvectors $\mathbf{x}_{n-r+1}, \ldots, \mathbf{x}_n$, then

$$\lambda_1 \leq \frac{\mathbf{x}^T A \mathbf{x}}{\mathbf{x}^T \mathbf{x}} \leq \lambda_{n-r}.$$

When $\mathbf{x}$ is orthogonal to $\mathbf{x}_1, \ldots, \mathbf{x}_{j-1}$, we have $u_1 = 0$, $u_2 = 0, \ldots,$ $u_{j-1} = 0$, so that

$$\mathbf{x}^T A \mathbf{x} = \lambda_j u_j^2 + \cdots + \lambda_n u_n^2 \geq \lambda_j(u_j^2 + \cdots + u_n^2) = \lambda_j \mathbf{x}^T \mathbf{x}.$$

When $\mathbf{x}$ is orthogonal to $\mathbf{x}_{n-r+1}, \ldots, \mathbf{x}_n$, we have $u_{n-r+1} = 0, \ldots,$ $u_n = 0$. In this event

$$\mathbf{x}^T A \mathbf{x} = \lambda_1 u_1^2 + \cdots + \lambda_{n-r} u_{n-r}^2 \leq \lambda_{n-r}(u_1^2 + \cdots + u_{n-r}^2) = \lambda_{n-r} \mathbf{x}^T \mathbf{x}.$$

From these inequalities we conclude that the last two parts of Property 3 hold.

Keeping the order of the eigenvalues and eigenvectors of A described above, we have as a consequence of this result the following:

COROLLARY. The kth eigenvalue λ_k of A is the

1. maximum of the Rayleigh quotient $R(\mathbf{x})$ of A on the subspace generated by the first k eigenvectors $\mathbf{x}_1, \mathbf{x}_2, \ldots, \mathbf{x}_k$.
2. minimum of $R(\mathbf{x})$ on the subspace generated by the last $n - k + 1$ eigenvectors $\mathbf{x}_k, \mathbf{x}_{k+1}, \ldots, \mathbf{x}_n$.

As a further property of $R(\mathbf{x})$ we have:

PROPERTY 4. Let μ_{min} and μ_{max} be respectively the minimum and maximum of the Rayleigh quotient $R(\mathbf{x})$ on a linear subspace $\mathbf{S}$ of dimension k. We have

$$\mu_{min} \leq \lambda_{n-k+1}, \qquad \mu_{max} \geq \lambda_k,$$

where λ_{n-k+1} and λ_k are eigenvalues of A ordered as described above.

Since $\mathbf{S}$ is of dimension k, there is a vector $\mathbf{x}$ in $\mathbf{S}$ orthogonal to the $k - 1$ eigenvectors $\mathbf{x}_1, \mathbf{x}_2, \ldots, \mathbf{x}_{k-1}$. It follows from Property 3 that we have

$$\lambda_k \leq R(\mathbf{x}) \qquad \text{as well as} \qquad R(\mathbf{x}) \leq \mu_{max}.$$

Hence $\lambda_k \leq \mu_{max}$.

Similarly, there is a vector $\mathbf{y}$ in $\mathbf{S}$ orthogonal to the $k - 1$ eigenvectors $\mathbf{x}_{n-k+2}, \ldots, \mathbf{x}_n$. By Property 3 again

$$R(\mathbf{y}) \leq \lambda_{n-k+1} \qquad \text{as well as} \qquad \mu_{min} \leq R(\mathbf{y}).$$

It follows that $\mu_{min} \leq \lambda_{n-k+1}$, as was to be proved.

Exercises

1. **(a)** Find the eigenvalues and corresponding eigenvectors for each of the following matrices.

$$A_1 = \begin{bmatrix} 2 & 0 \\ 0 & 3 \end{bmatrix}, \qquad A_2 = \begin{bmatrix} 5 & -2 \\ -2 & 2 \end{bmatrix}, \qquad A_3 = \begin{bmatrix} 8 & 2 \\ 2 & 5 \end{bmatrix}.$$

   (b) In each case express the Rayleigh quotient $R(\mathbf{x})$ in terms of the components x and y of a vector $\mathbf{x} = (x, y)$.

   (c) Compute $R(\mathbf{x})$ for each of the eigenvectors found in part (a). Verify that in each case, $R(\mathbf{x})$ is the eigenvalue corresponding to the eigenvector $\mathbf{x}$.

2. Let A be the matrix

$$A = \begin{bmatrix} a & b \\ b & c \end{bmatrix}.$$

   (a) With $\mathbf{x} = (x, y)$, express the Rayleigh quotient $R(\mathbf{x})$ of A in terms of x and y.

   (b) Show that the gradient $\mathbf{g} = A\mathbf{x} - R(\mathbf{x})\mathbf{x}$ of $R(\mathbf{x})$ at $\mathbf{x}$ is expressible in the form

$$\mathbf{g} = \frac{bx^2 - (a - c)xy - by^2}{x^2 + y^2} \begin{bmatrix} -y \\ x \end{bmatrix}.$$

   (c) Conclude that an eigenvector $\mathbf{x} = (x, y)$ of A can be found by a nonzero solution (x, y) of the equation

$$bx^2 - (a - c)xy - by^2 = 0.$$

   (d) Apply these results to the case in which A is the matrix A_2 in Exercise 1.

3. Show that the Rayleigh quotient has the property that

$$R(c\mathbf{x}) = R(\mathbf{x})$$

   for every nonzero scalar c.

4. Let $\lambda_1 \leq \lambda_2 \leq \cdots \leq \lambda_n$ be the eigenvalues of a symmetric matrix A. With the help of Property 4, establish the following.

   (a) *Minimax Principle*. The kth eigenvalue λ_k of A is the minimum of the maximum of the Rayleigh quotient $R(\mathbf{x})$ of A on k-dimensional linear subspaces of $\mathbf{R}^n$.

   (b) *Maximin Principle*. The kth eigenvalue λ_k of A is the maximum of the minimum of $R(\mathbf{x})$ on $(n - k + 1)$-dimensional linear subspaces of $\mathbf{R}^n$.

5. Observe that the definition of the Rayleigh quotient does not require A to be symmetric. Show that Property 1 holds even if A is not symmetric.

6. Continuing with Exercise 5, let A and B be the matrices

$$A = \begin{bmatrix} 1 & 2 \\ 0 & -1 \end{bmatrix}, \qquad B = \frac{1}{2}(A + A^T) = \begin{bmatrix} 1 & 1 \\ 1 & -1 \end{bmatrix}.$$

   (a) Show that A and B have the same Rayleigh quotient $R(\mathbf{x})$.

(b) Show that the maximum of $R(\mathbf{x})$ is $\sqrt{2}$, the maximum eigenvalue of B. This exceeds the maximum eigenvalue 1 of A.

(c) Conclude that when A is not symmetric, the maximum of the Rayleigh quotient may exceed the maximum eigenvalue of A.

7. Let A be a symmetric matrix. Suppose that the inequality

$$\mathbf{x}^T A \mathbf{x} \le 0$$

holds for all nonzero vectors $\mathbf{x}$ in a k-dimensional linear subspace of $\mathbf{R}^n$. Show that A has at least k negative eigenvalues. *Hint:* Use the result given in Exercise 4.

7.3   QUADRATIC FUNCTIONS AND ELLIPSOIDS

In analytic geometry it was seen that the equation

$$x^2 + y^2 = r^2$$

represents a circle of radius r whose center is at the origin. The circle of radius r whose center is at the point (x_0, y_0) is given by the equation

$$(x - x_0)^2 + (y - y_0)^2 = r^2.$$

Similarly, in Section 7.1, we saw that the equation

$$ax^2 + 2bxy + cy^2 = k \qquad (a > 0,\, ac - b^2 > 0,\, k > 0)$$

is an equation of an ellipse in the xy-plane. Its center is at the origin. If we move the ellipse so that its center is at the point (x_0, y_0), while keeping its orientation, its equation becomes

$$a(x - x_0)^2 + 2b(x - x_0)(y - y_0) + c(y - y_0)^2 = k \qquad (k > 0).$$

Using matrices, let A be a positive definite symmetric $n \times n$-matrix. In Section 7.1 we saw that the set of points $\mathbf{x}$ satisfying the equation

$$\mathbf{x}^T A \mathbf{x} = k \qquad (k > 0)$$

is an ellipse when $n = 2$, is a two-dimensional ellipsoid when $n = 3$, is a three-dimensional ellipsoid when $n = 4$, and, in general, is an $(n - 1)$-dimensional ellipsoid. In each case its center is at the origin $\mathbf{x}_0 = \mathbf{0}$. When we move this ellipsoid so that its center is at an arbitrary fixed point $\mathbf{x}_0$, its equation takes the form

$$(\mathbf{x} - \mathbf{x}_0)^T A (\mathbf{x} - \mathbf{x}_0) = k \qquad (k > 0).$$

Expanding and using the relation $\mathbf{x}^T A \mathbf{x}_0 = \mathbf{x}_0^T A \mathbf{x}$, our equation becomes

$$\mathbf{x}^T A \mathbf{x} - 2\mathbf{x}_0^T A \mathbf{x} + \mathbf{x}_0^T A \mathbf{x}_0 = k \qquad (k > 0).$$

This equation in $\mathbf{x}$ is of the form

$$F(\mathbf{x}) = \mathbf{x}^T A\mathbf{x} - 2\mathbf{h}^T\mathbf{x} = \alpha \qquad (\alpha > \alpha_0),$$

where

$$\mathbf{h} = A\mathbf{x}_0, \qquad \alpha = k - \mathbf{x}_0^T A\mathbf{x}_0, \qquad \alpha_0 = -\mathbf{x}_0^T A\mathbf{x}_0 = F(\mathbf{x}_0).$$

From these relations, we draw the following conclusions:

PROPOSITION 1. Let A be a positive definite symmetric $n \times n$-matrix and let $F(\mathbf{x})$ be the *quadratic function*

$$F(\mathbf{x}) = \mathbf{x}^T A\mathbf{x} - 2\mathbf{h}^T\mathbf{x},$$

where $\mathbf{h}$ is a fixed vector. Then

(a) The function $F(\mathbf{x})$ has its minimum value α_0 at the solution $\mathbf{x} = \mathbf{x}_0$ of the equation

$$A\mathbf{x} = \mathbf{h}.$$

(b) For each $\alpha > \alpha_0$, the ellipsoid defined by the equation

$$F(\mathbf{x}) = \alpha$$

has its center at the solution $\mathbf{x} = \mathbf{x}_0$ of the equation

$$A\mathbf{x} = \mathbf{h}.$$

This result tells us that we can find the solution of $A\mathbf{x} = \mathbf{h}$ by finding the minimizer $\mathbf{x}_0$ of $F(\mathbf{x})$ in any manner. It also tells that we can solve $A\mathbf{x} = \mathbf{h}$ by finding the common center of the ellipsoids defined by the equation $F(\mathbf{x}) = \alpha$. These facts can be used in designing special methods for solving the linear equation $A\mathbf{x} = \mathbf{h}$.

■ **EXAMPLE 1** _____

Consider the case in which

$$A = \begin{bmatrix} 4 & -5 \\ -5 & 9 \end{bmatrix}, \qquad \mathbf{h} = \begin{bmatrix} -12 \\ 26 \end{bmatrix}, \qquad \mathbf{x} = \begin{bmatrix} x \\ y \end{bmatrix}.$$

In this case

$$F(\mathbf{x}) = \mathbf{x}^T A\mathbf{x} - 2\mathbf{h}^T\mathbf{x} = 4x^2 - 10xy + 9y^2 + 24x - 52y.$$

The equation $A\mathbf{x} = \mathbf{h}$ in coordinate form is

$$\begin{aligned} 4x - 5y &= -12 \\ -5x + 9y &= 26. \end{aligned}$$

Solving, it is found that the solution is $\mathbf{x}_0 = (2, 4)$. Moreover, $F(\mathbf{x}_0) = -80$. We have

$$F(\mathbf{x}) - F(\mathbf{x}_0) = 4x^2 - 10xy + 9y^2 + 24x - 52y + 80$$
$$= 4(x - 2)^2 - 10(x - 2)(y - 4) + 9(y - 4)^2.$$

Since the right member is positive unless $\mathbf{x} = \mathbf{x}_0 = (2, 4)$ it follows that $\mathbf{x}_0$ minimizes $F(\mathbf{x})$, as stated in Proposition 1. Moreover, the equation

$$4x^2 - 10xy + 9y^2 + 24x - 52y = \alpha \qquad (\alpha > -80)$$

with α as a parameter, defines a one-parameter family of similar ellipses having the point $\mathbf{x}_0 = (2, 4)$ as their common center.

Continuing with the general case, we note that the quadratic function

$$F(\mathbf{x}) = \mathbf{x}^T A \mathbf{x} - 2\mathbf{h}^T \mathbf{x}$$

has the following property.

PROPOSITION 2. Let $\mathbf{x}_1$ be a fixed point. We have the relation

$$F(\mathbf{x}) - F(\mathbf{x}_1) = (\mathbf{x} - \mathbf{x}_1)^T A(\mathbf{x} - \mathbf{x}_1) - 2\mathbf{r}_1^T(\mathbf{x} - \mathbf{x}_1),$$

where $\mathbf{r}_1 = \mathbf{h} - A\mathbf{x}_1 = A(\mathbf{x}_0 - \mathbf{x}_1)$.

To establish this result we use the relations

$$F(\mathbf{x}) = \mathbf{x}^T A \mathbf{x} - 2\mathbf{h}^T \mathbf{x} = \mathbf{x}^T A \mathbf{x} - 2\mathbf{x}^T A \mathbf{x}_1 - 2\mathbf{r}_1^T \mathbf{x}$$
$$F(\mathbf{x}_1) = \mathbf{x}_1^T A \mathbf{x}_1 - 2\mathbf{h}^T \mathbf{x}_1 = \mathbf{x}_1^T A \mathbf{x}_1 - 2\mathbf{x}_1^T A \mathbf{x}_1 - 2\mathbf{r}_1^T \mathbf{x}_1.$$

Subtracting, we find that

$$F(\mathbf{x}) - F(\mathbf{x}_1) = \mathbf{x}^T A \mathbf{x} - 2\mathbf{x}^T A \mathbf{x}_1 + \mathbf{x}_1^T A \mathbf{x}_1 - 2\mathbf{r}_1^T(\mathbf{x} - \mathbf{x}_1).$$

Collecting terms, we obtain the relation described in Proposition 2.

When a point $\mathbf{x}$ satisfies the equation

$$\mathbf{r}_1^T(\mathbf{x} - \mathbf{x}_1) = 0,$$

we have, by Proposition 2, the relation

$$F(\mathbf{x}) - F(\mathbf{x}_1) = (\mathbf{x} - \mathbf{x}_1)^T A(\mathbf{x} - \mathbf{x}_1) > 0$$

unless $\mathbf{x} = \mathbf{x}_1$. This gives the following:

COROLLARY. Let **T** be the set of points $\mathbf{x}$ satisfying the equation

$$\mathbf{r}_1^T(\mathbf{x} - \mathbf{x}_1) = 0.$$

The inequality

$$F(\mathbf{x}) > F(\mathbf{x}_1)$$

holds for all points $\mathbf{x}$ in **T** different from $\mathbf{x}_1$.

The vector $\mathbf{r}_1$ has a name. In the calculus of functions of n variables the vector

$$\mathbf{g} = A\mathbf{x} - \mathbf{h}$$

is the **gradient** of the function

$$\tfrac{1}{2}F(\mathbf{x}) = \tfrac{1}{2}\mathbf{x}^T A\mathbf{x} - \mathbf{h}^T\mathbf{x}$$

at $\mathbf{x}$. It points in the direction of steepest ascent of this function at $\mathbf{x}$. Its negative

$$\mathbf{r} = \mathbf{h} - A\mathbf{x}$$

is the **negative gradient** of $\tfrac{1}{2}F(\mathbf{x})$ at $\mathbf{x}$ and points in the direction of steepest descent. For our purposes the scale factor $\tfrac{1}{2}$ of F is irrelevant. Accordingly, we call $\mathbf{g}$ not "the" but "a" gradient of F and we call $\mathbf{r}$ "a" negative gradient of F. The vector $\mathbf{g}$ is in the direction of steepest ascent of F at $\mathbf{x}$ and $\mathbf{r}$ is in the direction of steepest descent of F at $\mathbf{x}$. The subscript 1 on $\mathbf{r}_1$ signifies that $\mathbf{r}_1$ is a negative gradient of F at the point $\mathbf{x}_1$. Observe that $\mathbf{r}_1 = \mathbf{0}$ only if $\mathbf{x}_1$ is the minimizer $\mathbf{x}_0$ of F.

The set $\mathbf{T}$ described in our corollary also has a name. If $\mathbf{x}_1 = \mathbf{x}_0$, then $\mathbf{r}_1 = \mathbf{0}$ and $\mathbf{T}$ is the whole space $\mathbf{R}^n$. Otherwise, $\mathbf{T}$ is a **hyperplane** through $\mathbf{x}_1$. It consists of all points $\mathbf{x}$ such that the vector $\mathbf{x} - \mathbf{x}_1$ is orthogonal to $\mathbf{r}_1$. The vector $\mathbf{r}_1$ is a normal of $\mathbf{T}$. If we move the origin to the point $\mathbf{x}_1$, the hyperplane $\mathbf{T}$ becomes an $(n - 1)$-dimensional linear space comprised of all vectors $\mathbf{y} = \mathbf{x} - \mathbf{x}_1$ orthogonal to $\mathbf{r}_1$. Inasmuch as $F(\mathbf{x}) > F(\mathbf{x}_1)$ on $\mathbf{T}$ except at $\mathbf{x}_1$, it follows that $\mathbf{T}$ has only the point $\mathbf{x}_1$ in common with the ellipsoid $\mathbf{E}$ defined by the equation

$$F(\mathbf{x}) = F(\mathbf{x}_1).$$

The hyperplane $\mathbf{T}$ is therefore the **tangent** of $\mathbf{E}$ at $\mathbf{x}_1$. The term **tangent space** is also used for $\mathbf{T}$. The vector $\mathbf{r}_1$ at $\mathbf{x}_1$ points inward and is orthogonal to the tangent $\mathbf{T}$. It is called an **inner normal** of $\mathbf{E}$ at $\mathbf{x}_1$. The situation for the case $n = 2$ is illustrated in Figure 1. In this case $\mathbf{E}$ is an ellipse through $\mathbf{x}_1$ whose

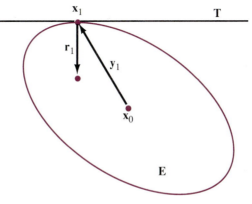

FIGURE 1

center is at $\mathbf{x}_0$. The line $\mathbf{T}$ is the tangent to $\mathbf{E}$ at $\mathbf{x}_1$ and $\mathbf{r}_1$ is an inner normal of $\mathbf{E}$ at $\mathbf{x}_1$. The vector

$$\mathbf{y}_1 = \mathbf{x}_1 - \mathbf{x}_0$$

joining the point $\mathbf{x}_1$ to the center $\mathbf{x}_0$ of $\mathbf{E}$ can be viewed to be a *radius* of $\mathbf{E}$. It cannot be determined explicitly until we have found the center $\mathbf{x}_0$. It is connected to $\mathbf{r}_1$ by the relation

$$\mathbf{r}_1 = \mathbf{h} - A\mathbf{x}_1 = A(\mathbf{x}_0 - \mathbf{x}_1) = -A\mathbf{y}_1.$$

Summarizing, we have the following proposition illustrated by Figure 1.

PROPOSITION 3. Let $\mathbf{x}_1$ be a point other than the minimizer $\mathbf{x}_0$ of F. The set of points $\mathbf{x}$ satisfying the equation

$$F(\mathbf{x}) = F(\mathbf{x}_1)$$

is an ellipsoid $\mathbf{E}$ through $\mathbf{x}_1$ having its center at $\mathbf{x}_0$. The vector

$$\mathbf{r}_1 = \mathbf{h} - A\mathbf{x}_1$$

is an inner normal of $\mathbf{E}$ at $\mathbf{x}_1$. The tangent $\mathbf{T}$ to $\mathbf{E}$ at $\mathbf{x}_1$ is the set of points $\mathbf{x}$ satisfying the equation

$$\mathbf{r}_1^T(\mathbf{x} - \mathbf{x}_1) = 0.$$

This equation can be rewritten in the form

$$\mathbf{x}_1^T A\mathbf{x} - \mathbf{h}^T(\mathbf{x} + \mathbf{x}_1) = F(\mathbf{x}_1).$$

The second equation for $\mathbf{T}$ can be obtained by writing the first equation in the form

$$(\mathbf{h} - A\mathbf{x}_1)^T (\mathbf{x} - \mathbf{x}_1) = 0,$$

expanding, and collecting terms. In the two-dimensional case in which

$$A = \begin{bmatrix} a & b \\ b & c \end{bmatrix}, \qquad \mathbf{h} = \begin{bmatrix} d \\ e \end{bmatrix}, \qquad \mathbf{x} = \begin{bmatrix} x \\ y \end{bmatrix},$$

the equation

$$F(\mathbf{x}) = \alpha$$

of $\mathbf{E}$, written in the coordinate form, becomes

$$ax^2 + 2bxy + cy^2 - 2dx - 2ey = \alpha.$$

The second equation in Proposition 3 for the tangent $\mathbf{T}$ to $\mathbf{E}$ at a point $\mathbf{x}_1 = (x_1, y_1)$ on $\mathbf{E}$ takes the form

$$ax_1x + b(x_1y + y_1x) + cy_1y - d(x + x_1) - e(y + y_1) = \alpha.$$

Notice that this equation is obtained by writing the equation for **E** in the form

$$axx + b(xy + yx) + cyy - d(x + x) - e(y + y) = \alpha$$

and appropriately putting the subscript 1 on x's and y's. Using matrices we find that the inner normal

$$\mathbf{r}_1 = \mathbf{h} - A\mathbf{x}_1$$

of **E** at $\mathbf{x}_1 = (x_1, y_1)$ is

$$\mathbf{r}_1 = \begin{bmatrix} d \\ e \end{bmatrix} - \begin{bmatrix} a & b \\ b & c \end{bmatrix} \begin{bmatrix} x_1 \\ y_1 \end{bmatrix} = \begin{bmatrix} d - ax_1 - by_1 \\ e - bx_1 - cy_1 \end{bmatrix}.$$

Observe that the components of $\mathbf{r}_1$ are the negatives of the coefficients of x and y in our tangential equation for **T**.

■ **EXAMPLE 1 (continued)** ────────────────────────────────

To determine α so that the ellipse **E** defined by the equation

$$4x^2 - 10xy + 9y^2 + 24x - 52y = \alpha$$

passes through the point $\mathbf{x}_1 = (10, 10)$, we set $x = 10$ and $y = 10$ in this equation and find that $\alpha = 20$. Set $\alpha = 20$. Then the tangent **T** to **E** at $\mathbf{x}_1 = (10, 10)$ is given by the equation

$$40x - 5(10y + 10x) + 90y + 12(x + 10) - 26(y + 10) = 20.$$

Collecting like terms, our tangential equation becomes

$$2x + 14y = 160.$$

According to the remarks made above, an inner normal of **E** at $\mathbf{x}_1 = (10, 10)$ is $\mathbf{r}_1 = (-2, -14)$. As was seen earlier, the point $\mathbf{x}_0 = (2, 4)$ is the center of **E**. It follows that the "radius" of **E** at $\mathbf{x}_1$ is

$$\mathbf{y}_1 = \mathbf{x}_1 - \mathbf{x}_0 = (10, 10) - (2, 4) = (8, 6).$$

The point $\mathbf{x}_2 = (-6, -2) = \mathbf{x}_0 - \mathbf{y}_1$ lies on the line **L** through $\mathbf{x}_0$ and $\mathbf{x}_1$. It is a second point of intersection of **E** and **L**, as can be seen by substitution in the equation for **E**. The "radius" of **E** at $\mathbf{x}_2$ is

$$\mathbf{y}_2 = \mathbf{x}_2 - \mathbf{x}_0 = (-8, -6) = -\mathbf{y}_1.$$

Let A be the matrix associated with **E**. The vector

$$\mathbf{r}_2 = -A\mathbf{y}_2 = A\mathbf{y}_1 = -\mathbf{r}_1 = (2, 14)$$

is an inner normal of **E** at $\mathbf{x}_2$. The equation

$$-2x - 14y = 36$$

defines the tangent to **E** at $\mathbf{x}_2$. It is parallel to the tangent at $\mathbf{x}_1$.

Exercises

1. Find the centers of the ellipses defined by the equations

   (a) $4x^2 + 9y^2 = 72$ (b) $4x^2 + 9y^2 - 8x + 18y = 84$

   (c) $4x^2 - 10xy + 9y^2 = 12$ (d) $4x^2 - 10xy + 9y^2 + 18x + 8y = 82$

2. Show that the point $(3, 2)$ lies on each of the ellipses described in Exercise 1. Find the equation of the tangent to these ellipses at the point $(3, 2)$.

3. Verify that the equation

$$ax^2 + 2bxy + cy^2 + 2dx + 2ey = \alpha$$

   can be written in the matrix form

$$\mathbf{x}^T A \mathbf{x} - 2\mathbf{h}^T \mathbf{x} = \alpha,$$

   where

$$A = \begin{bmatrix} a & b \\ b & c \end{bmatrix}, \qquad \mathbf{x} = \begin{bmatrix} x \\ y \end{bmatrix}, \qquad \mathbf{h} = \begin{bmatrix} -d \\ -e \end{bmatrix}.$$

   Find A and $\mathbf{h}$ for each of the equations given in Exercise 1.

4. Consider the matrix and vectors

$$A = \begin{bmatrix} 5 & 1 & -1 \\ 1 & 4 & 1 \\ -1 & 1 & 9 \end{bmatrix}, \qquad \mathbf{h}_1 = \begin{bmatrix} 7 \\ 4 \\ -9 \end{bmatrix}, \qquad \mathbf{h}_2 = \begin{bmatrix} 16 \\ 12 \\ 6 \end{bmatrix}.$$

   Find the minimum point and the minimum value of

$$F(\mathbf{x}) = \mathbf{x}^T A \mathbf{x} - 2\mathbf{h}^T \mathbf{x}$$

   when (a) $\mathbf{h} = \mathbf{h}_1$ and (b) $\mathbf{h} = \mathbf{h}_2$.

5. Let A be a positive definite matrix. Let $\mathbf{x}_0$ be the minimum point of

$$F(\mathbf{x}) = \mathbf{x}^T A \mathbf{x} - 2\mathbf{h}^T \mathbf{x}.$$

   Show that $F(\mathbf{x}_0) = -\mathbf{h}^T \mathbf{x}_0$ is the minimum value of $F(\mathbf{x})$.

In Exercises 6 to 11 the reader is asked to establish certain properties of poles and polars relative to a fixed ellipse $\mathbf{E}$ in the xy-plane. We shall use vector notations. We select the origin to be the center of $\mathbf{E}$. Points in the xy-plane will be denoted by $\mathbf{z} = (x, y)$ with or without subscripts. By a suitable choice of a positive definite symmetric matrix A, the equation of our fixed ellipse $\mathbf{E}$ can be put in the form

$$F(\mathbf{z}) = \mathbf{z}^T A \mathbf{z} = 1.$$

Why? For each point $\mathbf{z}_1$ not at the origin, there is a unique line $\mathbf{P}_1$ defined by the equation

$$\mathbf{z}_1^T A \mathbf{z} = 1.$$

The line $\mathbf{P}_1$ is called the *polar line* of $\mathbf{z}_1$ and $\mathbf{z}_1$ is the *pole* of $\mathbf{P}_1$.

Establish the following properties of poles and polar lines:

6. If z_1 is on **E**, then P_1 is the tangent to **E** at z_1. *Hint:* The relation $z_1^T A z_1 = 1$ states that z_1 is on **E** and on the polar line P_1.

7. If z_2 is a point on P_1, then z_1 is a point on the polar line P_2 of z_2. *Hint:* Interpret the equation $z_1^T A z_2 = 1$ in two ways.

8. If z_2 and z_3 are distinct points on P_1, then z_1 is the point of intersection of the polar lines P_2 and P_3 of z_2 and z_3. *Hint:* Use the result given in Exercise 7.

9. Suppose that the point z_1 is exterior to the ellipse **E**. Show that the polar line P_1 of z_1 can be constructed in the manner shown in Figure 2. Construct the tangents P_2 and P_3 to **E** which pass through the point z_1. Let z_2 and z_3 be, respectively, the points of tangency of P_2 and P_3 to **E**. The line P_1 through z_2 and z_3 is the polar line of z_1. Why? This construction can be reversed. Let P_1 be a line that cuts **E** in two points z_2 and z_3 and does not pass through the center of **E**. Construct the tangents (polar lines) P_2 and P_3 to **E** at the points z_2 and z_3, respectively. The intersection z_1 of P_2 and P_3 is the pole of P_1.

10. Let z_1 be an interior point of **E** apart from its center. The polar line P_1 of z_1 can be obtained by the construction shown in Figure 3. Find the poles z_2 and z_3 of two distinct lines P_2 and P_3 through z_1. The line P_1 through the points z_2 and z_3 is the polar line of z_1.

11. Let L_1 be a line through z_1 and the center **0** of **E**, as shown schematically in Figure 4. This line is given parametrically by the equation $z = t z_1$. Show that:

   (a) The polar lines

   $$t z_1^T A z = 1$$

   of points $t z_1$ on L_1 are parallel to the polar line P_1 of z_1 and hence also to the line

   $$z_1^T A z = 0.$$

   The lines are said to be *conjugate* to L_1.

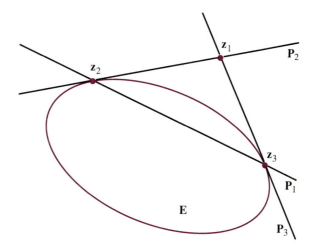

FIGURE 2

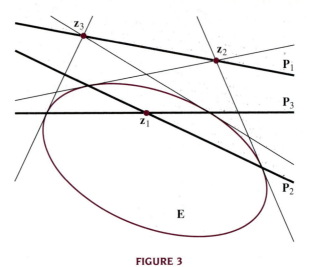

FIGURE 3

(b) The lines L_1 and P_1 intersect in the point

$$\mathbf{z}_{11} = t_1 \mathbf{z}_1 \qquad \text{with} \qquad t_1 = \frac{1}{F(\mathbf{z}_1)} \, .$$

The point $\mathbf{z}_{11}$ is called the *reciprocal* of $\mathbf{z}_1$ relative to **E**. The reciprocal of $\mathbf{z}_{11}$ is $\mathbf{z}_1$.

(c) $F(\mathbf{z}_1)F(\mathbf{z}_{11}) = 1$.

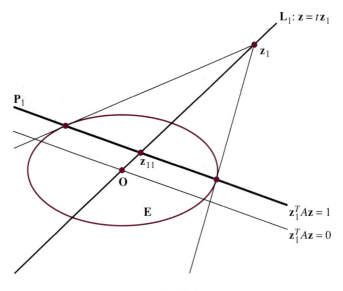

FIGURE 4

(d) The polar line $\mathbf{P}_1$ is tangent at $\mathbf{z}_{11}$ to the ellipse

$$F(\mathbf{z}) = F(\mathbf{z}_{11}).$$

(e) The point $\mathbf{z}_{11}$ is the minimizer of $F(\mathbf{z})$ on $\mathbf{P}_1$.

(f) Suppose that $\mathbf{z}_1$ is exterior to $\mathbf{E}$. Then, as shown in Figure 4, its polar line $\mathbf{P}_1$ cuts $\mathbf{E}$ in two points $\mathbf{z}_2$ and $\mathbf{z}_3$. Show that

$$\mathbf{z}_{11} = \tfrac{1}{2}(\mathbf{z}_2 + \mathbf{z}_3).$$

Hence $\mathbf{z}_{11}$ is the midpoint of the chord of $\mathbf{E}$ joining $\mathbf{z}_2$ to $\mathbf{z}_3$. *Hint:* Set $\mathbf{z}_4 = \tfrac{1}{2}(\mathbf{z}_2 + \mathbf{z}_3)$. Since $\mathbf{z}_2^T A \mathbf{z}_2 = \mathbf{z}_3^T A \mathbf{z}_3 = 1$, we have $(\mathbf{z}_2 - \mathbf{z}_3)^T A \mathbf{z}_4 = 0$ as well as $(\mathbf{z}_2 - \mathbf{z}_3)^T A \mathbf{z}_{11} = 0$. This is possible only if $\mathbf{z}_4 = \mathbf{z}_{11}$.

Remark: These results can be extended to higher-dimensional cases. Then $\mathbf{E}$ is an ellipsoid and *polar lines* become *polar planes*.

8

Solving Linear
Equations and
Finding Eigenvalues

8.1 AN ALTERNATIVE ELIMINATION PROCESS

There are many methods for finding solutions of a linear equation

$$Ax = h.$$

where A is a nonsingular $n \times n$-matrix and $\mathbf{h}$ is a vector. Why are there so many? This is because there is no method that is best for all situations. The most widely used method is some version of the Gaussian elimination method which we presented in Section 2.3. There are situations when it is useful to use the conjugate gradient method given in Section 8.3. It has the advantage that under certain circumstances a good estimate of the solution is obtained early in the computation. There are also situations in which the Gauss–Seidel method or some modification thereof appears to be the most convenient to use. These methods will be described in Section 8.4.

In this section we describe a widely used alternative to the elimination method given in Chapter 2. Consider a square matrix

$$\mathbf{A} = [a_{ij}] \qquad (i, j = 1, 2, \ldots, n).$$

For each integer $k \le n$, we have the principal submatrix

$$A_k = [a_{ij}] \qquad (i, j = 1, 2, \ldots, k)$$

and the principal minor

$$\Delta_k = \det [a_{ij}] = \det A_k \qquad (i, j = 1, 2, \ldots, k).$$

Accordingly, we have

$$\Delta_1 = a_{11}, \qquad \Delta_2 = \begin{vmatrix} a_{11} & a_{12} \\ a_{21} & a_{22} \end{vmatrix}, \qquad \Delta_3 = \begin{vmatrix} a_{11} & a_{12} & a_{13} \\ a_{21} & a_{22} & a_{23} \\ a_{31} & a_{32} & a_{33} \end{vmatrix}, \quad \ldots.$$

For reference, we shall call these minors the **primary principal minors** of A. The corresponding principal submatrices of A will be called the **primary principal submatrices** of A. As an example, consider the matrix

$$A = \begin{bmatrix} 2 & 1 & -1 & 1 \\ 4 & 0 & -1 & 1 \\ 6 & 1 & -1 & 3 \\ 6 & -1 & 0 & 1 \end{bmatrix}.$$

Its primary principal minors are

$$\Delta_1 = 2, \quad \Delta_2 = \begin{vmatrix} 2 & 1 \\ 4 & 0 \end{vmatrix} = -4, \quad \Delta_3 = \begin{vmatrix} 2 & 1 & -1 \\ 4 & 0 & -1 \\ 6 & 1 & -1 \end{vmatrix} = -4, \quad \Delta_4 = \det A = 4,$$

as one readily verifies. The corresponding primary principal submatrices of A are

$$A_1 = [2], \qquad A_2 = \begin{bmatrix} 2 & 1 \\ 4 & 0 \end{bmatrix}, \qquad A_3 = \begin{bmatrix} 2 & 1 & -1 \\ 4 & 0 & -1 \\ 6 & 1 & -1 \end{bmatrix}, \qquad A_4 = A.$$

We shall make use of the following result.

> The values of the primary principal minors of a square matrix A are unaltered when a multiple of one row is added to a row that follows it. Such a row operation will be called a *forward* row operation of the third kind.

This follows because under such a row operation a primary principal minor is either unchanged or else a multiple of one of its rows is added to another. In either event its value is unchanged.

Of course, a *backward* row operation of the third kind is one in which a multiple of a row is added to a row that precedes it. Accordingly, there are two types of elementary row operations of the third kind, a forward one and a backward one. Our elimination procedure, described below, is comprised first using forward row operations, then using backward row operations, and finally using row operations of the second kind. Note that no interchanges of rows are used. This procedure is effective only if the primary principal minors of A are nonzero. A modification of this procedure is effective when A is nonsingular and it possesses zero primary principal minors.

When successive *forward* row operations are applied to the identity matrix I, we obtain a lower triangular matrix L whose main diagonal entries are all 1's. We shall call a lower triangular matrix having 1's as its main diagonal entries, a *normalized lower triangular matrix*. Every normalized lower triangular matrix L can be obtained from I in the manner described above. When we form the product LA we are adding to each row of A a linear combination of the rows that precede it. Accordingly, we consider L to be a *generalized forward row operator*, when operating on a matrix on the left. The first two of the following lower triangular matrices

$$L_1 = \begin{bmatrix} 1 & 0 & 0 \\ 2 & 1 & 0 \\ 3 & -4 & 1 \end{bmatrix}, \qquad L_2 = \begin{bmatrix} 1 & 0 & 0 \\ -1 & 1 & 0 \\ 1 & -1 & 1 \end{bmatrix}, \qquad L_3 = \begin{bmatrix} 2 & 0 & 0 \\ 5 & -1 & 0 \\ 6 & -7 & 3 \end{bmatrix}$$

are normalized but the last one is not.

Similarly, a *normalized upper triangular matrix* U is one whose main diagonal entries are all 1's. It is the transpose of a normalized lower triangular matrix. In the elimination procedure described below, the upper triangular matrices that we encounter usually are not normalized.

Consider now the linear equation

$$Ax = h,$$

where A is a matrix having no zero primary principal minor. To discuss more fully what happens, we shall use the enlarged augmented matrix

$$M = [A \ \vdots \ h \ \vdots \ I]$$

instead of the usual augmented matrix $[A \ \vdots \ h]$. The matrix I is, of course, the identity matrix. By successive *forward* row operations of the third kind we transform M into a matrix M_1 of the form

$$M_1 = [LA \ \vdots \ Lh \ \vdots \ L] = [U \ \vdots \ h_1 \ \vdots \ L],$$

where $U = LA$ is an upper triangular matrix and L is a *normalized* lower triangular matrix. We then use successive *backward* row operations to transform M_1 into a matrix of the form

$$M_2 = [VLA \ \vdots \ VLh \ \vdots \ VL] = [D \ \vdots \ h_1 \ \vdots \ N],$$

where $D = VLA$ is a diagonal matrix and V is a normalized upper triangular matrix whose values will not concern us. The next step is to use elementary operations of the second kind to transform M_2 into the matrix

$$M_3 = [I \ \vdots \ x_0 \ \vdots \ A^{-1}],$$

where $x_0 = A^{-1}h$ solves the equation $Ax = h$.

In the course of our discussion we have established the following result.

Suppose that no primary principal minor of A is zero. Then there is a normalized lower triangular matrix L such that the matrix $U = LA$ is an upper triangular matrix. Hence, with $L_1 = L^{-1}$, we see that our matrix A is the product $A = L_1U$ of a lower triangular matrix L_1 and an upper triangular matrix U.

The decomposition of A into the product $A = L_1U$ of a lower triangular matrix L and an upper triangular matrix U is called a ***Cholesky decomposition*** of A.

Observe that the upper triangular matrix U has the same primary principal minors $\Delta_1, \Delta_2, \ldots, \Delta_n$ as the matrix A. It follows that the primary principal minors of A can be computed from those of U. For the case $n = 4$, U is of the form

$$U = \begin{bmatrix} d_1 & * & * & * \\ 0 & d_2 & * & * \\ 0 & 0 & d_3 & * \\ 0 & 0 & 0 & d_4 \end{bmatrix}.$$

Accordingly, we have, for this case,

$$\Delta_1 = d_1, \qquad \Delta_2 = d_1 d_2, \qquad \Delta_3 = d_1 d_2 d_3, \qquad \Delta_4 = d_1 d_2 d_3 d_4,$$

so that

$$d_1 = \Delta_1, \qquad d_k = \frac{\Delta_k}{\Delta_{k-1}} \quad \text{for } k > 1.$$

The extension of these formulas for a general $n \times n$-matrix A is immediate. We now seek to solve the equation

$$A\mathbf{x} = \mathbf{h}$$

for the case in which

$$A = \begin{bmatrix} 2 & 1 & -1 \\ 4 & 0 & -1 \\ 6 & 1 & -1 \end{bmatrix}, \qquad \mathbf{h} = \begin{bmatrix} 2 \\ 3 \\ 6 \end{bmatrix}.$$

We have selected $\mathbf{h}$ so that the solution is $\mathbf{x}_0 = (1, 1, 1)$. In order to find this solution, we form the augmented matrix

$$M = \begin{bmatrix} 2 & 1 & -1 & \vdots & 2 & \vdots & 1 & 0 & 0 \\ 4 & 0 & -1 & \vdots & 3 & \vdots & 0 & 1 & 0 \\ 6 & 1 & -1 & \vdots & 6 & \vdots & 0 & 0 & 1 \end{bmatrix} = [A \; \vdots \; \mathbf{h} \; \vdots \; I].$$

We now carry out the forward elimination.

Forward Elimination

To do so in a compact fashion we write the rows of M as the top rows in boxes as shown below. We write the modifications in each row directly below the row that is changed. The numbers on the left is the order in which these changes are carried out. At each stage the bottom row in each box is the row of the matrix with which we are concerned. The arrows indicate the rows of the original matrix M.

$$\rightarrow \quad \begin{array}{ccccccc} 2 & 1 & -1 & \vdots & 2 & \vdots & 1 & 0 & 0 \end{array}$$

$$\rightarrow \quad \begin{array}{ccccccc} 4 & 0 & -1 & \vdots & 3 & \vdots & 0 & 1 & 0 \end{array}$$

1 $\quad \begin{array}{ccccccc} 0 & -2 & 1 & \vdots & -1 & \vdots & -2 & 1 & 0 \end{array}$ **Add** -2(row 1) to row 2

$$\rightarrow \quad \begin{array}{ccccccc} 6 & 1 & -1 & \vdots & 6 & \vdots & 0 & 0 & 1 \end{array}$$

2 $\quad \begin{array}{ccccccc} 0 & -2 & 2 & \vdots & 0 & \vdots & -3 & 0 & 1 \end{array}$ **Add** -3(row 1) to row 3

3 $\quad \begin{array}{ccccccc} 0 & 0 & 1 & \vdots & 1 & \vdots & -1 & -1 & 1 \end{array}$ **Add** $-$(row 2) to row 3

The statements on the right explain how this row was obtained.

The result obtained from our forward elimination is the matrix

$$M_1 = \begin{bmatrix} 2 & 1 & -1 & \vdots & 2 & \vdots & 1 & 0 & 0 \\ 0 & -2 & 1 & \vdots & -1 & \vdots & -2 & 1 & 0 \\ 0 & 0 & 1 & \vdots & 1 & \vdots & -1 & -1 & 1 \end{bmatrix} = [LA \quad \vdots \quad L\mathbf{h} \quad \vdots \quad L].$$

We next carry out the backward elimination.

Backward Elimination

We carry out backward elimination on M_1 as follows:

$$\begin{array}{cccccccc} & 2 & 1 & -1 & \vdots & 2 & \vdots & 1 & 0 & 0 \end{array}$$

5 $\quad \begin{array}{cccccccc} 2 & 1 & 0 & \vdots & 3 & \vdots & 0 & -1 & 1 \end{array}$ **Add** row 3 to row 1

6 $\quad \begin{array}{cccccccc} 2 & 0 & 0 & \vdots & 2 & \vdots & -1/2 & 0 & 1/2 \end{array}$ **Add** $(1/2)$(row 2) to row 1

$$\begin{array}{cccccccc} 0 & -2 & 1 & \vdots & -1 & \vdots & -2 & 1 & 0 \end{array}$$

4 $\quad \begin{array}{cccccccc} 0 & -2 & 0 & \vdots & -2 & \vdots & -1 & 2 & -1 \end{array}$ **Add** $-$(row 3) to row 2

$$\begin{array}{cccccccc} 0 & 0 & 1 & \vdots & 1 & \vdots & -1 & -1 & 1 \end{array}$$

The matrix obtained by the backward elimination on M_1 is

$$M_2 = \begin{bmatrix} 2 & 0 & 0 & \vdots & 2 & \vdots & -1/2 & 0 & 1/2 \\ 0 & -2 & 0 & \vdots & -2 & \vdots & -1 & 2 & -1 \\ 0 & 0 & 1 & \vdots & 1 & \vdots & -1 & -1 & 1 \end{bmatrix} = [D \quad \vdots \quad N\mathbf{h} \quad \vdots \quad N].$$

We now scale the rows of M_2 so the first nonzero entry in each row is 1. To do so, we divide row 1 by 2; divide row 2 by -2. This yields the final matrix

$$M_3 = \begin{bmatrix} 1 & 0 & 0 & \vdots & 1 & \vdots & -1/4 & 0 & 1/4 \\ 0 & 1 & 0 & \vdots & 1 & \vdots & 1/2 & -1 & 1/2 \\ 0 & 0 & 1 & \vdots & 1 & \vdots & -1 & -1 & 1 \end{bmatrix} = [I \quad \vdots \quad \mathbf{x}_0 \quad \vdots \quad A^{-1}].$$

The vector $x_0 = (1, 1, 1)$, given by the middle column, is the solution of our equation $Ax = h$. The last three columns of M_3 are the columns of A^{-1}.

Referring to the matrix M_1, we see that L is the first of the matrices

$$L = \begin{bmatrix} 1 & 0 & 0 \\ -2 & 1 & 0 \\ -1 & -1 & 1 \end{bmatrix}, \quad L_1 = \begin{bmatrix} 1 & 0 & 0 \\ 2 & 1 & 0 \\ 3 & 1 & 1 \end{bmatrix}, \quad U = LA = \begin{bmatrix} 2 & 1 & -1 \\ 0 & -2 & 1 \\ 0 & 0 & 1 \end{bmatrix}.$$

The last matrix $U = LA$ is the upper triangular matrix appearing in M_1. The middle matrix L_1 is the inverse of L, which we computed so that we could verify that

$$L_1 U = \begin{bmatrix} 1 & 0 & 0 \\ 2 & 1 & 0 \\ 3 & 1 & 1 \end{bmatrix} \begin{bmatrix} 2 & 1 & -1 \\ 0 & -2 & 1 \\ 0 & 0 & 1 \end{bmatrix} = \begin{bmatrix} 2 & 1 & -1 \\ 4 & 0 & -1 \\ 6 & 1 & -1 \end{bmatrix} = A.$$

The primary principal minors of U and hence of A are

$$\Delta_1 = 2, \qquad \Delta_2 = -4, \qquad \Delta_3 = -4.$$

The elimination routine exhibited above can be compacted in a single "blocked" matrix as follows:

$\rightarrow$	2	1	-1	2		1	0	0	
5	2	1	0	3		0	-1	1	Add row 3 to row 1
6	2	0	0	2		$-1/2$	0	1/2	Add (1/2)(row 3) to row 1
7	**1**	**0**	**0**	**1**		**$-1/4$**	**0**	**1/4**	divided row 1 by 2
$\rightarrow$	4	0	-1	3		0	1	0	
1	0	-2	1	-2		-2	1	0	Add -2(row 1) to row 2
4	0	-2	0	-2		-1	2	-1	Add -(row 3) to row 2
8	**0**	**1**	**0**	**1**		**1/2**	**-1**	**1/2**	Divide row 2 by -2
$\rightarrow$	6	1	-1	6		0	0	1	
2	0	-2	2	0		-3	0	1	Add -3(row 1) to row 3
3	**0**	**0**	**1**	**1**		**-1**	**-1**	**1**	Add -(row 2) to row 3

The boldfaced rows are the rows of the final matrix M_3 given above. The numbers on the left indicate the order in which the operations 1 through 8 were performed. The statements on the right explain how that row was obtained.

Modification of the Preceding Elimination Routine

Consider the problem of solving the linear equation

$$A\mathbf{x} = \mathbf{h}$$

for the case in which

$$A = \begin{bmatrix} 0 & -1 & 1 \\ 3 & 6 & 1 \\ 2 & 4 & -2 \end{bmatrix}, \qquad \mathbf{h} = \begin{bmatrix} 3 \\ 2 \\ -4 \end{bmatrix}.$$

The primary principal minors of A are

$$\Delta_1 = 0, \qquad \Delta_2 = \begin{vmatrix} 0 & -1 \\ 3 & 6 \end{vmatrix} = 3, \qquad \Delta_3 = \det A = -8.$$

Since one of them, Δ_1, is zero, the procedure used above cannot be used without modification. However, because A is nonsingular, we can obtain an effective routine by using interchange of rows in the forward elimination part of the routine. We present this modification in the following routine, which we present in compact form using the conventions described above. Since we are not concerned with finding the inverse of A, we use the augmented matrix

$$M = [A \ \vdots \ \mathbf{h}]$$

instead of the enlarged augmented matrix $[A \ \vdots \ \mathbf{h} \ \vdots \ I]$ used earlier. The arrows indicate the rows of M.

$\rightarrow$	0	-1	1	$\vdots$	3	
1	2	4	-2	$\vdots$	-4	Interchange rows 1 and 3
5	2	4	0	$\vdots$	0	Add (1/2)(row 3) to row 1
6	2	0	0	$\vdots$	4	Add 4(row 2) to row 1
7	1	0	0	$\vdots$	2	Divide row 1 by 2
$\rightarrow$	3	6	1	$\vdots$	2	
2	0	0	4	$\vdots$	8	Add $-(3/2)$(row 1) to row 2
3	0	-1	1	$\vdots$	3	Interchange rows 2 and 3
4	0	-1	0	$\vdots$	1	Add $-(1/4)$(row 3) to row 2
8	0	1	0	$\vdots$	-1	Divide rows 2 by -1
$\rightarrow$	2	4	-2	$\vdots$	-4	
1	0	-1	1	$\vdots$	3	Interchange rows 1 and 3
3	0	0	4	$\vdots$	8	Interchange rows 2 and 3
9	0	0	1	$\vdots$	2	Divide row 3 by 4

The solution to our equation $A\mathbf{x} = \mathbf{h}$ is given by the boldface numbers in the last column. It is $\mathbf{x}_0 = (2, -1, 2)$.

The procedure just described can also be broken up into a forward part, a backward part, and a scaling part. The forward part is

$$
\begin{array}{c}
\rightarrow \\
\mathbf{1} \\
\\
\rightarrow \\
2 \\
\\
\\
\rightarrow \\
1 \\
\\
\end{array}
\quad
\begin{array}{|ccc:c|l}
0 & -1 & 1 & 3 & \\
2 & 4 & -2 & -4 & \text{Interchange rows 1 and 3} \\ \hline
3 & 6 & 1 & 2 & \text{Add } -(3/2)(\text{row 1}) \text{ to row 2} \\
0 & 0 & 4 & 8 & \\
0 & -1 & 1 & 3 & \text{Interchange rows 2 and 3} \\ \hline
2 & 4 & -2 & -4 & \\
0 & -1 & 1 & 3 & \text{Interchange rows 1 and 3} \\
0 & 0 & 4 & 8 & \text{Interchange rows 2 and 3} \\
\end{array}
$$

The remaining steps form the backward part and the scaling part. Observe that the forward part transforms the augmented matrix

$$
M = [A \mid \mathbf{h}] = \begin{bmatrix} 0 & -1 & 1 & \vdots & 3 \\ 3 & 6 & 1 & \vdots & 2 \\ 2 & 4 & -2 & \vdots & -4 \end{bmatrix}
$$

$$
\text{into } M_1 = [U \mid \mathbf{k}] = \begin{bmatrix} 2 & 4 & -2 & \vdots & -4 \\ 0 & -1 & 1 & \vdots & 3 \\ 0 & 0 & 4 & \vdots & 8 \end{bmatrix}
$$

by interchanging rows and by the use of a forward row operation of the third kind. In this procedure we have transformed the equation

$$A\mathbf{x} = \mathbf{h} \quad \text{into an equation} \quad U\mathbf{x} = \mathbf{k}$$

where U is upper triangular. Our equation $U\mathbf{x} = \mathbf{k}$ in component form becomes

$$
\begin{array}{rcrcrcr}
2x & + & 4y & - & 2z & = & -4 \\
& & -y & + & z & = & 3 \\
& & & & 4z & = & 8.
\end{array}
$$

By backward substitution we find that $z = 2$, $y = -1$, $x = 2$, as before. Of course, performing this backward substitution is equivalent (but not the same) as the backward and scaling parts of the elimination procedure given above. In this case there is little difference between the two methods. For large systems, it may be preferable to use backward substitution on $U\mathbf{x} = \mathbf{k}$

because it involves fewer arithmetic operations and so is subject to fewer round-off errors.

In the first example we could stop at the **u** end of the forward elimination procedure and form the equation

$$U\mathbf{x} = \mathbf{k}$$

with

$$U = L\mathbf{k} = \begin{bmatrix} 2 & 1 & -1 \\ 0 & -2 & 1 \\ 0 & 0 & 1 \end{bmatrix}, \qquad \mathbf{k} = L\mathbf{h} = \begin{bmatrix} 2 \\ -1 \\ 1 \end{bmatrix}.$$

In component form, with $\mathbf{x} = (x, y, z)$ we have

$$2x + y - z = 2$$
$$-2 + z = -1$$
$$z = 1.$$

Using backward substitution, we have, as before, $z = 1$, $y = 1$, and $x = 1$.

Exercises

1. In each of the following cases solve the equation $A\mathbf{x} = \mathbf{h}$ by the elimination method described in this section. Check your answer by substitution. In each case find the normalized lower triangular matrix L such that $U = LA$ is upper triangular. Also exhibit U, and solve the equation $U\mathbf{x} = \mathbf{k} = L\mathbf{h}$ by backward substitution.

   (a) $A = \begin{bmatrix} 2 & -2 & 4 \\ -2 & 3 & -5 \\ 4 & -5 & 11 \end{bmatrix}$, $\mathbf{h} = \begin{bmatrix} 4 \\ -4 \\ 10 \end{bmatrix}$

   (b) $A = \begin{bmatrix} 1 & 1 & 1 \\ 1 & 4 & -1 \\ 1 & -1 & 3 \end{bmatrix}$, $\mathbf{h} = \begin{bmatrix} 3 \\ 4 \\ 3 \end{bmatrix}$

   (c) $A = \begin{bmatrix} 2 & 0 & 0 \\ -2 & 3 & 0 \\ 4 & -5 & 11 \end{bmatrix}$, $\mathbf{h} = \begin{bmatrix} 2 \\ 1 \\ 10 \end{bmatrix}$

   (d) $A = \begin{bmatrix} 1 & -2 & 4 \\ 0 & 1 & -5 \\ 0 & 0 & 1 \end{bmatrix}$, $\mathbf{h} = \begin{bmatrix} 3 \\ -4 \\ 1 \end{bmatrix}$

   (e) $A = \begin{bmatrix} 1 & 2 & -1 & 1 \\ 2 & 5 & 0 & 2 \\ -1 & 0 & 6 & 0 \\ 1 & 2 & 0 & 3 \end{bmatrix}$, $\mathbf{h} = \begin{bmatrix} 0 \\ 2 \\ -1 \\ 1 \end{bmatrix}$

2. Let L be a normalized lower triangular $n \times n$-matrix.

   (a) Show that $B = I - L$ has the property that $B^n = 0$.

   (b) Show that $L^{-1} = I + B + B^2 + \cdots + B^{n-1}$.

   (c) Find the inverses of the matrices L obtained in Exercise 1. In each case verify that $A = L^{-1}U$.

3. Let D be a diagonal matrix whose main diagonal entries are the main diagonal entries of an upper triangular matrix U. Show that the matrices $U_1 = D^{-1}U$ and $U_2 = UD^{-1}$ are normalized upper triangular matrices. Show further that if D_1 and D_2 are diagonal matrices such that $D_1D_2 = D^{-1}$, then $U_3 = D_1UD_2$ is a normalized upper triangular matrix. Apply these results to the matrix

$$U = \begin{bmatrix} 2 & -8 & -32 \\ 0 & 4 & 16 \\ 0 & 0 & 8 \end{bmatrix}.$$

4. Referring to Exercise 3, what are the analogous results for lower triangular matrices? Give an example.

5. Let A be a nonsingular matrix. Show that the primary principal minors of A are nonzero in each of the following cases:

   (a) A is a diagonal matrix.

   (b) A is lower triangular.

   (c) A is upper triangular.

   (d) $A = LU$, where L is lower triangular and U is upper triangular.

6. Show that the primary principal minors of a nonsingular matrix A are all nonzero if and only if it is expressible as a product $A = LU$ of a lower triangular matrix L and an upper triangular matrix U.

7. Let A be a matrix whose primary principal minors are all nonzero. Show that A is expressible in the form

$$A = LDU$$

where

   (a) L is a lower triangular matrix whose main diagonal entries are all 1's.

   (b) D is a diagonal matrix.

   (c) U is an upper triangular matrix whose main diagonal entries are all 1's.

   Show that these matrices are unique.

8. Suppose that one of the primary principal minors of A, say Δ_k, is zero. Show that there is a nonzero vector $\mathbf{x}$ such that

$$\mathbf{x}^T A \mathbf{x} = 0.$$

   Conclude that if the inequality

$$\mathbf{x}^T A \mathbf{x} > 0$$

   holds for all nonzero vectors $\mathbf{x}$, then the primary principal minors of A are all nonzero. *Hint:* Let A_k be the primary principal submatrix of A such that $\Delta_k = \det A_k = 0$. Choose a nonzero k-dimensional vector $\mathbf{y}$ such that $A_k\mathbf{y} = \mathbf{0}$. The vector $\mathbf{x} = (\mathbf{y}, 0)$ has

$$\mathbf{x}^T A \mathbf{x} = \mathbf{y}^T A_k \mathbf{y} = \mathbf{0}.$$

8.2 ELIMINATION IN THE SYMMETRIC CASE

We continue to study the problem of computing the solution $\mathbf{x}$ of the linear equation

$$A\mathbf{x} = \mathbf{h}$$

in which A is an $n \times n$-matrix whose primary principal minors are all nonzero. We now make the additional assumption that A is symmetric. Of course, we can solve this equation by the method described in the preceding section. However, in this section we give a variant of this elimination routine. In this variant we again use the enlarged augmented matrix

$$M = [A \mid \mathbf{h} \mid I].$$

As before, we use forward row operations to obtain the matrix

$$M_1 = [U \mid \mathbf{c} \mid L],$$

where L is a ***normalized*** lower triangular matrix,
    $U = LA$ is upper triangular with $d_1, d_2, \ldots, d_n$ as its main diagonal entries,
    $\mathbf{c} = L\mathbf{h}$ is a column vector with entries $c_1, c_2, \ldots, c_n$.

We now form the column vector $\mathbf{a}$ whose entries are

$$a_1 = \frac{c_1}{d_1}, \qquad a_2 = \frac{c_2}{d_2}, \qquad \ldots, \qquad a_n = \frac{c_n}{d_n}.$$

As we shall see presently, the solution $\mathbf{x}_0$ to the equation $A\mathbf{x} = \mathbf{h}$ is then given by the formula

$$\mathbf{x}_0 = L^T\mathbf{a}.$$

Schematically, for the case $n = 3$, we have

$$M_1 = \begin{bmatrix} d_1 & * & * & \vdots & c_1 & \vdots & 1 & 0 & 0 \\ 0 & d_2 & * & \vdots & c_2 & \vdots & e & 1 & 0 \\ 0 & 0 & d_3 & \vdots & c_3 & \vdots & f & g & 1 \end{bmatrix} = [U \mid \mathbf{c} \mid L],$$

$$\mathbf{x}_0 = \begin{bmatrix} 1 & e & f \\ 0 & 1 & g \\ 0 & 0 & 1 \end{bmatrix}\begin{bmatrix} c_1/d_1 \\ c_2/d_2 \\ c_3/d_3 \end{bmatrix}$$

with $\mathbf{x}_0$ as the solution of $A\mathbf{x} = \mathbf{h}$.

■ **EXAMPLE 1** _____

When
$$A = \begin{bmatrix} 2 & -2 & 4 \\ -2 & 3 & -5 \\ 4 & -5 & 11 \end{bmatrix}, \quad h = \begin{bmatrix} 4 \\ -4 \\ 10 \end{bmatrix},$$

we form the augmented matrix $M = [A \vdots h \vdots I]$ whose rows are designated by arrows in the scheme shown below. We then perform the forward elimination and obtain a matrix M_1 whose rows are given by the boldfaced rows in this scheme. We get

→	**2**	**−2**	**4**	┊ **4**	┊	**1**	**0**	**0**	
→	−2	3	−5	┊ −4	┊	0	1	0	
1	**0**	**1**	**−1**	┊ **0**	┊	**1**	**1**	**0**	Add row 1 to row 2
→	4	−5	11	┊ 10	┊	0	0	1	
2	0	−1	3	┊ 2	┊	−2	0	1	Add −2(row 1) to row 3
3	**0**	**0**	**2**	┊ **2**	┊	**−1**	**1**	**1**	Add row 2 from row 3

The matrix M_1 given by the boldface numbers is

$$M_1 = \begin{bmatrix} 2 & -2 & 4 & \vdots & 4 & \vdots & 1 & 0 & 0 \\ 0 & 1 & -1 & \vdots & 0 & \vdots & 1 & 1 & 0 \\ 0 & 0 & 2 & \vdots & 2 & \vdots & -1 & 1 & 1 \end{bmatrix}.$$

We have $d_1 = 2$, $d_2 = 1$, $d_3 = 2$, $c_1 = 4$, $c_2 = 0$, and $c_3 = 2$, so that the solution of $Ax = h$ is

$$x_0 = \begin{bmatrix} 1 & 1 & -1 \\ 0 & 1 & 1 \\ 0 & 0 & 1 \end{bmatrix} \begin{bmatrix} 4/2 \\ 0/1 \\ 2/2 \end{bmatrix} = \begin{bmatrix} 1 \\ 1 \\ 1 \end{bmatrix}.$$

The procedure used above is based on the following result.

Let A be a *symmetric* matrix whose primary principal minors are nonzero.
Let L be a *normalized* lower triangular matrix such that $U = LA$ is upper triangular. Then the matrix

$$D = UL^T = LAL^T$$

is a diagonal matrix. Moreover, the inverse of A is given by the formula

$$A^{-1} = L^T D^{-1} L.$$

Observe that D is upper triangular because it is the product of two upper triangular matrices, namely, U and L^T. It is also symmetric because A is symmetric and $D = LAL^T$. A matrix having these two properties has zeros below the main diagonal and above the main diagonal and is a diagonal matrix. Note it is unnecessary to form the product UL^T to find D. This is because the main diagonal entries of D are the same as the main diagonal entries of U. Solving the equation $LAL^T = D$ for A, we find that

$$A = L^{-1}D(L^T)^{-1}, \qquad A^{-1} = L^T D^{-1}L.$$

The solution $\mathbf{x}_0$ of $A\mathbf{x} = \mathbf{h}$ is therefore

$$\mathbf{x}_0 = A^{-1}\mathbf{h} = L^T D^{-1}L\mathbf{h} = L^T\mathbf{a}, \qquad \mathbf{a} = D^{-1}\mathbf{c}, \qquad \mathbf{c} = L\mathbf{h}.$$

This is the formula for $\mathbf{x}_0$ that we used earlier.

Observe that the relation $LAL^T = D$ tells us that A is diagonalized by the upper triangular matrix L^T. It follows that, under the transformation,

$$\mathbf{x} = L^T\mathbf{y},$$

we have

$$\mathbf{x}^T A\mathbf{x} = \mathbf{y}^T LAL^T\mathbf{y} = \mathbf{y}^T D\mathbf{y} = d_1 y_1^2 + d_2 y_2^2 + \cdots + d_n y_n^2.$$

The verification of the last equality is like that for the case $n = 3$. When $n = 3$ we have

$$\mathbf{y}^T D\mathbf{y} = \begin{bmatrix} y_1 & y_2 & y_3 \end{bmatrix}\begin{bmatrix} d_1 & 0 & 0 \\ 0 & d_1 & 0 \\ 0 & 0 & d_3 \end{bmatrix}\begin{bmatrix} y_1 \\ y_2 \\ y_3 \end{bmatrix} = d_1 y_1^2 + d_2 y_2^2 + d_3 y_3^2.$$

As we have said earlier, a symmetric matrix A is said to be ***positive definite*** if

$$\mathbf{x}^T A\mathbf{x} > 0 \qquad \text{unless} \qquad \mathbf{x} = \mathbf{0}.$$

It follows from the relation

$$\mathbf{x}^T A\mathbf{x} = d_1 y_1^2 + d_2 y_2^2 + \cdots + d_n y_n^2 \qquad \text{with} \qquad \mathbf{x} = L^T\mathbf{y}$$

just established that A is positive definite if and only if the numbers d_1, d_2, ..., d_n are all positive. Because the primary principal minors Δ_1, Δ_2, ..., Δ_n of A satisfy the relations

$$\Delta_1 = d_1, \qquad \Delta_k = d_k\Delta_{k-1} \qquad (k > 1),$$

we have the following result.

> A symmetric matrix A is positive definite if and only if its primary principal minors are all positive.

Using this criterion we see that the matrix

$$A = \begin{bmatrix} 2 & -2 & 4 \\ -2 & 3 & -5 \\ 4 & -5 & 11 \end{bmatrix}$$

is positive definite because its primary principal minors are the positive numbers

$$\Delta_1 = 2, \qquad \Delta_2 = 2, \qquad \Delta_3 = 4.$$

The results described above can be put in another form expressed in terms of the **conjugacy relation**

$$\mathbf{p}^T A \mathbf{q} = 0$$

between two vectors $\mathbf{p}$ and $\mathbf{q}$. When this relation holds we say that $\mathbf{p}$ is **conjugate to** $\mathbf{q}$ and that $\mathbf{q}$ is **conjugate to** $\mathbf{p}$. This terminology is one that is used in geometrical considerations. The term **A-orthogonal** is sometimes used in place of the term *conjugate*. In our discussion of conjugacy we shall assume that our matrix A is positive definite.

Let $\mathbf{p}_1, \mathbf{p}_2, \ldots, \mathbf{p}_n$ be the column vectors of the normalized upper triangular matrix L^T appearing in the relation $LAL^T = D$ described above. When $n = 3$, the product $LAL^T = D$ takes the form

$$\begin{bmatrix} \mathbf{p}_1^T \\ \mathbf{p}_2^T \\ \mathbf{p}_3^T \end{bmatrix} A [\mathbf{p}_1 \quad \mathbf{p}_2 \quad \mathbf{p}_3] = \begin{bmatrix} \mathbf{p}_1^T A \mathbf{p}_1 & \mathbf{p}_1^T A \mathbf{p}_2 & \mathbf{p}_1^T A \mathbf{p}_3 \\ \mathbf{p}_2^T A \mathbf{p}_1 & \mathbf{p}_2^T A \mathbf{p}_2 & \mathbf{p}_2^T A \mathbf{p}_3 \\ \mathbf{p}_3^T A \mathbf{p}_1 & \mathbf{p}_3^T A \mathbf{p}_2 & \mathbf{p}_3^T A \mathbf{p}_3 \end{bmatrix} = \begin{bmatrix} d_1 & 0 & 0 \\ 0 & d_2 & 0 \\ 0 & 0 & d_3 \end{bmatrix}.$$

From this equation, we see that the relations

$$\mathbf{p}_j^T A \mathbf{p}_k = 0; \quad j \neq k, \qquad \mathbf{p}_k^T A \mathbf{p}_k = d_k > 0,$$

hold when $n = 3$. They also hold in the general n-dimensional case. Vectors $\mathbf{p}_1, \mathbf{p}_2, \ldots, \mathbf{p}_n$ of this type are said to be **mutually conjugate**. In the example given above in which

$$A = \begin{bmatrix} 2 & -2 & 4 \\ -2 & 3 & -5 \\ 4 & -5 & 11 \end{bmatrix}, \qquad \mathbf{h} = \begin{bmatrix} 4 \\ -4 \\ 10 \end{bmatrix}, \qquad L^T = \begin{bmatrix} 1 & 1 & -1 \\ 0 & 1 & 1 \\ 0 & 0 & 1 \end{bmatrix}$$

the columns of L^T are

$$\mathbf{p}_1 = \begin{bmatrix} 1 \\ 0 \\ 0 \end{bmatrix}, \qquad \mathbf{p}_2 = \begin{bmatrix} 1 \\ 1 \\ 0 \end{bmatrix}, \qquad \mathbf{p}_3 = \begin{bmatrix} -1 \\ 1 \\ 1 \end{bmatrix}.$$

It is easily verified numerically that these vectors are mutually conjugate with respect to our 3×3-matrix A. We have

$$d_1 = \mathbf{p}_1^T A \mathbf{p}_1 = 2, \qquad d_2 = \mathbf{p}_2^T A \mathbf{p}_2 = 1, \qquad d_3 = \mathbf{p}_3^T A \mathbf{p}_3 = 2$$

as the main diagonal entries of $D = LAL^T$. Moreover,

$$c_1 = \mathbf{p}_1^T \mathbf{h} = 4, \qquad c_2 = \mathbf{p}_2^T \mathbf{h} = 0, \qquad c_3 = \mathbf{p}_3^T \mathbf{h} = 2$$

are the components of the vector $\mathbf{c} = L\mathbf{h}$ used earlier in our computations. The ratios

$$a_1 = \frac{c_1}{d_1} = 2, \qquad a_2 = \frac{c_2}{d_2} = 0, \qquad a_3 = \frac{c_3}{d_3} = 1$$

have the property that the vector

$$\mathbf{x}_0 = a_1\mathbf{p}_1 + a_2\mathbf{p}_2 + a_3\mathbf{p}_3 = 2\begin{bmatrix} 1 \\ 0 \\ 0 \end{bmatrix} + 0\begin{bmatrix} 1 \\ 1 \\ 0 \end{bmatrix} + 1\begin{bmatrix} -1 \\ 1 \\ 1 \end{bmatrix} = \begin{bmatrix} 1 \\ 1 \\ 1 \end{bmatrix}$$

solves our equation $A\mathbf{x} = \mathbf{h}$. Frequently, the vector $\mathbf{x}_0$ is computed by the algorithm

$$\mathbf{x}_1 = \mathbf{0}, \qquad \mathbf{x}_{k+1} = \mathbf{x}_k + a_k\mathbf{p}_k \qquad \text{for } k = 1, 2, 3$$

so that $\mathbf{x}_4$ is the solution $\mathbf{x}_0$ of our problem. When this computation is made, we have

$$\mathbf{x}_1 = \mathbf{0}, \qquad \mathbf{x}_2 = \begin{bmatrix} 2 \\ 0 \\ 0 \end{bmatrix}, \qquad \mathbf{x}_3 = \mathbf{x}_2, \qquad \mathbf{x}_4 = \begin{bmatrix} 1 \\ 1 \\ 1 \end{bmatrix} = \mathbf{x}_0.$$

There are many ways to construct a set of mutually conjugate vectors $\mathbf{p}_1, \mathbf{p}_2, \ldots, \mathbf{p}_n$ for a positive definite symmetric matrix A. One method will be described in the next section. Another method is to proceed as follows:

Select a nonsingular matrix W and set $B = W^T A W$. By the forward elimination routine construct a normalized matrix L such that the matrix

$$U = LB = LW^T A W$$

is an upper triangular matrix. The matrix $P = WL^T$ has the property that $D = P^T A P$ is a diagonal matrix. Accordingly, the columns $\mathbf{p}_1$, $\mathbf{p}_2, \ldots, \mathbf{p}_n$ of P are mutually conjugate relative to A.

■ **EXAMPLE 2**

Consider the matrices

$$A = \begin{bmatrix} 2 & -2 & 4 \\ -2 & 3 & -5 \\ 4 & -5 & 11 \end{bmatrix}, \quad W = \begin{bmatrix} 1 & 0 & 1 \\ -1 & 1 & 1 \\ 0 & 1 & 0 \end{bmatrix}, \quad B = W^T A W = \begin{bmatrix} 9 & 4 & -1 \\ 4 & 4 & 0 \\ -1 & 0 & 1 \end{bmatrix}.$$

By the forward elimination method for B we obtain the matrices

$$U = LB = \begin{bmatrix} 9 & 4 & -1 \\ 0 & 20/9 & 4/9 \\ 0 & 0 & 4/5 \end{bmatrix}, \quad L = \begin{bmatrix} 1 & 0 & 0 \\ -4/9 & 1 & 0 \\ 1/5 & -1/5 & 1 \end{bmatrix}.$$

Then the matrices

$$P = WL^T = \begin{bmatrix} 1 & -4/9 & 6/5 \\ -1 & 13/9 & 3/5 \\ 0 & 1 & -1/5 \end{bmatrix}, \quad AP = \begin{bmatrix} 4 & 2/9 & 2/5 \\ -5 & 2/9 & 2/5 \\ 9 & 2 & -2/5 \end{bmatrix}$$

have the property that

$$D = P^T(AP) = \begin{bmatrix} 9 & 0 & 0 \\ 0 & 20/9 & 0 \\ 0 & 0 & 2/5 \end{bmatrix}$$

is a diagonal matrix. It follows that the columns

$$\mathbf{p}_1 = \begin{bmatrix} 1 \\ -1 \\ 0 \end{bmatrix}, \quad \mathbf{p}_2 = \frac{1}{9}\begin{bmatrix} -4 \\ 13 \\ 9 \end{bmatrix}, \quad \mathbf{p}_3 = \frac{1}{5}\begin{bmatrix} 6 \\ 3 \\ -1 \end{bmatrix}$$

of P are mutually conjugate relative to A.

The procedure just described transforms the column vectors $\mathbf{w}_1, \ldots, \mathbf{w}_n$ of a nonsingular matrix W into mutually conjugate vectors $\mathbf{p}_1, \ldots, \mathbf{p}_n$. An analysis of the operations used will show that our procedure is equivalent to a Gram–Schmidt process with $\mathbf{p}^T A \mathbf{q}$ as its inner product. We leave the verification of this fact as an exercise.

Exercises

1. Solve $A\mathbf{x} = \mathbf{h}$ in each of the following cases by the method described at the beginning of this section. In each case find the associated normalized lower triangular matrix L and find $D = LAL^T = UL^T$. Check your solution of $A\mathbf{x} = \mathbf{h}$. Use the augmented matrix $M = [A \mid \mathbf{h} \mid I]$.

(a) $A = \begin{bmatrix} 1 & -1 & 2 \\ -1 & 3 & -4 \\ 2 & -4 & 9 \end{bmatrix}$, $\quad \mathbf{h} = \begin{bmatrix} 3 \\ -1 \\ 7 \end{bmatrix}$

(b) $A = \begin{bmatrix} 1 & -1 & 2 \\ -1 & 0 & -1 \\ 2 & -1 & 4 \end{bmatrix}$, $\quad \mathbf{h} = \begin{bmatrix} 2 \\ -2 \\ 5 \end{bmatrix}$

(c) $A = \begin{bmatrix} 3 & 1 & 1 \\ 1 & 5 & 1 \\ 1 & 1 & 3 \end{bmatrix}$, $\quad \mathbf{h} = \begin{bmatrix} 3 \\ -3 \\ 1 \end{bmatrix}$

(d) $A = \begin{bmatrix} 5 & 1 & -1 \\ 1 & 5 & 1 \\ -1 & 1 & 5 \end{bmatrix}$, $\mathbf{h} = \begin{bmatrix} 1 \\ 1 \\ 0 \end{bmatrix}$

(e) $A = \begin{bmatrix} 1 & 2 & -1 & 1 \\ 2 & 5 & 0 & 2 \\ -1 & 0 & 6 & 0 \\ 1 & 2 & 0 & 3 \end{bmatrix}$, $\mathbf{h} = \begin{bmatrix} 0 \\ 2 \\ -1 \\ 1 \end{bmatrix}$

2. Let a, b, and c be nonzero numbers and set

$$A = \begin{bmatrix} a & -a & 2a \\ -a & a+b & -2a-b \\ 2a & -2a-b & 4a+b+c \end{bmatrix}.$$

(a) By use of the forward elimination method applied to the augmented matrix $M = [A \ \ I]$, obtain a normalized lower triangular matrix L such that $U = LA$ is upper triangular. Find $D = LAL^T$. Show that A is positive definite if and only if the numbers a, b, and c are all positive.

(b) Verify that the matrices A in parts (a) and (b) of Exercise 1 are of this type. Find the values of a, b, and c in each case.

8.3   A CONJUGATE GRADIENT ROUTINE

In addition to the various elimination routines for solving a linear equation

$$A\mathbf{x} = \mathbf{h}$$

there is another finite step method that has been found to be useful when A is a positive definite symmetric $n \times n$-matrix. It is called the **conjugate gradient method** or simply the **cg-method**. Its name follows from the fact that it can be viewed to be a generalized steepest-descent method for minimizing the quadratic function

$$F(\mathbf{x}) = \mathbf{x}^T A \mathbf{x} - 2\mathbf{h}^T \mathbf{x}$$

which we studied in Chapter 7. It is an iterative method which has the desirable property that frequently a good estimate of the solution $\mathbf{x}_0$ is obtained early in the computations. In fact, if the eigenvalues of A are clustered about m values, a good estimate of the solution is obtained after m steps. In addition, it has associated with it an algorithm for finding the characteristic polynomial for A or a factor thereof. We shall illustrate some of these properties in examples. Their proofs can be found in the reference below.

As was seen in the preceding section the solution $\mathbf{x}_0$ of $A\mathbf{x} = \mathbf{h}$ can be expressed as a linear combination of mutually conjugate vectors $\mathbf{p}_1, \mathbf{p}_2, \ldots,$ $\mathbf{p}_n$. A method of generating mutually conjugate vectors was described in the preceding section. An alternative method is given by the following algorithm. In this algorithm, A is a positive definite symmetric matrix.

METHOD OF CONJUGATE GRADIENTS (cg-method) for solving

$$Ax = h.$$

Initial Step. Select a point $\mathbf{x}_1$ and compute

(1a) $\qquad \mathbf{r}_1 = \mathbf{h} - A\mathbf{x}_1, \qquad \mathbf{p}_1 = \mathbf{r}_1.$

Iterative Steps. Successively, for $k = 1, 2, \ldots$, proceed as follows: having found $\mathbf{x}_k, \mathbf{r}_k, \mathbf{p}_k$, compute

(1b) $\qquad A\mathbf{p}_k, \qquad d_k = \mathbf{p}_k^T A\mathbf{p}_k, \qquad c_k = \mathbf{p}_k^T \mathbf{r}_k, \qquad a_k = \dfrac{c_k}{d_k},$

(1c) $\qquad \mathbf{x}_{k+1} = \mathbf{x}_k + a_k\mathbf{p}_k, \qquad \mathbf{r}_{k+1} = \mathbf{r}_k - a_k A\mathbf{p}_k,$

(1d) $\qquad e_k = -\mathbf{p}_k^T A\mathbf{r}_{k+1}, \qquad b_k = \dfrac{e_k}{d_k},$

(1e) $\qquad \mathbf{p}_{k+1} = \mathbf{r}_{k+1} + b_k\mathbf{p}_k.$

Termination. Terminate at the mth step if $\mathbf{r}_{m+1} = \mathbf{0}$. Then $m \leq n$ and $\mathbf{x}_0 = \mathbf{x}_{m+1}$ is the solution of $A\mathbf{x} = \mathbf{h}$. The scalars c_k and b_k are also given by the formulas

(2) $\qquad c_k = \|\mathbf{r}_k\|^2, \qquad b_k = \dfrac{c_{k+1}}{c_k}.$

It should be noted that the vector $\mathbf{r}_k$ is the residual

$$\mathbf{r}_k = \mathbf{h} - A\mathbf{x}_k.$$

This follows by induction. It holds for $k = 1$, by definition. If it holds for an integer k, it holds for the next integer $k + 1$, as can be seen by the computation

$$\mathbf{h} - A\mathbf{x}_{k+1} = \mathbf{h} - A(\mathbf{x}_k + a_k\mathbf{p}_k) = \mathbf{r}_k - a_k A\mathbf{p}_k = \mathbf{r}_{k+1}.$$

This completes the induction.

The cg-method has the following important properties.

The vectors $\mathbf{p}_1, \mathbf{p}_2, \mathbf{p}_3, \ldots$ are mutually conjugate and the residuals $\mathbf{r}_1, \mathbf{r}_2, \mathbf{r}_3, \ldots$ are mutually orthogonal. That is,

(3) $\qquad \mathbf{p}_j^T A\mathbf{p}_k = 0; \quad j \neq k, \qquad \mathbf{r}_j^T \mathbf{r}_k = 0; \quad j \neq k.$

The proof of these and other properties of the cg-method is given in the Supplementary Proofs at the end of the section.

It is interesting and useful to note that there exist polynomials $R_k(\lambda)$ and $P_k(\lambda)$ such that

$$\mathbf{r}_k = R_k(A)\mathbf{r}_1, \qquad \mathbf{p}_k = P_k(A)\mathbf{r}_1 \qquad (k = 1, 2, \ldots, m + 1).$$

These polynomials are generated by the algorithm

$$R_1 = P_1 = 1, \qquad R_{k+1} = R_k - a_k\lambda P_k, \qquad P_{k+1} = R_{k+1} + b_k P_k,$$

where the a_k's and b_k's are the coefficients generated by the cg-algorithm. Here, we have $P_{m+1} = R_{m+1}$ because $b_m = 0$. It can be shown that the polynomial $R_{m+1}(\lambda)$ is a factor of the characteristic polynomial of A. Thus there is a connection between the cg-method and eigenvalue theory. Consider the following

■ **EXAMPLE 1** _____

In this example we use the cg-method to find the solution $\mathbf{x}_0$ of $A\mathbf{x} = \mathbf{h}$ for the case in which

$$A = \begin{bmatrix} 1 & 1 & 1 \\ 1 & 4 & -1 \\ 1 & -1 & 3 \end{bmatrix}, \qquad \mathbf{h} = \begin{bmatrix} 3 \\ 4 \\ 3 \end{bmatrix}, \qquad \mathbf{x}_1 = \begin{bmatrix} 3 \\ 0 \\ 0 \end{bmatrix}.$$

The vectors $\mathbf{x}_k$, $\mathbf{r}_k$, $\mathbf{p}_k$, and $\mathbf{s}_k = A\mathbf{p}_k$ and scalars a_k, b_k, c_k, and d_k generated by this method are listed in the following table:

$\mathbf{x}_1$	$\mathbf{r}_1$	$\mathbf{p}_1$	$\mathbf{s}_1$	$\mathbf{x}_2$	$\mathbf{r}_2$	$\mathbf{p}_2$	$\mathbf{s}_2$	$\mathbf{x}_3$	$\mathbf{r}_3$	$\mathbf{p}_3$	$\mathbf{s}_3$	$\mathbf{x}_4$	$\mathbf{r}_4$
3	0	0	1	3	$-1/4$	$-1/4$	1/8	5/2	$-1/2$	$-3/2$	$-1/2$	1	0
0	1	1	4	1/4	0	1/8	0	1/2	0	1/2	0	1	0
0	0	0	-1	0	1/4	1/4	3/8	1/2	$-1/2$	1/2	$-1/2$	1	0

$d_1 = 4$	$c_1 = 1$		$d_2 = 1/16$	$c_2 = 1/8$		$d_3 = 1/2$	$c_3 = 1/2$
$a_1 = 1/4$	$b_1 = 1/8$		$a_2 = 2$	$b_2 = 4$		$a_3 = 1$	$b_4 = 0$

The solution to our problem is $\mathbf{x}_0 = \mathbf{x}_4 = (1, 1, 1)$. It is easily verified that the vectors $\mathbf{p}_1$, $\mathbf{p}_2$, and $\mathbf{p}_3$ are mutually conjugate and that the residuals $\mathbf{r}_1$, $\mathbf{r}_2$, and $\mathbf{r}_3$ are mutually orthogonal. The corresponding polynomials $R_k(\lambda)$ and $P_k(\lambda)$ are

$$R_1(\lambda) = P_1(\lambda) = 1, \qquad R_2(\lambda) = 1 - \frac{\lambda}{4}, \qquad P_2(\lambda) = \frac{9}{8} - \frac{\lambda}{4},$$

$$R_3(\lambda) = \frac{2 - 5\lambda + \lambda^2}{2}, \qquad P_3(\lambda) = \frac{11 - 7\lambda + \lambda^2}{2},$$

$$R_4(\lambda) = \frac{2 - 16\lambda + 8\lambda^2 - \lambda^3}{2}.$$

The polynomial $2R_4(\lambda)$ is the characteristic polynomial of A.

■ **EXAMPLE 2**

In this case we solve $A\mathbf{x} = \mathbf{h}$ by the cg-method for the situation in which

$$A = \begin{bmatrix} 5 & 1 & -1 \\ 1 & 5 & 1 \\ -1 & 1 & 5 \end{bmatrix}, \qquad \mathbf{h} = \begin{bmatrix} 1 \\ 1 \\ 0 \end{bmatrix}, \qquad \mathbf{x}_1 = \mathbf{0}, \qquad \mathbf{p}_1 = \mathbf{r}_1 = \mathbf{h}.$$

In this event $A\mathbf{p}_1 = A\mathbf{h} = 6\mathbf{h}$, $d_1 = 12$, $c_1 = 2$, $a_1 = \frac{1}{6}$, $\mathbf{x}_2 = \mathbf{x}_1 + a_1\mathbf{p}_1 = \frac{1}{6}\mathbf{h}$, $\mathbf{r}_2 = \mathbf{r}_1 - a_1 A\mathbf{p}_1 = \mathbf{h} - \frac{1}{6}(6\mathbf{h}) = \mathbf{0}$. The solution is $\mathbf{x}_0 = \mathbf{x}_2 = \frac{1}{6}\mathbf{h}$. It was obtained by one step of the cg-method. The polynomial $R_2(\lambda) = 1 - \lambda/6$ is a factor of the characteristic polynomial $\Delta(\lambda) = (3 - \lambda)(6 - \lambda)^2$ of A.

This illustrates the following result.

> Let $\mathbf{h}$ be an eigenvector of A with λ as the corresponding eigenvalue. The cg-method, with $\mathbf{x}_1 = \mathbf{0}$, gives the solution $\mathbf{x}_0 = (1/\lambda)\mathbf{h}$ of $A\mathbf{x} = \mathbf{h}$ in one step.

■ **EXAMPLE 3**

Solve the equation $A\mathbf{x} = \mathbf{h}$ by the cg-method for the case in which

$$A = \begin{bmatrix} 5 & 1 & -1 \\ 1 & 5 & 1 \\ -1 & 1 & 5 \end{bmatrix}, \qquad \mathbf{h} = \begin{bmatrix} 1 \\ 0 \\ 0 \end{bmatrix}, \qquad \mathbf{y} = \begin{bmatrix} 1 \\ -1 \\ 1 \end{bmatrix}, \qquad \mathbf{z} = \begin{bmatrix} 2 \\ 1 \\ -1 \end{bmatrix}.$$

The vectors $\mathbf{y}$ and $\mathbf{z}$ are eigenvectors of A. The vectors and scalars generated by the cg-method in solving $A\mathbf{x} = \mathbf{h}$ with $\mathbf{x}_1 = \mathbf{0}$ are given in the following table.

$\mathbf{x}_1$	$\mathbf{r}_1$	$\mathbf{p}_1$	$\mathbf{s}_1$	$\mathbf{x}_2$	$\mathbf{r}_2$	$\mathbf{p}_2$	$\mathbf{s}_2$	$\mathbf{x}_3$	$\mathbf{r}_3$
0	1	1	5	1/5	0	2/25	0	4/18	0
0	0	0	1	0	−1/5	−1/5	−18/25	−1/18	0
0	0	0	−1	0	1/5	1/5	18/25	1/18	0

$d_1 = 5$ $c_1 = 1$ $d_2 = 36/125$ $c_2 = 10/125$

$a_1 = 1/5$ $b_1 = 2/25$ $a_3 = 5/18$

The solution $\mathbf{x}_0 = \mathbf{x}_3 = (1/18)(4, -1, 1)$ was obtained in two steps. This happened because the vector $\mathbf{h}$ is a linear combination

$$\mathbf{h} = \tfrac{1}{3}\mathbf{y} + \tfrac{1}{3}\mathbf{z}$$

of two eigenvectors **y** and **z** of A corresponding to the distinct eigenvalues $\lambda = 3$ and $\mu = 6$ of A. In this case the polynomial $R_3(\lambda) = (3 - \lambda)$ $(6 - \lambda)/18$ and is a factor of the characteristic polynomial $\Delta(\lambda) = (3 - \lambda)(6 - \lambda)^2$.

This illustrates the following property of the cg-method.

> Suppose that **h** is a linear combination of m eigenvectors of A corresponding to m distinct eigenvalues of A. Then the cg-algorithm, with $x_1 = 0$, solves $Ax = h$ in at most m steps.

It also illustrates the following property of the cg-method.

> Suppose that the matrix A has exactly m distinct eigenvalues. Then the cg-method solves $Ax = h$ in at most m steps.

The matrix A appearing in Examples 2 and 3 has 3, 6, and 6 as its three eigenvalues. Accordingly, it has just two distinct eigenvalues, 3 and 6. It follows that it takes at most two steps of the cg-method to solve $Ax = h$.

In general, when A is an $n \times n$-matrix, it usually has n distinct eigenvalues and it normally takes n steps of the cg-method to obtain the solution, assuming that exact arithmetic is performed. When numbers are rounded off in our computations, this may not be the case. In using the cg-method on a computer, it is essential that high-precision arithmetic be used. This is true for any algorithm involving matrix multiplications.

One of the important properties of the cg-method is that in large systems, one frequently gets good estimates of the solution early in the computations. For a detailed development of the cg-method, the reader is referred to the book *Conjugate Direction Methods in Optimization* by Magnus R. Hestenes (Springer-Verlag, New York, 1980). References to the original papers by Steifel and Hestenes are found in this book. Extensions of the cg-method are also given and include the following two algorithms for the nonsymmetric case.

The cg-method can be extended to the case in which A is an arbitrary square nonsingular matrix. This follows because the equation

$$A^T A x = A^T h$$

has the same solution as $Ax = h$. Applying cg-algorithm (1) to this modified equation, we obtain a cg-method defined by the following equations:

(3a) x_1 arbitrary, $r_1 = h - Ax_1$, $g_1 = A^T r_1$, $p_1 = g_1$,

(3b) $\mathbf{s}_k = A\mathbf{p}_k,$ $d_k = \|\mathbf{s}_k\|^2,$ $c_k = \mathbf{p}_k^T\mathbf{g}_k$ or $\|\mathbf{g}_k\|^2,$ $a_k = \dfrac{c_k}{d_k},$

(3c) $\mathbf{x}_{k+1} = \mathbf{x}_k + a_k\mathbf{p}_k,$ $\mathbf{r}_{k+1} = \mathbf{r}_k - a_k A\mathbf{p}_k,$

(3d) $\mathbf{g}_{k+1} = A^T\mathbf{r}_{k+1},$ $b_k = \dfrac{\|\mathbf{g}_{k+1}\|^2}{\|\mathbf{g}_k\|^2},$

(3e) $\mathbf{p}_{k+1} = \mathbf{g}_{k+1} + b_k\mathbf{p}_k.$

The derivation of this algorithm from algorithm (1) will be left as an exercise. Observe that in this algorithm we do not explicitly form the matrix product A^TA. This algorithm remains valid when A is an $m \times n$-matrix of rank n. In this case the final vector $\mathbf{x}_{k+1}$ is a least square solution of $A\mathbf{x} = \mathbf{h}$. The algorithm can be modified so as to give a least square solution of $A\mathbf{x} = \mathbf{h}$ for a general $m \times n$-matrix A.

There is a second way to apply the cg-algorithm to the nonsymmetric case. In this case we set $\mathbf{x} = A^T\mathbf{y}$ in the equation $A\mathbf{x} = \mathbf{h}$, thereby giving us the equation

$$AA^T\mathbf{y} = \mathbf{h}$$

having a positive definite symmetric AA^T. Applying the cg-algorithm to this case and using the relation $\mathbf{x} = A^T\mathbf{y}$ we obtain the following algorithm.

(4a) $\mathbf{x}_1 = \mathbf{0},$ $\mathbf{r}_1 = \mathbf{h},$ $\mathbf{p}_1 = A^T\mathbf{r}_1,$

(4b) $s_k = A\mathbf{P}_k,$ $d_k = \|\mathbf{p}_k\|^2,$ $c_k = \|\mathbf{r}_k\|^2,$ $a_k = \dfrac{c_k}{d_k},$

(4c) $\mathbf{x}_{k+1} = \mathbf{x}_k + a_k\mathbf{p}_k,$ $\mathbf{r}_{k+1} = \mathbf{r}_k - a_k s_k,$

(4d) $\mathbf{p}_{k+1} = A^T\mathbf{r}_{k+1} + b_k\mathbf{p}_k$ with $b_k = \dfrac{\|\mathbf{r}_{k+1}\|^2}{\|\mathbf{r}_k\|^2}.$

The derivation of this algorithm from algorithm (1) will be left as an exercise. In this algorithm we do not form the product AA^T explicitly. It remains valid for a $m \times n$-matrix A of rank m.

The cg-method can be viewed to be a generalization of a well-known method called the **gradient method** or **method of steepest descent**. This algorithm can be obtained from the cg-algorithm (1a)–(1e) by setting $b_k = 0$ in each step. The defining relations are then

$\mathbf{x}_1$ arbitrary, $\mathbf{r}_1 = \mathbf{h} - A\mathbf{x}_1,$

$\mathbf{x}_{k+1} = \mathbf{x}_k + a_k\mathbf{r}_k,$ $\mathbf{r}_{k+1} = \mathbf{r}_k - a_k A\mathbf{r}_k,$

$a_k = \dfrac{c_k}{d_k},$ $c_k = \|\mathbf{r}_k\|^2,$ $d_k = \mathbf{r}_k^T A\mathbf{r}_k.$

This algorithm is normally nonterminating and is particularly useful when the ratio of the largest eigenvalue to the smallest eigenvalue is close to 1.

As remarked above, the cg-method is a terminating method when exact computations are made. However, when the cg-method is put in a computer it normally becomes a nonterminating method. This follows because of roundoff of errors, the residual $\mathbf{r}_k = \mathbf{h} - A\mathbf{x}_k$ is not zero. As remarked above, to ensure the effectiveness of the cg-method, high-precision arithmetic should be used. Even so, $\mathbf{x}_{n+1}$ is sometimes a better estimate than $\mathbf{x}_n$ of the solution $\mathbf{x}_0$ of $A\mathbf{x} = \mathbf{h}$. When $\mathbf{r}_n$ (or $\mathbf{r}_{n+1}$) is not sufficiently small for termination, restart the algorithm, with the last computed estimate as the initial vector $\mathbf{x}_1$. Restarting after n steps is more effective than continuing with the algorithm. There are many ways to improve the effectiveness of the cg-method, but we shall not discuss them here. The cg-method together with some auxiliary computation has become one of the standard tools for solving linear partial differential equations. It is also one of the standard tools in quadratic and nonquadratic optimization problems.

Exercises

1. Solve $A\mathbf{x} = \mathbf{h}$ by the cg-method for the cases given in Exercise 1 of Section 8.2. Construct a chart as in Example 1 of this section. In these cases select the initial point $\mathbf{x}_1$ as follows: (a)$\mathbf{x}_1 = (3, 0, 0)$ (b) $\mathbf{x}_1 = (2, 0, 0)$; (c) $\mathbf{x}_1 = (1, 0, 0)$; (d) $\mathbf{x}_1 = 0$; (e) $\mathbf{x}_1 = (1, 0, 0, 0)$. In each case find the associated polynomials $R_k(\lambda)$ and $P_k(\lambda)$.

2. Associate with a positive definite symmetric matrix A the functions

$$F(\mathbf{x}) = \tfrac{1}{2}\mathbf{x}^T A\mathbf{x} - \mathbf{h}^T\mathbf{x},$$

$$G(\mathbf{x}) = A\mathbf{x} - \mathbf{h} = \text{gradient of } F(\mathbf{x}).$$

Show that the cg-algorithm can be restated as follows:
   Initial Step. Choose a point $\mathbf{x}_1$ and set $\mathbf{r}_1 = -G(\mathbf{x}_1)$.
   Iterative Steps. For $k = 1, 2, 3, \ldots$:
(1) Find the minimizer $\mathbf{x}_{k+1} = \mathbf{x}_k + a_k\mathbf{p}_k$ on the line

$$\mathbf{x} = \mathbf{x}_k + t\mathbf{p}_k.$$

(2) Compute

$$\mathbf{p}_{k+1} = -G(\mathbf{x}_{k+1}) + \frac{\|G(\mathbf{x}_{k+1})\|^2}{\|G(\mathbf{x}_k)\|^2}\,\mathbf{p}_k.$$

   Termination. Terminate when $G(\mathbf{x}_{m+1}) = \mathbf{0}$. Then $\mathbf{x}_{m+1}$ is the minimizer of F and the solution $\mathbf{x}_0$ of $A\mathbf{x} = \mathbf{h}$.
   This algorithm can be applied to a general function $F(x)$ with $G(x)$ as its gradient. In the general case the method is nonterminating. Experience has shown that convergence is improved if one restarts the algorithm after $m \geq n$ steps.

Supplementary Proofs

Use the defining relations (1a) to (1e) of the cg-algorithm to establish the following properties of the cg-method.

▬ **P1.** Verify that the scalars a_k and b_k in the cg-algorithm are determined by the relations

$$\mathbf{p}_k^T \mathbf{r}_{k+1} = 0, \qquad \mathbf{p}_k^T A \mathbf{p}_{k+1} = 0.$$

We have

$$\mathbf{p}_k^T \mathbf{r}_{k+1} = \mathbf{p}_k^T (\mathbf{r}_k - a_k A \mathbf{p}_k) = c_k - a_k d_k = 0,$$

$$\mathbf{p}_k^T A \mathbf{p}_{k+1} = \mathbf{p}_k^T A \mathbf{r}_{k+1} + b_k \mathbf{p}_k^T A \mathbf{p}_k = -e_k + b_k d_k = 0. \qquad ▬$$

▬ **P2.** Establish the following relations

$$c_k = \|\mathbf{r}_k\|^2, \qquad d_k = \mathbf{p}_k^T A \mathbf{r}_k = \mathbf{r}_k^T A \mathbf{p}_k, \qquad \mathbf{r}_k^T \mathbf{r}_{k+1} = 0,$$

$$c_{k+1} = \mathbf{p}_{k+1}^T \mathbf{r}_k, \qquad b_k = \frac{c_{k+1}}{c_k}.$$

We have, for $k > 1$,

$$c_k = \mathbf{p}_k^T \mathbf{r}_k = (\mathbf{r}_k + b_{k-1} \mathbf{p}_{k-1})^T \mathbf{r}_k = \mathbf{r}_k^T \mathbf{r}_k = \|\mathbf{r}_k\|^2,$$

$$d_k = \mathbf{p}_k^T A \mathbf{p}_k = \mathbf{p}_k^T A (\mathbf{r}_k + b_{k-1} \mathbf{p}_{k-1}) = \mathbf{p}_k^T A \mathbf{r}_k.$$

These equations also hold when $k = 1$ since $\mathbf{p}_1 = \mathbf{r}_1$. Also,

$$\mathbf{r}_k^T \mathbf{r}_{k+1} = \mathbf{r}_k^T (\mathbf{r}_k - a_k A \mathbf{p}_k) = c_k - a_k d_k = 0,$$

$$c_{k+1} = \mathbf{p}_{k+1}^T \mathbf{r}_{k+1} = \mathbf{p}_{k+1}^T (\mathbf{r}_k - a_k A \mathbf{p}_k) = \mathbf{p}_{k+1}^T \mathbf{r}_k,$$

$$c_{k+1} = \mathbf{p}_{k+1}^T \mathbf{r}_k = (\mathbf{r}_{k+1} + b_k \mathbf{p}_k)^T \mathbf{r}_k = b_k c_k, \qquad b_k = \frac{c_{k+1}}{c_k}. \qquad ▬$$

▬ **P3.** Show that

$$\frac{\mathbf{p}_k}{c_k} = \frac{\mathbf{r}_k}{c_k} + \frac{\mathbf{r}_{k-1}}{c_{k-1}} + \cdots + \frac{\mathbf{r}_1}{c_1}.$$

This follows because $b_k = c_{k+1}/c_k$ so that

$$\frac{\mathbf{p}_{k+1}}{c_{k+1}} = \frac{\mathbf{r}_{k+1}}{c_{k+1}} + \frac{\mathbf{p}_k}{c_k}$$

$$= \frac{\mathbf{r}_{k+1}}{c_{k+1}} + \frac{\mathbf{r}_k}{c_k} + \cdots + \frac{\mathbf{r}_1}{c_1}. \qquad ▬$$

P4. Show that

(i) $$\mathbf{r}_j^T \mathbf{r}_{k+1} = 0 \, (j < k),$$

(ii) $$\mathbf{p}_j^T A \mathbf{p}_{k+1} = 0 \, (j < k).$$

The proof will be made by induction. This result holds when $k = 1$. Suppose that it holds when $k < i$. We shall show that it holds when $k = i$, that is, we shall show that

(iii) $$\mathbf{r}_j^T \mathbf{r}_{i+1} = 0,$$

(iv) $$\mathbf{p}_j^T A \mathbf{p}_{i+1} = 0.$$

If $\mathbf{r}_{i+1} = \mathbf{0}$, the algorithm is finished and there is nothing additional to prove. Otherwise, both $\mathbf{r}_{i+1}$ and $\mathbf{p}_{i+1}$ are nonzero and we proceed as follows. We already know, by P1 and P2, that equations (iii) and (iv) hold when $j = i$. Hence assume that $j < i$. In this case we have the relation

(v) $$\mathbf{r}_j^T A \mathbf{p}_i = 0.$$

This follows because, by (ii) with $k = i$,

$$\mathbf{r}_j^T A \mathbf{p}_i = (\mathbf{p}_j + b_{j-1}\mathbf{p}_{j-1})^T A \mathbf{p}_i = 0 - 0 = 0$$

when $j > 1$. It also holds when $j = 1$ since $\mathbf{r}_1 = \mathbf{p}_1$. Using this result we see that (iii) holds because

$$\mathbf{r}_j^T \mathbf{r}_{i+1} = \mathbf{r}_j^T (\mathbf{r}_i - a_i A \mathbf{p}_i) = 0 - 0 = 0.$$

We observe next that

(vi) $$\mathbf{r}_j^T \mathbf{p}_{i+1} = c_{i+1} \qquad (j \le i)$$

because, by the orthogonality relations of the $\mathbf{r}_k$'s,

$$\mathbf{r}_j^T \mathbf{p}_{i+1} = \mathbf{r}_j^T [c_{i+1} \frac{\mathbf{r}_{i+1}}{c_{i+1}} + \cdots + \frac{\mathbf{r}_j}{c_j} + \cdots + \frac{\mathbf{r}_1}{c_1} = \frac{c_{i+1} \|\mathbf{r}_j\|^2}{c_j} = c_{i+1}.$$

From this result we see that

$$a_j \mathbf{p}_j^T A \mathbf{p}_{i+1} = (\mathbf{r}_j - \mathbf{r}_{j+1})^T \mathbf{p}_{i+1} = c_{i+1} - c_{i+1} = 0.$$

Since $a_j > 0$, we have

$$\mathbf{p}_j^T A \mathbf{p}_{i+1} = 0$$

and our induction is complete.

P5. Establish the relations

$$\mathbf{p}_j^T \mathbf{r}_k = 0 \quad (j < k), \qquad \mathbf{p}_j^T \mathbf{r}_k = c_k \quad (j \ge k),$$

$$\mathbf{r}_k^T A \mathbf{p}_k = d_k, \qquad \mathbf{r}_{k+1}^T A \mathbf{p}_k = -b_k d_k, \qquad \mathbf{r}_j^T A \mathbf{p}_k = 0 \quad \text{otherwise,}$$

$$\mathbf{r}_k^T A \mathbf{r}_k = d_k + b_{k-1}^2 d_{k-1} \quad (k > 1), \qquad \mathbf{r}_1^T A \mathbf{r}_1 = d_1,$$

$$\mathbf{r}_k^T A \mathbf{r}_{k+1} = -b_k d_k, \qquad \mathbf{r}_j^T A \mathbf{r}_k = 0 \quad \text{if } |j - k| > 1.$$

The verification of these relations will be left to the reader. ■

■ **P6.** Establish the following relations, which are satisfied by the residuals $\mathbf{r}_1, \mathbf{r}_2, \ldots$ in the cg-algorithms

$$\mathbf{r}_2 = \mathbf{r}_1 - a_1 A \mathbf{r}_1,$$

$$\mathbf{r}_{k+1} = (1 + \tau_k)\mathbf{r}_k - a_k A \mathbf{r}_k - \tau_k \mathbf{r}_{k-1}, \qquad \tau_k = \frac{a_k b_{k-1}}{a_{k-1}}.$$

Verify that the corresponding estimates $\mathbf{x}_1, \mathbf{x}_2, \ldots$ are

$$\mathbf{x}_2 = \mathbf{x}_1 + a_1 \mathbf{r}_1,$$

$$\mathbf{x}_{k+1} = \mathbf{x}_k + a_k \mathbf{r}_k + \tau_k(\mathbf{x}_k - \mathbf{x}_{k-1}).$$

The relation between the $\mathbf{r}$'s is obtained by eliminating $A\mathbf{p}_{k-1}$ and $A\mathbf{p}_k$ from the relations

$$\mathbf{r}_k = \mathbf{r}_{k-1} - a_{k-1} A\mathbf{p}_{k-1},$$
$$A\mathbf{p}_k = A\mathbf{r}_k + b_{k-1} A\mathbf{p}_{k-1},$$
$$\mathbf{r}_{k+1} = \mathbf{r}_k - a_k A\mathbf{p}_k$$

■

■ **P7.** Continuing with Exercise P6, set $\alpha_1 = a_1$ and

$$\alpha_k = \frac{a_k}{1 + \tau_k}, \qquad \beta_k = \frac{\tau_k}{1 + \tau_k} \quad (k > 1).$$

Show that the cg-algorithm is equivalent to the following algorithm. *Initial Step.* Choose $\mathbf{x}_1$ arbitrarily and compute

$$\mathbf{x}_2 = \mathbf{x}_1 + \alpha_1 \mathbf{r}_1, \qquad \mathbf{r}_2 = \mathbf{r}_1 - \alpha_1 \mathbf{r}_1, \qquad \alpha_1 = \mathbf{r}_1^T A \mathbf{r}_1$$

Iterative Steps. For $k = 2, 3, \ldots$, compute

$$\alpha_k = \frac{\|\mathbf{r}_k\|^2}{\mathbf{r}_k^T A \mathbf{r}_k}, \qquad \beta_k = -\frac{\mathbf{r}_{k-1}^T A \mathbf{r}_k}{\|\mathbf{r}_{k-1}\|^2},$$

$$\mathbf{x}_{k+1} = \frac{\mathbf{x}_k + \alpha_k \mathbf{r}_k - \beta_k \mathbf{x}_{k-1}}{1 - \beta_k},$$

$$\mathbf{r}_{k+1} = \frac{\mathbf{r}_k - \alpha_k A \mathbf{r}_k - \beta_k \mathbf{r}_{k-1}}{1 - \beta_k}$$

Termination. Terminate when $\mathbf{r}_{m+1} = \mathbf{0}$. Then $\mathbf{x}_0 = \mathbf{x}_{m+1}$ is the solution of $A\mathbf{x} = \mathbf{h}$. Show that the scalars α_k and β_k are determined by the relations

$$\mathbf{r}_k^T \mathbf{r}_{k+1} = 0, \qquad \mathbf{r}_{k-1}^T \mathbf{r}_{k+1} = 0.$$

This algorithm is sometimes called **gradient partan**. ■

8.4 GAUSS–SEIDEL METHODS

In the preceding pages we have discussed algorithms for solving the linear equation $A\mathbf{x} = \mathbf{h}$ that are finite in character in the sense that the solution is obtained in a finite number of steps under perfect computations. There are additional algorithms that are infinite in character in that normally the solution cannot be obtained in a finite number of steps but have the property that a "good" approximation is found after a reasonable number of steps. In this section we describe a particular algorithm, called a Gauss–Seidel algorithm. This algorithm and variations thereof are used extensively in applications. It is simple to program on a computer and is a self-correcting procedure.

We begin with a geometrical description of the routine in the two-dimensional case. We are given two lines, $\mathbf{L}_1$ and $\mathbf{L}_2$ and seek to find their point of intersection. We consider four cases, as shown in Figures 1 to 4. Consider first the situation in Figure 1. Starting with a point 1 we move in the x-direction until we hit a point 2 on the line $\mathbf{L}_1$. Next we move in the y-direction until we hit a point 3 on $\mathbf{L}_2$. Then we move in the x-direction to a point 4 on $\mathbf{L}_1$, followed by a motion in the y-direction to a point 5 on $\mathbf{L}_2$. Proceeding in this manner we obtain successive points on $\mathbf{L}_1$ and $\mathbf{L}_2$ which "converage" to the point of intersection of $\mathbf{L}_1$ and $\mathbf{L}_2$. In this manner we can obtain a good estimate of this intersection after a reasonable number of steps. Using this method in the situation shown in Figure 2, we obtain a path that approaches the solution but winds around the point of intersection.

Although the procedure just described is effective in a large number of important cases, it is not effective in every situation. In Figure 3 we wind around the solution on the sides of a rectangle and get nowhere. In Figure 4 we move away from the point of intersection of these lines. At the end of this section we describe modifications of the method which are effective in these situations.

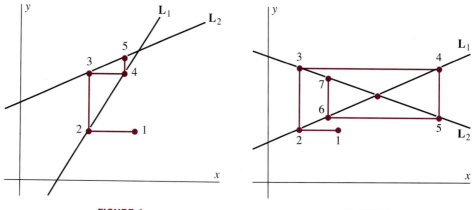

FIGURE 1   FIGURE 2

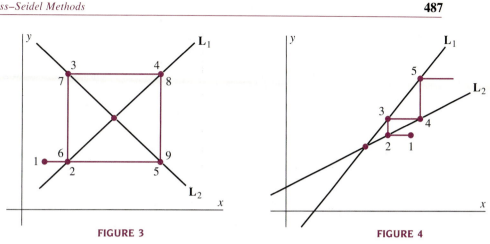

FIGURE 3   **FIGURE 4**

We now turn to an analytical description of the Gauss–Seidel method for solving a linear equation

$$\mathbf{A}\mathbf{x} = \mathbf{h}.$$

We begin with the two-dimension case. We consider the simple pair of equations

$$2x - y = 3$$
$$-x + 2y = 6$$

whose solution is $x = 4$ and $y = 5$. To solve these equations we proceed as follows.

1. Let (x_1, y_1) be an initial estimate of the solution

2. Set $y = y_1$ in the first equation, solve for x and get an estimate (x_2, y_1) that satisfies the first equation.

3. Set $x = x_2$ in the second equation, solve for y and get an estimate (x_2, y_2) that satisfies the second equation.

4. Repeat the process with (x_2, y_2) playing the role of (x_1, y_1).

When we choose $x_1 = 0$ and $y_1 = 0$, we obtain the following five estimates of the solution of our pair of linear equations.

Step	x	y
1	1.5	0
	1.5	3.75
2	3.375	3.75
	3.375	4.6875
3	3.84374	4.6875
	3.84375	4.921875
4	3.9609375	4.921875
	3.9609375	4.98046875
5	3.990234375	4.9951171875
	3.99755854373	4.998779271875

It is clear that as the procedure proceeds our estimates approach the solution $(x, y) = (4, 5)$ of our equations.

Let us analyze further the procedure we employed. Starting with an estimate (x_1, y_1) of the solution, we obtain two new estimates, $(x_1 + u, y_1)$ and $(x_1 + u, y_1 + v)$, such that

$$2(x_1 + u) - \quad y_1 \quad = 3$$
$$-(x_1 + u) + 2(y_1 + v) = 6.$$

These equations can be rewritten as

$$2u \quad\quad = 3 - (2x_1 - y_1)$$
$$-u + 2v = 6 - (-x_1 + 2y_1).$$

In matrix form these equations become

$$\begin{bmatrix} 2 & 0 \\ -1 & 2 \end{bmatrix} \begin{bmatrix} u \\ v \end{bmatrix} = \begin{bmatrix} 3 \\ 6 \end{bmatrix} - \begin{bmatrix} 2 & -1 \\ -1 & 2 \end{bmatrix} \begin{bmatrix} x_1 \\ y_1 \end{bmatrix}.$$

This matrix equation is of the form

$$\mathbf{Bu} = \mathbf{h} - A\mathbf{x}_1,$$

where

$$A = \begin{bmatrix} 2 & -1 \\ -1 & 2 \end{bmatrix}, \quad B = \begin{bmatrix} 2 & 0 \\ -1 & 2 \end{bmatrix}, \quad \mathbf{u} = \begin{bmatrix} u \\ v \end{bmatrix}, \quad \mathbf{x}_1 = \begin{bmatrix} x_1 \\ y_1 \end{bmatrix}, \quad \mathbf{h} = \begin{bmatrix} 3 \\ 6 \end{bmatrix}.$$

Observe how B is related to A. We have

$$\mathbf{u} = B^{-1}(\mathbf{h} - A\mathbf{x}_1),$$

so that our iteration becomes

$$\mathbf{x}_{k+1} = \mathbf{x}_k + \mathbf{u} = \mathbf{x}_k + B^{-1}\mathbf{r}_k, \quad \mathbf{r}_k = \mathbf{h} - A\mathbf{x}_k.$$

This iteration is valid when A is $n \times n$ and B is obtained from A by replacing the entries above the main diagonal by zeros. We verify this for the case $n = 3$. In this event we seek to solve the equations

$$a_1 x + b_1 y + c_1 z = h_1$$
$$a_2 x + b_2 y + c_2 z = h_2$$
$$a_3 x + b_3 y + c_3 z = h_3.$$

Our iteration takes the form:

1. Select and estimate (x_1, y_1, z_1) of the solution.

2. Form the equations

$$a_1(x_1 + u) + b_1\, y_1 \quad\quad\quad + c_1 z_1 = h_1$$
$$a_2(x_1 + u) + b_2(y_1 + v) \quad\quad + c_1 z_1 = h_2$$
$$a_3(x_1 + u) + b_2(y_1 + v) + c_1(z_1 + w) = h_3$$

3. Solve the first equation for u, then solve the second for v, and finally solve the third equation for w. The last estimate $(x_2, y_2, z_2) = (x_1 + u, y_1 + v, z_1 + w)$ becomes the new estimate (x_1, y_1, z_1) and the process is repeated.

By the argument used in the two-dimensional case, this procedure yields the iteration

$$\mathbf{x}_{k+1} = \mathbf{x}_k + B^{-1}\mathbf{r}_k, \qquad \mathbf{r}_k = \mathbf{h} - A\mathbf{x}_k,$$

where

$$A = \begin{bmatrix} a_1 & b_1 & c_1 \\ a_2 & b_2 & c_2 \\ a_3 & b_3 & c_3 \end{bmatrix}, \qquad B = \begin{bmatrix} a_1 & 0 & 0 \\ a_2 & b_2 & 0 \\ a_3 & b_3 & c_3 \end{bmatrix}.$$

We call B the **lower triangular part** of A.

The method just described is called the **forward Gauss–Seidel method**. There is a **backward Gauss–Seidel method** which proceeds as follows for the case $n = 3$.

Select an estimate (x_1, y_1, z_1) of the solution. Form the equations

$$\begin{aligned} a_1(x_1 + u) + b_1(y_1 + v) + c_1(z_1 + w) &= h_1 \\ a_2\,x_1 \quad\;\; + b_2(y_1 + v) + c_1(z_1 + w) &= h_2 \\ a_3\,x_1 \quad\;\; + b_2 y_1 \quad\;\; + c_1(z_1 + w) &= h_3. \end{aligned}$$

Solve the last equation for w, then solve the second for v, and finally solve the first for u. The last estimate, $(x_2, y_2, z_2) = (x_1 + u, y_1 + v, z_1 + w)$, becomes the new estimate, (x_1, y_1, z_1), and the process is repeated.

By the argument used in the preceding case, this procedure yields the iteration

$$\mathbf{x}_{k+1} = \mathbf{x}_k + C^{-1}\mathbf{r}_k, \qquad \mathbf{r}_k = \mathbf{h} - A\mathbf{x}_k,$$

where

$$A = \begin{bmatrix} a_1 & b_1 & c_1 \\ a_2 & b_2 & c_2 \\ a_3 & b_3 & c_3 \end{bmatrix}, \qquad C = \begin{bmatrix} a_1 & b_1 & c_1 \\ 0 & b_2 & c_2 \\ 0 & 0 & c_3 \end{bmatrix}.$$

We call C the **upper triangular part** of A.

There is one other form of the Gauss–Seidel method that is in use. It consists of alternating the forward and backward routines. Having obtained the kth estimate $\mathbf{x}_k$, we obtain the next estimate $\mathbf{x}_{k+1}$ by the formulas

$$\begin{aligned} \mathbf{y}_k &= \mathbf{x}_k + B^{-1}(\mathbf{h} - A\mathbf{x}_k) \\ \mathbf{x}_{k+1} &= \mathbf{y}_k + C^{-1}(\mathbf{h} - A\mathbf{y}_k). \end{aligned}$$

These formulas can be combined to form a single formula,

$$\mathbf{x}_{k+1} = \mathbf{x}_k + C^{-1}DB^{-1}(\mathbf{h} - A\mathbf{x}_k).$$

where D is the diagonal matrix whose main diagonal entries are those of A. We call D the **diagonal part** of A. To derive this formula we note that

$$\mathbf{x}_{k+1} - \mathbf{x}_k = \mathbf{x}_{k+1} - \mathbf{y}_k + \mathbf{y}_k - \mathbf{x}_k$$
$$= C^{-1}[\mathbf{h} - A\mathbf{x}_k - AB^{-1}(\mathbf{h} - A\mathbf{x}_k)] + B^{-1}(\mathbf{h} - A\mathbf{x}_k)$$
$$= C^{-1}[B - A + C]B^{-1}(\mathbf{h} - A\mathbf{x}_k)$$
$$= C^{-1}DB^{-1}(\mathbf{h} - A\mathbf{x}_k),$$

where $D = B + C - A$ is the diagonal matrix whose main diagonal entries are those of A.

The three iterations are of the form

$$\mathbf{x}_{k+1} = \mathbf{x}_k + H\mathbf{r}_k, \qquad \mathbf{r}_k = \mathbf{h} - A\mathbf{x}_k,$$

where $H = B^{-1}$ in the forward routine, $H = C^{-1}$ in the backward routine, and $H = C^{-1}DB^{-1}$ in the combined routine. These iterations are effective when A is a positive definite symmetric matrix or when A has certain other properties which we shall not describe.

However, in many cases, the effectiveness of these methods can be enhanced by the introduction of a relaxation factor β. To see intuitively that this might be effective, let us look at the two-dimensional case shown in Figure 5. We seek to find the intersection of two lines $\mathbf{L}_1$ and $\mathbf{L}_2$. Starting at a point 1 we proceed in the x-direction until we hit the line $\mathbf{L}_1$ and go slightly past to a point 2. We then move in the y-direction to a point 3 just past the line $\mathbf{L}_2$. We then repeat this process and move in the x-direction to a point 4 beyond the line $\mathbf{L}_1$. Then we go in the y-direction to a point 5 just past the line $\mathbf{L}_2$. Repeating this process in a suitable manner we approach the intersection more rapidly than by the process described in the beginning of this section.

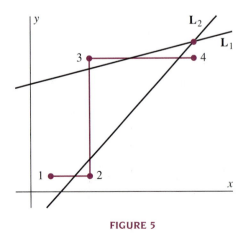

FIGURE 5

To describe this procedure analytically, we consider the case $n = 3$. Using the notations introduced above, the forward relaxed routine proceeds as follows:

1. Select a number β close to 1 (usually larger than 1).
2. Select an estimate (x_1, y_1, z_1) of the solution.
3. Form the equations

$$a_1(x_1 + u) + b_1 y_1 \qquad\qquad + c_1 z_1 \qquad = h_1$$
$$a_2(x_1 + \beta u) + b_2(y_1 + v) + c_1 z_1 \qquad = h_2$$
$$a_3(x_1 + \beta u) + b_2(y_1 + \beta v) + c_1(z_1 + w) = h_3.$$

4. Solve the first equation for u, then solve the second equation for v, and finally solve the third equation for w. The last estimate,

$$(x_2, y_2, z_2) = (x_1 + \beta u, y_1 + \beta v, z_1 + \beta w),$$

becomes the new estimate (x_1, y_1, z_1) and the process is repeated.

By the argument used in the earlier case, this procedure yields the iteration

$$\mathbf{x}_{k+1} = \mathbf{x}_k + \beta(D + \beta L)^{-1}\mathbf{r}_k, \qquad \mathbf{r}_k = \mathbf{h} - A\mathbf{x}_k,$$

where D and L are the first two of the following matrices:

$$D = \begin{bmatrix} a_1 & 0 & 0 \\ 0 & b_2 & 0 \\ 0 & 0 & c_3 \end{bmatrix}, \qquad L = \begin{bmatrix} 0 & 0 & 0 \\ a_2 & 0 & 0 \\ a_3 & b_3 & 0 \end{bmatrix}, \qquad U = \begin{bmatrix} 0 & b_1 & c_1 \\ 0 & 0 & c_2 \\ 0 & 0 & 0 \end{bmatrix}.$$

Normally, β is restricted to the interval $0 < \beta < 2$. When $\beta > 1$ we have **overrelaxation** and the algorithm is called the **SO-routine** (successive overrelaxation routine). When $\beta < 1$ we have **underrelaxation**.

Similarly, the relaxed backward routine is of the form

$$\mathbf{x}_{k+1} = \mathbf{x}_k + \beta(D + \beta U)^{-1}\mathbf{r}_k.$$

The relaxed combined routine takes the form

$$\mathbf{x}_{k+1} = \mathbf{x}_k + \beta(2 - \beta)(D + \beta U)^{-1}D(D + \beta L)^{-1}\mathbf{r}_k.$$

The last routine with $\beta > 1$ is called the **SSO-routine**. The SO- and SSO-routines are popular and practical routines for solving special linear equations $A\mathbf{x} = \mathbf{h}$ which arise in applications. These algorithms have been used effectively in obtaining numerical solutions of partial differential equations. We shall not pursue the topic of relaxation further.

There is a simple extension of the Gauss–Seidel method for solving a linear equation

$$A\mathbf{x} = \mathbf{h}.$$

In the two-dimensional case we can replace the motions in the *x*-direction and *y*-direction by motions in two other fixed directions, say a **u**-direction and a **v**-direction. In Figure 6 we show how the algorithm proceeds when **u** and **v** are the normals of the lines L_1 and L_2 whose intersection is sought. Let U be the matrix whose columns are the vectors **u** and **v**. Then the method of obtaining corrections in the **u**- and **v**-directions with $x_1 = 0$ is equivalent to applying the standard Gauss–Seidel method to the equation

$$AU\mathbf{y} = \mathbf{h} \qquad (\mathbf{x} = U\mathbf{y}).$$

Similarly, in the *n*-dimensional case we can move in the directions of the column vectors of a nonsingular $n \times n$-*matrix* U instead of in the directions of the coordinate axes given by the column vectors of the identity matrix I. Of particular importance is the choice $U = A^T$, in which case we are solving the equation

$$AA^T\mathbf{y} = \mathbf{h}$$

by the Gauss–Seidel routine. In this case we are moving successively in the directions of the normals to the hyperplanes associated with the equation $A\mathbf{x} = \mathbf{h}$. For example, in the case $n = 3$, these normals are the normals

$$\mathbf{n}_1 = (a_1, b_1, c_1), \qquad \mathbf{n}_2 = (a_2, b_2, c_2), \qquad \mathbf{n}_3 = (a_3, b_3, c_3)$$

of the 2-planes defined by the equations

$$a_1 x + b_1 y + c_1 z = h_1$$
$$a_2 x + b_2 y + c_2 z = h_2$$
$$a_3 x + b_3 y + c_3 z = h_3.$$

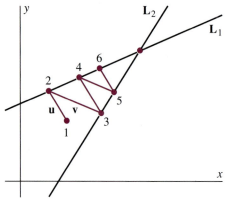

FIGURE 6

The method of using normals is often called a ***Kaczmarc routine***. It has the property that it always works, although in some cases convergence might be slow.

There is another variation of the Gauss–Seidel method. In the two-dimensional case it consists of replacing the lines L_1 and L_2 in the figures given above by two other lines M_1 and M_2 having the same point of intersection. When the lines L_1 and L_2 are given by the equations

$$ax + by - h = 0$$

$$cx + dy - k = 0,$$

then M_1 and M_2 are given by equations of the form

$$p(ax + by - h) + q(cx + dy - k) = 0$$

$$r(ax + by - h) + s(cx + dy - k) = 0.$$

In matrix notation we have replaced

$$A\mathbf{x} = \mathbf{h} \quad \text{by} \quad VA\mathbf{x} = V\mathbf{h},$$

where

$$A = \begin{bmatrix} a & b \\ c & d \end{bmatrix}, \qquad V = \begin{bmatrix} p & q \\ r & s \end{bmatrix}, \qquad \mathbf{x} = \begin{bmatrix} x \\ y \end{bmatrix}, \qquad \mathbf{h} = \begin{bmatrix} h \\ k \end{bmatrix}.$$

Thus, before applying the Gauss–Seidel routine, it is sometimes desirable to precondition our equations by replacing the equation $A\mathbf{x} = \mathbf{h}$ by $VA\mathbf{x} = V\mathbf{h}$ or more generally by $VAU\mathbf{y} = V\mathbf{h}$. This illustrates the variety of methods that are generated by the Gauss–Seidel routine. We shall not pursue these ideas further.

Exercises

1. Perform a few steps of the Gauss–Seidel routine in each of the following cases. Start at the origin.

   (a) $2x - y = 5$
       $x + 2y = 10$

   (b) $5x + y = 13$
       $x + 10y = 32$

   (c) $2x - y = 4$
       $2x + y = 12$

   (d) $x + y = 6$
       $2x - y = 3$

   In which cases is the algorithm effective?

2. For parts (a), (c), and (d) of Exercise 1, instead of correcting the x- and y-directions, make the corrections in the directions of the normals to the lines defined by the equations.

3. Consider the following systems of equations:

(1) x $= 3$ (2) $2x - y + 3z = 4$
     $2x + 3y$ $= 12$ $- 5y + 2z = -3$
     $5x - y - 4z = 9$ $7z = 7$

Solve the first system by the forward Gauss–Seidel method and the second system by the backward Gauss–Seidel method. Conclude that in these instances, the method used is equivalent to a standard elimination method.

4. Let $F(\mathbf{x})$ be the quadratic function

$$F(\mathbf{x}) = \mathbf{x}^T A \mathbf{x} - 2\mathbf{h}^T \mathbf{x}.$$

Here A is a positive definite symmetric matrix. Select m vectors $\mathbf{u}_1, \mathbf{u}_2, \ldots, \mathbf{u}_m$ which span $\mathbf{R}^n$. They need not be linearly independent. Consider the following iteration for solving $A\mathbf{x} = \mathbf{h}$. Select an initial point $\mathbf{x}_1$. After having obtained the kth estimate $\mathbf{x}_k$, find the minimizer

$$\mathbf{x}_{k+1} = \mathbf{x}_k + a_k \mathbf{u}_k$$

of $F(\mathbf{x})$ on the line

$$\mathbf{x} = \mathbf{x}_k + t\mathbf{u}_k.$$

Having obtained the point $\mathbf{x}_{m+1}$, repeat the algorithm with $\mathbf{x}_{m+1}$ as the initial point $\mathbf{x}_1$. It can be shown that the estimates obtained by these repetitions converge to the solution $\mathbf{x}_0$ of $A\mathbf{x} = \mathbf{h}$. Show that

(a) If $m = n$ and $\mathbf{u}_1, \ldots, \mathbf{u}_n$ are the unit coordinate vectors

$$\mathbf{e}_1 = (1, 0, \ldots, 0), \quad \mathbf{e}_2 = (0, 1, \ldots, 0), \quad \ldots, \quad \mathbf{e}_n = (0, \ldots, 0, 1),$$

our algorithm is the forward Gauss–Seidel algorithm.

(b) If $m = n$ and $\mathbf{u}_1, \ldots, \mathbf{u}_n$ are the vectors $\mathbf{e}_n, \ldots, \mathbf{e}_1$ in reverse order, we have the backward Gauss–Seidel routine.

(c) If $m = 2n - 1$ and $\mathbf{u}_1, \ldots, \mathbf{u}_m$ are the vectors

$$\mathbf{e}_1, \ldots, \mathbf{e}_{n-1}, \mathbf{e}_n, \mathbf{e}_{n-1}, \ldots, \mathbf{e}_1$$

our routine is the combined Gauss–Seidel routine.

8.5   THE POWER METHOD FOR FINDING A DOMINANT EIGENVALUE

In our examples on eigenvalues of a square matrix A, we normally obtained the eigenvalues of A by first finding its characteristic polynomial $p(\lambda)$ and then solving the equation $p(\lambda) = 0$. This is usually easy to do when A is a small matrix. However, when A is a large matrix, it is sometimes difficult to compute its characteristic polynomial $p(\lambda)$ accurately. Also even after $p(\lambda)$ has been found, it is not always a simple matter to solve the equation $p(\lambda) = 0$.

In many applications all that is needed is the magnitude of the dominant eigenvalue of A. In this case it is unnecessary to compute the characteristic polynomial for A. Instead, one can use a method known as the ***power method***. An eigenvalue of A is a ***dominant eigenvalue*** of A if its absolute value exceeds the absolute value of every other eigenvalue of A. A corresponding eigenvector will be called a ***dominant eigenvector***.

Before illustrating the power method, let us observe that if $\mathbf{y}$ is an eigenvector of A corresponding to an eigenvalue of λ, that is, if

$$A\mathbf{y} = \lambda\mathbf{y},$$

then for every integer k,

$$A^k\mathbf{y} = \lambda^k\mathbf{y}.$$

This equation has two interpretations. It states that $\mathbf{y}$ is an eigenvector of A^k. It also states that $A^k\mathbf{y}$ is a multiple of the eigenvector $\mathbf{y}$ of A and so is also an eigenvector of A. It turns out that for any fixed nonzero vector $\mathbf{y}$ and a large integer k, the kth power $\mathbf{y}_k = A^k\mathbf{y}$ is a good estimate of an eigenvector of A. Usually, this eigenvector is the dominant eigenvector of A. Moreover, the Rayleigh quotient

$$\frac{\mathbf{y}_k^T A \mathbf{y}_k}{\mathbf{y}_k^T \mathbf{y}_k}$$

is a good estimate of the corresponding eigenvalue of A.

We now illustrate the power method for a simple case in which the eigenvalues and eigenvectors are known. In our computations we scale our estimates of an eigenvector. We can do so because if $\mathbf{y}$ is an estimate of an eigenvector $\mathbf{x}$, then $a\mathbf{y}$ is an estimate of the eigenvector $a\mathbf{x}$.

■ **EXAMPLE 1** _____

Consider the case in which A and its eigenvalues and eigenvectors are

$$A = \begin{bmatrix} 5 & -5 \\ -4 & 6 \end{bmatrix}, \qquad \lambda_1 = 10, \qquad \mathbf{x}_1 = \begin{bmatrix} 1 \\ -1 \end{bmatrix}, \qquad \lambda_2 = 1, \qquad \mathbf{x}_2 = \begin{bmatrix} 1 \\ 8 \end{bmatrix}.$$

Here λ_1 is the dominant eigenvalue and $\mathbf{x}_1$ is a corresponding dominant eigenvector normalized so that its maximum component is equal to 1 in absolute value. Starting with a vector $\mathbf{y}_1$, we compute normalized vectors $\mathbf{y}_2, \mathbf{y}_3, \ldots$ as follows:

$$\mathbf{y}_2 = \frac{A\mathbf{y}_1}{c_1}, \quad \mathbf{y}_3 = \frac{A\mathbf{y}_2}{c_2}, \quad \ldots, \quad \mathbf{y}_{k+1} = \frac{A\mathbf{y}_k}{c_k}, \quad \ldots.$$

Here c_k is the maximum component in $A\mathbf{y}_k$. The results are given in the following table:

y_1	Ay_1	y_2	Ay_2	y_3	Ay_3	y_4	Ay_4	y_5	Ay_5
1	5	1	9	1	9.9	1	9.99	1	9.999
0	-4	$-.8$	-8.8	$-.98$	-9.88	$-.998$	-9.988	$-.9998$	-9.9988
$c_1 = 5$		$c_2 = 9$		$c_3 = 9.9$		$c_4 = 9.99$		$c_5 = 9.999$	

The next estimate is $y_6 = (1, -.99998)$. In computing the vectors $y_3, y_4, \ldots$ we have rounded up the second component. It is clear that as k increases, the vector y_k becomes a better and better estimate of the dominant eigenvector x_1. Also, c_k tends to the dominant eigenvalue $\lambda_1 = 10$. The dominant eigenvalue is also estimated by the Rayleigh quotient,

$$\frac{y_5^T A y_5}{y_5^T y_5} = 10.179 \ldots .$$

Observe that this Rayleigh quotient is an overestimate of the dominant eigenvalue of A. When A is symmetric, a Rayleigh quotient cannot exceed its dominant eigenvalue.

This illustrates the following:

POWER METHOD FOR FINDING AN EIGENVALUE OF A MATRIX A

1. Select an arbitrary vector y_1.

2. For $k = 1, 2, \ldots$, compute

$$y_{k+1} = \frac{A y_k}{c_k},$$

where c_k is a scale factor chosen according to some rule.

3. At the kth step the Rayleigh quotient

$$R(y_k) = \frac{y_k^T A y_k}{y_k^T y_k}$$

is an estimate of the eigenvalue we are seeking.

It should be noted that

$$y_{k+1} = \frac{A^k y_1}{b_k}, \qquad b_k = c_1 c_2 \cdots c_k.$$

Thus y_{k+1} is determined by the kth power of A. Because of this, the method is called the *power method*.

The Rayleigh quotient need not be computed at each step. We compute it only when we wish to have an estimate of the eigenvalue sought. Normally, this eigenvalue is the dominant eigenvalue of A. In rare cases it could be another eigenvalue, but because of round-off errors this rarely happens unless we put other restrictions on y_k.

There are various rules for choosing the scale factor c_k, such as:

1. Choose $c_k = 1$ at each step. In this case y_k could become very long or very short according as the dominant eigenvalue is greater or less than 1.

2. Choose c_k to be the maximum component of Ay_k as was done in Example 1. In this case c_k or its negative becomes an estimate of the dominant eigenvalue. In addition, there are only small changes in the length of y_k as k progresses.

3. Use choices 1 and 2 intermittently. For example, use choice 1 for several steps and then switch to choice 2 for awhile, then go back to choice 1, and so on.

4. Choose c_k to be a power of 2 such that the maximum component of y_{k+1} has absolute value between $\frac{1}{2}$ and 1. This is easy to do on a computer when floating-point arithmetic is used.

Of course, there are other choices of c_k that could be used. For example, we could choose c_k to be the length of Ay_k. Then y_{k+1} is a unit vector. However, this choice increases the amount of computation at each step, so we do not recommend its use.

To see why the power method works let us examine the case when A is a symmetric 3×3-matrix. The matrix A has three eigenvalues λ_1, λ_2, and λ_3 with x_1, x_2, and x_3 as corresponding eigenvectors. We suppose that these eigenvalues have the property that

$$|\lambda_1| > |\lambda_2| \geq |\lambda_3|.$$

Then λ_1 is a dominant eigenvalue of A. Moreover, the dominant eigenvector x_1 is orthogonal to the eigenvectors x_2 and x_3. Also x_3 can be chosen to be orthogonal to x_2 if it does not have this property already, as is the case when λ_2 and λ_3 are distinct. Since the eigenvectors x_1, x_2, and x_3 form a basis for $\mathbf{R}^3$, a vector y in $\mathbf{R}^3$ is expressible in the form

$$y = ax_1 + bx_2 + cx_3.$$

Suppose that the coefficient a is not zero. Let us use the power method iteration

$$y_{k+1} = Ay_k \quad \text{with} \quad y_1 = y$$

and no scaling. We shall show that, for large values of k,

$$\mathbf{y}_{k+1} = A^k\mathbf{y} = \lambda_1^k(a\mathbf{x}_1 + \mathbf{e}_k),$$

where $\mathbf{e}_k$ is a very short vector. This means that $\mathbf{y}_{k+1}$ is a good approximation of the dominant eigenvector $\lambda_1^k a\mathbf{x}_1$. To see this, recall that because

$$A^k\mathbf{x}_1 = \lambda_1^k\mathbf{x}_1, \qquad A^k\mathbf{x}_2 = \lambda_2^k\mathbf{x}_2, \qquad A^k\mathbf{x}_3 = \lambda_3^k\mathbf{x}_3,$$

we have

$$A^k\mathbf{y} = aA^k\mathbf{x}_1 + bA^k\mathbf{x}_2 + cA^k\mathbf{x}_3 = a\lambda_1^k\mathbf{x}_1 + b\lambda_2^k\mathbf{x}_2 + c\lambda_3^k\mathbf{x}_3.$$

This can be rewritten in the form

$$A^k\mathbf{y} = \lambda_1^k(a\mathbf{x}_1 + b\mu_2^k\mathbf{x}_2 + c\mu_3^k\mathbf{x}_3) = \lambda_1^k(a\mathbf{x}_1 + \mathbf{e}_k),$$

where μ_2 and μ_3 are the ratios

$$\mu_2 = \frac{\lambda_2}{\lambda_1}, \qquad \mu_3 = \frac{\lambda_3}{\lambda_1}$$

and $\mathbf{e}_k$ is the error vector

$$\mathbf{e}_k = b\mu_2^k\mathbf{x}_2 + c\mu_3^k\mathbf{x}_3.$$

Because λ_1 is a dominant eigenvalue, we have the inequalities

$$1 > |\mu_2| \geq |\mu_3|.$$

It follows that for large values of k, the values μ_2^k and μ_3^k are very small. For example,

$$\text{when} \quad \mu_2 = .1, \quad \text{then} \quad \mu_2^{10} = .0000000001,$$

and μ_3^{10} is smaller still. On the other hand, when μ_2 is close to 1, say

$$\mu_2 = .999, \quad \text{then} \quad \mu_2^{10} = .99\ldots \quad \text{and} \quad \mu_2^{100} = .9\ldots.$$

In this case k has to be very large before μ_2^k is very small. It follows that the rate of "convergence" depends on the size of the ratio μ_2. Even when μ_2 is close to 1, the Rayleigh quotient of the iterates gives a good estimate of λ_1 early on.

Suppose next that $a = 0$, so that our initial vector $\mathbf{y}$ is a linear combination

$$\mathbf{y} = b\mathbf{x}_2 + c\mathbf{x}_3$$

of the eigenvectors $\mathbf{x}_2$ and $\mathbf{x}_3$. Suppose further that b is not zero and that $|\lambda_2| > |\lambda_3|$. Then by the argument given above, the vector $A^k\mathbf{y}$ will approximate the eigenvector $\lambda_2^k b\mathbf{x}_2$ for large values of k. This holds under the assumption of perfect computations. However, on a computer, round-off errors usually occur, so that the iterates $\mathbf{y}_k$ will also become dependent on $\mathbf{x}_1$ and the vector $\mathbf{y}_k$ will eventually approach a multiple of a dominant eigenvector $\mathbf{x}_1$. There is a way to stop this if one has a reasonable estimate z_1 of x_1. One has only to introduce additional steps that orthogonalize y_k to z_1.

The power method can be modified by replacing A by $B = A - \sigma I$, where σ is a fixed number. Then λ is an eigenvalue of A if and only if $\lambda - \sigma$ is an eigenvalue of B. For example, if $n = 4$ and 10, 5, 1, -4 are the eigenvalues of A and $\sigma = 9$, then 1, -4, -8, and -13 are the eigenvalues of B. Note that $\mu = -13$ is the dominant eigenvalue of B. If in this case we apply the power method to B, we will obtain an estimate μ_k of the eigenvalue $\mu = -13$ of B. The number $\mu_k + \sigma = \mu_k + 9$ is an estimate of the least eigenvalue $\lambda = -4$ of A. Thus by "shifting" the eigenvalues in this manner we can obtain the least as well as the largest eigenvalues of A. There are other variations of the power method that will enable us to find all the eigenvalues of a symmetric matrix A, but we shall not pursue them here.

One other variation should be mentioned. Suppose that we choose a number σ and apply the power method to the inverse $(A - \sigma I)^{-1}$ of $B = A - \sigma I$. Then we will obtain an estimate of the dominant eigenvalue of B^{-1}. Writing this estimate in the form $1/(\lambda - \sigma)$, it can be shown that λ is an estimate to the eigenvalue of A closest to σ. The difficulty with this procedure is that it requires the computation of the inverse of the matrix B. The method just described is called the **inverse power method**.

Exercises

1. Find the eigenvalues and corresponding eigenvectors of the following matrices. Then apply the power method with $y_1 = (1, 1)$ as the initial vector. Verify that the power method yields a good approximation of the dominant eigenvalue, except perhaps where y_1 is already an eigenvector.

   (a) $\begin{bmatrix} 1 & 0 \\ 1 & 5 \end{bmatrix}$ (b) $\begin{bmatrix} 4 & 2 \\ 2 & 4 \end{bmatrix}$ (c) $\begin{bmatrix} 1 & 8 \\ 8 & -11 \end{bmatrix}$

2. Show that if A is a matrix of rank 1, then Ay is either zero or else an eigenvector of A.

3. Let λ_1, λ_2, and λ_3 be the eigenvalues of a symmetric 3×3-matrix A and let x_1, x_2, and x_3 be the corresponding eigenvectors. Construct the matrices

$$E = \frac{x_1 x_1^T}{x_1^T x_1}, \qquad E_2 = \frac{x_2 x_2^T}{x_2^T x_2}, \qquad E_3 = \frac{x_3 x_3^T}{x_3^T x_3}.$$

   (a) What are the ranks of E_1, E_2, and E_3?

   (b) Show that 0, λ_2, and λ_3 are the eigenvalues of $B = A - \lambda_1 E_1$. What are corresponding eigenvectors of B? (The matrix B is said to be obtained from A by deflation.)

(c) What are the eigenvalues of $C = B - \lambda_2 E_2$?

(d) Show that $C = \lambda_3 E_3$. Conclude that

$$A = \lambda_1 E_1 + \lambda_2 E_2 + \lambda_3 E_3.$$

(e) What are the corresponding results for an $n \times n$ symmetric matrix?

4. Apply the results given in Exercise 3 to the situation in which

$$A = \begin{bmatrix} 3 & -1 & 1 \\ -1 & 5 & -1 \\ 1 & -1 & 3 \end{bmatrix}, \quad \mathbf{x}_1 = \begin{bmatrix} 1 \\ 1 \\ 1 \end{bmatrix}, \quad \mathbf{x}_2 = \begin{bmatrix} 1 \\ 0 \\ -1 \end{bmatrix}, \quad \mathbf{x}_3 = \begin{bmatrix} 1 \\ -2 \\ 1 \end{bmatrix}.$$

What are the corresponding eigenvalues λ_1, λ_2, and λ_3?

5. When the power method is applied to the matrix A in Exercise 4 with $\mathbf{y}_1 = (1, 0, 0)$ as the initial vector, which eigenvalue will you obtain? After deflating A sending this eigenvalue to zero in the deflated matrix B, apply the power method to B. Which eigenvalue will you get?

6. Extend the power method for a symmetric matrix A as follows:

   (1) Apply the power method to A to obtain a dominant eigenvector $\mathbf{x}_1$ and the corresponding eigenvalue λ_1.

   (2) Scale $\mathbf{x}_1$ to be a unit vector and set $B = A - \lambda_1 \mathbf{x}_1 \mathbf{x}_1^T$.

   (3) Repeat (1) and (2) with A replaced by B, obtaining thereby a second unit eigenvector $\mathbf{x}_2$ and eigenvalue λ_2 of A and a second deflated matrix $C = B - \lambda_2 \mathbf{x}_2 \mathbf{x}_2^T$.

   (4) Repeat (1) and (2) with A replaced by C. Proceeding in this manner, one will obtain all the eigenvalues of A. However, there are some numerical difficulties in this procedure because round-off errors will interfere. Normally, this procedure will work well for finding the first two or three dominant eigenvalues.

Appendix I
Polynomials

Our purpose in this section is to review the concept of polynomials and to describe some of their properties. Recall that a polynomial $f(t)$ in t can be viewed as a linear combination

$$f(t) = a_0 + a_1 t + a_2 t^2 + \cdots + a_n t^n$$

of the powers $t^0 = 1, t, t^2, \ldots, t^n$ of a variable t. Here n is any integer. If the coefficients $a_0, a_1, \ldots, a_n$ are all zero, then $f(t)$ is the *zero polynomial*. If a_n is not zero, $f(t)$ is said to be of degree n. In general, the **degree of $f(t)$ is the largest integer k for which the coefficient a_k of t^k is not zero.** The zero polynomial has no degree. The polynomials

$$2 - t, \quad 2 - t + 0t^2, \quad t - t^5, \quad 1 - t^{12}, \quad 2 - 3t^3 + t^6$$

are of degrees 1, 1, 5, 12, and 6, respectively. Note that we normally omit writing the powers of t having zero coefficients. A polynomial can be written in many ways. For example, the expressions

$$8 + 6t + t^2, \quad t^2 + 6t + 8, \quad 6t + 8 + t^2, \quad (4 + t)(2 + t)$$

are different representations of the same polynomial. In our discussion of polynomials we write a polynomial $f(t)$ in t either as a sum

$$f(t) = a_0 + a_1 t + a_2 t^2 + \cdots + a_n t^n$$

of ascending powers of t or else as the sum

$$f(t) = c_0 t^n + c_1 t^{n-1} + \cdots + c_{n-1} t + c_n,$$

of descending powers of the t, whichever is convenient in context. Of course, we can use a variable other than t to describe a polynomial. For example,

$$2s^3 + s + 5, \quad 6z^4 - 7z^3 + 2z, \quad 5q^2 - 33q + 1,$$

$$2u + 3, \quad 10w^7 + 16w^5 - 6w^3, \quad 4\lambda^3 - 7\lambda + 3$$

are, respectively, polynomials in s, z, q, u, w, and λ (Greek lowercase lambda). A polynomial such as $f(t) = 5$ is a **constant polynomial**. It is of degree zero. A **linear polynomial** is one of the form $at + b$. If a is not zero, the polynomial $at^2 + bt + c$ is said to be **quadratic**. A polynomial of degree 3 is called a **cubic**. A polynomial is a **quartic** if it is of degree 4. And so on.

In elementary algebra we learned how to add two polynomials and how to multiply two polynomials. A polynomial $g(t)$ is said to be a **factor** of a polynomial $f(t)$ if there is a polynomial $q(t)$ such that

$$f(t) = g(t)q(t).$$

The degree of $f(t)$ is the sum of the degrees of $g(t)$ and $q(t)$. We say that $g(t)$

divides $f(t)$. Normally, when we divide $f(t)$ by $g(t)$, there is a remainder. In particular, when we divide a polynomial $f(t)$ of degree n by the linear polynomial $t - c$, we obtain a result of the form

$$f(t) = (t - c)q(t) + r,$$

where $q(t)$ is the **quotient** and r is a constant, called the **remainder**. For example, in the numerical case in which

$$f(t) = t^3 - 9t^2 + 6t + 5 = (t - 2)(t^2 - 7t - 8) - 11$$

we have

$$q(t) = t^2 - 7t - 8, \qquad r = -11.$$

Observe that if we set $t = 2$, then $f(2) = -11$, the remainder. In the general case given above, we have

$$f(c) = (c - c)q(c) + r = r.$$

Consequently, when we divide $f(t)$ by $t - c$ the remainder is $f(c)$. This gives us the following:

REMAINDER THEOREM. When a polynomial $f(t)$ of degree $n > 0$ is divided by $t - c$, we obtain a relation of the form

$$f(t) = (t - c)q(t) + f(c),$$

where $f(c)$ is the remainder and the quotient $q(t)$ is a polynomial of degree $n - 1$.

The result just given enables us to give a simple method for computing $f(c)$ and $q(t)$. This method is called ***synthetic division***. For the case $n = 3$ we have

$$f(t) = a_0 t^3 + a_1 t^2 + a_2 t + a_3 = (t - c)(b_0 t^2 + b_1 t + b_2) + f(c).$$

Equating the coefficients of like powers of t, we find that

$$b_0 = a_0, \qquad b_1 = a_1 + cb_0, \qquad b_2 = a_2 + cb_1, \qquad f(c) = a_3 + cb_2.$$

These equations can be put in the form

a_0	a_1	a_2	a_3	$\lfloor c$
	cb_0	cb_1	cb_2	
b_0	b_1	b_2	$f(c)$	

Each entry below the line is the sum of the two entries directly above it. The values b_0, b_1, b_2, and $f(c)$ are computed successively. For example, when $f(t) = t^3 - 9t^2 + 6t + 5$ and $c = 2$, we make the computations

$$
\begin{array}{cccc|}
1 & -9 & 6 & 5 \quad \underline{2} \\
& 2(1) & 2(-7) & 2(-8) \\
\hline
1 & -7 & -8 & -11 = f(2)
\end{array}
$$

We have $q(t) = t^2 - 7t - 8$ and $f(2) = -11$. Similarly, when $f(t) = 2t^4 - 7t^2 - t + 4$ and $c = -3$, we have

$$
\begin{array}{ccccc|}
2 & 0 & -7 & -1 & 4 \quad \underline{-3} \\
& -6 & 18 & -33 & 102 \\
\hline
2 & -6 & 11 & -34 & 106 = f(-3)
\end{array}
$$

so that $q(t) = 2t^3 - 6t^2 + 11t - 34$ and $f(-3) = 106$.

A real or complex number $t = c$ will be called a **zero** or a **root** of a polynomial $f(t)$ if $f(c) = 0$. It can be shown that every polynomial $f(t)$ of degree $n > 0$ has at least one root. This result is known as the **fundamental theorem of algebra**. If $t = c$ is a root of $f(t)$, then by the remainder theorem, we have

$$f(t) = (t - c)q(t) + f(c) = (t - c)q(t),$$

so that $t - c$ is a factor of $f(t)$. This gives us the following:

> **FACTOR THEOREM.** A number $t = c$ is a root of a polynomial $f(t)$ of degree $n > 0$ if and only if $t - c$ is a factor of $f(t)$.

For example, $t = -1$ is a zero of the polynomial

$$f(t) = t^3 + t^2 - 4t - 4,$$

as can be seen by substitution or by the division

$$
\begin{array}{cccc|}
1 & 1 & -4 & -4 \quad \underline{-1} \\
& -1 & 0 & 4 \\
\hline
1 & 0 & -4 & 0 = f(-1)
\end{array}
$$

The quotient is $q(t) = t^2 - 4 = (t - 2)(t + 2)$. Hence $f(t)$ has the factorization

$$f(t) = (t + 1)(t - 2)(t + 2).$$

The numbers -1, 2, and -2 are the zeros of $f(t)$. They are distinct. However, the polynomial

$$f(t) = (2t - 2)^3 = 8(t - 1)(t - 1)(t - 1)$$

has three coincident zeros: 1, 1, and 1. The polynomial

$$f(t) = t^2 - 6t + 25 = (t - 3 - 4i)(t - 3 + 4i)$$

has the complex numbers $3 + 4i$ and $3 - 4i$ as its zeros. It has no real zeros. These examples illustrate the following result.

A polynomial $f(t)$ of degree $n > 0$ has n zeros $t_1, t_2, \ldots, t_n$, not necessarily distinct. Accordingly, $f(t)$ can be expressed in the factored form

$$f(t) = h(t - t_1)(t - t_2) \cdots (t - t_n)$$

where h is a nonzero constant. Alternatively, we have

$$f(t) = k(t_1 - t)(t_2 - t) \cdots (t_n - t), \qquad k = (-1)^n h.$$

The proof in the general case is like that for the case $n = 3$. We know that $f(t)$ has one zero, say $t = t_1$. So we can write $f(t)$ in the factored form

$$f(t) = (t - t_1)f_1(t)$$

where $f_1(t)$ is of degree $n - 1 = 2$. The polynomial $f_1(t)$ has a zero t_2 so that

$$f_1(t) = (t - t_2)f_2(t),$$

where $f_2(t)$ is of degree $n - 2 = 1$. Since $f_2(t)$ is linear it can be written in the form $f_2(t) = h(t - t_3)$. Consequently,

$$f(t) = h(t - t_1)(t - t_2)(t - t_3),$$

as was to be shown.

As noted in an example given above, a real polynomial can have complex roots. In fact, for a real polynomial, complex roots occur in pairs, as stated in the following result.

Let $f(t)$ be a real polynomial of degree $n > 0$. If a complex number $t = a + bi$ is a root of $f(t)$, so is its conjugate $\bar{t} = a - bi$.

The proof in the general case is like that for the case $n = 4$. Accordingly, suppose that $t = a + bi$ is a solution of the equation

$$f(t) = a_0 t^4 + a_1 t^3 + a_2 t^2 + a_3 t + a_4 = 0.$$

Then, by conjugation, we see that $\bar{t} = a - bi$ satisfies the equation

$$\overline{f(t)} = f(\bar{t}) = a_0 \bar{t}^4 + a_1 \bar{t}^3 + a_2 \bar{t}^2 + a_3 \bar{t} + a_4 = 0$$

and so is a root of $f(t)$.

As an illustration, consider the polynomial

$$f(t) = t^4 + t^2 + 1.$$

It can be verified that the complex numbers

$$w = \cos\frac{\pi}{3} + i \sin\frac{\pi}{3}, \qquad \overline{w} = \cos\frac{\pi}{3} - i \sin\frac{\pi}{3},$$

$$w^2 = \cos\frac{2\pi}{3} + i \sin\frac{2\pi}{3}, \qquad \overline{w}^2 = \cos\frac{2\pi}{3} - i \sin\frac{2\pi}{3}$$

are roots of $f(t)$. In this verification, it is convenient to use *DeMoivre's theorem*, which states that

$$(\cos \theta + i \sin \theta)^m = \cos m\theta + i \sin m\theta$$

for every integer m. We also use the fact that

$$\cos\frac{\pi}{3} = -\cos\frac{2\pi}{3} = \frac{1}{2}.$$

Observe that because the conjugate $a - bi$ of a complex zero $a + bi$ of a real polynomial $f(t)$ is also a zero of $f(t)$, the real polynomial

$$p(t) = (t - a - bi)(t - a + bi) = (t - a)^2 + b^2$$

divides $f(t)$. Because b is not zero, the polynomial $p(t)$ is **irreducible** in the sense that it has no real linear factors. A real linear polynomial is also irreducible. From these results we see that

---

A real polynomial $f(t)$ of degree $n > 0$ is the product

$$f(t) = kp_1(t)p_2(t) \cdots p_m(t)$$

of powers of real irreducible factors of degree 1 or 2.

For example, if 2, 2, 5, 5, 5, $1 + i$, $1 - i$, $1 + i$, and $1 - i$ are the roots of $f(t)$, then $f(t)$ is expressible in the form

$$f(t) = k(t - 2)^2(t - 5)^3(t^2 - 2t + 2)^2 = kp_1(t)p_2(t)p_3(t),$$

where

$$p_1(t) = (t - 2)^2, \qquad p_2(t) = (t - 5)^3, \qquad p_3(t) = (t^2 - 2t + 2)^2.$$

In this event, we say that $t = 2$ is a root of multiplicity 2. So also are the roots $t = 1 + i$ and $1 - i$. The root $t = 5$ is a root of multiplicity 3. In general, $t = c$ is a root of $f(t)$ of **multiplicity** m if $(t - c)^m$ divides $f(t)$ but $(t - c)^{m+1}$ does not.

In many applications we encounter polynomials with integers as coefficients. These often have integers or rational numbers as roots. To obtain a root of this type we make use of the following result.

Let

$$f(t) = a_0 t^n + a_1 t^{n-1} + \cdots + a_{n-1} t + a_n$$

be a polynomial whose coefficients are integers. If an integer $t = c$ is a root of $f(t)$, then c divides a_n. More generally, suppose that the quotient $g = c/d$ of two integers c and d is a root of $f(t)$. If c and d have no factor in common other than ± 1, then c divides a_n and d divides a_0.

For example, when $f(t)$ is a cubic

$$f(t) = a_0 t^3 + a_1 t^2 + a_2 t + a_3$$

with integer coefficients and $x = c/d$ is a rational root in lowest terms, then

$$d^3 f\left(\frac{c}{d}\right) = a_0 c^3 + a_1 c^2 d + a_2 c d^2 + a_3 d^3 = 0.$$

Since c divides the first three terms, it must divide the last term and so divides a_3 because c does not divide d^3 unless $c = \pm 1$. Similarly, d divides a_0. The proof in the general case can be made in the same manner.

As an illustration of how this result can be used, consider the polynomial

$$f(t) = 2t^4 + 9t^3 + 2t^2 - 20t - 15.$$

By the result just obtained, the possible rational roots are

$$\pm 1, \quad \pm 1/2, \quad \pm 3, \quad \pm 3/2, \quad \pm 5, \quad \pm 5/2, \quad \pm 15, \quad \pm 15/2.$$

By inspection, it is seen that $t = -1$ is a root. Divide $f(t)$ by $t + 1$ as follows:

2	9	2	-20	-15	$\underline{\vert -1}$
	-2	-7	5	15	
2	7	-5	-15	0	

Consequently, $f(t) = (t + 1)q(t)$ with $q(t) = 2t^3 + 7t^2 - 5t - 15$. The remaining roots of $f(t)$ are roots of $q(t)$. The possible rational roots of $q(t)$ are the same as those of $f(t)$. Testing with $t = 3/2$,

2	7	-5	-15	$\underline{\vert 3/2}$
	3	15	15	
2	10	10	0	

we see that $t = 3/2$ is a root. Hence

$$f(t) = (t + 1)(t - 3/2)(2t^2 + 10t + 10) = (t + 1)(2t - 3)(t^2 + 5t + 5).$$

The remaining roots of $f(t)$ are the roots of $t^2 + 5t + 5$. By the quadratic formula, it is seen that these roots are $(-5 + \sqrt{5})/2$ and $-(5 + \sqrt{5})/2$.

The following rule gives us information on the number of positive roots of a real polynomial $f(t)$. The number of negative roots of $f(t)$ is equal to the number of positive roots of $f(-t)$. The proof of this rule can be found in standard textbooks on the theory of equations.

DESCARTES' RULE OF SIGNS. The number of positive real roots of a real polynomial

$$f(t) = a_0 t^n + a_1 t^{n-1} + \cdots + a_{n-1}t + a_n$$

is either equal to the number of variations of signs in its coefficients or is less than that number by a positive even number. A root of multiplicity m is counted as m roots.

The number of variations of signs in the polynomial

$$f(t) = 2t^4 + 9t^3 + 2t^2 - 20t - 15$$

is one, so $f(t)$ has one positive root, as noted above. Also because

$$f(-t) = 2t^4 - 9t^3 + 2t^2 + 20t - 15$$

has three variations in signs, $f(t)$ has either 3 or 1 negative roots. We found earlier that $f(t)$ has three negative roots.

The polynomial $t^4 - 3t^3 + t + 1$ has either two or no positive roots; the exact amount is not determined by our rule of signs. However, $3t^3 - t - 1$ has exactly one positive root.

There is one other method of locating roots:

Suppose that $a < b$ and that $f(a)f(b) < 0$ for a real polynomial $f(t)$. Then there is at least one root of $f(t)$ on the interval $a < t < b$.

Intuitively, this result follows because the graph of $f(t)$ must cross the t-axis between a and b. We shall not give a more precise proof of this result.

This result can be used to approximate roots of a polynomial $f(t)$. Consider the cubic

$$f(t) = t^3 + t - 3.$$

We have $f(1) = -1$ and $f(2) = 7$ so that a root lies between 1 and 2. It should be a little larger than 1. Setting $t = 1.2$ and $t = 1.3$ in turn we find, by synthetic division, that $f(1.2) = -0.072$, $f(1.3) = 0.497$. We now know that a root lies between 1.2 and 1.3. Trying $t = 1.21$ and $t = 1.22$, we find that $f(1.21) < 0$ and $f(1.22) > 0$ so that a root lies between 1.21 and 1.22. Proceeding in this manner we can estimate this root as close as we wish. A method of this kind is called ***Horner's method***.

Continuing, we begin by recalling the following property of division of polynomials.

DIVISION ALGORITHM. Let $f(t)$ and $g(t)$ be polynomials with $g(t)$ not zero. There exists a unique polynomial $q(t)$, called the ***quotient***, and a unique polynomial $r(t)$, called the ***remainder***, such that

$$f(t) = g(t)q(t) + r(t),$$

where $r(t)$ is either zero or the degree of $r(t)$ is less than the degree of $g(t)$.

If $f(t)$ is zero or if $f(t)$ is of lower degree than $g(t)$, then $q(t) = 0$ and $r(t) = f(t)$. If the degree of $g(t)$ does not exceed that of $f(t)$, the polynomials $q(t)$ and $r(t)$ can be obtained by a long division routine. For example, when

$$f(t) = t^5 + 5t^3 + 6t^2 + 1$$

and $g(t) = t^2 + 2t + 1$, we proceed by long division in the following familiar fashion:

$$
\begin{array}{r}
t^3 - 2t^2 + 8t - 8 \\
t^2 + 2t + 1 \enclose{longdiv}{t^5 + 0t^4 + 5t^3 + 6t^2 + 0t + 1} \\
\underline{t^5 + 2t^4 + t^3} \\
-2t^4 + 4t^3 + 6t^2 \\
\underline{-2t^4 - 4t^3 - 2t^2} \\
8t^3 + 8t^2 + 0t \\
\underline{8t^3 + 16t^2 + 8t} \\
-8t^2 - 8t + 1 \\
\underline{-8t^2 - 16t - 8} \\
8t + 9
\end{array}
$$

to obtain the relation

$$t^5 + 5t^3 + 6t^2 + 1 = (t^2 + 2t + 1)(t^3 - 2t^2 + 8t - 8) + (8t + 9),$$

where $a(t) = t^3 - 2t^2 + 8t - 8$ is the quotient and $r(t) = 8t + 9$ is the remainder.

Let $f(t)$ and $g(t)$ be polynomials, not both zero. A polynomial that divides

$f(t)$ and $g(t)$ is called a ***common divisor*** of $f(t)$ and $g(t)$. A polynomial $d(t)$ will be called a ***greatest common divisor*** (GCD) of $f(t)$ and $g(t)$ if:

1. $d(t)$ is a common divisor of $f(t)$ and $g(t)$.

2. Every common divisor of $f(t)$ and $g(t)$ divides $d(t)$.

It should be noted that if $d(t)$ is a GCD of $f(t)$ and $g(t)$, then so is $cd(t)$ for every nonzero scalar c.

To show that every pair of polynomials $f(t)$ and $g(t)$, not both zero, have a GCD, let **S** be the class of all polynomials of the form $m(t)f(t) + n(t)g(t)$, where $m(t)$ and $n(t)$ are arbitrary polynomials. This class contains some nonzero polynomials. Let

$$d(t) = p(t)f(t) + q(t)g(t)$$

be a polynomial of **S** of lowest degree. We shall show that $d(t)$ is a GCD of $f(t)$ and $g(t)$. To see that $d(t)$ is a divisor of $f(t)$, we use the division algorithm and write $f(t)$ in the form

$$f(t) = e(t)d(t) + r(t),$$

where $r(t)$ is either zero or is of degree less than that of $d(t)$. This relation can be put in the form

$$r(t) = f(t) - e(t)[p(t)f(t) + q(t)g(t)]$$
$$= [1 - e(t)p(t)]f(t) + [-e(t)q(t)]g(t)$$

so that $r(t)$ is in **S**. Since $d(t)$ is a polynomial of lowest degree in **S**, we must have $r(t) = 0$. Hence $d(t)$ divides $f(t)$. By the same argument it is seen that $d(t)$ divides $g(t)$. If $d_1(t)$ is a second common divisor of $f(t)$ and $g(t)$, we can write $f(t)$ and $g(t)$ in the form $f(t) = f_1(t)d_1(t)$ and $g(t) = g_1(t)d_1(t)$. We then have the relation

$$d(t) = p(t)f_1(t)d_1(t) + q(t)g_1(t)d_1(t),$$

from which we conclude that $d_1(t)$ divides $d(t)$. The polynomial $d(t)$ is therefore a GCD of $f(t)$ and $g(t)$, as was to be shown.

If $d(t)$ and $d_1(t)$ are two GCDs of $f(t)$ and $g(t)$, each must divide the other and so must be of the same degree. This is possible only if $d_1(t) = cd(t)$, where c is a nonzero scalar.

Two polynomials $f(t)$ and $g(t)$ are said to be ***relatively prime*** if they are not both zero and have a GCD of degree zero. In this event $d(t) = 1$ is a GCD of $f(t)$ and $g(t)$. It follows that ***two polynomials $f(t)$ and $g(t)$ are relatively prime if and only if there exist polynomials $p(t)$ and $q(t)$ such that***

$$1 = p(t)f(t) + q(t)g(t).$$

Exercises

1. What are the degrees of the following polynomials?

$$f(t) = t^5 + 2t^4 + 3t^3 + 3t^2 + 3t + 3, \qquad g(t) = t^2 + t + 1,$$

$$q(t) = t^3 + t^2 + t + 1, \qquad r(t) = t + 2.$$

   Show that $f(t) = q(t)g(t) + r(t)$.

2. Establish the following properties of sums and products of two polynomials $f(t)$ and $g(t)$ of degrees m and n, respectively.

   (a) If m and n are not equal, the degree of $s(t) = f(t) + g(t)$ is the larger of m and n.

   (b) If $m = n$, then, if $s(t) = f(t) + g(t)$ is not the zero polynomial, the degree of $s(t)$ cannot exceed m.

   (c) The degree of the product $p(t) = f(t)g(t)$ is $m + n$.

3. Use synthetic division to find the values of the following polynomials at $t = 2$.

   (a) $2t^4 - 3t^3 + 2t^2 - t + 5$ (b) $t^3 + t^2 - 5$

   (c) $t^5 - 2t^3 - t + 6$ (d) $2t^3 - t$

4. The following polynomials have rational roots. Find them.

   (a) $2t^2 - t - 6$ (b) $2s^4 + s^3 - 6s^2$

   (c) $10\lambda^3 - \lambda^2 - 8\lambda + 3$ (d) $18 - 29\lambda + 12\lambda^2 - \lambda^3$

5. By the use of Descartes' rule of signs determine how many positive roots the polynomials given in Exercise 3 can have. Also determine the number of negative roots they can have.

6. Under what conditions on the real numbers a, b, and c will the quadratic $at^2 + 2bt + c$ have complex roots. What does this say about b when $a = c = 1$?

7. Show that the determinant

$$p(\lambda) = \begin{vmatrix} 1 - \lambda & 2 \\ 1 & 3 - \lambda \end{vmatrix}$$

   is a polynomial of degree 2 in λ. Find its roots.

8. Show that the determinant

$$p(\lambda) = \begin{vmatrix} 5 - \lambda & 0 & 8 \\ -1 & 2 - \lambda & 4 \\ 2 & 0 & 3 - \lambda \end{vmatrix}$$

   is a polynomial of degree 3 in λ. Find its roots.

9. Set $w = \cos(2\pi/3) + i \sin(2\pi/3)$. Show that the numbers 1, w, and w^2 are roots of the equation $t^3 = 1$ and so are cube roots of unity.

10. Set $w = \cos(2\pi/n) + i \sin(2\pi/n)$. Show that the numbers $1, w, w^2, \ldots, w^{n-1}$ are the nth roots of unity. *Hint:* Use DeMoivre's theorem.

11. Show that if $f(t)$ is a polynomial of degree n, then for every number c, $t - c$ divides $g(t) = f(t) - f(c)$.

Remark: There is a standard method for solving the equation $f(t) = 0$, known as ***Newton's method.*** It involves the concept of the derivative $f'(t)$ of $f(t)$, a basic concept in calculus. Starting with a well-chosen initial estimate t_1 of the solution, successive "improved" estimates $t_2, t_3, \ldots$ are obtained by the formula

$$t_{k+1} = t_k - \frac{f(t_k)}{f'(t_k)}.$$

For example, when $f(t) = t^3 + t - 3$, $f'(t) = 3t^2 + 1$. In this event our iteration becomes

$$t_{k+1} = t_k - \frac{t_k^3 + t_k - 3}{3t_k^2 + 1} = \frac{2t_k^3 + 3}{3t_k^2 + 1}.$$

Choosing $t_1 = 2$, we have $t_2 = 19/13 = 1.46 \ldots$. Using $t_2 = 1.46$, we have $t_3 = 1.214 \ldots$. Using $t_3 = 1.214$, we find that $t_4 = 1.2134119 \ldots$. Because $f(1.2134119) = 0.000001 \ldots$, we accept $t = 1.2134119$ as our estimate of the solution. An improved estimate can be obtained by performing another step of Newton's algorithm.

Answers
to
Selected Exercises

CHAPTER 1

Section 1.0

1. (a) $I = \begin{bmatrix} 2 & 3 \\ 5 & 8 \end{bmatrix}$; $II = \begin{bmatrix} 9 & 1 \\ -1 & 1 \end{bmatrix}$; $III = \begin{bmatrix} 2 & 3 \\ 6 & 9 \end{bmatrix}$, $IV = \begin{bmatrix} 1 & 7 \\ 12 & -3 \end{bmatrix}$;

$V = \begin{bmatrix} 4 & 8 \\ 2 & -5 \end{bmatrix}$; $VI = \begin{bmatrix} -9 & 2 \\ 36 & -8 \end{bmatrix}$.

(b) $I = \begin{bmatrix} 7 \\ 4 \end{bmatrix}$; $II = \begin{bmatrix} 19 \\ 9 \end{bmatrix}$; $III = \begin{bmatrix} 7 \\ 20 \end{bmatrix}$; $IV = \begin{bmatrix} 8 \\ 4 \end{bmatrix}$; $V = \begin{bmatrix} 9 \\ 2 \end{bmatrix}$; $VI = \begin{bmatrix} 3 \\ 1 \end{bmatrix}$.

(c) $I = \left[\begin{array}{cc:c} 2 & 3 & 7 \\ 5 & 8 & 4 \end{array}\right]$; $II = \left[\begin{array}{cc:c} 9 & 1 & 19 \\ -1 & 1 & 9 \end{array}\right]$; $III = \left[\begin{array}{cc:c} 2 & 3 & 7 \\ 6 & 9 & 20 \end{array}\right]$;

$IV = \left[\begin{array}{cc:c} 1 & 7 & 8 \\ 12 & -3 & 4 \end{array}\right]$; $V = \left[\begin{array}{cc:c} 4 & 8 & 9 \\ 2 & -5 & 2 \end{array}\right]$; $VI = \left[\begin{array}{cc:c} -9 & 2 & 3 \\ 36 & -8 & 1 \end{array}\right]$.

2. (c) $I = \left[\begin{array}{ccc:c} 2 & 5 & 1 & 1 \\ 1 & -1 & 0 & 1 \\ 1 & 1 & 1 & 2 \end{array}\right]$; $II = \left[\begin{array}{ccc:c} 1 & 1 & -1 & 2 \\ 1 & 0 & 1 & 1 \\ 0 & 1 & -1 & 0 \end{array}\right]$;

$III = \left[\begin{array}{ccc:c} 1 & 1 & 1 & 1 \\ 1 & -1 & -1 & 4 \\ 3 & 1 & 1 & 2 \end{array}\right]$; $IV = \left[\begin{array}{ccc:c} 1 & 1 & 1 & 1 \\ 1 & -1 & 1 & 4 \\ 2 & 3 & -1 & 2 \end{array}\right]$;

$V = \left[\begin{array}{ccc:c} 1 & 2 & 3 & 1 \\ -1 & 9 & 4 & 2 \\ 2 & 1 & -3 & 3 \end{array}\right]$; $VI = \left[\begin{array}{ccc:c} 1 & 1 & 0 & 0 \\ 1 & -1 & 1 & 1 \\ 6 & -4 & -1 & 2 \end{array}\right]$.

3. (a) $2x - 3y = 4$ **(b)** $x + 3y = 5$
  $6x + 9y = -3$ $2x \quad\quad = -6$

(c) $2x + 3y = -2$ **(d)** $3x \qquad\;\;\; = -6$

       $x \qquad\; = 8$ $y - 2z = 5$

Section 1.1

1. A is 2×3, B is 2×4, C is 3×2, D is 3×4.

    E is 3×1, F is 1×5, G is 8×9, H is 3×4.

2. A: $\begin{bmatrix} 5 \\ -1 \end{bmatrix}$; B: $\begin{bmatrix} 6 \\ 7 \end{bmatrix}$; C: $\begin{bmatrix} -1 \\ -2 \\ -3 \end{bmatrix}$; F: $[1]$; H: $\begin{bmatrix} k_1 \\ k_2 \\ k_3 \end{bmatrix}$; D: $[3 \quad 6 \quad 0 \quad 8]$; E: $[7]$;

    G: $[g_{31} \quad g_{32} \quad \cdots \quad g_{39}]$; H: $[h_3 \quad k_3 \quad m_3 \quad n_3]$.

    A: $R_1 = [2 \quad 5 \quad 6]$, $R_2 = [0 \quad -1 \quad 7]$; $C_1 = \begin{bmatrix} 2 \\ 0 \end{bmatrix}$, $C_2 = \begin{bmatrix} 5 \\ -1 \end{bmatrix}$, $C_3 = \begin{bmatrix} 6 \\ 7 \end{bmatrix}$.

    B: $R_1 = [4 \quad 6 \quad 8 \quad 7]$, $R_2 = [5 \quad 7 \quad 9 \quad 1]$; $C_1 = \begin{bmatrix} 4 \\ 5 \end{bmatrix}$, $C_2 = \begin{bmatrix} 7 \\ 7 \end{bmatrix}$,

       $C_3 = \begin{bmatrix} 8 \\ 9 \end{bmatrix}$, $C_4 = \begin{bmatrix} 7 \\ 1 \end{bmatrix}$.

    C: $R_1 = [1 \quad -1]$, $R_2 = [2 \quad -2]$, $R_3 = [3 \quad -3]$; $C_1 = \begin{bmatrix} 1 \\ 2 \\ 3 \end{bmatrix}$, $C_2 = \begin{bmatrix} -1 \\ -2 \\ -3 \end{bmatrix}$.

    D: $R_1 = [1 \quad 4 \quad 5 \quad -6]$, $R_2 = [-2 \quad 5 \quad -4 \quad 7]$, $R_3 = [3 \quad 6 \quad 0 \quad 8]$.

       $C_1 = \begin{bmatrix} 1 \\ -2 \\ 3 \end{bmatrix}$, $C_2 = \begin{bmatrix} 4 \\ 5 \\ 6 \end{bmatrix}$, $C_3 = \begin{bmatrix} 5 \\ -4 \\ 0 \end{bmatrix}$, $C_4 = \begin{bmatrix} -6 \\ 7 \\ 8 \end{bmatrix}$.

3. (a) $\begin{bmatrix} 2 & 5 & 16 \\ 0 & -1 & 7 \end{bmatrix}$;   **(b)** $\begin{bmatrix} 2 & 10 & 6 \\ 0 & -2 & 7 \end{bmatrix}$;   **(c)** $\begin{bmatrix} 1 & -1 \\ 2 & -2 \\ 2 & -4 \end{bmatrix}$;   **(d)** $\begin{bmatrix} 1 & -1 \\ 2 & -2 \\ -3 & 3 \end{bmatrix}$;

    (e) $\begin{bmatrix} 1 & 0 \\ 2 & 0 \\ 3 & 0 \end{bmatrix}$;   **(f)** $\begin{bmatrix} 1 & -1 & 5 \\ 2 & -2 & 6 \\ 3 & -3 & 7 \end{bmatrix}$.

4. $d_{12} = 4$, $d_{23} = -4$, $d_{34} = 8$, $d_{32} = 6$, $d_{33} = 0$.

5. (a) 19;   **(b)** 21;   **(c)** 6;   **(d)** 18;   **(e)** 17;   **(f)** 0.

6. (a) -4;   **(b)** 6;   **(c)** π.

7. (a) $b = 0$;   **(b)** $a = 0$, $b = 0$, $c = 1$;   **(c)** 4;   **(d)** $b = -4$;   **(e)** $b = 1$.

8. $A = \begin{bmatrix} 5 & 8 \\ 7 & 10 \\ 9 & 12 \end{bmatrix}$.

9. $B = \begin{bmatrix} 12 & 27 & 50 & 81 \\ 30 & 48 & 74 & 108 \end{bmatrix}$.

10. $C = \begin{bmatrix} -1 + \sqrt{10} & 1 + 2\sqrt{10} & -1 + 3\sqrt{10} & 1 + 4\sqrt{10} \\ -1 + 2\sqrt{10} & 1 + 4\sqrt{10} & -1 + 6\sqrt{10} & 1 + 8\sqrt{10} \\ -1 + 3\sqrt{10} & 1 + 6\sqrt{10} & -1 + 9\sqrt{10} & 1 + 12\sqrt{10} \end{bmatrix}.$

11. **(a)** $\begin{bmatrix} 8 + 3a_{11} & 8 + 3a_{12} \\ 8 + 3a_{21} & 8 + 3a_{12} \\ 8 + 3a_{31} & 8 + 3a_{32} \end{bmatrix}$; **(b)** $\begin{bmatrix} b_{11} & b_{12} & b_{13} \\ b_{21} & b_{22} & b_{23} \end{bmatrix}$; **(c)** $\begin{bmatrix} 4 & 4 & 4 \\ 9 & 9 & 9 \\ 16 & 16 & 16 \\ 25 & 25 & 25 \end{bmatrix}$;

(d) $\begin{bmatrix} 1 & 2 & 3 \\ 1 & 2 & 3 \\ 1 & 2 & 3 \end{bmatrix}$.

12. $S_2 = \begin{bmatrix} 1 & -1 \\ -1 & 1 \end{bmatrix}$, $S_3 = \begin{bmatrix} 1 & -1 & 1 \\ -1 & 1 & -1 \\ 1 & -1 & 1 \end{bmatrix}$; $S_4 = \begin{bmatrix} 1 & -1 & 1 & -1 \\ -1 & 1 & -1 & 1 \\ 1 & -1 & 1 & -1 \\ -1 & 1 & -1 & 1 \end{bmatrix}$

13. $I_3 = \begin{bmatrix} 1 & 0 & 0 \\ 0 & 1 & 0 \\ 0 & 0 & 1 \end{bmatrix} = [\delta_{ij}] \; (i, j = 1, 2, 3).$

$[\delta_{ij}] = \begin{bmatrix} 1 & 0 & 0 & 0 & 0 \\ 0 & 1 & 0 & 0 & 0 \\ 0 & 0 & 1 & 0 & 0 \end{bmatrix} = [I_3 \mid \theta]$ where $\theta = \begin{bmatrix} 0 & 0 \\ 0 & 0 \\ 0 & 0 \end{bmatrix}$ is the 3×2-zero

matrix $(i = 1, 2, 3; j = 1, 2, 3, 4, 5)$.

14. A: $[2 \quad -1 \quad 7]$, B: $[4 \quad 7 \quad 5 \quad 7]$, C: $[1 \quad 1]$, D: $[1 \quad -2 \quad 7]$, E: $[52]$,
F: $[a_{11} \quad a_{22} \quad a_{33}]$.

15. **(a)** 8; **(b)** 23; **(c)** $\begin{bmatrix} 0 & -1 \\ -1 & 0 \end{bmatrix}$; **(d)** $\begin{bmatrix} -6 & -1 & 5 \\ 2 & -9 & 6 \\ 3 & -3 & 0 \end{bmatrix}$;

(e) $\begin{bmatrix} 2 - x & 5 & 6 \\ 0 & -1 - x & 7 \\ 5 & 6 & 7 - x \end{bmatrix}$, $8 - 3x = 0$, $x = 8/3$.

16. **(a)** 1/4, 1/6.

(b) 2nd row, 4th column; 3rd row, 3rd column; 4th row, 2nd column.

(c) Use associative and commutative laws to show that $\dfrac{1}{i + j - 1} = \dfrac{1}{j + i - 1}$.

(d) $[1 \quad 1/3 \quad 1/5 \quad 1/7]$. **(e)** $H_3 = \begin{bmatrix} 1 & 1/2 & 1/3 \\ 1/2 & 1/3 & 1/4 \\ 1/3 & 1/4 & 1/5 \end{bmatrix}$.

(f) $H_5 = \begin{bmatrix} 1 & 1/2 & 1/3 & 1/4 & 1/5 \\ 1/2 & 1/3 & 1/4 & 1/5 & 1/6 \\ 1/3 & 1/4 & 1/5 & 1/6 & 1/7 \\ 1/4 & 1/5 & 1/6 & 1/7 & 1/8 \\ 1/5 & 1/6 & 1/7 & 1/8 & 1/9 \end{bmatrix}$.

17. (a) [5 10 15];   **(b)** [−1 1 −1 1 −1];   **(c)** [2 5 10 17];

  (d) [−k − 1 k + 1 −k − 1 k + 1 −k − 1];   **(e)** [0 0 0 · · · 0].

18. $V = \text{diag}\{\pi, \sqrt{10}\};$ $W = \text{diag}\{14\frac{1}{2}\};$ $X = \text{diag}\{-1, 1, -1\};$ $Y = \text{diag}\{2, 3, 5, 0\}.$

19. (a) $M = \begin{bmatrix} 1 & 4 & 1 & 1 & 1 \\ 2 & 5 & 1 & 0 & 1 \\ 3 & 6 & 0 & 0 & 1 \end{bmatrix};$ **(b)** $M = \begin{bmatrix} 1 & 4 & 0 & 1 & 1 \\ 2 & 5 & 1 & 0 & 1 \\ 3 & 6 & 0 & 0 & 1 \end{bmatrix};$

  (c) $\begin{bmatrix} 1 & 4 & 0 & 1 & 1 \\ 2 & 5 & 1 & 0 & 1 \\ 3 & 6 & 0 & 0 & 1 \end{bmatrix}.$

20. (a) $M = \begin{bmatrix} 1 & -1 & 1 & 0 \\ 2 & 3 & -1 & 4 \\ 0 & 0 & 1 & 0 \\ 0 & 0 & 0 & 1 \end{bmatrix};$ **(b)** $\begin{bmatrix} 1 & -1 & 1 \\ 2 & 3 & -1 \\ 0 & 1 & 5 \end{bmatrix};$ **(c)** $\begin{bmatrix} 1 & -1 & 1 & 0 \\ 2 & 3 & -1 & 4 \\ 0 & 0 & 1 & 0 \\ 0 & 0 & 0 & 1 \end{bmatrix};$

  (d) $\begin{bmatrix} 1 & 1 & -1 \\ 2 & 3 & -1 \\ 0 & 0 & 1 \\ 0 & 0 & 0 \end{bmatrix}.$

Section 1.2

1. $A + C = C + A = \begin{bmatrix} 7 & -2 \\ 1 & 4 \end{bmatrix},$ $A - C = \begin{bmatrix} -3 & 2 \\ -3 & -2 \end{bmatrix},$ $C - A = \begin{bmatrix} 3 & -2 \\ 3 & 2 \end{bmatrix};$

  $E + B = B + E = [5 \; 6 \; 7],$ $B - E = [-5 \; -6 \; -7],$ $E - B = [5 \; 6 \; 7];$

  $D + F = F + D = \begin{bmatrix} 1 & 2 \\ 2 & 1 \\ 1 & -3 \end{bmatrix},$ $D - F = \begin{bmatrix} 1 & -2 \\ -2 & 1 \\ 1 & 3 \end{bmatrix},$ $F - D = \begin{bmatrix} -1 & 2 \\ 2 & -1 \\ -1 & -3 \end{bmatrix}.$

2. $5A = \begin{bmatrix} 10 & 0 \\ -5 & 5 \end{bmatrix};$ $4B = [0 \; 0 \; 0];$ $-3C = \begin{bmatrix} -15 & 6 \\ -6 & -9 \end{bmatrix};$ $7D = \begin{bmatrix} 7 & 0 \\ 0 & 7 \\ 7 & 0 \end{bmatrix};$

  $2E = [10 \; 12 \; 14];$ $-6F = \begin{bmatrix} 0 & -12 \\ -12 & 0 \\ 0 & 18 \end{bmatrix}.$

3. $5A + 3C = \begin{bmatrix} 25 & -6 \\ 1 & 14 \end{bmatrix};$ $4B + 2E = [10 \; 12 \; 14];$ $7D - 6F = \begin{bmatrix} 7 & -12 \\ -12 & 7 \\ 7 & 18 \end{bmatrix}.$

4. $H + L = L + H = \begin{bmatrix} 0 & 0 & 0 \\ 0 & 0 & 0 \end{bmatrix}$, $H - L = \begin{bmatrix} 2 & -2 & 2 \\ -2 & 2 & -2 \end{bmatrix}$,

$L - H = \begin{bmatrix} -2 & 2 & -2 \\ 2 & -2 & 2 \end{bmatrix}$, $J + M = M + J = [13 + 3\sqrt{42}]$,

$J - M = [13 - 3\sqrt{42}]$, $M - J = [3\sqrt{42} - 13]$;

$K + N = N + K = \begin{bmatrix} 1 & -1 \\ -1 & 1 \end{bmatrix}$, $K - N = \begin{bmatrix} \cos^2\theta - \sin^2\theta & \cos^2\theta - \sin^2\theta \\ \cos^2\theta - \sin^2\theta & \cos^2\theta - \sin^2\theta \end{bmatrix}$,

$N - K = \begin{bmatrix} \sin^2\theta - \cos^2\theta & \sin^2\theta - \cos^2\theta \\ \sin^2\theta - \cos^2\theta & \sin^2\theta - \cos^2\theta \end{bmatrix}$.

5. (a) $\begin{bmatrix} 2\pi/3 & 32/3 & 28 \\ 10/3 & 4/3 & 2 \\ 1/3 & 2/3 & 1 \end{bmatrix}$; **(b)** $\begin{bmatrix} -7 & 7 & -7 \\ 7 & -7 & 7 \end{bmatrix}$; **(c)** $[13\sqrt[3]{42}]$;

(d) $\begin{bmatrix} 1 & -\tan^2\theta \\ -\tan^2\theta & 1 \end{bmatrix}$; **(e)** $\begin{bmatrix} -\sqrt{10} & \sqrt{10} & -\sqrt{10} \\ \sqrt{10} & -\sqrt{10} & \sqrt{10} \end{bmatrix}$; **(f)** $[0]$;

(g) $\begin{bmatrix} 1 & -\cot^2\theta \\ -\cot^2\theta & 1 \end{bmatrix}$.

6. (a) $A - 3I = \begin{bmatrix} 0 & -1 \\ 5 & -1 \end{bmatrix}$, $B - 4I = \begin{bmatrix} -3 & 3 & 7 \\ 0 & 0 & 6 \\ 0 & 0 & 1 \end{bmatrix}$, $C - 7I = [-3]$.

(b) $2 \times 2 : 3 \times 3 : 1 \times 1$.

7. (a) $\begin{bmatrix} -5 & 6 \\ -11 & -4 \end{bmatrix}$; **(b)** $\begin{bmatrix} 25 & -6 \\ 1 & 14 \end{bmatrix}$; **(c)** $[20 \quad 24 \quad 28]$; **(d)** $\begin{bmatrix} 8 & 10 \\ 10 & 8 \\ 8 & -15 \end{bmatrix}$.

8. (a) $\begin{bmatrix} 3 & -2 \\ 3 & 2 \end{bmatrix}$; **(b)** $\begin{bmatrix} -3 & 2 \\ -3 & -2 \end{bmatrix}$; **(c)** $\begin{bmatrix} -2 & 0 \\ 1 & -1 \end{bmatrix}$; **(d)** $\begin{bmatrix} -1 & 2 \\ 2 & -1 \\ -1 & -3 \end{bmatrix}$.

(e) $\begin{bmatrix} -5/3 & 14/3 \\ 14/3 & -5/3 \\ -5/3 & -7 \end{bmatrix}$; **(f)** $[135/16 \quad 81/8 \quad 189/16]$;

(g) $\begin{bmatrix} -21/\sqrt{\pi} & 21/\sqrt{\pi} & -21/\sqrt{\pi} \\ 21/\sqrt{\pi} & -21/\sqrt{\pi} & 21/\sqrt{\pi} \end{bmatrix}$; **(h)** $[55/2 \quad 33 \quad 77/2]$.

9. (a) $bA + C$; **(b)** $1/b[A(1 - a) + C(1 - a - b)]$; **(c)** $-2C$;

(d) $1/(b + 1)[D(-a - b - 1) + aF]$, $b \neq -1$; **(e)** $1/5F(1 + a)$

10. No.

11. Yes, yes.

12. X is a linear combination of I and J since

$$\begin{bmatrix} 3 & -2 \\ 2 & 3 \end{bmatrix} = 3\begin{bmatrix} 1 & 0 \\ 0 & 1 \end{bmatrix} + 2\begin{bmatrix} 0 & -1 \\ 1 & 0 \end{bmatrix}.$$

Section 1.3

1. $AB = \begin{bmatrix} 1 & 1 & 1 \\ -1 & 1 & 1 \end{bmatrix}$, $BC = \begin{bmatrix} 4 & 0 & 0 \\ 2 & 0 & 1 \end{bmatrix}$, $YA = \begin{bmatrix} 0 & 2 \\ 3 & 1 \\ 3 & 3 \end{bmatrix}$,

$YB = \begin{bmatrix} 1 & 1 & 1 \\ 2 & -1 & 2 \\ 3 & 0 & 3 \end{bmatrix}$, $CY = \begin{bmatrix} 1 & 4 \\ 5 & 2 \\ 3 & 0 \end{bmatrix}$, $XZ = \begin{bmatrix} 1 & -1 & 1 & 2 \\ -2 & 2 & -2 & -4 \\ 1 & 1 & 1 & 2 \end{bmatrix}$,

$BX = \begin{bmatrix} 2 \\ -2 \end{bmatrix}$, $BY = \begin{bmatrix} 4 & 1 \\ 2 & -1 \end{bmatrix}$, $CX = \begin{bmatrix} 5 \\ 3 \\ -1 \end{bmatrix}$.

2. $DF = \begin{bmatrix} 9 & 9 \\ 5 & 5 \\ 1 & 1 \end{bmatrix}$, $DV = \begin{bmatrix} 44 \\ 26 \\ 8 \end{bmatrix}$, $EU = \begin{bmatrix} 18 & 16 \end{bmatrix}$, $FV = \begin{bmatrix} 9 \\ 9 \end{bmatrix}$,

$UF = \begin{bmatrix} 18 & 18 \\ 1 & 1 \\ 14 & 14 \\ 0 & 0 \\ 3 & 3 \end{bmatrix}$, $UV = \begin{bmatrix} 81 \\ 1 \\ 70 \\ 0 \\ 10 \end{bmatrix}$, $VE = \begin{bmatrix} 8 & -8 & 8 & -8 & 8 \\ 1 & -1 & 1 & 1 & 1 \end{bmatrix}$,

$VW = \begin{bmatrix} 16 & 40 \\ 2 & 5 \end{bmatrix}$, $WF = \begin{bmatrix} 7 & 7 \end{bmatrix}$, $WV = \begin{bmatrix} 21 \end{bmatrix}$.

3. $IA = \begin{bmatrix} 1 & 1 \\ -1 & 1 \end{bmatrix}$, $AI = \begin{bmatrix} 1 & 1 \\ -1 & 1 \end{bmatrix}$, $A^2 = \begin{bmatrix} 0 & 2 \\ -2 & 0 \end{bmatrix}$, $A^3 = \begin{bmatrix} -2 & 2 \\ -2 & -2 \end{bmatrix}$,

$B = \begin{bmatrix} 0 & 0 \\ 0 & 0 \end{bmatrix}$.

4. $(AB)C = A(BC) = \begin{bmatrix} 1 & 5 & 2 & 2 \\ -3 & -5 & 2 & 0 \end{bmatrix}$.

5. $A(B + C) = AB + AC = \begin{bmatrix} 3 & 1 & 4 \\ 1 & -1 & 2 \end{bmatrix}$.

6. $AB = \begin{bmatrix} 1 & 1 & 1 \\ -1 & 1 & -1 \end{bmatrix}$, $AC = \begin{bmatrix} 2 & 0 & 3 \\ 2 & -2 & 3 \end{bmatrix}$, $2B - 3C = \begin{bmatrix} 2 & -3 & 2 \\ -6 & 5 & -9 \end{bmatrix}$,

$2AB - 3AC = \begin{bmatrix} -4 & 2 & -7 \\ -8 & 8 & -11 \end{bmatrix}$.

7. $A^2 = 2A - I = \begin{bmatrix} 1 & 2 \\ 0 & 1 \end{bmatrix}$, $A^3 = 3A - 2I = \begin{bmatrix} 1 & 3 \\ 0 & 1 \end{bmatrix}$, $A^4 = 4A - 3I = \begin{bmatrix} 1 & 4 \\ 0 & 1 \end{bmatrix}$.

9. **(a)** $AB = \begin{bmatrix} 8 & 10 \\ 2 & 2 \end{bmatrix}$, $BA = \begin{bmatrix} -1 & 7 \\ -1 & 11 \end{bmatrix}$.

(b) $AB = \begin{bmatrix} 21 & 20 \\ 15 & 16 \end{bmatrix}$, $BA = \begin{bmatrix} 10 & 9 & 8 \\ 2 & 3 & 4 \\ 24 & 24 & 24 \end{bmatrix}$.

(c) $AB = \begin{bmatrix} 14 & 340 \\ 28 & 508 \\ 1 & 60 \end{bmatrix}$, BA cannot be found.

(d) $AB = \begin{bmatrix} 19 & 8 \\ 1 & 2 \end{bmatrix}$, $BA = \begin{bmatrix} 16 & 25 \\ 2 & 5 \end{bmatrix}$.

10. (a) $AB = \begin{bmatrix} 0 \\ 0 \end{bmatrix}$; **(b)** $AB = \begin{bmatrix} 0 \\ 0 \end{bmatrix}$; **(c)** $AB = [0]$; **(d)** $AB = \begin{bmatrix} 0 & 0 \\ 0 & 0 \end{bmatrix}$.

12. (a) $X = \dfrac{1}{b}(BA - AB)$; **(b)** $X = I$; **(c)** $X = AB$.

13. (a) $XY = \begin{bmatrix} 3 & 4 & 5 & 6 \\ 0 & 0 & 0 & 0 \\ 3 & 4 & 5 & 6 \\ 6 & 8 & 10 & 12 \end{bmatrix}$; $YX = [20]$.

(b) $XY = \begin{bmatrix} 2 & 4 \\ 3 & 6 \\ 5 & 10 \\ 8 & 16 \end{bmatrix}$; YX cannot be found.

(c) $XY = \begin{bmatrix} 0 & 0 & 0 & 0 & 0 \\ 0 & 2 & 6 & 3 & 5 \end{bmatrix}$; YX cannot be found.

14. $CD = \begin{bmatrix} 12 & -8 & 0 \\ 4 & 0 & 6 \\ 2 & 4 & 0 \end{bmatrix}$, $CD = \begin{bmatrix} 12 & -4 & 0 \\ 8 & 0 & 4 \\ 6 & 6 & 0 \end{bmatrix}$.

15. $AD = \begin{bmatrix} a_{11}d_{11} & a_{12}d_{22} & a_{13}d_{33} \\ a_{21}d_{11} & a_{22}d_{22} & a_{23}d_{33} \end{bmatrix}$,

$DB = \begin{bmatrix} d_{11}b_{11} & d_{11}b_{12} & d_{11}b_{13} & d_{11}b_{14} \\ d_{22}b_{21} & d_{22}b_{22} & d_{22}b_{23} & d_{22}b_{24} \\ d_{33}b_{31} & d_{33}b_{32} & d_{33}b_{33} & d_{33}b_{34} \end{bmatrix}$,

$UV = \begin{bmatrix} d_1 e_1 & d_1 f + a e_2 & d_1 h + a g + c e_3 \\ 0 & d_2 e_2 & d_2 g + b e_3 \\ 0 & 0 & d_3 e_3 \end{bmatrix}$.

17. $x = 1$, $y = 1$; $x = 7$, $y = -4$.

Section 1.4

1. $U^T = \begin{bmatrix} 3 & 5 \\ 2 & 4 \end{bmatrix}$, $V^T = \begin{bmatrix} 1 & 2 \\ 5 & 3 \\ 3 & 5 \\ 6 & 8 \end{bmatrix}$, $W^T = [\pi \quad \sqrt{10} \quad 3.14 \quad 22/7]$,

$X^T = \begin{bmatrix} 6 & 0 \\ 28 & 1 \\ 496 & 0 \end{bmatrix}$, $Y^T = \begin{bmatrix} x_1 & y_1 & z_1 \\ x_2 & y_2 & z_2 \\ x_3 & y_3 & z_3 \\ x_4 & y_4 & z_4 \end{bmatrix}$, $Z^T = [14]$.

2. $A^T = \begin{bmatrix} 1 & -1 \\ 1 & 2 \end{bmatrix}$, $B^T = \begin{bmatrix} 0 & 1 \\ 0 & 1 \\ 0 & 1 \end{bmatrix}$, $C^T = \begin{bmatrix} 1 & -1 & 1 \\ -1 & 2 & 3 \\ 1 & 3 & 4 \end{bmatrix}$, $P = \begin{bmatrix} 1 & 1 & 1 \\ 2 & 2 & 2 \end{bmatrix}$,

$Q = P^T = \begin{bmatrix} 1 & 2 \\ 1 & 2 \\ 1 & 2 \end{bmatrix}$, $V = W^T = \begin{bmatrix} 0 & 0 & 0 \\ 1 & 4 & 8 \end{bmatrix}$, $W = V^T = \begin{bmatrix} 0 & 1 \\ 0 & 4 \\ 0 & 8 \end{bmatrix}$.

3. L, N, and T are all symmetric.

4. **(a)** By inspection, we see that the matrices

$$AR = \begin{bmatrix} 5 & 4 & 3 \\ 4 & 3 & 2 \\ 3 & 2 & 1 \end{bmatrix}, \quad RA = \begin{bmatrix} 1 & 2 & 3 \\ 2 & 3 & 4 \\ 3 & 4 & 5 \end{bmatrix}$$

are symmetric.

(b) By inspection, we see that the matrices

$$AR = \begin{bmatrix} e & d & c \\ d & c & b \\ c & b & a \end{bmatrix}, \quad RA = \begin{bmatrix} a & b & c \\ b & c & d \\ c & d & e \end{bmatrix}$$

are symmetric.

5. **(a)** $\begin{bmatrix} 4 & 4 \\ 3 & 6 \end{bmatrix}$, **(b)** $\begin{bmatrix} 6 & 1 \\ 28 & 2 \\ 496 & 1 \end{bmatrix}$, **(c)** $\begin{bmatrix} 7 & 2 \\ 1 & 11 \end{bmatrix}$, **(d)** $\begin{bmatrix} 5 & 0 & 5 \\ 0 & 6 & 4 \\ 5 & 4 & 8 \end{bmatrix}$,

(e) $\begin{bmatrix} 5 & 7 \\ 4 & 10 \end{bmatrix}$, **(f)** $\begin{bmatrix} -1 & 19 \\ -2 & 16 \end{bmatrix}$, **(g)** $\begin{bmatrix} 8 & 7 \\ 6 & 6 \end{bmatrix}$, **(h)** $\begin{bmatrix} 20 & 19 \\ 20 & 19 \\ 20 & 19 \end{bmatrix}$.

6. **(i)** **(a)** $2A + 3B = \begin{bmatrix} 6 & 13 \\ 13 & 4 \end{bmatrix}$, **(b)** $2A + 3B = \begin{bmatrix} 11 & 13 \\ 13 & 19 \end{bmatrix}$,

(c) $2A + 3B = \begin{bmatrix} 2 & 10 & 10 \\ 10 & 0 & 8 \\ 10 & 8 & 4 \end{bmatrix}$.

(ii) $AB = \begin{bmatrix} 5 & 3 \\ 2 & 5 \end{bmatrix}$, $\qquad AB = \begin{bmatrix} 9 & 9 \\ 21 & 21 \end{bmatrix}$, $\qquad AB = \begin{bmatrix} 4 & 12 & 4 \\ 0 & 6 & 0 \\ 2 & 14 & 2 \end{bmatrix}$.

7. (i) $AB = \begin{bmatrix} 10 & 6 \\ 6 & 4 \end{bmatrix}$ is symmetric. $\qquad$ (ii) $BA = \begin{bmatrix} 10 & 6 \\ 6 & 4 \end{bmatrix}$ is symmetric.

8. (a) $p(A) = \begin{bmatrix} 43 & 35 \\ 35 & 36 \end{bmatrix}$ symmetric; $\qquad$ (b) $q(B) = \begin{bmatrix} 6 & 0 \\ 0 & 6 \end{bmatrix}$ symmetric;

$\qquad$ (c) $r(B) = \begin{bmatrix} 4 & 4 \\ 4 & 4 \end{bmatrix}$ symmetric; $\qquad$ (d) $s(A, B) = \begin{bmatrix} 45 & 31 \\ 29 & 40 \end{bmatrix}$.

9. $\bar{A} = \begin{bmatrix} 5 & 3-i \\ 3+i & 6 \end{bmatrix}$, $A^* = \begin{bmatrix} 5 & 3+i \\ 3-i & 6 \end{bmatrix}$, $\quad A$ is Hermitian.

$\bar{B} = \begin{bmatrix} 1-2i \\ 1+2i \end{bmatrix}$, $B^* = [1-2i \quad 1+2i]$.

$\bar{C} = \begin{bmatrix} 3+i & 6 \\ 2 & 5+5i \\ -3i & -i \end{bmatrix}$, $C^* = \begin{bmatrix} 3+i & 2 & -3i \\ 6 & 5+5i & -i \end{bmatrix}$.

$\bar{D} = \begin{bmatrix} -i & -2i & -3i \\ 1 & 2 & 3 \end{bmatrix}$, $D^* = \begin{bmatrix} -i & 1 \\ -2i & 2 \\ -3i & 3 \end{bmatrix}$.

$\bar{E} = \begin{bmatrix} 1 & -i & 1-2i \\ i & 2 & 3+4i \\ 1+2i & 3-4i & 3 \end{bmatrix}$, $E^* = \begin{bmatrix} 1 & i & 1+2i \\ -i & 2 & 3-4i \\ 1-2i & 3+4i & 3 \end{bmatrix}$, $\quad E$ is Hermitian.

10. No, because diagonal elements remain in their same positions.

11. $A = \begin{bmatrix} 5 & 3 \\ 3 & 6 \end{bmatrix} + i\begin{bmatrix} 0 & 1 \\ -1 & 0 \end{bmatrix}$, $\quad B = \begin{bmatrix} 1 \\ 1 \end{bmatrix} + i\begin{bmatrix} 2 \\ -2 \end{bmatrix}$, $\quad C = \begin{bmatrix} 3 & 6 \\ 2 & 5 \\ 0 & 0 \end{bmatrix} + i\begin{bmatrix} -1 & 0 \\ 0 & -5 \\ 3 & 1 \end{bmatrix}$,

$D = \begin{bmatrix} 0 & 0 & 0 \\ 1 & 2 & 3 \end{bmatrix} + i\begin{bmatrix} 1 & 2 & 3 \\ 0 & 0 & 0 \end{bmatrix}$, $\quad E = \begin{bmatrix} 1 & 0 & 1 \\ 0 & 2 & 3 \\ 1 & 3 & 3 \end{bmatrix} + i\begin{bmatrix} 0 & 1 & 2 \\ -1 & 0 & -4 \\ -2 & 4 & 0 \end{bmatrix}$.

13. $C = C^T = \begin{bmatrix} 29 & 57 \\ 57 & 170 \end{bmatrix}$, $\quad D = D^T = \begin{bmatrix} 65 & -42 \\ -42 & 53 \end{bmatrix}$.

14. (a) P, Q, R, S, T are all the same as I with various rows or columns permuted.

$\qquad$ (b) $P, Q,$ and T are symmetric.

15. $x = -1/3$ so $D = \begin{bmatrix} 6 & 0 \\ 0 & 20/3 \end{bmatrix}$.

16. $x = -\dfrac{b}{a}$, $D = \begin{bmatrix} a & 0 \\ 0 & c - b^2/a \end{bmatrix}$.

17. $x = -\dfrac{1}{2}\left(\dfrac{c}{b} + 1\right)$, $D = \begin{bmatrix} -1 & 0 \\ 0 & -1-2b \end{bmatrix}$.

Section 1.5

1. **(a)** $\begin{bmatrix} 1 & 2 \\ 1 & 3 \end{bmatrix}\begin{bmatrix} x \\ y \end{bmatrix} = \begin{bmatrix} -2 \\ -6 \end{bmatrix}$, $\begin{bmatrix} 1 \\ 1 \end{bmatrix}x + \begin{bmatrix} 2 \\ 3 \end{bmatrix}y$,

   $x = 6, y = -4$.

   (b) $\begin{bmatrix} 1 & 0 & 1 \\ 0 & 1 & 1 \\ 1 & 0 & 2 \end{bmatrix}\begin{bmatrix} x \\ y \\ z \end{bmatrix} = \begin{bmatrix} 7 \\ 8 \\ 12 \end{bmatrix}$, $\begin{bmatrix} 1 \\ 0 \\ 1 \end{bmatrix}x + \begin{bmatrix} 0 \\ 1 \\ 0 \end{bmatrix}y + \begin{bmatrix} 1 \\ 1 \\ 2 \end{bmatrix}z = \begin{bmatrix} 7 \\ 8 \\ 12 \end{bmatrix}$,

   $x = 2, y = 3, z = 5$.

   (c) $\begin{bmatrix} 1 & 1 & 1 \\ 2 & -1 & 1 \\ 3 & 2 & -3 \end{bmatrix}\begin{bmatrix} x \\ y \\ z \end{bmatrix} = \begin{bmatrix} 4 \\ 8 \\ -5 \end{bmatrix}$, $\begin{bmatrix} 1 \\ 2 \\ 3 \end{bmatrix}x + \begin{bmatrix} 1 \\ -1 \\ 2 \end{bmatrix}y + \begin{bmatrix} 1 \\ 1 \\ -3 \end{bmatrix}z = \begin{bmatrix} 4 \\ 8 \\ -5 \end{bmatrix}$,

   $x = 2, y = -1, z = 3$.

   (d) $\begin{bmatrix} 1 & 5 & 6 \\ 1 & 2 & 3 \\ 0 & 3 & 3 \\ 0 & 0 & 1 \end{bmatrix}\begin{bmatrix} x \\ y \\ z \end{bmatrix} = \begin{bmatrix} -12 \\ -6 \\ -6 \\ -1 \end{bmatrix}$, $\begin{bmatrix} 1 \\ 1 \\ 0 \\ 0 \end{bmatrix}x + \begin{bmatrix} 5 \\ 2 \\ 3 \\ 0 \end{bmatrix}y + \begin{bmatrix} -6 \\ 3 \\ 3 \\ 1 \end{bmatrix}z = \begin{bmatrix} -12 \\ -6 \\ -6 \\ -1 \end{bmatrix}$,

   $x = -1, y = -1, z = -1$.

   (e) $= \begin{bmatrix} 2 & 3 & 5 \\ 1 & 6 & -4 \end{bmatrix}\begin{bmatrix} x_1 \\ x_2 \\ x_3 \end{bmatrix} = \begin{bmatrix} 8 \\ 13 \end{bmatrix}$, $\begin{bmatrix} 2 \\ 1 \end{bmatrix}x_1 + \begin{bmatrix} 3 \\ 6 \end{bmatrix}x_2 + \begin{bmatrix} 5 \\ -4 \end{bmatrix}x_3 = \begin{bmatrix} 8 \\ 13 \end{bmatrix}$,

   $x = 1 - (14/3)a, \quad y = 2 + (13/9)a, \quad z = a$.

   (f) $\begin{bmatrix} 1 & -2 & 3 \\ 1 & 2 & 0 \\ 3 & 2 & 0 \\ 5 & 2 & 0 \end{bmatrix}\begin{bmatrix} x_1 \\ x_2 \\ x_3 \end{bmatrix} = \begin{bmatrix} -10 \\ -3 \\ -1 \\ -9 \end{bmatrix}$, $\begin{bmatrix} 1 \\ 1 \\ 3 \\ 5 \end{bmatrix}x_1 + \begin{bmatrix} -2 \\ 2 \\ 2 \\ 2 \end{bmatrix}x_2 + \begin{bmatrix} 3 \\ 0 \\ 0 \\ 0 \end{bmatrix}x_3 = \begin{bmatrix} -10 \\ -3 \\ -1 \\ -9 \end{bmatrix}$.

   No solution.

   (g) $[13 \quad -2 \quad 11 \quad -51]\begin{bmatrix} x_1 \\ x_2 \\ x_3 \\ x_4 \end{bmatrix} = [10]$,

   $x_1 = 1/13(10 + 2a - 11b + 5c); \quad x_2 = a; \quad x_3 = b; \quad x_4 = c$.

2. **(a)** $\begin{bmatrix} 1 \\ -1 \end{bmatrix} + \begin{bmatrix} 1 \\ 1 \end{bmatrix} = \begin{bmatrix} 2 \\ 0 \end{bmatrix}$;   **(b)** $\begin{bmatrix} 1 \\ 2 \\ -1 \end{bmatrix} + 2\begin{bmatrix} 0 \\ 1 \\ 0 \end{bmatrix} = \begin{bmatrix} 1 \\ 4 \\ -1 \end{bmatrix}$;

(c) $0\begin{bmatrix} 2 \\ 5 \\ 5 \end{bmatrix} + 0\begin{bmatrix} 1 \\ 0 \\ 0 \end{bmatrix} + \begin{bmatrix} 0 \\ 1 \\ 0 \end{bmatrix} = \begin{bmatrix} 0 \\ 1 \\ 0 \end{bmatrix}$;   (d) $2\begin{bmatrix} 2 \\ 4 \\ -1 \\ 0 \end{bmatrix} = \begin{bmatrix} 4 \\ 8 \\ -2 \\ 0 \end{bmatrix}$.

3. (a) $X = \begin{bmatrix} 1 \\ 1 \\ -1 \end{bmatrix}$;   (b) $X = \begin{bmatrix} 1 \\ 2 \\ -1 \end{bmatrix}$;   (c) $\begin{bmatrix} 0 \\ 0 \\ 1 \\ -1 \end{bmatrix}$;   (d) $\begin{bmatrix} 2 \\ 0 \\ -1 \end{bmatrix}$.

4. Columns of A, D are linearly independent.
Columns of B, C, E are linearly dependent.

5. Rows of A, D, C, E linearly independent.
Rows of B are linearly dependent.

10. (a) $(2)(-2) - (4)(-1) = -4 + 4 = 0 \Rightarrow$ linearly dependent.

(b) $(5)(3) - (1)(7) = 15 - 7 = 8 \Rightarrow$ linearly independent.

(c) $(3)(8) - (4)(6) = 24 - 24 = 0 \Rightarrow$ linearly dependent.

Section 1.6

1. $\|\mathbf{x}\| = 5$, $\|\mathbf{y}\| = 13$
$\|\mathbf{z}\| = \sqrt{2}$, $\|\mathbf{u}\| = 13$
$\|\mathbf{v}\| = \sqrt{5}$ $\|\mathbf{w}\| = 2$

2. $\mathbf{x} = (3/5, -4/5)$, $\mathbf{y} = (-12/13, 5/13)$, $\mathbf{z} = (-1/\sqrt{2}, 1/\sqrt{2})$,
$\mathbf{u} = (3/13, -4/13, 12/13)$, $\mathbf{y} = (0, 2/\sqrt{5}, 1/\sqrt{5})$, $\mathbf{w} = (1/2, 1/2, 1/2, 1/2)$.

4. (a) $\mathbf{x} \cdot \mathbf{y} = 3$,   (b) $\mathbf{x} \cdot \mathbf{y} = 0$,   (c) $\mathbf{x} \cdot \mathbf{y} = 2$,   (d) $\mathbf{x} \cdot \mathbf{y} = 0$,
$\cos \theta = 3/\sqrt{10}$ $\cos \theta = 0$ $\cos \theta = 2/\sqrt{42}$ $\cos \theta = 0$

5. (a) $k = -2$;   (b) $k = 9$;   (c) k can be anything.

10. (b) $\mathbf{h} = \left(\dfrac{-165}{169}, \dfrac{396}{169} \right)$.

CHAPTER 2

Section 2.1

2. (a) $\begin{bmatrix} 2/5 & -1 \\ -1 & 13/5 \end{bmatrix}$;   (b) $\begin{bmatrix} 4/37 & -9/37 \\ 27/37 & -70/37 \end{bmatrix}$;   (c) $\begin{bmatrix} -1/6 & -1/2 & -1/3 \\ -3/10 & -7/10 & -1/2 \\ 1/4 & 1/2 & 1/4 \end{bmatrix}$;

(d) $\begin{bmatrix} 1/3 & -1 & -2/3 \\ 2/3 & -9 & -19/3 \\ 7/3 & 5/3 & 5/3 \end{bmatrix}$;   (e) $\begin{bmatrix} 29 & -75 \\ -75 & 194 \end{bmatrix}$.

3. (a) $A_1 X = 0$ only if $X = \begin{bmatrix} x \\ y \end{bmatrix} = \begin{bmatrix} 0 \\ 0 \end{bmatrix}$;   (b) $B_1 X = 0$ only if $X = \begin{bmatrix} x \\ y \end{bmatrix} = \begin{bmatrix} 0 \\ 0 \end{bmatrix}$

4. (a) $X = \begin{bmatrix} x_1 \\ x_2 \end{bmatrix} = \begin{bmatrix} 26 \\ -67 \end{bmatrix};$ **(b)** $X = \begin{bmatrix} x_1 \\ x_2 \end{bmatrix} = \begin{bmatrix} 2/37 \\ -5/37 \end{bmatrix},$

(c) $X = \begin{bmatrix} x_1 \\ x_2 \\ x_3 \end{bmatrix} = \begin{bmatrix} -39 \\ -96 \\ 24 \end{bmatrix};$ **(d)** $X = \begin{bmatrix} x_1 \\ x_2 \end{bmatrix} = \begin{bmatrix} -11 \\ -4 \end{bmatrix}.$

5. (a) $b = -1/2,$ **(b)** $b = 1/3,$ **(c)** $b = 1/2,$ **(d)** $b = -4.$

6. (a) $X = \begin{bmatrix} x_1 \\ -x_1 \end{bmatrix},$ $x_1 \in R;$ **(b)** $X = \begin{bmatrix} -3/2x_2 \\ x_2 \end{bmatrix},$ $x_2 \in R;$

(c) $X = \begin{bmatrix} x_1 \\ -x_1 \\ 0 \end{bmatrix},$ $x_1 \in R;$ **(d)** $X = \begin{bmatrix} -3x_3 \\ -4x_3 \\ x_3 \\ 0 \end{bmatrix},$ $x_3 \in R.$

7. (a) A is nonsingular; **(b)** B is singular; **(c)** C is singular; **(d)** D is nonsingular; **(e)** E is singular.

8. (a) $A^{-1} = \begin{bmatrix} 5 & -2 \\ -2 & 1 \end{bmatrix},$ **(b)** $B^{-1} = \begin{bmatrix} 4/26 & 2/26 \\ -7/26 & 3/26 \end{bmatrix},$

(c) C is not invertible because det $C = 0,$

(d) $D^{-1} = \begin{bmatrix} -1/3 & 0 \\ 1/9 & 1/6 \end{bmatrix},$

(e) E is not invertible because det $E = 0.$

9. (b) $A^{-3} = (A^{-1})^3,$ $A^{-5} = (A^{-1})^5.$

12. $A^{-1} = -A^2 - 3A + 5.$

13. (a) No solutions.

(b) Infinitely many solutions; $\mathbf{x} = \begin{bmatrix} x_1 \\ -3 - x_1 \end{bmatrix}.$

(c) Infinitely many solutions; $\mathbf{x} = \begin{bmatrix} x_1 \\ 3 + 2x_1 \end{bmatrix}.$

(d) No solutions.

14. (a) One solution; **(b)** No solution; **(c)** No solution.

19. $A^{-1} = \dfrac{1}{a^2 + b^2} \begin{bmatrix} a & -b \\ b & a \end{bmatrix}.$

(a) $c = \dfrac{1}{\det A};$ **(b)** $a = 4,$ $b = 4,$ $c = 25$

$a = 12,$ $b = -5,$ $c = 169$

21. (a) B is $2 \times 2,$ C is $4 \times 4,$ D is $8 \times 8.$

22. **(a)** follows because if a matrix is invertible so also is its transpose. **(b)** follows from **(a)** because a matrix is invertible if and only if it is nonsingular. **(c)** follows because $X = 0$ is the only solution of $MX = 0$ if and only if the columns of M are linearly independent. **(d)** follows from **(c)**. By **(b)**, **(e)** and **(f)** are equivalent to **(c)** and **(d)** for A^T. **(g)** and **(h)** follow from the preceding statements.

23. **(a)** If A is singular, its p rows are linearly dependent. Consequently the first p rows of M are linearly dependent. It follows that the rows of M are linearly dependent and hence that M is singular. **(b)** If B is singular, its q columns are linearly dependent. Consequently the last q columns of M are linearly dependent. It follows that the columns of M are linearly dependent and hence that M is singular. **(c)** If M is nonsingular, then neither A nor B can be singular by **(a)** and **(b)**. The formula for M^{-1} given in **(d)** can be verified by showing that it gives us the relation $MM^{-1} = I$.

Section 2.2

1. A, elementary; B, elementary; C, not elementary; D, elementary;
E, elementary; F, elementary; G, not elementary; H, not elementary;
I, not elementary.

2. $A^{-1} = \begin{bmatrix} 1 & -4 \\ 0 & 1 \end{bmatrix}$, $B^{-1} = \begin{bmatrix} 0 & 1 \\ 1 & 0 \end{bmatrix}$, $E^{-1} = \begin{bmatrix} 1 & 0 & 0 \\ 0 & -1/4 & 0 \\ 0 & 0 & 1 \end{bmatrix}$, $F^{-1} = \begin{bmatrix} 0 & 1 & 0 \\ 1 & 0 & 0 \\ 0 & 0 & 1 \end{bmatrix}$.

3. $C = \begin{bmatrix} 0 & 1 \\ 1 & 0 \end{bmatrix} \begin{bmatrix} 3 & 0 \\ 0 & 1 \end{bmatrix}$, $C^{-1} = \begin{bmatrix} 0 & 1/3 \\ 1 & 0 \end{bmatrix}$,

D is singular; D^{-1} doesn't exist.

$G = \begin{bmatrix} 3 & 0 & 0 \\ 0 & 1 & 0 \\ 0 & 0 & 1 \end{bmatrix} \begin{bmatrix} 1 & 0 & 0 \\ 0 & 1 & 0 \\ -2 & 0 & 1 \end{bmatrix}$, $G^{-1} = \begin{bmatrix} 1/3 & 0 & 0 \\ 0 & 1 & 0 \\ 2/3 & 0 & 1 \end{bmatrix}$,

$H = \begin{bmatrix} 0 & 1 & 0 \\ 1 & 0 & 0 \\ 0 & 0 & 1 \end{bmatrix} \begin{bmatrix} 1 & 0 & 0 \\ 0 & 0 & 1 \\ 0 & 1 & 0 \end{bmatrix}$, $H^{-1} = \begin{bmatrix} 0 & 1 & 0 \\ 0 & 0 & 1 \\ 1 & 0 & 0 \end{bmatrix}$,

$I = \begin{bmatrix} 6 & 0 & 0 \\ 0 & 1 & 0 \\ 0 & 0 & 1 \end{bmatrix} \begin{bmatrix} 1 & 0 & 0 \\ 0 & 4 & 0 \\ 0 & 0 & 1 \end{bmatrix}$, $I^{-1} = \begin{bmatrix} 1/6 & 0 & 0 \\ 0 & 1/4 & 0 \\ 0 & 0 & 1 \end{bmatrix}$.

4. **(a)** $A = \begin{bmatrix} 1/4 & 0 \\ 0 & 1 \end{bmatrix} \begin{bmatrix} 1 & 0 \\ -1/2 & 1 \end{bmatrix}$, $B = \begin{bmatrix} 1 & 2/3 \\ 0 & 1 \end{bmatrix} \begin{bmatrix} 1 & 0 \\ 0 & 3 \end{bmatrix}$,

$C = \begin{bmatrix} 0 & 1 \\ 1 & 0 \end{bmatrix} \begin{bmatrix} 4 & 0 \\ 0 & 1 \end{bmatrix} \begin{bmatrix} 1 & 0 \\ 2 & 1 \end{bmatrix}$,

$D = \begin{bmatrix} -2/7 & 0 \\ 0 & 1 \end{bmatrix} \begin{bmatrix} 1 & -1 \\ 0 & 1 \end{bmatrix} \begin{bmatrix} 1 & 0 \\ -6 & 1 \end{bmatrix} \begin{bmatrix} 1 & 0 \\ 0 & 7 \end{bmatrix}$,

$$E = \begin{bmatrix} 1 & 0 & 0 \\ 0 & 4 & 0 \\ 0 & 0 & 1 \end{bmatrix} \begin{bmatrix} 1 & 0 & 0 \\ 0 & 1 & 0 \\ 3 & 0 & 1 \end{bmatrix},$$

(b) $A^{-1} = \begin{bmatrix} 1/4 & 0 \\ -1/2 & 1 \end{bmatrix}$, $B^{-1} = \begin{bmatrix} 1 & -2/3 \\ 0 & 1/3 \end{bmatrix}$, $C^{-1} = \begin{bmatrix} 0 & 1/4 \\ 1 & -1/2 \end{bmatrix}$,

$$D^{-1} = \begin{bmatrix} -7/2 & 1 \\ -3 & 1 \end{bmatrix}, \quad E^{-1} = \begin{bmatrix} 1 & 0 & 0 \\ 0 & 1/4 & 0 \\ -3 & 0 & 1 \end{bmatrix}.$$

5. (a) $X = \begin{bmatrix} 3/4 \\ 5/2 \end{bmatrix}$;   **(b)** $X = \begin{bmatrix} -2/3 \\ -2/3 \end{bmatrix}$;   **(c)** $X = \begin{bmatrix} 1/4 \\ -1/2 \end{bmatrix}$;

(d) $X = \begin{bmatrix} -7/2(\pi) + \sqrt{10} \\ -3\pi + \sqrt{10} \end{bmatrix}$;   **(e)** $X = \begin{bmatrix} 2 \\ 0 \\ -7 \end{bmatrix}$.

7. No.

8. (a) $\begin{bmatrix} 1 & 1 \\ 1 & 0 \end{bmatrix}$;   **(b)** $\begin{bmatrix} 1 & 6/5 & 3/5 \\ 0 & 4 & 1 \\ 0 & 1 & 0 \end{bmatrix}$;   **(c)** $\begin{bmatrix} 1 & 6 & 2 \\ 7 & -4 & 0 \\ 5 & -1 & 3 \end{bmatrix}$.

9. (a) $\begin{bmatrix} 17 & 2 \\ 6 & -4 \end{bmatrix}$;   **(b)** $\begin{bmatrix} 23/4 & 2 \\ 3 & 4 \end{bmatrix}$;   **(c)** $\begin{bmatrix} 5 & 2 & -1 \\ 6 & 3 & 0 \\ -4 & -2 & 1 \end{bmatrix}$.

10. Is not possible because the rows/columns are multiples of one another.

11. (a) $Q_1 = \begin{bmatrix} 0 & 1 & 0 \\ 1 & 0 & 0 \\ 0 & 0 & 1 \end{bmatrix} \begin{bmatrix} 1 & 0 & 0 \\ 0 & 0 & 1 \\ 0 & 1 & 0 \end{bmatrix}$; $Q_2 = \begin{bmatrix} 0 & 1 & 0 \\ 1 & 0 & 0 \\ 0 & 0 & 1 \end{bmatrix} \begin{bmatrix} 0 & 0 & 1 \\ 0 & 1 & 0 \\ 1 & 0 & 0 \end{bmatrix}$.

(b) $Q_1^T = Q_1^{-1} = \begin{bmatrix} 0 & 1 & 0 \\ 0 & 0 & 1 \\ 1 & 0 & 0 \end{bmatrix}$; $Q_2^T = Q_2^{-1} = \begin{bmatrix} 0 & 0 & 1 \\ 1 & 0 & 0 \\ 0 & 1 & 0 \end{bmatrix}$.

13. $D^{-1} = \begin{bmatrix} 1/2 & 0 & 0 \\ 0 & -1/3 & 0 \\ 0 & 0 & 1/5 \end{bmatrix}$, $X^{-1} = \begin{bmatrix} 1 & -2 & 0 \\ 0 & 1 & 0 \\ 0 & -3 & 1 \end{bmatrix}$, $Y^{-1} = \begin{bmatrix} 1 & 2/5 & 0 \\ 0 & -1/5 & 0 \\ 0 & 3/5 & 1 \end{bmatrix}$;

$$P^{-1} = \begin{bmatrix} 0 & 1 & 0 \\ 0 & 0 & 1 \\ 1 & 0 & 0 \end{bmatrix}, \quad Q^{-1} = \begin{bmatrix} 1 & 0 & -7 \\ 0 & 1 & 0 \\ 0 & 0 & 1 \end{bmatrix}, \quad R^{-1} = \begin{bmatrix} 1 & 0 & 0 \\ 2/5 & -1/5 & 3/5 \\ 0 & 0 & 1 \end{bmatrix};$$

$$U^{-1} = \begin{bmatrix} 1 & 2/3 & 0 \\ 0 & -1/3 & 0 \\ 0 & 0 & 1 \end{bmatrix}, \quad V^{-1} = \begin{bmatrix} 1 & 0 & 0 \\ 0 & 1 & 0 \\ 0 & -3 & 1 \end{bmatrix}, \quad W^{-1} = \begin{bmatrix} 1/2 & 1 & 0 \\ -1/2 & 0 & 1 \\ 1/2 & 0 & 0 \end{bmatrix}.$$

Section 2.3

1. $A^{-1} = \begin{bmatrix} 0 & -1 \\ 1 & 4 \end{bmatrix}$, $B^{-1} = \begin{bmatrix} -1/2 & 3/2 \\ 1 & -2 \end{bmatrix}$,

$C^{-1} = \begin{bmatrix} 1 & 1 & 0 \\ 1 & 2 & 0 \\ 0 & 1 & 1 \end{bmatrix}$, $D^{-1} = \begin{bmatrix} 1/12 & -1/8 & 2/3 \\ 1/4 & 1/8 & -1 \\ -1/12 & 1/8 & 1/3 \end{bmatrix}$,

$E^{-1} = \begin{bmatrix} 0 & -1/2 & 0 & 1/2 \\ 1/2 & 0 & -1/2 & 0 \\ 1/2 & 0 & 0 & -1/2 \\ 0 & 1/2 & 1/2 & 0 \end{bmatrix}$.

2. $A^{-1} = \begin{bmatrix} 1 & 1 & -1 \\ -1 & -1/2 & 3/2 \\ 0 & -1/2 & 1/2 \end{bmatrix}$, $B^{-1} = \begin{bmatrix} -1/2 & -3/2 & 1 \\ -3/4 & -5/4 & 3/2 \\ 1/4 & 3/4 & -1/2 \end{bmatrix}$, C is not invertible.

3. (a) $A = P_{12}S_{12}(-1)T_{12}(4)$,
$\quad B = T_{12}(1/2)S_2(-1/2)T_{21}(3)S_1(4)$,
$\quad C = P_{13}T_{12}(-1)T_{13}(2)P_{23}T_{32}(-2)T_{31}(1)T_{21}(-1)$,
$\quad D = P_{13}T_{13}(4)S_2(2)T_{23}(3)S_3(-12)T_{31}(3)T_{32}(1)$,
$\quad E = T_{12}(-1)T_{13}(1)T_{14}(1)T_{32}(-1)P_{23}P_{34}S_2(-2)S_3(-2)S_4(2)T_{41}(1)T_{31}(1)T_2(1)$.

(b) $A^{-1} = T_{12}(-4)S_{12}(1)P_{12}$,
$\quad C^{-1} = T_{21}(1)T_{32}(2)T_{31}(-1)P_{23}T_{13}(-2)T_{12}(1)P_{13}$,
$\quad D^{-1} = T_{32}(-1)T_{31}(-3)S_3(-1/12)T_{23}(-3)S_2(1/2)T_{13}(-4)P_{13}$
$\quad E^{-1} = T_{21}(-1)T_{31}(-1)T_{41}(-1)S_4(1/2)S_3(-1/2)S_2(-1/2)P_{34}P_{23}T_{32}(1)T_{14}(-1)T_{13}(-1)T_{12}(1)$

4. (a) $X = \begin{bmatrix} -3 \\ 14 \end{bmatrix}$, **(b)** $X = \begin{bmatrix} 7/2 \\ -4 \end{bmatrix}$, **(c)** $X = \begin{bmatrix} 0 \\ -1 \\ 1 \end{bmatrix}$,

(d) $X = \begin{bmatrix} 37/24 \\ -45/24 \\ 11/24 \end{bmatrix}$, **(e)** $X = \begin{bmatrix} 1/2 \\ 1/2 \\ 0 \\ 0 \end{bmatrix}$.

Section 2.4

1. (a) Yes; **(b)** no; **(c)** yes; **(d)** no; **(e)** yes; **(f)** no.

2. (a) $\begin{bmatrix} 1 & -1 & -1 \\ 0 & 2 & -10 \end{bmatrix} \begin{bmatrix} x \\ y \\ z \end{bmatrix} = \begin{bmatrix} 2 \\ 1 \end{bmatrix}$, $\begin{bmatrix} x \\ y \\ z \end{bmatrix} = \begin{bmatrix} 5/2 + 6b \\ 1/2 + 5b \\ b \end{bmatrix}$, $b \in R$.

(b) $\begin{bmatrix} 2 & 0 & 2 & -4 \\ 0 & 1 & 2 & -1 \\ 0 & 2 & 5 & -4 \end{bmatrix} \begin{bmatrix} x \\ y \\ z \\ w \end{bmatrix} = \begin{bmatrix} -12 \\ 3 \\ 9 \end{bmatrix}$, $\begin{bmatrix} x \\ y \\ z \\ w \end{bmatrix} = \begin{bmatrix} -9 \\ -3 - 3b \\ 3 + 2b \\ b \end{bmatrix}$, $b \in R$.

(c) $\begin{bmatrix} 3 & 6 & -12 \\ 2 & 4 & -8 \end{bmatrix} \begin{bmatrix} x \\ y \\ z \end{bmatrix} = \begin{bmatrix} 9 \\ 6 \end{bmatrix}$, $\quad \begin{bmatrix} x \\ y \\ z \end{bmatrix} = \begin{bmatrix} 3 - 2a + 4b \\ a \\ b \end{bmatrix}$, $\quad a, b \in R$.

(d) $\begin{bmatrix} -2 & 0 & -4 & 0 \\ 1 & 0 & 2 & 3 \\ 3 & 0 & 6 & 1 \end{bmatrix} \begin{bmatrix} x \\ y \\ z \\ w \end{bmatrix} = \begin{bmatrix} -6 \\ -3 \\ 9 \end{bmatrix}$; there are no solutions.

3. (b) $\begin{bmatrix} 1 & 0 & \vdots & 3 \\ 0 & 1 & \vdots & 5 \end{bmatrix}$, (d) $\begin{bmatrix} 1 & 0 & 0 & \vdots & -3 \\ 0 & 1 & 0 & \vdots & 4 \\ 0 & 0 & 0 & \vdots & 0 \end{bmatrix}$, (f) $\begin{bmatrix} 1 & 0 & 0 & 5 \\ 0 & 1 & 0 & 7 \\ 0 & 0 & 1 & 4 \\ 0 & 0 & 0 & 0 \end{bmatrix}$

4. (a) $\begin{bmatrix} x_1 \\ x_2 \\ x_3 \\ x_4 \end{bmatrix} = \begin{bmatrix} 4 - a - 2b \\ b \\ -1 + 2a \\ a \end{bmatrix}$, (b) $\begin{bmatrix} x_1 \\ x_2 \\ x_3 \\ x_4 \\ x_5 \end{bmatrix} = \begin{bmatrix} 7 - a - b \\ 1 + b - 2a \\ a \\ 2 + 2b \\ b \end{bmatrix}$.

CHAPTER 3

Section 3.1

1. $11, 33, 0, a^2 + b^2, 1, 1, 0.$
2. $-27, 27, 27.$
3. $-27, 27, 27.$
4. $81, 0, 0, 45, 5, 4.$
5. $63, 6.$
8. $54, 9, -36.$

Section 3.2

2. $3, 1, 0, -40, -300.$
3. $abcd.$
8. (a) $x = -3, -1, 1;$　(b) $x = 1$ or $x = -2;$　(c) $1, 2, 9;$　(d) $1, 3, 9.$
9. (a) $(b - a)(c - a)(d - a)(c - b)(d - b)(d - c);$　(b) $16, -768$
10. $y = \dfrac{(y_2 - y_1)}{(x_2 - x_1)} x + \dfrac{(x_2 y_1 - x_1 y_2)}{(x_2 - x_1)};$　$y = x + 1.$
12. (a) $9/2;$　(b) $17;$　(c) $13.$
15. (a) $x^2 + y^2 - 5x - 5y = 0;$　(b) $x^2 + y^2 - 8x - 8y + 14 = 0;$
  (c) $x^2 + y^2 - 2x + 2y - 48 = 0.$
18. $13, -18, 1, -1, 1.$

Section 3.3

1. (b) $(\text{adj } A)A = A(\text{adj } A) = (\det A)I = \begin{bmatrix} ad - bc & 0 \\ 0 & ad - bc \end{bmatrix}$.

2. $\text{adj } A = \begin{bmatrix} bc & 0 & 0 \\ 0 & ac & 0 \\ 0 & 0 & ab \end{bmatrix}$; $\text{adj } B = \begin{bmatrix} -1 & 0 & 0 \\ 0 & 0 & -1 \\ 0 & -1 & 0 \end{bmatrix}$;

$\text{adj } C = \begin{bmatrix} 1 & 0 & 0 \\ -a & 1 & 0 \\ -b & 0 & 1 \end{bmatrix}$; $\text{adj } D = \begin{bmatrix} 4 & -4 & -4 & 4 \\ 4 & 4 & -4 & -4 \\ 4 & -4 & 4 & -4 \\ 4 & 4 & 4 & 4 \end{bmatrix}$;

$\text{adj } E = \begin{bmatrix} 66 & -24 & 11 & -6 \\ -24 & 9 & -4 & 2 \\ 11 & -4 & 2 & -1 \\ -6 & 2 & -1 & 1 \end{bmatrix}$; $\text{adj } F = \begin{bmatrix} 39 & -52 & 0 & 0 \\ -26 & 39 & 0 & 0 \\ -8 & 7 & 2 & -1 \\ 17 & -23 & -1 & 7 \end{bmatrix}$;

$\text{adj } G = \begin{bmatrix} -2 & 8 & 13 & -17 \\ -4 & -10 & -13 & 57 \\ 0 & 0 & 13 & -39 \\ 0 & 0 & -13 & 13 \end{bmatrix}$; $\text{adj } H = \begin{bmatrix} 2 & 3 & 0 & 0 \\ 5 & 8 & 0 & 0 \\ 0 & 0 & 3 & -2 \\ 0 & 0 & -4 & 3 \end{bmatrix}$.

7. (a) $x = 1, y = -1$; **(b)** $x = 124/17, y = -26/17$;

(c) $x_1 = 10/3, x_2 = -7, x_3 = -28/3$;

(d) $x = 2, \quad y = 2, \quad z = -3$.

8. $x = \dfrac{(h_1 - a_{13}k)a_{22} - (h_2 - a_{23}k)a_{12}}{a_{11}a_{22} - a_{21}a_{12}}$;

$y = \dfrac{a_{11}(h_2 - a_{23}k) - a_{21}(h_1 - a_{13}k)}{a_{11}a_{22} - a_{21}a_{12}}$

assuming all a's, h's, and k's are constants, and $a_{11}a_{22} - a_{21}a_{12} \neq 0$.

Section 3.4

1. A: $r = 2$, B: $r = 1$, C: $r = 2$
D: $r = 2$, E: $r = 1$, F: $r = 3$
G: $r = 3$, H: $r = 4$, K: $r = 4$

2. A: **(a)** $\begin{bmatrix} 2 & 1 \\ 1 & 0 \end{bmatrix}$; **(b)** $\begin{bmatrix} 2 \\ 1 \end{bmatrix}, \begin{bmatrix} 1 \\ 0 \end{bmatrix}$; **(c)** $[2 \quad 1], [1 \quad 0]$.

B: **(a)** $[2]$; **(b)** $\begin{bmatrix} 2 \\ 4 \end{bmatrix}$, **(c)** $[2 \quad 1]$.

C: **(a)** $\begin{bmatrix} 1 & 0 \\ 0 & 1 \end{bmatrix}$; **(b)** $\begin{bmatrix} 1 \\ 0 \end{bmatrix}, \begin{bmatrix} 0 \\ 1 \end{bmatrix}$; **(c)** $[1 \quad 0 \quad 3], [0 \quad 1 \quad 1]$.

D: **(a)** $\begin{bmatrix} 1 & 0 \\ 0 & 1 \end{bmatrix}$;   **(b)** $\begin{bmatrix} 1 \\ 0 \\ 1 \end{bmatrix}, \begin{bmatrix} 0 \\ 1 \\ 1 \end{bmatrix}$;   **(c)** [1   0   3], [0   1   1].

E: **(a)** [1];   **(b)** $\begin{bmatrix} 1 \\ -2 \\ 3 \end{bmatrix}$;   **(c)** [1   1 -2].

F: **(a)** $\begin{bmatrix} 1 & 1 & 0 \\ 0 & 1 & 2 \\ 0 & 0 & 1 \end{bmatrix}$;   **(b)** $\begin{bmatrix} 1 \\ 0 \\ 0 \end{bmatrix}, \begin{bmatrix} 1 \\ 1 \\ 0 \end{bmatrix}, \begin{bmatrix} 0 \\ 2 \\ 1 \end{bmatrix}$;

    (c) [1 1 0], [0 1 2], [0 0 1].

G: **(a)** $\begin{bmatrix} 3 & 4 & 1 \\ 2 & 3 & 0 \\ 5 & 6 & 7 \end{bmatrix}$;   **(b)** $\begin{bmatrix} 3 \\ 2 \\ 5 \\ 0 \end{bmatrix}, \begin{bmatrix} 4 \\ 3 \\ 6 \\ 1 \end{bmatrix}, \begin{bmatrix} 1 \\ 0 \\ 7 \\ -6 \end{bmatrix}$.

    (c) [3   4   1   1], [2   3   0   2], [5   6   7   1].

H: **(a)** H **(b)** take 4 columns of H;   **(c)** 4 rows of H.

K: **(a)** K;   **(b)** 4 columns of K;   **(c)** 4 rows of K.

CHAPTER 4

Section 4.1

1. H: $(5 - \lambda)^2 - 16$. 1, 9; K: $6 - 5\lambda + \lambda^2$; 2, 3; L: $(4 - \lambda)^2 + 9$; $4 \pm 3i$;
   M: $(12 - \lambda)^2 + 25$; $12 \pm 5\sqrt{2}\,i$; N: $[16 - (5 - \lambda)^2]\lambda$; 1, 9, 0;
   P: $1 - \lambda^3$; 1, $(1/2)(-1 \pm i\sqrt{3})$; Q: $-9 + 9\lambda + \lambda^2 - \lambda^3$; 1, 3, -3;
   R: $(3 - 4\lambda + \lambda^2)(4 - 4\lambda + \lambda^2)$; 1, 3, 2, 2; S: $(5 - \lambda)(1 - \lambda^3)$; 5, 1, $(1/2)(-1 \pm i\sqrt{3})$;
   T: $(4 - \lambda^2)(9 - \lambda^2)$; 2, -2, 3, -3.

4. $A = \begin{bmatrix} 0 & -1 & 0 & 0 \\ 0 & 0 & -1 & 0 \\ 0 & 0 & 0 & -1 \\ 4 & 0 & -5 & 0 \end{bmatrix}$.

5. U: $1 - \lambda^3$; V: $-1 - \lambda^3$; W: $-1 - \lambda^3$; X: $-1 + \lambda^4$; Y: $1 + \lambda^4$;
   Z: $-1 + \lambda^4$.

6. **(a)** -1, 1;   **(b)** $-1/2 + i\sqrt{3}/2$, $-1/2 - i\sqrt{3}/2$, 1;   **(c)** i, -1, $-i$, 1;

   (d) $w_1 = \cos 2\pi/5 + i \sin 2\pi/5$, w_1^2, w_1^3, w_1^4, 1.

7. $C = \begin{bmatrix} 0 & 5/2 & 0 & -1/2 \\ 5/2 & 0 & -1/2 & 0 \\ 0 & -1/2 & 0 & 5/2 \\ -1/2 & 0 & 5/2 & 0 \end{bmatrix}$.   Matrices A and C have same eigenvalues,

   $\lambda_1 = 2$, $\lambda_2 = -2$, $\lambda_3 = 3$, $\lambda_4 = -3$.

8. $a = 16$; $b = 16$; $c = 8$; $d = 4$.

9. (a) $-3, -1, 1$;   **(b)** 4, 6, 8;   **(c)** $2 + c, 4 + c, 6c$;   **(d)** $-2, -4, -6$;
   (e) 4, 8, 12;   **(f)** $2c, 4c, 6c$;   **(g)** 7, 11, 15.

10. (a) ± 1;   **(b)** $\pm 1, \pm 3$;   **(c)** $\pm 1, \pm 2, \pm 4, \pm 8$;   **(d)** $\pm 1, \pm 11$.

Section 4.2

1. H: For $\lambda = 1$, $(1, -1)$; for $\lambda = 9$, $(1, 1)$
   K: For $\lambda = 2$, $(1, 1)$; for $\lambda = 3$, $(2, 3)$.
   L and M: none
   N: for $\lambda = 1$, $(-8, 8, 5)$; for $\lambda = 9$, $(0, 0, 1)$.
   P: for $\lambda = 1$, $(1, -1, 1)$.
   Q: for $\lambda = 1$, $(1, 1, -1)$; for $\lambda = 3$, $(1, -3, 9)$; for $\lambda = -3$, $(1, 3, 9)$.
   R: for $\lambda = 1$, $(1, 1, 22, -99)$; for $\lambda = 3$, $(1, 3, 94, -211)$; for $\lambda = 2$, $(0, 0, 1, -3)$.
   S: for $\lambda = 1$, $(4, 4, 4, -9)$; for $\lambda = 5$, $(0, 0, 0, 1)$.
   T: for $\lambda = 2$, $(1, 0, 0, 1)$; for $\lambda = -2$, $(1, 0, 0, -1)$; for $\lambda = 3$, $(0, 1, 1, 0)$;
   for $\lambda = -3$, $(0, 1, -1, 0)$.

2. $(1, 1, 1, 1)$, $(-1, 1, -1, 1)$, $(1, 1, -1, -1)$, and $(1, -1, -1, 1)$

3. $C\mathbf{y} = CB\mathbf{x} = BAB^{-1}B\mathbf{x} = BA\mathbf{x} = B\lambda\mathbf{x} = \lambda B\mathbf{x} = \lambda\mathbf{y}$.

7. (a) By computation, $\mathbf{u}_1^T\mathbf{v}_2 = 0$ and $\mathbf{u}_2^T\mathbf{v}_1 = 0$.

   (b), (c) By **(a)** $A\mathbf{u}_1 = (\mathbf{u}_1(\mathbf{v}_1^T\mathbf{u}_1)) + \mathbf{0} = \lambda_1\mathbf{u}_1$ with $\lambda_1 = \mathbf{v}_1^T\mathbf{u}_1$ as the corresponding eigenvalue of A.
   Also, by **(a)**, $A\mathbf{u}_2 = \mathbf{0} + \mathbf{u}_2(\mathbf{v}_2^T\mathbf{u}_2) = \lambda_2\mathbf{u}_2$ with $\lambda_2 = \mathbf{v}_2^T\mathbf{u}_2$ as an eigenvalue.

   (d), (e) Apply conclusion **(b)**, **(c)** to A^T.

   (f) $A\mathbf{u}_3 = \mathbf{0} = 0\mathbf{u}_3$, $A^T\mathbf{v}_3 = \mathbf{0} = 0\mathbf{v}_3$.

8. The answers given in Exercise 7 hold here also.

9. (a) $A\mathbf{x} = 0\mathbf{x}$;   **(b)** Use **(a)**;   **(c)** $AY = AXD = 0D = 0$.

10. (a) Since A is nonsingular its eigenvalues are nonzero. $\mathbf{y} = A\mathbf{x} = \lambda\mathbf{x}$ is a nonzero multiple of $\mathbf{x}$ and so is an eigenvector of A with λ as the corresponding eigenvalue. $\mathbf{z} = A^2\mathbf{x} = A\mathbf{y} = \lambda\mathbf{y} = \lambda^2\mathbf{x}$ and so is also an eigenvector of A. Likewise $A^k\mathbf{x}$ is an eigenvector of A.

   (b) The columns of $Y = XD$ are nonzero multiples of the columns of X and so are eigenvectors of A. The columns of $Z = AX$ are nonzero multiples of the columns of A and so are eigenvectors of A. The same is true for the columns of A^kX.

11. (a) By computation.

   (b) This holds because the main diagonal entries of an upper triangular matrix are its eigenvalues.

   (c) For A, $\mathbf{x}_1 = (1, 0, 0, 0)$;   for B, $\mathbf{x}_1$ and $\mathbf{x}_2 = (0, 1, 0, 0)$;   for C, $\mathbf{x}_1$, $\mathbf{x}_2$ and $\mathbf{x}_3 = (0, 0, 1, 0)$.

   (d) $\mathbf{x}_1$ is a common eigenvector.

(e) x_2 is an eigenvector of B which is not an eigenvector of A.

(f) x_3 has this property.

(g) All nonzero vectors.

12. **(a)** By computation.

(b) The main diagonal entries of an upper triangular matrix are its eigenvalues.

(c) The nonzero multiples of $x_1 = (1, 0, 0)$ are their eigenvectors.

14. **(a)** follows from the result given in Exercise 7.

(b) is a restatement of **(a)**.

(c) If $p(\lambda)$ and $q(\lambda)$ are relatively prime, then (see Appendix) there are polynomials $f(\lambda)$ and $g(\lambda)$ such that

$$1 = f(\lambda)p(\lambda) + g(\lambda)q(\lambda).$$

Then

$$I = f(A)p(A) + g(A)q(A) = g(A)q(A) \qquad \text{since } p(A) = 0.$$

Consequently $q(A)$ is invertible with $g(A)$ as its inverse. It follows that $q(A)$ is nonsingular.

(d) Suppose that λ_1 is a common zero of $p(\lambda)$ and $q(\lambda)$. Since $p(\lambda_1) = 0$, λ_1 is an eigenvalue of A. Let x_1 be the corresponding eigenvector of A. Then $Ax_1 = \lambda_1 x_1$ and, by **(a)**, $q(A)x_1 = q(\lambda_1)x_1 = 0$ (because $q(\lambda_1) = 0$). Hence $q(A)$ is singular.

15. **(e)** If $r(\lambda)$ is a GCD of $p(\lambda)$ and $q(\lambda)$, then (see Appendix) there exist polynomials $f(\lambda)$ and $g(\lambda)$ such that

$$r(\lambda) = f(\lambda)p(\lambda) + g(\lambda)q(\lambda).$$

Hence

$$r(A) = f(A)p(A) + g(A)q(A) = g(A)q(A) = 0$$

when $g(A) = 0$.

(f) If $q(A) = 0$, the degree of $q(\lambda)$ is not less than the degree of $m(\lambda)$. By the division algorithm, there is a polynomial $c(\lambda)$ and a polynomial $d(\lambda)$ of degree less than that of $m(\lambda)$ such that

$$q(\lambda) = c(\lambda)m(\lambda) + d(\lambda).$$

Consequently, if $q(A) = 0$, then

$$0 = q(A) = c(A)m(A) + d(A) = d(A).$$

But the degree of $d(\lambda)$ is less than that of $m(\lambda)$. By our choice of $m(\lambda)$ this is possibly only if $d(\lambda) = 0$ for all λ. It follows that $q(\lambda) = c(\lambda)m(\lambda)$ and hence that $m(\lambda)$ divides $q(\lambda)$.

Section 4.3

1. Since A is symmetric, by Property 7 it is diagonalizable.

2. The eigenvalues are: A: $9, -1$; B: $11, 1$; C: $11, 1$; D: $25, -1$; E: $3, 9, -1$; F: $3, 9, -1$; G: $7, 2, -2$; H: $-1, 1, 9, 11$; K: $-1, 1, 9, 25$; L: $3, 9, -1, 7, 2, -2$.

3. This is because the main diagonal entries of these matrices are its eigenvalues. Since they are distinct these matrices are diagonalizable.

5. $\lambda_1 = 2$, $\lambda_2 = 0$, $\lambda_3 = 1$. $4, 0$, and 1. $2^n, 0$, and 1.

7. $A = A^T$; $\lambda_1 = \sqrt{2}$, $\lambda_2 = -\sqrt{2}$.
   $\mathbf{x}_1 = (1 + \sqrt{2}, 1)$ and $\mathbf{x}_2 = (1 - \sqrt{2}, 1)$.
   $$\begin{bmatrix} \dfrac{1 + \sqrt{2}}{\sqrt{4 + 2\sqrt{2}}} & \dfrac{1 - \sqrt{2}}{\sqrt{4 - 2\sqrt{2}}} \\ \dfrac{1}{\sqrt{4 + 2\sqrt{2}}} & \dfrac{1}{\sqrt{4 - 2\sqrt{2}}} \end{bmatrix} = X = \begin{bmatrix} \dfrac{X_1}{\|X_1\|} & \dfrac{X_2}{\|X_2\|} \end{bmatrix}.$$

8. **(a)** $\lambda_1 = 0$, $\lambda_2 = -\sqrt{2}$, $\lambda_3 = \sqrt{2}$;
   (b) $\mathbf{x}_1 = (0, 1, 1)$, $\mathbf{x}_2 = (\sqrt{2}, -1, 1)$, and $\mathbf{x}_3 = (\sqrt{2}, 1, -1)$.
   (c) $X = \begin{bmatrix} 0 & \sqrt{2} & \sqrt{2} \\ 1 & -1 & 1 \\ 1 & 1 & -1 \end{bmatrix}$; $Y = \begin{bmatrix} 0 & \sqrt{2}/2 & \sqrt{2}/2 \\ 1/\sqrt{2} & -1/2 & 1/2 \\ 1/\sqrt{2} & 1/2 & -1/2 \end{bmatrix}.$

9. $D = D^T = (X^{-1}AX)^T = (AX)^T(X^{-1})^T$
   $\qquad = X^T A^T (X^T)^{-1} = ((X^T)^{-1})^{-1} A^T (X^T)^{-1} = Y^{-1} A^T Y.$

Section 4.4

1. U: $\lambda = 6$, multiplicity 2.
   $E = \begin{bmatrix} 6 & 1 \\ 0 & 6 \end{bmatrix}.$
   V: $\lambda_1 = 4$ and $\lambda_2 = 6$, $\mathbf{x}_1 = (1, 1)$ and $\mathbf{x}_2 = (-1, 1)$. Canonical form D.
   $\begin{bmatrix} 6 & 0 \\ 0 & 4 \end{bmatrix}.$
   $\begin{bmatrix} 6 & 0 \\ 0 & 4 \end{bmatrix}.$
   W: $\lambda = 6$, multiplicity 2, $\mathbf{x} = (2, 2)$. Canonical form E.
   $\begin{bmatrix} 6 & 1 \\ 0 & 6 \end{bmatrix}.$
   X: $\lambda_1 = 10$ and $\lambda_2 = -5$, $\mathbf{x}_1 = (1, 0)$ and $\mathbf{x}_2 = (-1/5, 1)$. Canonical type D.
   $\begin{bmatrix} 10 & 0 \\ 0 & -5 \end{bmatrix}.$ U and W are similar by the transitive property.

2. $X^{-1}PX = \begin{bmatrix} 1 & 0 \\ 0 & 1/4 \end{bmatrix} \begin{bmatrix} 0 & 1 \\ 0 & 0 \end{bmatrix} \begin{bmatrix} 1 & 0 \\ 0 & 4 \end{bmatrix} = \begin{bmatrix} 0 & 4 \\ 0 & 0 \end{bmatrix} = Q.$

$X^{-1}RX = \begin{bmatrix} 1 & 0 \\ 0 & 1/4 \end{bmatrix} \begin{bmatrix} 5 & 1 \\ 0 & 5 \end{bmatrix} \begin{bmatrix} 1 & 0 \\ 0 & 4 \end{bmatrix} = \begin{bmatrix} 5 & 4 \\ 0 & 5 \end{bmatrix} = S.$

3. $\begin{bmatrix} 2 & -3 \\ 3 & 2 \end{bmatrix}$

4. $\begin{bmatrix} 2 & 0 \\ 0 & 1 \end{bmatrix}, \begin{bmatrix} 1 & 0 \\ 0 & 2 \end{bmatrix}$

6. Because they do not have the same eigenvalues.

7. Choose nonsingular matrices X and Y so that $D = X^{-1}AX$ and $E = Y^{-1}BY$. The main diagonals of D and E are the same except possibly for their order. By Property 4, D and E are similar. Hence A and B are similar.

8. Because they all have 2 and 3 as distinct eigenvalues.

10. K: $\lambda_1 = 1$, $\lambda_2 = -1$, and $\lambda_3 = 0$. $\mathbf{x}_1 = (1, 0, 0)$, $\mathbf{x}_2 = (1, -1, 0)$, and $\mathbf{x}_3 = (7, -2, -1)$.
$$\begin{bmatrix} 1 & 0 & 0 \\ 0 & -1 & 0 \\ 0 & 0 & 0 \end{bmatrix} = D.$$
L: $\lambda_1 = \lambda_2 = 2$ and $\lambda_3 = 1$. $\mathbf{x}_1 = \mathbf{x}_2 = (1, 0, 0)$ and $\mathbf{x}_3 = (0, 1, 0)$.
$$\begin{bmatrix} 2 & 1 & 0 \\ 0 & 2 & 0 \\ 0 & 0 & 1 \end{bmatrix} = E.$$
M: $\lambda = 1$ multiplicity 3. $\mathbf{x} = (0, 0, 1)$.
$$\begin{bmatrix} 1 & 1 & 0 \\ 0 & 1 & 1 \\ 0 & 0 & 1 \end{bmatrix} = F.$$
N: $\lambda_1 = 1$, $\lambda_2 = 1 + i$ and $\lambda_3 = 1 - i$. $\mathbf{x}_1 = (1, -1, 1)$, $\mathbf{x}_2 = (1, i, 0)$ and $\mathbf{x}_3 = (1, -i, 0)$.
$$\begin{bmatrix} 1 & 1 & 0 \\ -1 & 1 & 0 \\ 0 & 0 & 1 \end{bmatrix} = G.$$

11. $D = \begin{bmatrix} 2 & 0 & 0 \\ 0 & 2 & 0 \\ 0 & 0 & 2 \end{bmatrix}$; $E = \begin{bmatrix} 2 & 1 & 0 \\ 0 & 2 & 0 \\ 0 & 0 & 2 \end{bmatrix}$; and $F = \begin{bmatrix} 2 & 1 & 0 \\ 0 & 2 & 1 \\ 0 & 0 & 2 \end{bmatrix}$.

15. The first conclusion holds because if $A\mathbf{y} = \lambda\mathbf{y}$ then

$$A\mathbf{x} = AP\mathbf{y} = PB\mathbf{y} = \lambda P\mathbf{y} = \lambda\mathbf{x}.$$

Because of the nonsingularity of P, the vectors $\mathbf{y}_1 = P^{-1}\mathbf{x}_1, \ldots, \mathbf{y}_k = P^{-1}\mathbf{x}_k$ are linearly independent if and only if $\mathbf{x}_1 = P\mathbf{y}_1, \ldots, \mathbf{x}_k = P\mathbf{y}_k$ are linearly independent. Hence the second conclusion holds.

16. $A = 1,\quad B = 2,\quad C = 3,\quad D = 2.$

Section 4.5

1. A: $|5 - \lambda| \le 1,\quad |9 - \lambda| \le 2,$
B: $|5 - \lambda| \le 4,\quad |-5 - \lambda| \le 4,$
C: $|5 - \lambda| \le 4,$
D: $|1 - \lambda| \le 1,$
E: $|3 - \lambda| \le 2,\quad |5 - \lambda| \le 2,\quad |3 - \lambda| \le 2,$
F: $|3 - \lambda| \le 1,\quad |7 - \lambda| \le 2,\quad |\lambda| \le 1,$
G: $|9 - \lambda| \le 3,\quad |-5 - \lambda| \le 3,\quad |21 - \lambda| \le 11.$

2. A^T: $|9 - \lambda| \le 1,\quad |5 - \lambda| \le 2,$
B^T: $|5 - \lambda| \le 4,\quad |-5 - \lambda| \le 4,$
C^T: $|5 - \lambda| \le 4,$
D^T: $|1 - \lambda| \le 1,$
E^T: $|3 - \lambda| \le 2,\quad |5 - \lambda| \le 2,$
F^T: $|3 - \lambda| \le 1,\quad |7 - \lambda| \le 1,\quad |-\lambda| \le 2,$
G^T: $|9 - \lambda| \le 6,\quad |-5 - \lambda| \le 7,\quad |21 - \lambda| \le 4.$

3. H: $0 \le \lambda \le 4$; K: $-2 \le \lambda \le 6.$

4. For L, the disks are $|\lambda - 1| \le 2, |\lambda - 5| \le 1, |\lambda - 9| \le 1$. Since no two overlap, the eigenvalues are distinct and each disk contains a (real) eigenvalue. For M, these disks are $|\lambda - 2| \le 2, |\lambda - 5| \le 0, |\lambda - 9| \le 1$. They are pairwise nonoverlapping so that each contains an eigenvalue. This tells us that $\lambda = 5$ is an eigenvalue. For N, the disks $|\lambda - 5| \le 2, |\lambda - 1| \le 1, |\lambda + 3| \le 1$ are also pairwise nonoverlapping. So the eigenvalues are real and distinct.

Section 4.6

1. A_1: $\mathbf{y}_1 = \begin{bmatrix} e^{2t} \\ 0 \end{bmatrix},\quad \mathbf{y}_2 = \begin{bmatrix} 0 \\ e^{3t} \end{bmatrix},\quad \mathbf{y} = a\mathbf{y}_1 + b\mathbf{y}_2.$

A_2: $\mathbf{y}_1 = \begin{bmatrix} -2e^t \\ 1e^t \end{bmatrix},\quad \mathbf{y}_2 = \begin{bmatrix} 2 & e^{9t} \\ 1 & e^{9t} \end{bmatrix},\quad \mathbf{y} = \begin{bmatrix} 2ae^t + 2be^{9t} \\ ae^t + be^{9t} \end{bmatrix}.$

A_3: $\mathbf{y} = a\begin{bmatrix} te^{2t} \\ e^{2t} \end{bmatrix} + b\begin{bmatrix} e^{2t} \\ 0 \end{bmatrix}.$

A_4: $\mathbf{y} = \begin{bmatrix} a \\ b \end{bmatrix}.$

2. A_5: $\mathbf{y} = a\begin{bmatrix} e^{2t} \\ 0 \\ 0 \end{bmatrix} + b\begin{bmatrix} 8e^t \\ -2e^t \\ 1e^t \end{bmatrix} + c\begin{bmatrix} 4e^{9t} \\ 14e^{9t} \\ 7e^{9t} \end{bmatrix}.$

A_6: $\mathbf{y} = a\begin{bmatrix} te^{2t} \\ e^{2t} \\ 0 \end{bmatrix} + b\begin{bmatrix} e^{2t} \\ 0 \\ 0 \end{bmatrix} + c\begin{bmatrix} 0 \\ 0 \\ e^{3t} \end{bmatrix}.$

$$A_7: \quad y = a \begin{bmatrix} \frac{1}{2}t^2e^{2t} \\ te^{2t} \\ 0 \end{bmatrix} + b \begin{bmatrix} te^{2t} \\ e^{2t} \\ 0 \end{bmatrix} + c \begin{bmatrix} e^{2t} \\ 0 \\ 0 \end{bmatrix}.$$

CHAPTER 5

Section 5.1

1. A, C, F, are vector spaces.

9. Yes.

10. The class of differentiable functions is a subclass of the linear space of continuous functions. The class is a linear subspace, hence, a linear space in its own right.

11. Yes.

Section 5.2

1. A: $\{(-1, 1, 0), (-1, 0, 1)\}$, dim $A = 2$.

C: $\left\{ \begin{bmatrix} 1 & 0 \\ 0 & 1 \end{bmatrix}, \begin{bmatrix} 0 & 1 \\ 1 & 0 \end{bmatrix} \right\}$, dim $C = 2$.

F: $\{(1, 0)\}$, dim $F = 1$.

2. Yes, $\mathbf{x}, \mathbf{y}$ lie in the subspace and they form a basis.

3. The dimension is 4. $\left\{ \begin{bmatrix} 1 & 0 \\ 0 & -1 \end{bmatrix}, \begin{bmatrix} 0 & 1 \\ 0 & 0 \end{bmatrix}, \begin{bmatrix} 0 & 0 \\ 1 & 0 \end{bmatrix}, \begin{bmatrix} 0 & 0 \\ 0 & 1 \end{bmatrix} \right\}$ is a basis for this space. The subspace generated by $\left\{ \begin{bmatrix} 1 & 0 \\ 0 & 0 \end{bmatrix}, \begin{bmatrix} 0 & 1 \\ 0 & 0 \end{bmatrix} \right\}$; i.e. matrices of the form $\begin{bmatrix} a & b \\ 0 & 0 \end{bmatrix}$ is a two-dimensional subspace.

4. $\begin{bmatrix} 1 & 0 & 0 \\ 0 & 0 & 0 \end{bmatrix}, \begin{bmatrix} 0 & 1 & 0 \\ 0 & 0 & 0 \end{bmatrix}, \begin{bmatrix} 0 & 0 & 1 \\ 0 & 0 & 0 \end{bmatrix}, \begin{bmatrix} 0 & 0 & 0 \\ 1 & 0 & 0 \end{bmatrix}, \begin{bmatrix} 0 & 0 & 0 \\ 0 & 1 & 0 \end{bmatrix}, \begin{bmatrix} 0 & 0 & 0 \\ 0 & 0 & 1 \end{bmatrix}.$

The dimension is 6.

6. Yes. $\left\{ \begin{bmatrix} a & b & c \\ d & e & 0 \\ 0 & 0 & 0 \end{bmatrix} \middle| a, b, c, d, e \in \mathbf{R} \right\}$ is a subspace of dim 5.

7. Six.   Three.   Five.

The set of all upper triangular matrices of the form

$$\begin{bmatrix} 2a & 3a & 4a \\ 0 & 4a & 3a \\ 0 & 0 & 6a \end{bmatrix} = a \begin{bmatrix} 2 & 3 & 4 \\ 0 & 4 & 3 \\ 0 & 0 & 6 \end{bmatrix}.$$

8. $\{(-1, 2, 1, 0), (1, -4, 0, 1)\}$ is a basis for $\mathbf{N}$ of dimension two.

9. (a) $(1, 1)$ and $(0, 1)$;   (b) 2;   (c) $B = \begin{bmatrix} 1 & 0 \\ 1 & 1 \end{bmatrix}$;   (d) $(1, 0), (2, 3)$.

10. $A = \begin{bmatrix} 1 & 0 & 2 & 3 & 4 \\ 0 & 1 & 0 & 0 & 0 \\ 0 & 0 & 0 & 0 & 0 \end{bmatrix}$.

11. $\mathbf{u} = (1, 0, 0, 0, 0)$, $\mathbf{v} = (0, 1, 0, 0, 0)$, $\mathbf{w} = (0, 0, 1, 0, 0)$; $\mathbf{u}, \mathbf{v}, \mathbf{w}$ are linearly independent. Let $V = \{(0, 0, 0, a, b) \in \mathbf{R}^5\}$. Yes, the sum is direct. Yes, $\mathbf{u}, \mathbf{v}, \mathbf{w}, \mathbf{x}, \mathbf{y}$ are linearly independent.

Section 5.3

1. $\{(1, 0, -5), (0, 1, 2)\}$ and $\{(2, 5, 0), (-1, 0, 5)\}$ are bases for $5x - 2y + z = 0$.

$$\begin{bmatrix} 1 & 0 \\ 0 & 1 \\ -5 & 2 \end{bmatrix} \begin{bmatrix} 2 & -1 \\ 5 & 0 \end{bmatrix} = \begin{bmatrix} 2 & -1 \\ 5 & 0 \\ 0 & 5 \end{bmatrix}.$$

2. $\{(-1/2, 1)\}$ and $\{(1, -2)\}$ are bases for $2x + y = 0$.

$$\begin{bmatrix} -2 & 0 \\ 0 & -2 \end{bmatrix} \begin{bmatrix} -1/2 \\ 1 \end{bmatrix} = \begin{bmatrix} 1 \\ -2 \end{bmatrix}.$$

3. $(1, 6, 7)$ is a basis for the intersection of $\mathbf{P}$ and $\mathbf{Q}$.
$(1, 6, 7)$ and $(1, 0, 5)$ is a basis for $\mathbf{P}$.
$(1, 6, 7)$ and $(1, 0, 1)$ is a basis for $\mathbf{Q}$.

$$\begin{bmatrix} 1 & 1 & 1 \\ 6 & 0 & 0 \\ 7 & 5 & 1 \end{bmatrix} \text{ is a basis operator for } \mathbf{R}^3.$$

4. $\mathbf{N}$ has basis $\{(-2, 1, 0, 0, 0), (-1, 0, -1, 1, 0), (-3, 0, -1, 0, 1)\}$. Dimension of $\mathbf{N}$ is 3. $\{(1\ 2\ 0\ 1\ 3), (0\ 0\ 1\ 1\ 1)\}$ is a basis for $\mathbf{R}$. Dimension of $\mathbf{R}$ is 2.

5. $\{(1\ 0\ 1), (0\ 1\ 1)\}$, $\{(1, 1, 1), (1, -1, 0)\}$ are two bases for $\mathbf{C}$. dim $\mathbf{C} = 2$.

Section 5.4

1. (a) $2 = \langle \mathbf{f}, \mathbf{g} \rangle$; $-6 = \langle \mathbf{g}, \mathbf{h} \rangle$; $-2 = \langle \mathbf{h}, \mathbf{f} \rangle$.
(b) $\langle \mathbf{x}, \mathbf{y} \rangle = 8$; $\langle \mathbf{y}, \mathbf{z} \rangle = 0$; $\langle \mathbf{z}, \mathbf{x} \rangle = -4$.
(c) $\langle \mathbf{u}, \mathbf{v} \rangle = 1$.

2. (a) $(-1, 0, 1), (2, 2, 2), (4/3, -8/3, 4/3)$;
(b) $(2, -2, -4, 0), (3, 1, 1, 3) (2/15, 4/15, -1/15, -3/15)$;
(c) $(1, 0, 1, 0, 1), (2/3, -1, 2/3, 1, -4/3)$.

4. $B^{-1} = \dfrac{1}{\det B} \operatorname{adj} B = \begin{bmatrix} 1 & -b_{12} & b_{12}b_{23} - b_{13} \\ 0 & 1 & -b_{23} \\ 0 & 0 & 1 \end{bmatrix}$.

5. $F^T F = \begin{bmatrix} 2 & 4 & -4 \\ 4 & 10 & -2 \\ -4 & -2 & 30 \end{bmatrix}$, $\det F^T F = 16$. $G^T G = \begin{bmatrix} 2 & 0 & 0 \\ 0 & 2 & 0 \\ 0 & 0 & 4 \end{bmatrix}$, $\det G^T G = 16$.

CHAPTER 6

Section 6.1

1. T is a linear operator. $\begin{bmatrix} 1 & 1 \\ 1 & -1 \end{bmatrix}\begin{bmatrix} x \\ y \end{bmatrix} = \begin{bmatrix} x + y \\ x - y \end{bmatrix}$.

2. No, T is not a linear operator.

3. Yes, $\begin{bmatrix} 1 & 0 \\ 0 & 1 \end{bmatrix}\begin{bmatrix} x \\ y \end{bmatrix} = \begin{bmatrix} x \\ y \end{bmatrix}$.

4. Yes, $\begin{bmatrix} 0 & 1 \\ 0 & 0 \end{bmatrix}\begin{bmatrix} x \\ y \end{bmatrix} = \begin{bmatrix} y \\ 0 \end{bmatrix}$.

5. Yes, $\begin{bmatrix} 1 & 0 \\ 1 & 0 \\ 1 & 0 \end{bmatrix}\begin{bmatrix} x \\ y \end{bmatrix} = \begin{bmatrix} x \\ x \\ x \end{bmatrix}$.

6. Yes, $\begin{bmatrix} 1 & 1 \\ 1 & -1 \\ -1 & 1 \end{bmatrix}\begin{bmatrix} x \\ y \end{bmatrix} = \begin{bmatrix} x + y \\ x - y \\ y - x \end{bmatrix}$.

7. Yes, $\begin{bmatrix} 2 & 0 & 1 \\ 0 & 1 & -1 \\ 0 & 0 & 1 \end{bmatrix}\begin{bmatrix} x \\ y \\ z \end{bmatrix} = \begin{bmatrix} 2x + z \\ y - z \\ z \end{bmatrix}$.

8. Yes, $\begin{bmatrix} 0 & 0 & 2 \\ 0 & 2 & 0 \\ 2 & 0 & 0 \end{bmatrix}\begin{bmatrix} x \\ y \\ z \end{bmatrix} = 2\begin{bmatrix} z \\ y \\ x \end{bmatrix}$.

9. Yes, $\begin{bmatrix} 1 & 0 & 0 \\ 0 & 1 & 0 \end{bmatrix}\begin{bmatrix} x \\ y \\ z \end{bmatrix} = \begin{bmatrix} x \\ y \end{bmatrix}$.

10. Yes, $\begin{bmatrix} 2 & -1 & 0 \\ 0 & 1 & -3 \end{bmatrix}\begin{bmatrix} x \\ y \\ z \end{bmatrix} = \begin{bmatrix} 2x - y \\ y - 3z \end{bmatrix}$.

11. Yes, $\begin{bmatrix} 0 & 0 & 0 \\ 0 & 0 & 0 \end{bmatrix}\begin{bmatrix} x \\ y \\ z \end{bmatrix} = \begin{bmatrix} 0 \\ 0 \end{bmatrix}$.

12. No.

14. $A = \begin{bmatrix} 0 & 1 & 2 \\ 1 & -1 & 3 \end{bmatrix}$; $T(1, 1, 1) = (3, 3)$.

15. $A = \begin{bmatrix} 2 & 1 & -1 \\ 1 & 0 & 1 \end{bmatrix}$; $T(1, 2, 3) = (1, 4)$.

16. (a) and (c)

17. T is linear. No.

Section 6.2

1. (a) The null space has only **0** and the nullity is 0.

   (b) (0, 1) is a basis for the null space and the nullity is 1.

   (c) The null space has only **0** and the nullity is 0.

   (d) The null space has basis {(1, 0), (0, 1)} and the nullity is 2.

2. (a) Basis (1, −1, 0) and nullity is 1.

   (b) Only **0** and nullity 0.

   (c) {(1, 0, 1, 0, 1), (0, 0, 0, 1, 0)} is a basis for the null space and nullity is 2.

   (d) **0** is the only element for the null space and nullity is 0.

3. (a) 2;   (b) 1;   (c) 2;   (d) 0.

4. (a) 2;   (b) 3;   (c) 3;   (d) 3.

5. (a) $\begin{bmatrix} y \\ x \end{bmatrix}$;   (b) $\begin{bmatrix} x + 2y \\ -x + 3y \end{bmatrix}$;   (c) $\begin{bmatrix} 2x + y \\ -x + y + 2z \end{bmatrix}$.

   (d) $\begin{bmatrix} 4x - y + 2z \\ 5x - 3z \end{bmatrix}$;   (e) $\begin{bmatrix} 3u_1 + u_2 + 9u_4 \\ 4u_1 + u_2 + 3u_3 + 4u_4 \\ 5u_1 + 6u_3 + u_4 \end{bmatrix}$.

6. (a) $\begin{bmatrix} 0 & 1 \\ 1 & 0 \end{bmatrix}$;   (b) $\begin{bmatrix} 1 & 2 \\ -1 & 3 \end{bmatrix}$;   (c) $\begin{bmatrix} 2 & 1 & 0 \\ -1 & 1 & 2 \end{bmatrix}$;

   (d) $\begin{bmatrix} 4 & -1 & 2 \\ 5 & 0 & -3 \end{bmatrix}$;   (e) $\begin{bmatrix} 3 & 1 & 0 & 9 \\ 4 & 1 & 3 & 4 \\ 5 & 0 & 6 & 1 \end{bmatrix}$.

7. (a) **0** is the only element and nullity 0.

   (b) **0** is the only element and nullity 0.

   (c) (2, −4, 3) and nullity is 1.

   (d) The null space has basis (3, 22, 5) and nullity 1.

   (e) (−33, 18, 26, 9) is a basis for the null space and nullity is 1.

8. (a) 2;   (b) 2;   (c) 2;   (d) 2;   (e) 3

10. $\begin{bmatrix} 36 \\ 22 \\ 1 \end{bmatrix}$

11. (b) A^{-1}

14. The range of T is a subspace of $\mathbf{R}^m$ of dimension n.

15. $r = n - n = 0$. T is the zero transformation.

16. Range of T contains only the zero vector.

Section 6.3

1. **(c)** $R(A) = \text{span}\left\{\begin{bmatrix} 1 \\ 0 \end{bmatrix}, \begin{bmatrix} 1 \\ 1 \end{bmatrix}\right\}.$

2. $R(A) = IR^m.$

3. **(a)** $A\mathbf{x} = \mathbf{h}_1,\ A\mathbf{x} = \mathbf{h}_2$ are solvable.

   (b) $A\mathbf{x} = \mathbf{h}_1$ solvable, $A\mathbf{x} = \mathbf{h}_2$ is not solvable.

   (c) $A\mathbf{x} = \mathbf{h}_1$ is not solvable, $A\mathbf{x} = \mathbf{h}_2$ is solvable.

4. **(a)** $r(A) = $ number of non-zero rows of $F = 2.$ $q(A) = 4 - r(A) = 2.$

   (c) row-space of $A = $ column-space of $A^T = R(A^T).$

   (d) $F\mathbf{x} = 0 = \mathbf{x} = t\begin{bmatrix} -1 \\ 1 \\ 0 \\ 0 \end{bmatrix} + r\begin{bmatrix} -2 \\ 0 \\ -3 \\ 1 \end{bmatrix};\ \ r,\ t$ arbitrary.

6. **(a)** A basis for $R(A^T) = \{(1,\ -1,\ 2,\ 0,\ 2,\ 0),\ (0,\ 0,\ 0,\ 1,\ -1/2,\ 0),\ (0,\ 0,\ 0,\ 0,\ 0,\ 1)\}.$

8. **(a)** For $\mathbf{h}_1$: $\begin{bmatrix} 1 \\ 1 \end{bmatrix},$ for $\mathbf{h}_2$: $\begin{bmatrix} 1 \\ -1 \end{bmatrix}$ are least square solutions.

   (b) For $\mathbf{h}_1$: $\begin{bmatrix} 1 \\ 2 \\ 0 \\ 0 \end{bmatrix} + t\begin{bmatrix} 0 \\ 0 \\ 1 \\ 0 \end{bmatrix};$ for $\mathbf{h}_2$: $\begin{bmatrix} 4 \\ 0 \\ 0 \\ 1 \end{bmatrix} + s\begin{bmatrix} 0 \\ 0 \\ 1 \\ 0 \end{bmatrix}.$

   (c) for $\mathbf{h}_1$: $\begin{bmatrix} 1/2 \\ 0 \\ 0 \end{bmatrix},$ for $\mathbf{h}_2$: $\begin{bmatrix} 1 \\ -1 \\ -1 \end{bmatrix}.$

Section 6.4

1. **(a)** $E = \dfrac{1}{5}\begin{bmatrix} 1 & 2 \\ 2 & 4 \end{bmatrix},\ \ F = \dfrac{1}{5}\begin{bmatrix} 4 & -2 \\ -2 & 1 \end{bmatrix}.$

   (b) $E = \begin{bmatrix} 1 & 0 \\ 0 & 0 \end{bmatrix},\ \ F = \begin{bmatrix} 0 & 0 \\ 0 & 1 \end{bmatrix}.$

   (c) $E = \begin{bmatrix} 1 & 0 & 0 \\ 0 & 0 & 0 \\ 0 & 0 & 0 \end{bmatrix},\ \ F = \begin{bmatrix} 0 & 0 & 0 \\ 0 & 1 & 0 \\ 0 & 0 & 1 \end{bmatrix}.$

   (d) $E = \begin{bmatrix} 1 & 1 & 1 & 1 \\ 1 & 1 & 1 & 1 \\ 1 & 1 & 1 & 1 \\ 1 & 1 & 1 & 1 \end{bmatrix},\ \ F = \begin{bmatrix} 0 & -1 & -1 & -1 \\ -1 & 0 & -1 & -1 \\ -1 & -1 & 0 & -1 \\ -1 & -1 & -1 & 0 \end{bmatrix}.$

2. (a) $E = \dfrac{1}{3}\begin{bmatrix} 2 & -1 & 1 \\ -1 & 2 & 1 \\ 1 & 1 & 2 \end{bmatrix}$, $F = \begin{bmatrix} 1 & 1 & -1 \\ 1 & 1 & -1 \\ -1 & -1 & 1 \end{bmatrix}$.

(b) $E = \dfrac{1}{2}\begin{bmatrix} 1 & 0 & 1 \\ 0 & 2 & 0 \\ 1 & 0 & 1 \end{bmatrix}$, $F = \dfrac{1}{2}\begin{bmatrix} 1 & 0 & -1 \\ 0 & 0 & 0 \\ -1 & 0 & 0 \end{bmatrix}$.

(c) $E = \dfrac{1}{2}\begin{bmatrix} 1 & 0 & 1 & 0 \\ 0 & 1 & 0 & 1 \\ 1 & 0 & 1 & 0 \\ 0 & 1 & 0 & 1 \end{bmatrix}$, $F = \dfrac{1}{2}\begin{bmatrix} 1 & 0 & -1 & 0 \\ 0 & 1 & 0 & -1 \\ -1 & 0 & 1 & 0 \\ 0 & -1 & 0 & 1 \end{bmatrix}$.

3. (a) $E = \begin{bmatrix} 9/25 & 12/25 \\ 12/25 & 9/25 \end{bmatrix}$; **(b)** $\dfrac{1}{169}\begin{bmatrix} 9 & -12 & 36 \\ -12 & -16 & -48 \\ 36 & -48 & 144 \end{bmatrix}$.

Section 6.5

1. (a) $V = \begin{bmatrix} 0 & -1 \\ -1 & 0 \end{bmatrix}$; **(b)** $V = \begin{bmatrix} 0 & 0 & -1 \\ 0 & 1 & 0 \\ -1 & 0 & 0 \end{bmatrix}$

(c) $V = \begin{bmatrix} 1/2 & -1/2 & -1/2 & -1/2 \\ -1/2 & 1/2 & -1/2 & -1/2 \\ -1/2 & -1/2 & 1/2 & -1/2 \\ -1/2 & -1/2 & -1/2 & 1/2 \end{bmatrix}$.

2. (a) $V = \begin{bmatrix} 0 & 1 \\ 1 & 0 \end{bmatrix}$; **(b)** $V = \begin{bmatrix} 1/3 & 2/3 & -2/3 \\ 2/3 & 1/3 & 2/3 \\ -2/3 & 2/3 & 1/3 \end{bmatrix}$;

(c) $= \dfrac{1}{2}\begin{bmatrix} 1 & -1 & 1 & -1 \\ -1 & 1 & 1 & -1 \\ 1 & 1 & 1 & 1 \\ -1 & -1 & 1 & 1 \end{bmatrix}$.

3. $V_1 = \begin{bmatrix} 6/10 & 8/10 \\ 8/10 & -6/10 \end{bmatrix}$.

6. (a) $x_1 = 3$, $x_2 = 1$, $x_3 = 2$; **(b)** $x_1 = -2$, $x_2 = 0$, $x_3 = 2$.

7. $v_2 = (0, -1, 1, 2, 0) = (7, 2, 1, 2, 0) - (7, 3, 0, 0, 0)$; $v_2 = I - \dfrac{2v_2 v_2^T}{v_2^T v_2}$.

8. $v_3 = (3, 9, 18, 9, 18) - (3, 9, 27, 0, 0) = (0, 0, -9, 9, 18)$; $v_3 = I - \dfrac{2v_3 v_3^T}{v_3^T v_3}$.

9. $v_4 = (0, 27, 0, 3, 4) - (0, 27, 0, 5, 0) = (0, 0, 0, -2, 4)$.

10. P is a product of three orthogonal matrices, V_2, V_3, V_4.

Section 6.6

1. (a) h_1: L.S.E. $\begin{bmatrix} 1 & 1 \\ 1 & 1 \end{bmatrix} \bar{x} = \begin{bmatrix} 1 \\ 1 \end{bmatrix}$; L.S.S. $= \begin{bmatrix} r \\ 1 - r \end{bmatrix}$.

h_2: L.S.E. $\begin{bmatrix} 1 & 1 \\ 1 & 1 \end{bmatrix} \bar{x} = \begin{bmatrix} 4 \\ 4 \end{bmatrix}$; L.S.S. $= \begin{bmatrix} r \\ 4 - r \end{bmatrix}$.

h_3: L.S.E. $\begin{bmatrix} 1 & 1 \\ 1 & 1 \end{bmatrix} \bar{x} = \begin{bmatrix} 0 \\ 0 \end{bmatrix}$; L.S.S. $= \begin{bmatrix} r \\ -r \end{bmatrix}$.

(b) h_1: L.S.E. $\begin{bmatrix} 2 & 1 \\ 1 & 2 \end{bmatrix} \bar{x} = \begin{bmatrix} 1 \\ -1 \end{bmatrix}$; L.S.S. $= \begin{bmatrix} 1 \\ -1 \end{bmatrix}$.

h_2: L.S.E. $\begin{bmatrix} 2 & 1 \\ 1 & 2 \end{bmatrix} \bar{x} = \begin{bmatrix} 2 \\ 2 \end{bmatrix}$; L.S.S. $= \begin{bmatrix} 2/3 \\ 2/3 \end{bmatrix}$.

h_3: L.S.E. $\begin{bmatrix} 2 & 1 \\ 1 & 2 \end{bmatrix} \bar{x} = \begin{bmatrix} 0 \\ 0 \end{bmatrix}$; L.S.S. $= \begin{bmatrix} 0 \\ 0 \end{bmatrix}$.

(c) $h_1 =$ L.S.E. $\begin{bmatrix} 2 & 0 & 2 \\ 0 & 3 & 3 \\ 2 & 3 & 5 \end{bmatrix} \bar{x} = \begin{bmatrix} 2 \\ 1 \\ 3 \end{bmatrix}$; L.S.S. $= \begin{bmatrix} 1 - r \\ \frac{1}{3} - r \\ r \end{bmatrix}$

$h_2 =$ L.S.E. $\begin{bmatrix} 2 & 0 & 2 \\ 0 & 3 & 3 \\ 2 & 3 & 5 \end{bmatrix} \bar{x} = \begin{bmatrix} 4 \\ 6 \\ 10 \end{bmatrix}$; L.S.S. $= \begin{bmatrix} 2 - r \\ 2 - r \\ r \end{bmatrix}$

2. (a) For $h_1 = \begin{bmatrix} 1/2 \\ 1/2 \end{bmatrix}$; for $h_2 = \begin{bmatrix} 2 \\ 2 \end{bmatrix}$; for $h_3 = \begin{bmatrix} 0 \\ 0 \end{bmatrix}$.

(b) For h_1, h_2 all L.S.S. are orthogonal to nullspace of A.

(c) For $h_1 = \begin{bmatrix} 5/9 \\ -1/9 \\ 4/9 \end{bmatrix}$; for $h_2 = \begin{bmatrix} 2/3 \\ 2/3 \\ 4/3 \end{bmatrix}$.

3. (a) $\begin{bmatrix} t \\ 14 - t \end{bmatrix}$ is the least square solution with $h = 2h_1 + 3h_2$.

(b) $\begin{bmatrix} 4 \\ 0 \end{bmatrix}$ is the least square solution with $h = 2h_1 + 3h_2$.

(c) $\begin{bmatrix} 8 - t \\ \frac{20}{3} - t \\ t \end{bmatrix}$ is the least square solution with $h = 2h_1 + 3h_2$.

4. (a) $y = x - \dfrac{4}{5}$ (b) $y = \dfrac{13}{14}x - \dfrac{9}{14}$.

Section 6.7

2. $U^{-1} = \dfrac{1}{14}[1 \quad 2 \quad 3];$

$$V^{-1} = \frac{1}{2}\begin{bmatrix} 1 \\ 0 \\ 1 \\ 0 \end{bmatrix}, \quad W^{-1} = \frac{1}{3}[0 \quad 1 \quad 1 \quad 1], \quad Y^{-1} = \frac{1}{4}\begin{bmatrix} 1 \\ 1 \\ 1 \\ 1 \end{bmatrix}.$$

4. $A^{-1} = \dfrac{1}{28}\begin{bmatrix} 1 & 2 & 3 \\ 0 & 0 & 0 \\ 1 & 2 & 3 \\ 0 & 0 & 0 \end{bmatrix}, \quad B^{-1} = \dfrac{1}{6}\begin{bmatrix} 0 & 0 & 0 & 0 \\ 1 & 0 & 1 & 0 \\ 1 & 0 & 1 & 0 \\ 1 & 0 & 1 & 0 \end{bmatrix}.$

6. $(2A)^{-1} = \dfrac{1}{56}\begin{bmatrix} 1 & 2 & 3 \\ 0 & 0 & 0 \\ 1 & 2 & 3 \\ 0 & 0 & 0 \end{bmatrix}; \quad (5B)^{-1} = \dfrac{1}{30}\begin{bmatrix} 0 & 1 & 1 & 1 \\ 0 & 0 & 0 & 0 \\ 0 & 1 & 1 & 1 \\ 0 & 0 & 0 & 0 \end{bmatrix}.$

7. $A = \begin{bmatrix} 1 & 0 \\ 0 & 0 \end{bmatrix}, \quad B = \begin{bmatrix} 0 & 0 \\ 0 & 1 \end{bmatrix}.$

8. $P^{-1} = \begin{bmatrix} 1/6 & 1/3 & 1/6 \\ 1/3 & -1/3 & 1/3 \end{bmatrix}. \quad Q^{-1} = \begin{bmatrix} 1/2 & -1/6 \\ -1/2 & 5/4 \\ -0 & 1/3 \end{bmatrix}.$

9. $A^{-1} = \begin{bmatrix} 1/36 & 2/9 & 1/36 \\ 1/3 & -3/4 & 1/3 \\ 1/9 & -1/9 & 1/9 \end{bmatrix}.$

18. (a) $\begin{bmatrix} 1/4 & 0 & 1/4 \\ 1/3 & 1/3 & -1/3 \end{bmatrix};$ **(b)** $\begin{bmatrix} 1/2 & 0 & 1/2 \\ 0 & 1 & 0 \\ -1/2 & 0 & 1/2 \end{bmatrix};$

(c) $\begin{bmatrix} 1/3 & 1/6 & 1/6 \\ -1/3 & 1/3 & 1/3 \end{bmatrix};$ **(d)** $\begin{bmatrix} 1/3 & 1/6 & 1/6 \\ -1/3 & 1/3 & 1/3 \\ 0 & 1/2 & -1/2 \end{bmatrix}$

Section 6.8

1. (a) A_1: x-axis, y-axis; A_2: all lines through the origin; A_4: x − axis;
A_5: x-axis and the line $y = x$; A_6: the lines $y = x$ and $y = -x$; A_7: none;
A_8: y-axis.

(b) B_1: xy-, xz-, yz-planes and the coordinate axes; B_3: xy- and xz-planes, x-axis,
z-axis; B_4: x-axis, xy-plane; B_5: x-axis, xy-plane, the lines $y = 2x$, $z = 2y$, also
the 2-plane of points $(s + t, 2t, 4t)$.

4. A: $m(\lambda) = (\lambda - 2)$; B: $m(\lambda) = (\lambda - 2)^2$; C: $m(\lambda) = (\lambda - 2)^2$;
D: $m(\lambda) = (\lambda - 2)^3$; E: $m(\lambda) = (\lambda - 2)^4$.

CHAPTER 7

Section 7.1

1. (a) A_1: $4x^2 + 9y^2 = k$, ellipse;

A_2: $4x^2 - 9y^2 = k$, hyperbola;

A_3: $5x^2 + 8xy + 5y^2 = k$, ellipse;

A_4: $-5x^2 + 8xy - 5y^2 = k$, ellipse;

A_5: $x^2 + 4xy + y^2 = k$, hyperbola.

A_6: $2xy = k$, hyperbola

(b) A_1: coordinate axes.

A_2: coordinate axes.

A_3: $\bar{\mathbf{x}}_1 = \pm \begin{bmatrix} -1/\sqrt{2} \\ 1/\sqrt{2} \end{bmatrix}$, $\quad \bar{\mathbf{x}}_2 = \pm \begin{bmatrix} -1/\sqrt{2} \\ -1/\sqrt{2} \end{bmatrix}$;

A_4: $\bar{\mathbf{x}}_1 = \pm \begin{bmatrix} -1/\sqrt{2} \\ 1/\sqrt{2} \end{bmatrix}$, $\quad \bar{\mathbf{x}}_2 = \pm \begin{bmatrix} -1/\sqrt{2} \\ -1/\sqrt{2} \end{bmatrix}$;

A_5: $\bar{\mathbf{x}}_1 = \pm \begin{bmatrix} -1/\sqrt{2} \\ 1/\sqrt{2} \end{bmatrix}$, $\quad \bar{\mathbf{x}}_2 = \pm \begin{bmatrix} -1/\sqrt{2} \\ -1/\sqrt{2} \end{bmatrix}$;

A_6: $\bar{\mathbf{x}}_1 = \begin{bmatrix} -1/\sqrt{2} \\ 1/\sqrt{2} \end{bmatrix}$, $\quad \bar{\mathbf{x}}_2 = \pm \begin{bmatrix} -1/\sqrt{2} \\ -1/\sqrt{2} \end{bmatrix}$.

(c) A_1: $k = 97$; A_2: $k = -65$; A_3: $k = 113$; A_4: $k = -17$;

A_5: $k = 25$; A_6: $k = 12$

(d) A_1: positive definite; A_2: indefinite; A_3: positive definite;

A_4: negative definite; A_5: indefinite; A_6: indefinite.

(e) A_1: $(x, y) = (0, 0)$, i.e., origin; A_3: $x = 0$; A_2: the lines $y = \pm(2/3)x$;

A_4: $x = 0$; A_6: the coordinate axes.

2. (a) A_7: $x^2 + 4y^2 + 9z^2 = k$;

A_8: $x^2 + 4y^2 - 9z^2 = k$;

A_9: $3x^2 + 3y^2 + 3z^2 - 2xy - 2yz + 2xz = k$;

A_{10}: $x^2 - y^2 - z^2 = k$;

A_{11}: $y^2 + 2xz = k$;

A_{12}: $2x^2 + y^2 + 2z^2 + 2xz = k$.

(e) A_7, A_9, A_{12} are positive definite.

Section 7.2

1. (a) A_1: $\lambda_1 = 2$, $\lambda_2 = 3$; $\mathbf{x}_1 = \begin{bmatrix} 1 \\ 0 \end{bmatrix}$, $\mathbf{x}_2 = \begin{bmatrix} 0 \\ 1 \end{bmatrix}$.

A_2: $\lambda_1 = 1$, $\lambda_2 = 6$; $\mathbf{x}_1 = \begin{bmatrix} 1 \\ 2 \end{bmatrix}$, $\mathbf{x}_2 = \begin{bmatrix} 2 \\ -1 \end{bmatrix}$.

$$A_3: \lambda_1 = 4, \lambda_2 = 9; \quad \mathbf{x}_1 = \begin{bmatrix} 1 \\ -2 \end{bmatrix}, \quad \mathbf{x}_2 = \begin{bmatrix} 2 \\ 1 \end{bmatrix}.$$

(b) $A_1: R(x) = \dfrac{2x^2 + 3y^2}{x^2 + y^2}; \quad A_2: R(x) = \dfrac{5x^2 - 4xy + 2y^2}{x^2 + y^2}; \quad A_3: R(x) = \dfrac{8x^2 + 4xy + 5y^2}{x^2 + y^2}.$

Section 7.3

2. (a) $12x + 18y = 72$;   **(b)** $8x + 27y = 78$;   **(c)** $2x + 3y = 12$;

(d) $11x + 7y = 47$.

4. (a) Minimum point $\mathbf{x}_0 = (1, 1, -1)$;   minimum value $\alpha_0 = -20$.

CHAPTER 8

Because so many of the exercises in this chapter require verification of a computational algorithm and/or the use of a calculating device, the answers are not included.

Index